Visualizing

WEATHER
AND CLIMATE

Visualizing
WEATHER
AND CLIMATE

Bruce T. Anderson, PhD
Boston University

Alan Strahler, PhD
Boston University

WILEY

In collaboration with
THE NATIONAL GEOGRAPHIC SOCIETY

CREDITS

VP AND PUBLISHER Jay O'Callaghan

MANAGING DIRECTOR Helen McInnis

EXECUTIVE EDITOR Ryan Flahive

MANAGER OF PRODUCT DEVELOPMENT Nancy Perry

DEVELOPMENT EDITOR Charity Robey

ASSISTANT EDITOR Courtney Nelson

EDITORIAL ASSISTANT Erin Grattan

EXECUTIVE MARKETING MANAGER Jeffrey Rucker

MARKETING MANAGER Danielle Torio

MEDIA EDITORS Lynn Pearlman, Bridget O'Lavin

PRODUCTION MANAGER Micheline Frederick

Full Service Production Provided by
Camelot Editorial Services, LLC

CREATIVE DIRECTOR Harry Nolan

COVER DESIGNER Hope Miller

INTERIOR DESIGN Vertigo Design

PHOTO EDITOR Hilary Newman

PHOTO RESEARCHER Stacy Gold,
National Geographic Society

SENIOR ILLUSTRATION EDITOR Sandra Rigby

COVER CREDITS **Top photo:** © Bobby Model/NG Image
Collection **Bottom inset photos (left to right):** Courtesy
NOAA; © Peter Carsten/NG Image Collection; © SPL/Photo
Researchers; © Norbert Rossing/NG Image Collection;
© Corbis Digital Stock **Title page:** © Ira Block/NG
Image Collection

This book was set in New Baskerville by Preparé, Inc., and printed and bound by Quebecor World. The cover was printed by Phoenix Color.

To order books or for customer service please, call
1-800-CALL WILEY (225-5945).

ISBN-13   9780470147757
ISBN-10   047014775X

Printed in the United States of America
10 9 8 7 6 5 4 3 2 1

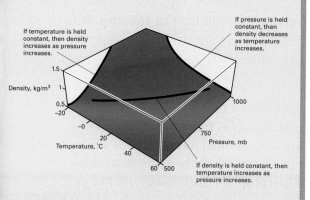

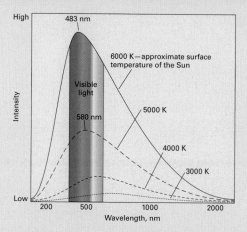

Visualizing *Weather and Climate* is designed to help your students learn effectively. Created in collaboration with the National Geographic Society and our Wiley Visualizing Consulting Editor, Professor Jan Plass of New York University, *Visualizing Weather and Climate* integrates rich visuals and media with text, to direct students' attention to important information. This approach represents complex processes, organizes related pieces of information, and integrates information into clear representations. Beautifully illustrated, *Visualizing Weather and Climate* gives your students the main concepts and applications of the discipline and also instills an appreciation and excitement about the richness of the subject.

Visuals, as used throughout this text, are instructional components that display facts, concepts, processes, or principles. The visuals include diagrams, graphs, maps, photographs, illustrations, schematics, animations, and videos, and together they create the foundation for the text.

Why should a textbook based on visuals be effective? Research shows that we learn better from integrated text and visuals than from either medium separately. Beginners in a subject benefit most from reading about the topic, attending class, and studying well-designed and integrated visuals. A visual, with good accompanying discussion, really can be worth a thousand words!

Well-designed visuals can also improve the efficiency with which a learner processes information. This process takes place as we integrate new information in our working memory with existing knowledge in our long-term memory.

Have you ever read a paragraph or a page in a book, stopped, and said to yourself: "I don't remember one thing I just read?" This may happen when your working memory has been overloaded, and the text you read was not successfully integrated into long-term memory. Visuals don't automatically solve the problem of overload, but well-designed visuals can reduce the number of elements that working memory must process, thus aiding learning.

You, as the instructor, facilitate your students' learning. Well-designed visuals, used in class, can help you in that effort. Here are six methods for using the visuals in the *Visualizing Weather and Climate* in classroom instruction.

1. **Assign students to study visuals in addition to reading the text.**

   It is important to make sure your students know that the visuals are just as essential as the text.

2. **Use visuals during class discussions or presentations.**

   By pointing out important information as the students look at the visuals during class discussions, you can help focus students' attention on key elements of the visuals and help them begin to organize the information and develop an integrated model of understanding.

3. **Use visuals to review content knowledge.**

   Students can review key concepts, principles, processes, vocabulary, and relationships displayed visually. Better understanding results when new information in working memory is linked to prior knowledge.

4. **Use visuals for assignments or when assessing learning.**

   Visuals can be used for comprehension activities or assessments. For example, students could be asked to identify examples of concepts portrayed in visuals. Visuals can be very useful for drawing inferences, for predicting, and for problem solving.

5. **Use visuals to situate learning in authentic contexts.**
   Learning is more meaningful when a learner can apply facts, concepts, and principles to realistic situations or examples. Visuals can provide that realistic context.

6. **Use visuals to encourage collaboration.**
   Collaborative groups often are required to practice interactive processes. These interactive, face-to-face processes provide the information needed to build a verbal mental model. Learners also benefit from collaboration in many situations such as decision making or problem solving.

Visualizing Weather and Climate not only aids student learning with extraordinary use of visuals, but it also offers an array of remarkable photos, media, and film from the National Geographic Society collections.

Additional information on learning tools and instructional design is provided electronically, including an *Instructor's Manual* that contains guidelines and suggestions on using the text and visuals most effectively. Other supplementary materials include a Test Bank with visuals used in assessment, PowerPoints, an Image Gallery to provide you with the same visuals used in the text, and Web-based learning materials for homework and assessment including images, video, and media resources from National Geographic.

This debut edition of *Visualizing Weather and Climate* offers students a valuable opportunity to identify and connect the central issues of weather, climate, and the atmosphere through a visual approach. As students explore the critical topics of weather and climate, their study of the role of the atmosphere must be interwoven with the behavior of the oceans, land surface, ecosystems, and human activity. *Visualizing Weather and Climate* reinforces these interacting components and, with its premier art program, vividly illustrates the overarching role that the atmosphere plays in influencing our planet's environment.

This book is intended to serve as an introductory text primarily for nonscience undergraduate students. The accessible format of *Visualizing Weather and Climate*, coupled with our assumption that students have little prior knowledge of atmospheric and environmental sciences, allows students to easily make the transition from jumping-off points in the early chapters to the more complex concepts they encounter later. With its highly visual presentation, which mirrors the very nature of weather and climate phenomena, this book is appropriate for use in one-semester atmospheric science courses offered by a variety of departments, including environmental studies, agriculture, geography, and geology.

ORGANIZATION

Visualizing Weather and Climate is organized around the premise that weather and climate are two interrelated and equally important influences on environmental processes and human activities. The book's first seven chapters lay the groundwork for creating an understanding of both weather and climate phenomena. Included topics are the forces and conditions that determine the state and composition of the atmosphere, the roles water and energy play in producing variations of the atmosphere in space and time, and the ways these variations influence the dynamics of the atmosphere and oceans.

Chapters 7, 8, 9, and 10 investigate in more detail specific weather phenomena and their influence on human society. The exploration starts with midlatitude weather systems and their relation to disturbances in the large-scale atmospheric circulations discussed in the opening chapters, with particular attention paid to the role variations in atmospheric water and energy play in producing the weather phenomena we experience at the surface. Next, tropical weather systems are investigated, with an eye toward describing the development, movement, and impacts of one of the most devastating natural phenomenon on the planet—tropical cyclones. From there, the section examines more local phenomena, including thunderstorms, lightning, and tornadoes, and the processes that give rise to these fearsome storms.

Chapters 11, 12, and 13 deal with climates of the Earth system, beginning with the processes responsible for producing variations in temperature and precipitation around the globe. The investigation continues by examining the climates of the world, with a specific eye toward the weather phenomena that give rise to the mean and seasonal patterns in temperature and precipitation that constitute these climates. Finally, the section explores the ways climate can vary. A particular emphasis is placed on both the various processes that can force changes in climates across the globe and the types of feedback that can either augment or reduce this forcing—a significant step forward compared with most discussions of climate variations.

Finally, Chapters 14, 15, and 16 focus specifically on human interaction with weather and climate. Those chapters investigate in detail the effects of weather and climate changes on human activities, along with the impacts of human activities on local weather and climate. The section then moves on to investigate how weather forecasts are made and how they can be used to mitigate some of the consequences of severe weather. In addition, a full chapter is dedicated to an investigation of the ways human activities may be influencing the global climate, how we can predict the scope of this influence, and how we might mitigate this influence in the future.

NATIONAL GEOGRAPHIC SOCIETY FACT-CHECKING

The National Geographic Society has performed an invaluable service in fact-checking *Visualizing Weather and Climate*. They have verified every fact in the book with two outside sources, to ensure that the text is accurate and up-to-date.

ILLUSTRATED BOOK TOUR

A number of pedagogical features using visuals have been developed specifically for *Visualizing Weather and Climate*. Presenting the highly varied and often technical concepts woven into a study of weather and climate phenomena raises challenges for reader and instructor alike. The **Illustrated Book Tour** on the following pages provides a guide to these diverse features.

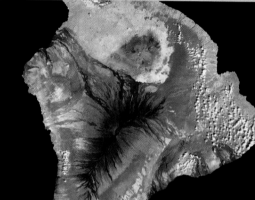

CHAPTER INTRODUCTIONS illustrate certain concepts in the chapter with concise stories about some of the world's most remarkable weather and climate phenomena. These narratives are featured alongside striking accompanying photographs. The chapter openers also include illustrated **CHAPTER OUTLINES** that use thumbnails of illustrations from the chapter to refer visually to the content.

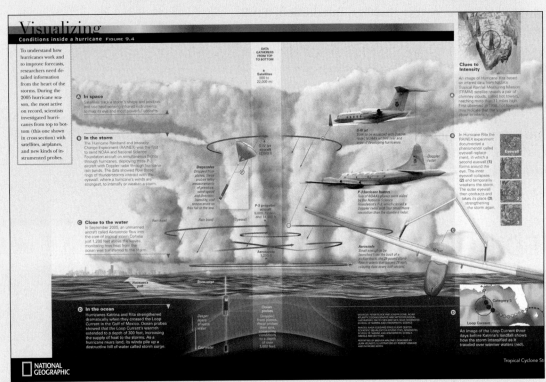

VISUALIZING features are specially designed, multipart visual spreads that focus on a key concept or topic in the chapter, exploring it in detail or in broader context using a combination of photos and figures.

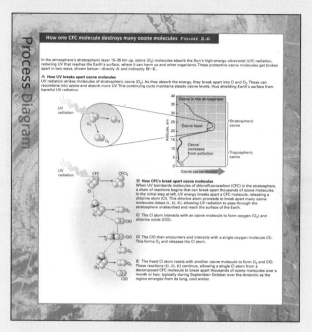

PROCESS DIAGRAMS present a series of figures or a combination of figures and photos that describe and depict a complex process, helping students to observe, follow, and understand the process.

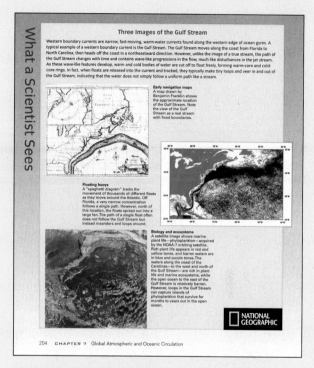

WHAT A SCIENTIST SEES are features that highlight a concept or phenomenon, using photos and figures that would stand out to a professional in the field, and helping students to develop observational skills.

WHAT IS HAPPENING IN THIS PICTURE? are end-of-chapter features that present students with a photograph that is relevant to chapter topics but illustrates a situation students are not likely to have encountered previously. The photograph is paired with questions designed to stimulate creative thinking.

OTHER PEDAGOGICAL FEATURES

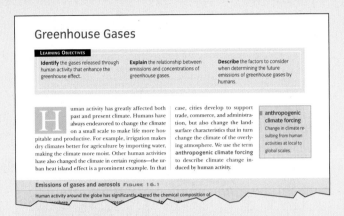

Greenhouse Gases

LEARNING OBJECTIVES

Identify the gases released through human activity that enhance the greenhouse effect.

Explain the relationship between emissions and concentrations of greenhouse gases.

Describe the factors to consider when determining the future emissions of greenhouse gases by humans.

Human activity has greatly affected both past and present climate. Humans have always endeavored to change the climate on a small scale to make life more hospitable and productive. For example, irrigation makes dry climates better for agriculture by importing water, making the climate more moist. Other human activities have also changed the climate in certain regions—the urban heat island effect is a prominent example. In that case, cities develop to support trade, commerce, and administration, but also change the land-surface characteristics that in turn change the climate of the overlying atmosphere. We use the term **anthropogenic climate forcing** to describe climate change induced by human activity.

■ **anthropogenic climate forcing** Change in climate resulting from human activities at local to global scales.

Emissions of gases and aerosols FIGURE 16.1

Human activity around the globe has significantly altered the chemical composition of

LEARNING OBJECTIVES at the beginning of each section indicate what the student must be able to do, to demonstrate mastery of the material.

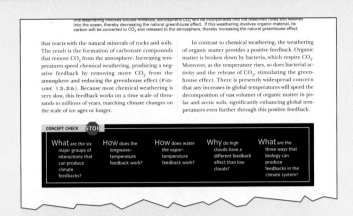

this weathering involves silicate minerals, atmospheric CO_2 will be incorporated into the dissolved rocks and washed into the ocean, thereby decreasing the natural greenhouse effect. If this weathering involves organic material, its carbon will be converted to CO_2 and released to the atmosphere, thereby increasing the natural greenhouse effect.

that reacts with the natural minerals of rocks and soils. The result is the formation of carbonate compounds that remove CO_2 from the atmosphere. Increasing temperatures speed chemical weathering, producing a negative feedback by removing more CO_2 from the atmosphere and reducing the greenhouse effect (FIGURE 13.26). Because most chemical weathering is very slow, this feedback works on a time scale of thousands to millions of years, matching climate changes on the scale of ice ages or longer.

In contrast to chemical weathering, the weathering of organic matter provides a positive feedback. Organic matter is broken down by bacteria, which respire CO_2. Moreover, as the temperature rises, so does bacterial activity and the release of CO_2, stimulating the greenhouse effect. There is presently widespread concern that any increases in global temperatures will speed the decomposition of vast volumes of organic matter in polar and arctic soils, significantly enhancing global temperatures even further through this positive feedback.

CONCEPT CHECK STOP

What are the six major groups of interactions that can produce climate feedbacks?

How does the longwave–temperature feedback work?

How does water vapor–temperature feedback work?

Why do high clouds have a different feedback effect than low clouds?

What are the three ways that biology can produce feedbacks in the climate system?

CONCEPT CHECK questions at the end of each section give students the opportunity to test their comprehension of the learning objectives.

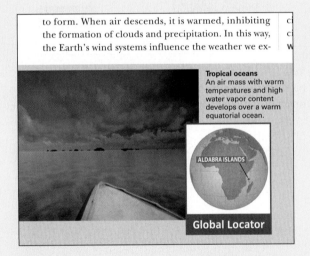

to form. When air descends, it is warmed, inhibiting the formation of clouds and precipitation. In this way, the Earth's wind systems influence the weather we ex-

Tropical oceans
An air mass with warm temperatures and high water vapor content develops over a warm equatorial ocean.

ALDABRA ISLANDS

Global Locator

GLOBAL LOCATOR MAPS, prepared specifically for this book by the National Geographic Society cartographers, accompany some photos. These locator maps help students visualize where the area depicted in the photo is situated on the Earth.

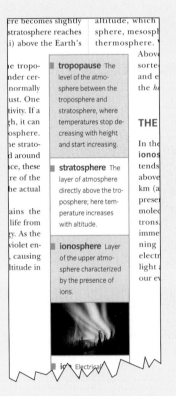

ere becomes slightly stratosphere reaches i) above the Earth's

e troponder cernormally ust. One tivity. If a gh, it can osphere. e stratod around ace, these re of the e actual

ains the life from y. As the violet en, causing ltitude in

■ **tropopause** The level of the atmosphere between the troposphere and stratosphere, where temperatures stop decreasing with height and start increasing.

■ **stratosphere** The layer of atmosphere directly above the troposphere; here temperature increases with altitude.

■ **ionosphere** Layer of the upper atmosphere characterized by the presence of ions.

altitude, which sphere, mesosph thermosphere.

Above sorte and e the h

THE

In the ionos tends above km (a prese molec trons. imme ning electr light our e

■ **ion** Electrical

MARGIN GLOSSARY TERMS (in green boldface) introduce each chapter's most important terms, often reinforced with a thumbnail photograph. The second most important terms appear in **black boldface** and are defined in the text; the third level of important terms appear in italics and are not formally defined, since they have usually been defined previously.

ILLUSTRATIONS AND PHOTOS support concepts covered in the text, elaborate on relevant issues, and add visual detail. Many of the photos originate from National Geographic's rich sources.

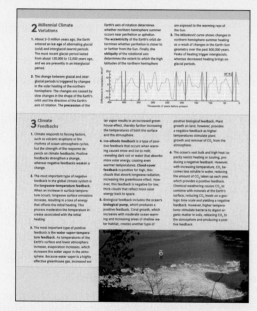

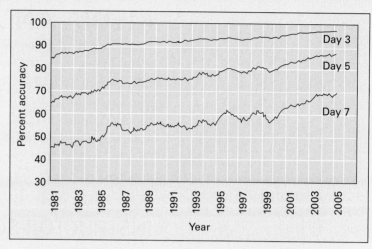

TABLES AND GRAPHS, with data sources cited at the end of the text, summarize and organize important information.

THE CHAPTER SUMMARY revisits each learning objective and redefines each margin glossary term, featured in boldface here and included in a list of **Key Terms**. Students are thus able to study vocabulary words in the context of related concepts. Each portion of the Chapter Summary is illustrated with a relevant photo from its respective chapter section.

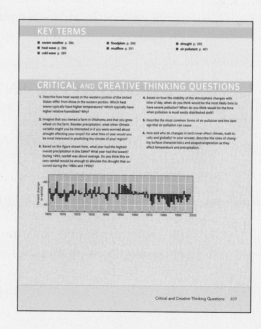

CRITICAL AND CREATIVE THINKING QUESTIONS encourage critical thinking and highlight each chapter's important concepts and applications.

MEDIA AND SUPPLEMENTS

Visualizing Weather and Climate is accompanied by a rich array of media and supplements that incorporate the visuals from the textbook extensively to form a pedagogically cohesive package. For example, a Process Diagram from the book appears in the Instructor's Manual with suggestions on using it as a PowerPoint in the classroom. That Process Diagram may also be the subject of a short video or an online animation; and it may also appear with questions in the Test Bank, as part of the chapter review, homework assignment, assessment questions, and other online features.

INSTRUCTOR SUPPLEMENTS

WileyPlus is a powerful online tool that provides instructors and students with an integrated suite of teaching and learning resources in one easy-to-use Web site.

These resources include:

VIDEOS

A rich collection of videos have been selected to accompany and enrich the text. Each chapter includes video clips, available online as digitized streaming video, that illustrate and expand on a concept or topic to aid student understanding. Accompanying each of the videos is contextualized commentary and questions that can further develop student understanding. The videos are available in **WileyPlus.**

POWERPOINT PRESENTATIONS AND IMAGE GALLERY

A complete set of highly visual PowerPoint presentations by Lawrence McGlinn of SUNY, New Paltz, is available online to enhance classroom presentations. Tailored to the text's topical coverage and learning objectives, these presentations are designed to convey key text concepts that are illustrated by embedded text art.

Image Gallery All photographs, figures, maps, and other visuals from the text are online and can be used as you wish in the classroom. These online electronic files allow you to easily incorporate them into your PowerPoint presentations as you choose, or to create your own overhead transparencies and handouts.

TEST BANK (AVAILABLE IN WILEYPLUS AND ELECTRONIC FORMAT)

The visuals from the textbook are also included in the Test Bank by Elizabeth J. Leppman of Walden University and Jonathon D. Little of Monroe Community College. The test items include multiple choice and essay questions testing a variety of comprehension levels. The test bank is available in two formats: online in MS Word files and as a Computerized Test Bank on a multiplatform CR-ROM. The easy-to-use test-generation program fully supports graphics, print tests, student answer sheets, and answer keys. The software's advanced features allow you to create an exam to your exact specifications.

INSTRUCTOR'S MANUAL (AVAILABLE IN ELECTRONIC FORMAT)

The manual begins with a special introduction on *Using Visuals in the Classroom*, prepared by Matthew Leavitt of the Arizona State University, in which he provides guidelines and suggestions on how to use the visuals in teaching the course. For each chapter, materials by James G. Duvall III of Contra Costa College include suggestions and directions for using Web-based learning modules in the classroom and for homework assignments, as well as over 50 creative ideas for in-class activities.

WEB-BASED LEARNING MODULES

A robust suite of multimedia learning resources has been designed for *Visualizing Weather and Climate*, focusing on and using the visuals from the book. Delivered via the Web, the content is organized into Tutorial animations. These animations visually support the learning of a difficult concept, process, or theory, many of them built around a specific feature such as a Process Diagram, Visualizing feature, or key visual in the chapter. The animations go beyond the content and visuals presented in the book, providing additional visual examples and descriptive narration.

ACKNOWLEDGMENTS

PROFESSIONAL FEEDBACK

Throughout the process of writing and developing this text and the visual pedagogy, we benefited from the comments and constructive criticism provided by the instructors and colleagues listed below. We offer our sincere appreciation to these individuals for their helpful reviews:

Al Armendariz,
Southern Methodist University;

Helen Cox,
California State University— Northridge;

Steven Emerman,
Simpson College - Indianola;

Martin Hackworth,
Idaho State University— Pocatello;

Lawrence Hipps,
Utah State University;

Cecil Keen,
Minnesota State University— Mankato;

Rodney Kubesh,
Saint Cloud State University;

Kevin Law,
Marshall University;

Scott Mandia,
Suffolk County Community College;

Jan Mikesell,
Gettysburg College;

Steven Newman,
Central Connecticut State University;

Kenneth Parsons,
Embry-Riddle Aeronautical University - Prescott;

David Porinchu,
Ohio State University— Columbus;

Debbie Schaum,
Embry-Riddle Aeronautical University—Daytona;

Justin Schoof,
Southern Illinois University— Carbondale;

SPECIAL THANKS

We are extremely grateful to the many members of the editorial and production staff at John Wiley and Sons who guided us through the challenging steps of developing this book. Their tireless enthusiasm, professional assistance, and endless patience smoothed the path as we found our way. We thank in particular: Ryan Flahive, who expertly launched and directed our process; Charity Robey, Developmental Editor; Nancy Perry, Manager, Product Development; Helen McInnis, Managing Director, Wiley Visualizing, who oversaw the concept of the book; Micheline Frederick, Production Manager, who stepped in whenever we needed expert advice; Jay O' Callaghan, Vice President and Publisher, who oversaw the entire project; and Jeffrey Rucker, Executive Marketing Manager for Wiley Visualizing, who adeptly represents the Visualizing imprint. We appreciate the efforts of Hilary Newman in obtaining some of our text photos, and Sandra Rigby, Senior Illustration Editor, for her expertise in managing the illustration program. We also wish to thank those who worked on the media and ancillary materials: Tom Kulesa and Lynn Pearlman, Senior Media Editors.

We wish to thank Stacy Gold, Research Editor and Account Executive at the National Geographic Image Collection, for her valuable expertise in selecting NGS photos. Many other individuals at National Geographic offered their expertise and assistance in developing this book: Francis Downey, Vice President and Publisher, and Richard Easby, Supervising Editor, National Geographic School Division; Mimi Dornack, Sales Manager, and Lori Franklin, Assistant Account Executive, National Geographic Image Collection; and Dierdre Bevington-Attardi, Project Manager, and Kevin Allen, Director of Map Services, National Geographic Maps.

PERSONAL ACKNOWLEDGMENTS

I welcome the opportunity to thank a number of people. Much of the time and effort that went into this text was supported by a Visiting Scientist appointment to the Grantham Institute for Climate Change, administered by Imperial College of Science, Technology, and Medicine in London. Many insightful and often scrutinizing comments and questions came from my colleagues and students at Boston University. Much of the love and encouragement came from my family—Dave, Gail, and Ian and most importantly my wife, Lara, to whom I dedicate this book.- BTA

ABOUT THE AUTHORS

Alan Strahler earned his Ph.D. degree in Geography from Johns Hopkins in 1969, and is presently Professor of Geography at Boston University. He has published over 250 articles in the refereed scientific literature, largely on the theory of remote sensing of vegetation, and has also contributed to the fields of plant geography, forest ecology, and quantitative methods. In 1993, he was awarded the Association of American Geographers/Remote Sensing Specialty Group Medal for Outstanding Contributions to Remote Sensing. With Arthur Strahler, he is a coauthor of seven textbook titles with eleven revised editions on physical geography and environmental science. He holds the honorary degree D.S.H.C. from the Université Catholique de Louvain, Belgium, and is a Fellow of the American Association for the Advancement of Science.

Bruce Anderson earned his Ph.D. degree in Ocean Sciences from Scripps Institution of Oceanography at University of California, San Diego in 1998, and is presently Professor of Geography and Environment at Boston University. He has published numerous articles in the refereed scientific literature, largely on regional impacts of climate variability, and has also contributed to the fields of atmospheric dynamics and hydrology, coupled ocean-atmosphere dynamics, and climate/vegetation interactions. In 2008, he was an inaugural Grantham Institute for Climate Change Visiting Fellow at Imperial College for Science, Technology and Medicine. He has also been a Royal Society Visiting Scientist, National Research Council Fellow and a NOAA Visiting Scientist Fellow. Presently he also serves as a Research Consultant for the Union of Concerned Scientists Northeast Climate Impacts Assessment (NECIA) project.

CONTENTS *in Brief*

COMPLETE TABLE OF CONTENTS

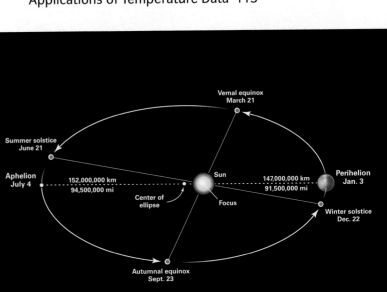

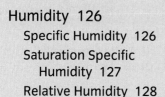
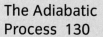

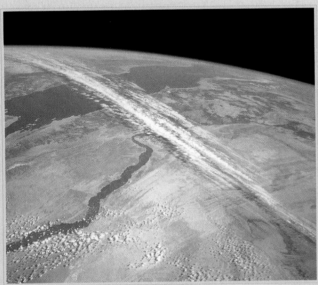

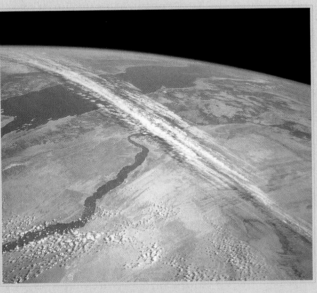

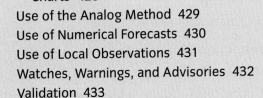

VISUALIZING FEATURES

Multipart visual presentations that focus on a key concept or topic in the chapter.

PROCESS DIAGRAMS

A series or combination of figures and photos that describe and depict a complex process

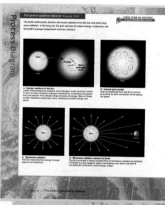

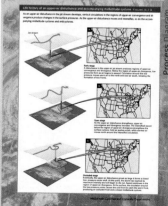

www.wileyplus.com

Wiley is committed to making your entire *WileyPLUS* experience productive & enjoyable by providing the help, resources, and personal support you & your students need, when you need it. It's all here: www.wileyplus.com –

TECHNICAL SUPPORT:

- ⊕ A fully searchable knowledge base of FAQs and help documentation, available 24/7
- ⊕ Live chat with a trained member of our support staff during business hours
- ⊕ A form to fill out and submit online to ask any question and get a quick response
- ⊕ **Instructor-only** phone line during business hours: 1.877.586.0192

FACULTY-LED TRAINING THROUGH THE WILEY FACULTY NETWORK:
Register online: www.wherefacultyconnect.com
Connect with your colleagues in a complimentary virtual seminar, with a personal mentor in your field, or at a live workshop to share best practices for teaching with technology.

1ST DAY OF CLASS...AND BEYOND!
Resources for You & Your Students Need to Get Started & Use WileyPLUS from the first day forward.

- ⊕ 2-Minute Tutorials on how to set up & maintain your *WileyPLUS* course
- ⊕ User guides, links to technical support & training options
- ⊕ *WileyPLUS for Dummies*: Instructors' quick reference guide to using *WileyPLUS*
- ⊕ Student tutorials & instruction on how to register, buy, and use *WileyPLUS*

YOUR WileyPLUS ACCOUNT MANAGER:
Your personal *WileyPLUS* connection for any assistance you need!

QuickStart

SET UP YOUR WileyPLUS COURSE IN MINUTES!
Select *WileyPLUS* courses with QuickStart contain pre-loaded assignments & presentations created by subject matter experts who are also experienced *WileyPLUS* users.

Interested? See–and Try *WileyPLUS* in action!
Details and Demo: www.wileyplus.com

Introducing Weather and Climate 1

In the last weeks of 1997 and the first weeks of 1998, extreme weather across the globe killed around 2100 people and caused $33 billion worth of property damage. Forest fires raged in Sumatra and Borneo, while large portions of Australia and the East Indies were plunged into drought. Torrential rains drenched Peru and Ecuador, and ice storms left four million people in Quebec and the northeastern United States stranded without power. This devastation was partly the result of climate changes induced by changes in the tropical Pacific known as El Niño.

An El Niño occurs every 3–7 years and lasts for 6–12 months. It was given the Spanish name El Niño (meaning "the little boy" or "Christ Child") by Peruvian fishermen whose anchovy harvests plummeted around Christmas. Under normal conditions in the tropical Pacific region, strong winds blow westward, "piling up" very warm ocean water in the western Pacific while cool, bottom water rises to the surface along the South American coast. During an El Niño, however, the westward winds weaken, and warm waters spread eastward across the Pacific. The subsequent changes in the overlying atmosphere set in motion by an El Niño can induce changes in weather around the globe, from droughts in Australia and Brazil to flooding in California and Chile. This series of changes is one of the most striking examples of the interaction between weather and climate and how the two can affect our lives here on the Earth.

Global Locator

As the El Niño of 2006–2007 began, fires similar to those of 1997–1998 swept across the vast Indonesian island of Borneo.

NATIONAL GEOGRAPHIC

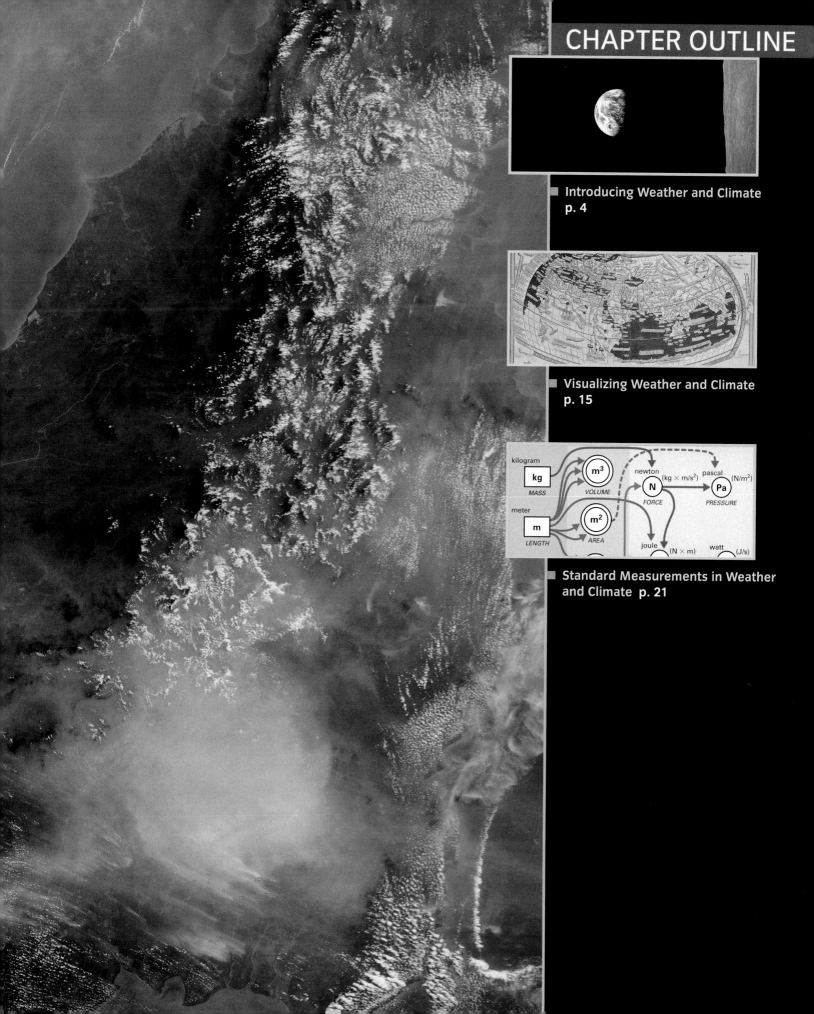

Introducing Weather and Climate

Define weather.

Define climate.

Explain different ways that weather and climate are related.

o gain a perspective of the topics covered in this text, it might be helpful to take a step back—way back. Imagine that you are an astronaut on a lunar base station in 2050 looking back toward the Earth (**FIGURE 1.1**). What do you see? From your lunar viewpoint, the first thing you notice is that the right-hand edge of the disk is in shadow, and you see the dramatic contrast between the day and night portions of the globe. The land and water masses of the illuminated side bask in the sunlight, and here the temperature of this portion of the Earth is slowly increasing. On the shadowed side of the planet, however, the heat of the oceans, atmosphere, and continents is continuously being lost—radiated into outer space both day and night. Without the offsetting heat provided by sunlight, the Earth's surface cools.

Over the next few hours, you notice the planet's slow rotation. The continent of Africa, seen in the figure, glides toward the edge of darkness, then disappears. Eventually South America emerges at the left side of the disk and slowly moves toward the center. Africa and South America receive the lion's share of the sunlight striking the Earth. These continents lie near the Equator, where they meet the strong, near-vertical rays of the Sun. North America and Europe, on the other hand, are shortchanged because they lie farther toward the pole and intercept the Sun's rays at a low angle. At the lower rim of the Earth's disk, the stark, white continent of Antarctica rotates slowly but remains sunlit, experiencing 24-hour daylight.

As the Earth days go by, you often gaze at your planet, marking the changes. Away from the Equator, huge swirls of clouds form, dissolve, and re-form in its atmosphere, making their way eastward across the disk. First one portion of a continent, and then another, is obscured. These swirls mark the passage of weather fronts and storms that are moved eastward by persistent winds high above the surface.

The Earth from space FIGURE 1.1

This image shows the Earth as seen from the surface of the Moon. White regions represent clouds and snow. Blue regions are oceans. The reddish-brown areas are the vast deserts of Africa. In this photo, only the left portion of the Earth receives sunlight; the right side of the Earth is experiencing night.

Near the Earth's Equator in Africa, you notice a band of patchy, persistent cloudiness bracketed north and south by reddish-brown areas of the Earth that are normally clear. The clouds result when warm, moist air is heated by excess solar energy and rises. As the air rises, it cools, and the moisture it contains condenses, forming clouds and producing rain. The lush, green landscape that is occasionally visible underneath the cloud belt seems to thrive on the warm temperature and abundant rainfall it receives. The reddish-brown areas are vast deserts. The air that rises over the Equator begins to descend in these regions, inhibiting cloud formation and rainfall. As the air descends toward the desert, it also warms. Showing the colors of rock and soil, the hot, dry deserts are barren of plants.

If you were to stay on the lunar base another six months, you would watch the slow changes of the planet with the seasons. By June, the Earth has changed its position with respect to the Sun. Antarctica has moved below the southern rim. The sunlit northern rim of the disk now includes northern Canada, Siberia, and the Arctic Ocean. The storm systems you saw buffeting the coastal regions of the United States and Europe have weakened and moved north. Now, however, a new set of storms, generated in the lower latitudes off western Africa, slowly migrate westward, occasionally strengthening as they go. A few of these may eventually grow into terrifying vortices of spiraling air hundreds of kilometers across—hurricanes.

In addition, during this time, a green wave of vegetation sweeps northward, up and across North America, Europe, and eastern Asia, following the warming temperatures of spring. The band of tropical cloudiness also moves northward, bringing along a green wave in Africa. Clearly, each region of the Earth has its unique climate, responding to the rhythm of the seasons in different ways.

What you have witnessed are the day-to-day and month-to-month changes in the circulation of the atmosphere across the globe. It is important to remember that the atmosphere is only one of the four spheres of the Earth, which include: the atmosphere, the lithosphere, the hydrosphere, and the biosphere (FIGURE 1.2).

The Earth's spheres FIGURE 1.2

There are four great realms, or spheres of the Earth: A the atmosphere, B the lithosphere, C the hydrosphere, and D the biosphere.

The **atmosphere** is a gaseous layer that surrounds the Earth. It receives heat and moisture from the surface and redistributes them, returning some heat and all the moisture to the surface. The atmosphere also supplies vital elements—carbon, hydrogen, oxygen, and nitrogen—that are needed to sustain life-forms.

The outermost solid layer of the Earth, or **lithosphere**, provides the platform for most earthly life-forms. The solid rock of the lithosphere bears a shallow layer of soil in which nutrient elements become available to organisms. The surface of the lithosphere is sculpted into landforms. These features—such as mountains, hills, and plains—provide varied habitats for plants, animals, and humans and influence local weather.

atmosphere Layers of gases surrounding the Earth and bound to it by the Earth's gravity.

lithosphere The solid portion of the Earth's surface, extending 100 km deep and comprising the ocean basins and continents.

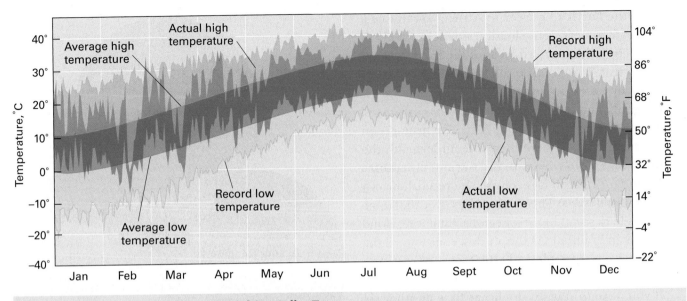

Time cycles of temperature at Wichita Falls, Texas FIGURE 1.3

This plot shows the daily temperatures at Wichita Falls, Texas, during 2006.

The liquid realm of the Earth is the **hydrosphere**, which is principally the mass of water in the world's oceans. It also includes solid ice in mountain and continental glaciers, which, like liquid water, is subject to flow under the influence of gravity. Within the atmosphere, water occurs as gaseous vapor, as well as liquid droplets and solid ice crystals that fall to the earth as *precipitation*. In the lithosphere, water is found in the uppermost layers in soils and in ground-water reservoirs.

The **biosphere** encompasses all living organisms of the Earth. Life-forms on the Earth utilize the gases of the atmosphere, the water of the hydrosphere, and the nutrients of the lithosphere, and so the biosphere is dependent on all three of the other great realms.

While our emphasis in this text will be on the atmosphere, we will often refer to the other three realms that interact with it.

You also may have noticed that many of the changes in the atmosphere exhibit different *time cycles*—rhythms in which processes change in a regular and repeatable fashion. For example, the annual revo-

■ **hydrosphere** Total water realm of the Earth's surface, including the oceans, surface waters of the lands, ground water, and water held in the atmosphere.

■ **biosphere** The network of all living organisms found on the Earth. Also that portion of the Earth in which life can exist.

lution of the Earth around the Sun, coupled with the tilt of the Earth's axis of rotation, generates a time cycle of incoming solar energy flow. We speak of this cycle as the *rhythm of the seasons*. The rotation of the Earth on its axis sets up the night-and-day cycle of darkness and light. The Moon, in its monthly orbit around the Earth, sets up its own time cycle. We see the lunar cycle in the timing and range of tides, with higher high tides and lower low tides ("spring tides") occurring both at full moon and at new moon.

The astronomical time cycles of Earth rotation and solar revolution—which are reflected in local temperatures, for instance (FIGURE 1.3)—will appear at several places throughout this text. Other time cycles with durations of a few years describe changes in the global interactions of the ocean with the atmosphere, which can have profound impacts upon temperatures and rainfall ranging from California to Australia. Still others, with durations of tens to hundreds of thousands of years, describe the alternate growth and shrinkage of the great ice sheets.

These time cycles (or *time scales*) also allow us to differentiate processes that affect the *weather* from those that affect *climate*. In general, we can think of weather as the changes that occur over time cycles of minutes to hours to days. Climate, on the other hand, is related to changes that occur over time cycles from months to years to millennia. While we will see that changes in weather and changes in climate can influence each other, for right now let us consider them separately.

WEATHER

When we looked down on the Earth from space, we identified two types of weather phenomena. In the midlatitudes, we saw inspiraling bands of clouds and clear skies—called *midlatitude* (or extratropical) *cyclones*; in other regions, the spiral direction is reversed and we find *midlatitude anticyclones*. In the lower latitudes, we saw smaller, more concentrated spirals of clouds and winds—called *tropical cyclones*. These are only two of the many weather phenomena you will encounter in this text.

What is weather? **Weather** is the state of the atmosphere at a particular place and time, typically described by temperature, moisture, wind, and precipitation conditions. Weather can change from minute to minute, but we usually speak of weather as the prevailing conditions of a day or two.

Another important concept is that of *weather systems*—recurring atmospheric circulation patterns and their associated weather changes. While any given change in weather is the result of a unique combination of interacting large-scale and small-scale processes, in this text we focus on weather systems, which produce similar types of weather changes. A few examples of the different types of weather systems you will encounter in this text (see **FIGURE 1.4**, on pages 8–9) include:

- *Midlatitude cyclones and anticyclones.* Large (hundreds of kilometers) inspirals and outspirals of surface air that repeatedly form, intensify, and

weather The state of the atmosphere at a particular location and time, usually determined by temperature, moisture, precipitation, and winds.

climate The average or prevailing weather for a given region, characterized by temperature, moisture, precipitation, and winds.

dissolve as they move slowly across the mid- and high-latitude regions of the globe

- *Fronts.* The boundary along which two large bodies of air with very different temperature and moisture characteristics interact, producing lifting and cooling of air leading to cloud formation and precipitation

- *Tropical cyclones.* Vast (approximately 100 km) inspiraling systems of very high winds and heavy rainfall that develop over very warm tropical oceans and move westward in lower latitudes, affecting the eastern coasts of most continents

- *Thunderstorms.* Storms spanning a few kilometers to a hundred kilometers in which intense vertical circulations develop, resulting in heavy precipitation, lightning and thunder, and sometimes severe winds

- *Tornadoes.* Small (from 100 m up to 2 km), intense vortices around which air spirals at speeds capable of causing total destruction of any structures and life-forms in their paths

CLIMATE

We use the term *weather* to describe how the state of the atmosphere changes from one hour to the next and from one day to the next. What about on longer time cycles—for instance, from one year to the next? In these cases, we refer to the state of the atmosphere in a given region as the **climate** of the region. While we can describe the state of the atmosphere using the same quantities—temperature, moisture, wind, and precipitation—there is one important difference. When we talk about what the climate of a region is like, we talk about its *average* characteristics and how these characteristics change over time. The average characteristics, however, are simply the average of all the different weather events that occurred at that location on a particular day (or month or even year).

Weather systems are recurring atmospheric patterns that produce similar types of weather with each occurrence. Shown here are some that we will cover in this book.

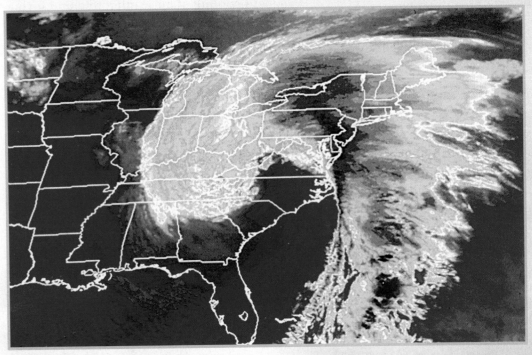

▲ Midlatitude cyclones

This satellite image shows a storm that struck the eastern U.S. in March of 1993. Note the characteristic "comma" shape and distinct regions of cloudy (yellow/black) and clear (blue) skies.

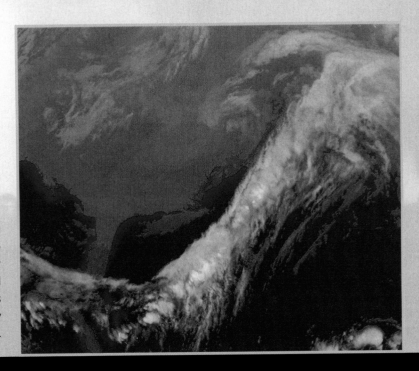

Fronts ▶

This satellite image shows a cold front pushing out across the Atlantic Ocean. The front represents a boundary between very cold, dry air to the west and warmer air to the east. It is marked by the sharp boundary between clear and cloudy skies.

NATIONAL GEOGRAPHIC

◀ Tropical cyclones
Hurricane Floyd sits poised off the coast of Florida. With winds peaking at 250 km/h (150 mph), it traveled parallel to the coast before making landfall in North Carolina on September 16, 1999.

Thunderstorms ▶
A massive thunderstorm threatens a house in Connecticut. These storms can bring heavy rain, high winds, lightning, and tornadoes.

◀ Tornadoes
An archetypical tornado descends from the base of a thunderstorm. The rapidly spinning winds kick up dust, seen at the bottom of the tornado. Although benign looking, the winds surrounding this tornado can be faster than 90 m/s (200 mph).

Thus when we talk about the climate of a region, we are really talking about what the average weather conditions are like in that region. For example, if we ask what the temperature in Juneau, Alaska, is in June, we are talking about the average temperature in June, measured over many years (usually around 30). On any given day in June in any given year, however, the actual temperature can be very different from the average temperature, as described in *What a Scientist Sees: Actual Temperatures and Temperature Departures.*

Actual Temperatures and Temperature Departures

Scientists install and record data from weather instruments, such as thermometers, in specific locations, such as this one in Juneau, Alaska. These thermometers will record the daily temperatures for a number of years before the average temperature of the location can be determined. Each actual measurement tells us something about the temperature on a given day. However, if scientists want to know whether it is colder or warmer than normal, they look at temperature departures.

To calculate a departure from normal—also called an *anomaly*—a scientist will subtract the average value from the actual value. This allows the scientist to identify which days (or years) were much warmer than normal and which were much cooler than normal. As seen here, on most days the actual temperature does not match the average temperature.

Weather station
A shelter containing thermometers and other weather instruments is installed in Juneau, Alaska. These instruments take measurements over many days and years, allowing scientists to determine the average temperature for a given day, as well as the temperature departures on a specific day.

Actual temperatures
The measured temperatures are shown by the colored line. The average temperatures for the given day are provided by the smooth line. When temperatures are higher than average, the curve is red; when they are lower, the curve is blue.

Temperature departures
The departures of measured temperatures from average temperatures are shown by the colored bars. When the value is close to 0, it means the temperature for that day was close to the average.

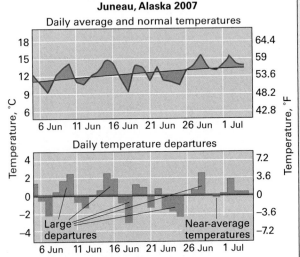

Juneau, Alaska 2007

Daily average and normal temperatures

Daily temperature departures

Large departures

Near-average temperatures

NATIONAL GEOGRAPHIC

Wheat fields FIGURE 1.5

Wheat prefers to grow in locations with relatively little moisture, warm conditions, and plenty of sunshine.

Why, then, do we concern ourselves with climate if there is no guarantee that the temperature (or rainfall or winds) on a given day will be the same as the average conditions? It is because many of the processes we as humans care about operate according to the climate of a region.

Consider a field of wheat in the plains of Kansas (FIGURE 1.5). Why does wheat grow well here? It is because of the climate. Wheat requires ample moisture in the spring but needs less as it matures and sets its seed—the grain that we harvest for food. Wheat also needs relatively warm conditions, but not too hot, and plenty of sunshine for photosynthesis and growth. It turns out that the Great Plains of the United States have just this type of climate, with cold, dry winters, but adequate rainfall and sunshine arriving in the summer.

Similarly, when engineers build an airport, they need to consider the average speed and direction of the wind so they can align the runways to enable airplanes to land and take off into the wind most frequently. When architects design a new building, they need to know the *heating degree days* and the *cooling degree days* for the region, because these indicate how often the heating and air conditioning in the building will need to be run. In the United States, both of these are determined by the average temperatures and how frequently they fall below or rise above 65°F.

Thus, while the weather can dictate how you dress on a particular day or whether you take an umbrella with you when you go out, the climate can dictate the types of clothes you have in your closet and whether your apartment has an air conditioner, a heater, or both.

Just as weather can vary over time, so can the climate of a given location. However, we do not refer to day-to-day climate changes. Instead, we refer to climate changes over the course of months, years, decades, and

El Niños

This satellite image shows the changes in precipitation associated with an El Niño—a change in sea-surface temperatures over the equatorial Pacific. Note that the changes in precipitation stretch eastward from the Pacific into South America and the Atlantic and westward into the Indian Ocean. They also extend northward and southward, producing changes in rainfall around the world.

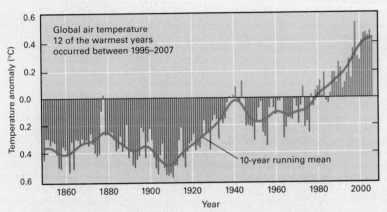

-1.2   -0.9   -0.6   -0.3   0.0   0.3   0.6   0.9   1.2

Dry   Precipitation anomaly   Wet

▲ **Seasons**

The annual bloom of cherry blossoms in Washington, D.C., heralds the arrival of spring, a period of warming temperatures and longer days that is part of the annual cycle.

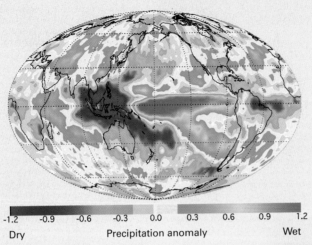

Global air temperature
12 of the warmest years occurred between 1995–2007

10-year running mean

▲ **Global Warming**

This graph shows that global average temperatures have been increasing since the early 1900s. This rise is coincident with increased emissions of heat-trapping gases like carbon dioxide and methane following the industrial revolution.

▲ **Ice Ages**

A massive glacier (the white area) retreats to the upper-right portion of this photo. Its initial location is marked by the edge of the rounded, gray landforms, which are silt and debris deposited by the glacier as it retreated. During the last glacial period, continental ice sheets covered much of the high- and midlatitude regions of the globe, then melted away quickly starting about 15,000 years ago.

Climate-change processes FIGURE 1.6

Climate can change because of many different processes. Here are some we will cover in this book.

even thousands of years. A few examples of the different climate-change processes that you will encounter in this text (**FIGURE 1.6**) include:

- *Seasons.* Cyclical change in the temperature and moisture characteristics of a given region associated with the orientation of the Earth during its annual passage around the Sun

- *El Niños.* Global-scale changes in winds, temperatures, and rainfall induced by the interaction of the atmosphere with the ocean across the tropical Pacific

- *Global warming.* Commonly used to refer to recent warming of the global-average temperature over the last 100 years, which coincides with significant increases in atmospheric concentrations of heat-trapping gases arising from humans' activities here on the Earth. Looking back farther, the global-average temperatures have both warmed and cooled on time-scales of decades to centuries.

- *Ice ages.* Very slow (over tens of thousands of years) changes in the global average temperature accompanied by the spread and retreat of massive glaciers of ice across the high- and midlatitude regions of the globe

WEATHER AND CLIMATE

So far we have seen that weather represents the day-to-day changes in the state of the atmosphere in a given region, and climate represents the *average* state of the atmosphere in a given region over a given time period. Are there any other connections between weather and climate? In fact there are quite a few. Climate really represents the average weather found in a given region, whether it is for a particular day of the year, a particular month, or even a particular century. Thus when we talk about a change in climate for a particular region, we are really referring to a change in the weather that affects that region.

As an example, let us consider a climate change you may be familiar with—namely, the change of seasons. Specifically, let us look at the change in weather that accompanies the change in seasons in Central Park in New York City (FIGURE 1.7). The first change in weather comes with the change in temperature—it is much colder in New York City in winter than in summer. While the change in temperatures is predominantly related to a decrease in the amount of *insolation*—incoming solar radiation—New York City re-

ceives during winter compared with summer, there are also weather-related reasons for it. During winter, New York City receives wave after wave of cold, dry air coming down from the high latitudes of Canada during *cold-air outbreaks*. These cold-air outbreaks, which can last from a few days to a week, are weather events whose occurrence dramatically drops the average temperature of New York City. Similarly, during summer, New York City receives warm, moist air coming up from the Atlantic Ocean near Florida. While this air brings hot, muggy weather to New York City for just a few days at a time, the constant influence of this air raises New York City's summer temperatures.

What are some other influences of weather on the climate in this region? For starters, the same change in prevailing *air masses*—cold, dry air in winter and warm, moist air in summer—changes the type of precipitation New York City receives. In winter, the drier air coming from Canada cannot support much precipitation, but when precipitation does occur, it occurs as snowfall. In summer, with lots of moisture in the air arriving from the south, precipitation can be much more intense and usually falls as rain. As a result, winters in New York City tend to be drier than summers.

Winter and summer in New York City FIGURE 1.7

The types of weather that a region such as New York City experiences change with the seasons. The result is a very different climate in winter than in summer.

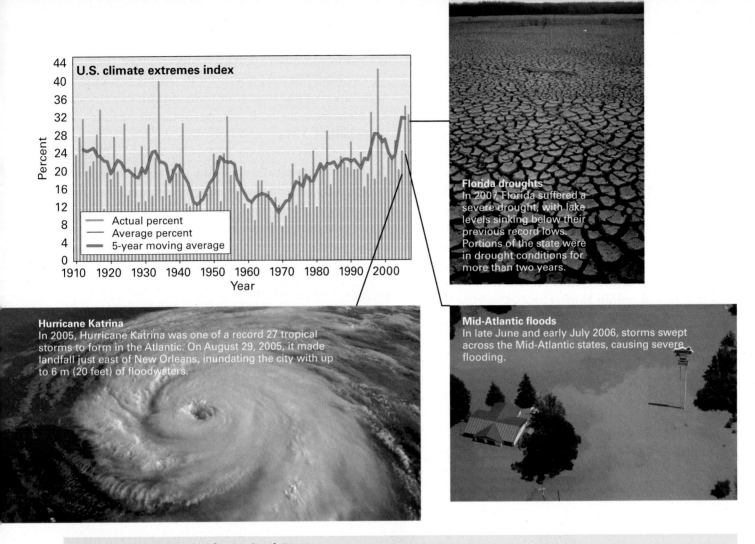

U.S. climate extremes index

Legend:
- Actual percent
- Average percent
- 5-year moving average

(Percent vs. Year, 1910–2000)

Florida droughts
In 2007, Florida suffered a severe drought, with lake levels sinking below their previous record lows. Portions of the state were in drought conditions for more than two years.

Hurricane Katrina
In 2005, Hurricane Katrina was one of a record 27 tropical storms to form in the Atlantic. On August 29, 2005, it made landfall just east of New Orleans, inundating the city with up to 6 m (20 feet) of floodwaters.

Mid-Atlantic floods
In late June and early July 2006, storms swept across the Mid-Atlantic states, causing severe flooding.

Climate extremes in the United States FIGURE 1.8

Extreme weather can change from year to year, as well as over longer time periods. This index shows changes in the number of extreme weather events—related to high and low temperatures, rainfall amounts, drought severity, and hurricane strength—affecting the United States during a given year.

Not only do changes in weather affect climate, changes in climate can also bring about changes in the weather. For instance, in winter, New York City experiences the passage of midlatitude cyclones. These weather systems tend to be much stronger in winter than in summer. In addition, they tend to be found further south, over the latitude of New York City, during winter. These large-scale storm systems, along with their accompanying fronts, can produce heavy precipitation over broad regions, coating whole states in snow. In summer, however, the predominant weather systems that influence New York City are fronts and thunderstorms. They tend to produce more localized precipita-

tion, which can be very heavy at times but also last for shorter periods of time. Hence, the type and intensity of weather events that affect New York City change with the seasons.

Similarly, the severity of weather events in the United States has changed over the last 35 years (FIGURE 1.8). During this time, the number and severity of hurricanes that have affected the United States have generally increased. The severity of *droughts*—prolonged periods with little to no rainfall—have also increased. At the same time, when precipitation does occur, it tends to be more intense under today's climate conditions than it was 30 years ago. Finally, the number of extremely warm days—those with temperatures higher than 90% of all recorded values—has also increased.

While the number of extreme weather events affecting the United States during any one year can be very different from the number in the next year, the slow but relatively steady change in the characteristics of weather events affecting the country over the last 35 years may indicate a change in the basic climate of the region. It is also apparent that these severe weather events—ranging from increased droughts that struck the farming regions of the Midwest in 2006 and 2007 to more frequent hurricanes like the one that struck New Orleans in 2005 to the historic floods that swept the Mid-Atlantic states from North Carolina to New York in 2006—significantly affect the livelihood, and lives, of the residents.

Given these changes, we then need to understand the causes. This is particularly important if we humans have a role in producing climate change through our industrial, agricultural, and leisure activities. Many of these activities—which involve the burning of coal, gas, and other fossil fuels—release heat-trapping gases to the atmosphere that subsequently lead to a warming of the global temperatures. In addition, increases in these gases may lead to a change in the characteristics and severity of weather that we experience, both today and well into the future.

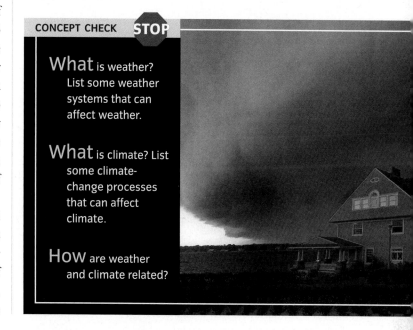

CONCEPT CHECK STOP

What is weather? List some weather systems that can affect weather.

What is climate? List some climate-change processes that can affect climate.

How are weather and climate related?

Visualizing Weather and Climate

LEARNING OBJECTIVES

Describe how graphs are made.   **Define** map projection.   **Explain** different ways that data can be presented on maps.

Scientists use a number of visualization tools to help describe and explain the phenomena they observe. For example, a satellite photo of a midlatitude cyclone makes it easy to see the structure of the cyclone and image its rotating motion. Drawings simplify and focus a picture for easier understanding. Diagrams can display relationships or help describe processes. A very important scientific visualization tool is the *graph* (also called a *plot*), which is used to display scientific data in a way that shows relationships. In addition, *maps* are used to show how phenomena are organized in space.

GRAPHS

You are probably already familiar with the use of graphs in everyday life. For example, your TV news program may show a daily graph of the Dow-Jones index, which measures the prices of certain industrial stocks. In this case, time is one set of data—or one *variable*—and the value of the Dow-Jones index is another set of data.

We can also plot the state of atmospheric variables with time. *What a Scientist Sees: Actual Temperatures and Temperature Departures* on page 10, shows a graph like this. In that case, there are actually two sets of data plotted with respect to time of year in June—the actual temperature and the average temperature.

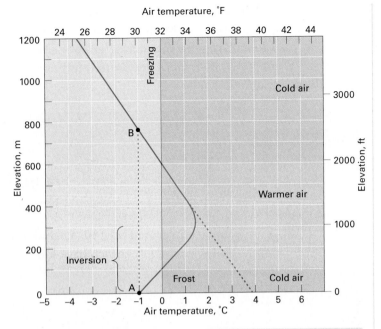

Temperature inversion FIGURE 1.9

While air temperature normally decreases with altitude (dashed line), in an inversion, temperature increases with altitude. In this example, the surface temperature is at −1°C (30°F), and temperature increases with altitude (solid line) for several hundred meters (1000 ft or so) above the ground. Temperature then decreases with altitude.

However, we do not simply have to plot data with time. We can plot any two sets of data. FIGURE 1.9 shows a graph of a *temperature inversion* in which temperature increases with height above the surface. In this case, we plot temperature as it is related to height above the surface at a particular location. Temperature is

given by the horizontal axis—the *x-axis*—while height above the ground is given by the vertical axis—the *y-axis*. Here we chose to plot height on the *y-axis* because we are used to thinking of height as a vertical direction. However, the information would be just the same if we switched around the axes. Plots can also come in different styles.

The plot shown here is called a *line graph,* because it presents the data as a continuous line. By presenting data in this way, we assume that the data vary smoothly as we move up through the atmosphere. Typically, atmospheric data are *discrete*—only available at select points in space or time. When we present these data as a line graph, we presume that there is a smooth transition of the data from one measurement to the next, again either in the intervening times or in the intervening locations.

If we cannot make this assumption, we typically plot the data as a bar graph. An example of a bar graph or bar plot is shown in FIGURE 1.10. This graph shows daily precipitation in Waco, Texas, during June 2007. Rainfall is typically intermittent with successive days of no rain followed by the passage of a storm, which brings rain for a few days. In this case, it is inappropriate to assume that the rainfall changes smoothly from one day to the next. Instead, we can only plot the data that we measure. Bar graphs maintain the discrete nature of these measurements.

Many times, however, whether one uses a bar graph or a line graph is more a matter of choice than it is based on any rigorous rules. Both types of plots will be used extensively throughout this text. In addition, other types of plots—pie charts and scatter plots are two examples—will also be used to present data.

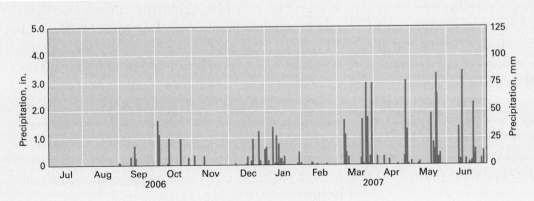

Precipitation in Waco, Texas FIGURE 1.10

This bar graph shows precipitation amounts for all days from July 2006 through June 2007. The data record the amount falling on a given day, which is often quite different from the amount falling the day before or the day after.

Mercator projection   In the **Merca-tor projection**, the meridians form a rectangular grid of straight vertical lines, while the parallels form straight horizontal lines (FIGURE 1.12A). The meridians are evenly spaced, but the spacing between parallels increases at higher latitude so that the spacing at 60° is double that at the Equator. As the map reaches closer to the poles, the spacing increases so much that the map must be cut off at some arbitrary parallel, such as 80° N. This change of scale artificially enlarges features near the poles relative to features at low latitudes.

Mercator projection Map projection of horizontal parallels and vertical meridians, with the space between parallels increasing poleward.

The Mercator projection has several special properties. Mercator's goal was to create a map that sailors could use to determine their course. A straight line drawn anywhere on his map gives you a line of constant compass direction. A navigator can simply draw a line between any two points on the map and measure the bearing, or direction angle of the line, with respect to a nearby meridian on the map. Because the meridian is a true north–south line, the angle will give the compass bearing to be followed. Once aimed in that compass direction, a ship or an airplane can be held to the same compass bearing to reach the final point or destination.

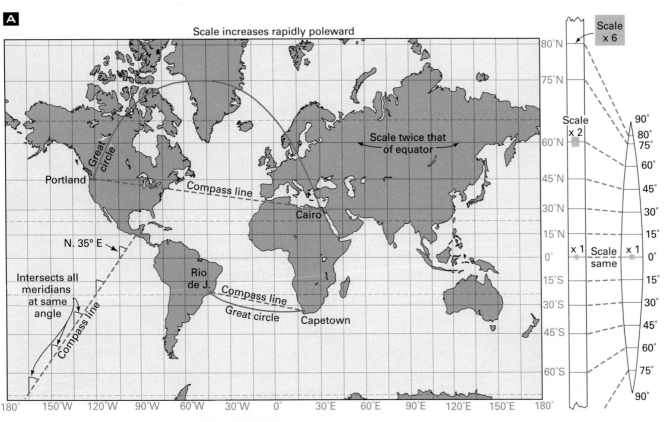

A

Scale increases rapidly poleward

The compass line connecting two locations, such as Portland and Cairo, shows the compass bearing of a course directly connecting them. However, the shortest distance between them lies on a great circle, which is the longer, curving line on this map projection.

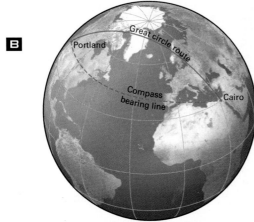

B

The true shortest distance between two points, drawn over the globe as the crow flies, will appear as a curved line on the Mercator projection.

The Mercator projection

FIGURE 1.12

Gerardus Mercator invented his navigator's map in 1569. It is a classic that has never gone out of style.

MAPS

Another way to present data is through the use of maps. The problem of just how to display the Earth, and data at various locations on the Earth, has puzzled *cartographers*, or mapmakers, throughout history (**FIGURE 1.11**). The oldest maps were limited by a lack of knowledge of the world rather than by difficulties caused by the Earth's curvature. They tended to represent political or religious views rather than geographic reality—ancient Greek maps from the 6th century B.C. show the world as an island, with Greece at its center, while medieval maps from the 14th century placed Jerusalem at the focus.

However, by the 15th century, ocean-faring explorers such as Columbus and Magellan were extending the reaches of the known world. These voyagers took mapmakers with them to record the new lands that they discovered, and navigation charts were highly valued. Mapmakers, who now had a great deal of information about the world to set down, were forced to tackle the difficulty of representing the curved surface of the Earth on a flat page.

For any map to be useful, we must know how position on the map relates to position on the Earth's three-dimensional surface. To establish this relation, we divide the globe into what is known as the *geographic grid*. This grid is made up of a system of imaginary circles, called parallels and meridians. Parallels are east-west lines running parallel to the Equator and are used to designate the *latitude* of a location. We define the Equator—the longest parallel—to be at 0° latitude with the North Pole at 90° north latitude (designated 90° N) and the South Pole at 90° south latitude (designated 90° S).

Meridians are north-south lines that join the North and South Poles and are used to designate the *longitude* of a location. Longitude is measured by angular distance from the prime meridian, which passes through Greenwich, England, and marks 0° longitude. Locations to the east of Greenwich have longitudes ranging from 0°–180° east longitude (designated 0°–180° E) and locations to the west of Greenwich have longitudes ranging from 0°–180° west longitude (designated 0°–180° W). Technically, 180° E is at the same longitude as 180° W, and is halfway around the world from Greenwich, England.

However, even with this geographic grid, we still cannot project a curved surface onto a flat sheet without some distortion. One of the earliest attempts to tackle the curvature problem for global-scale maps was made by the Belgian cartographer, Gerardus Mercator, in the 16th century, and it is still used today. A number of other systems, or **map projections**, have been developed to translate the curved surface of the Earth to a flat one. We will concentrate on the two most useful types, including Mercator's. Both have their own advantages and drawbacks.

> ### map projection
> A system of parallels and meridians representing the curvature of the Earth drawn on a flat surface.

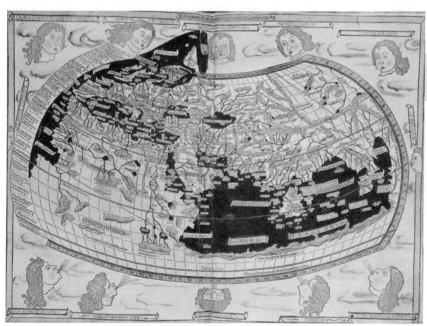

Ptolemy's map of the world
FIGURE 1.11

This atlas page shows a reproduction of a map of the world as it was known in ancient Greece.

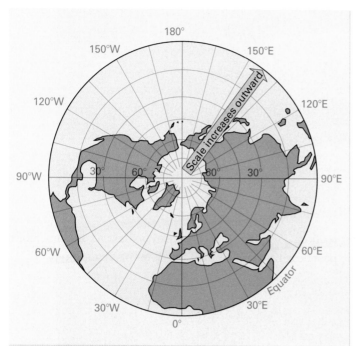

A polar projection FIGURE 1.13

This map centers on the North Pole. All longitude lines are straight lines radiating from the center point, and all latitude lines are concentric circles. The scale fraction increases in an outward direction, making shapes toward the edges of the map appear larger. Because the intersections of the parallels with the meridians always form true right angles, this projection shows the true shapes of all small areas.

However, this line does not necessarily follow the shortest actual distance between two points, which we can easily plot out on a globe (FIGURE 1.12B). We have to be careful—Mercator's map can make the shortest distance between two points appear much longer than the compass line joining them.

Because the Mercator projection shows the true compass direction of any straight line on the map, it is used to show many types of straight-line features. These include wind and ocean current flow lines and lines of equal values, such as lines of equal air temperature or equal air pressure. Thus, the Mercator projection is often chosen for maps of temperatures, winds, and pressures.

polar projection
Map projection centered on the Earth's North or South Pole.

Polar projection
Another map projection used throughout this text is the **polar projection**.

The polar projection (FIGURE 1.13) normally centers on either the North or the South Pole, and it is essential for visualizing weather maps of the polar regions. The map is usually cut off at lower latitudes to show only one hemisphere with the Equator forming the outer edge of the map.

In addition to deciding upon a map projection, it is necessary to decide how to present data given this projection. Many displays of data presented in map format are based on the concept of presenting data with equal values. A simple example is a topographic map, shown in FIGURE 1.14. Maps of this sort show a set of data plotted based on the values at various geographic locations. Here, the data are the elevations of the location above sea level. Running through regions with equal elevations are **isolines**—lines that connect data points

isolines Lines on a map that connect locations with equal values of a given variable.

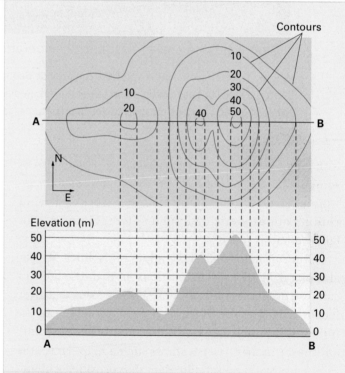

Contour map of topography FIGURE 1.14

This map shows lines, or contours, representing regions with constant elevation above sea level. The peak on the right-hand side is 55 m (181.5 ft) above sea level. As you move away from the peak, the number labeling each successive line decreases, indicating a decrease in elevation.

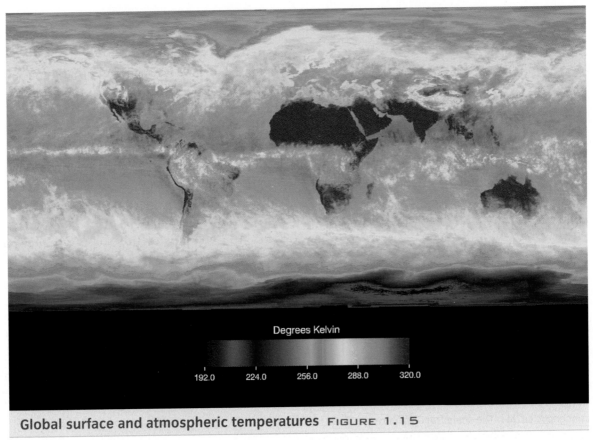

Global surface and atmospheric temperatures FIGURE 1.15

Because of their global view, satellites can provide data at locations all across the globe. Shown here is a satellite-based estimate of temperatures. Reds and oranges indicate warmer temperatures, while blues and purples indicate cooler temperatures, as shown by the color bar at the bottom.

with equal values. Isolines are also called *contours*, so we sometimes refer to these types of maps as *contour maps*. Isolines could also be drawn for other types of data, including atmospheric data such as temperatures, rainfall, even wind speed.

Contour maps only show values on isolines. We can also plot data values at all locations. In that case, we typically use different colors to represent the value at a given location. We can use as few as two colors, or we can use millions, in which case one color blends smoothly into another. FIGURE 1.15 shows such a map. Here, the values are based on satellite data that provide estimates of temperatures around the world. We can identify regions that have extremely high temperatures—for example, in the desert regions of Africa, the Middle East, and Australia. These are color-coded red and orange. As we move to higher latitudes, colors blend toward yellows and oranges, indicating cooler temperatures. In the very

highest latitudes of the northern and southern hemispheres, temperatures are extremely cold, indicated by the blue and purple colors.

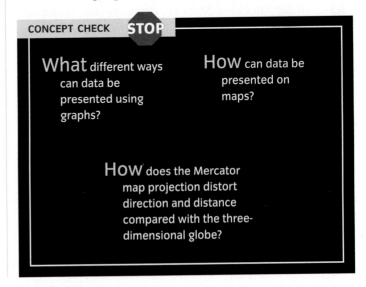

CONCEPT CHECK **STOP**

What different ways can data be presented using graphs?

How can data be presented on maps?

How does the Mercator map projection distort direction and distance compared with the three-dimensional globe?

Standard Measurements in Weather and Climate

To describe the atmosphere in which we live, we need to use a consistent set of measurement units. In studies of weather and climate, we use the metric units of the SI (Système International), which is the common scientific standard. This system uses meters, kilograms, and seconds as the three basic variables of distance, mass, and time (**FIGURE 1.16**).

For temperature, the SI system uses the basic unit of the kelvin (K) and recognizes degrees Celsius (°C) as a derived unit. Kelvin units are very similar to Celsius units in that a one-degree change in kelvins is the same as a one-degree change in Celsius degrees (which is a 9/5th degree change in Fahrenheit). However, the zero value is different for the two. Whereas in Celsius, zero degrees represents the temperature at which water freezes, zero kelvins, or *absolute zero*, represents the temperature at which all molecular motion ceases—it is the hypothetical (and unobtainable) coldest temperature any body can have. Presently, the lowest temperature ever measured is one half-billionth of a degree above zero K.

Why does the SI system use kelvin instead of degrees Celsius? It is principally because the Celsius scale allows for negative temperatures—in fact, any object with a temperature below the freezing point of water will have a negative value. Objects with these temperatures still have energy associated with molecular motions inside of them. Because technically **temperature** measures the amount of energy found in these molecular motions, having negative temperatures does not make physical sense. Instead, we measure the temperature associated with this molecular energy using kelvins. The other reason for converting from °C to K is that many of the mathematical formulas in the text require that the temperature be in K; putting in values with units °C will produce the wrong answer.

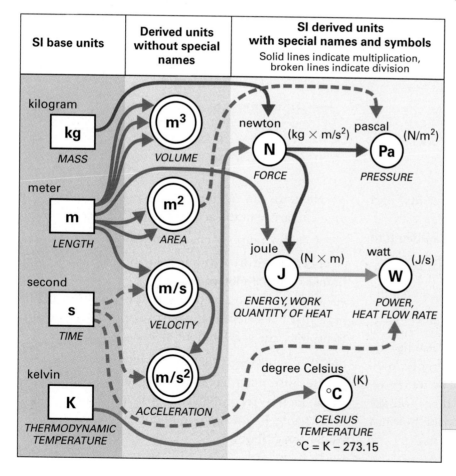

SI base units	Derived units without special names	SI derived units with special names and symbols

Solid lines indicate multiplication, broken lines indicate division

kilogram **kg** MASS

meter **m** LENGTH

second **s** TIME

kelvin **K** THERMODYNAMIC TEMPERATURE

m³ VOLUME

m² AREA

m/s VELOCITY

m/s² ACCELERATION

newton **N** (kg × m/s²) FORCE

pascal **Pa** (N/m²) PRESSURE

joule **J** (N × m) ENERGY, WORK QUANTITY OF HEAT

watt **W** (J/s) POWER, HEAT FLOW RATE

degree Celsius **°C** (K) CELSIUS TEMPERATURE °C = K − 273.15

temperature
A measure of the molecular energy within a given substance.

Metric units and their physical significance FIGURE 1.16

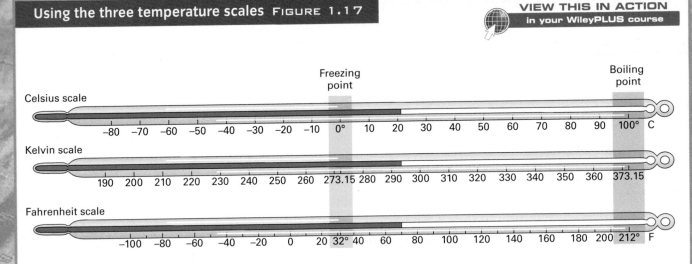

Using the three temperature scales FIGURE 1.17

VIEW THIS IN ACTION
in your WileyPLUS course

Celsius scale The basic unit of temperature used in this text is degrees Celsius, which is the metric unit for temperature and is abbreviated °C. The units are scaled so that 0°C is the temperature at which water freezes and 100°C is the temperature at which water boils, when at sea level. Sometimes this scale is also referred to as the centigrade scale.

Kelvin scale A second temperature scale is based on kelvin units, abbreviated K (there is no degree symbol "°" used with kelvin units; the name itself is in lower case while the abbreviation is in upper case). Kelvin units differ from Celsius units only by a constant factor of 273.15. Hence, the freezing point of water is 0°C = 273.15 K and the boiling point of water is 100°C = 373.15 K. A one-degree change in °C is the same as a one-degree change in K. To convert from °C to K, use the following equation: K = °C + 273.15.

Fahrenheit scale A third temperature scale is based upon the English units of Fahrenheit, which is abbreviated °F. Water freezes at 32°F and boils at 212°F. Notice that the difference between the freezing point and boiling point of water is greater in °F (212 − 32 = 180°F) compared with °C (100 − 0 = 100°C). Thus a change of 1°C is the same as a change of 180/100°F or 9/5°F = 1.8°F. To convert from temperatures in °C to °F, use the following equation: °F = (9/5 × °C) + 32. To convert from °F to °C, use the following equation: °C = 5/9 × (°F − 32).

The relationship between kelvin, Celsius, and Fahrenheit scales is discussed in FIGURE 1.17.

Now that we have the basic units, we can begin to define other physical quantities. The first is *velocity*, which represents the change in position over some period of time.

velocity = change in position / time

Velocity has units of meters/second. It also has the property of direction. Velocity without reference to direction is *speed*. In the atmospheric sciences, we use velocity to describe how fast and in what direction the winds are moving. We also use it to describe how fast ocean currents are moving.

Next, we can calculate the change in velocity with time and arrive at the acceleration of an object.

acceleration = change in velocity / time

Because we are dividing velocity by time, acceleration has units of meters/second^2.

Once we have acceleration, if we multiply by the mass that is accelerating, we arrive at the force needed to produce the acceleration:

force = mass × acceleration

Because mass has units of kg, force has units of kg × (meters/seconds^2). In metric units, force has another name, *newtons* (N), in honor of the famous scientist Sir Isaac Newton.

Process Diagram

While forces make objects, such as air molecules, accelerate and hence can change the speed and direction of the winds, in weather and climate studies we are usually more interested in the force applied over a given area, which is called *pressure*.

$$\text{pressure} = \text{force} / \text{area}$$

The units of pressure are kg / (meters × seconds²). As with forces, in metric units they have another name—*pascals* (Pa). However, when pressure in the atmosphere is measured, values are usually given in *millibars*. One millibar is the same as 100 pascals, so 1 millibar is the same as 1 *hectopascal* (hPa). It is becoming more common to use hectopascals for pressure measurements because they are more closely linked with the SI system on which all other atmospheric variables, such as temperature and wind speed, are based.

Throughout this text, we will discuss different measurements used to describe the atmosphere. Some typical ones we have discussed so far are *temperature, wind speed and direction,* and *pressure*. Another one that we will encounter is *specific humidity*—a measure of how much moisture is contained within a given mass of air. These make up the standard atmospheric measurements taken all over the world. We will also introduce measurements related to the amount of *energy* an object has, which is found by multiplying the force by a change in position.

$$\text{energy} = \text{force} \times \text{change in position}$$

Energy has units of N × meters, which is referred to as *joules* (J). If we then calculate how much energy is entering or leaving an object or a region over a given period of time, we find the energy flow, which is called *power*.

$$\text{power} = \text{change in energy} / \text{time}$$

Power has units of J/s and is also given the name *watts* (W).

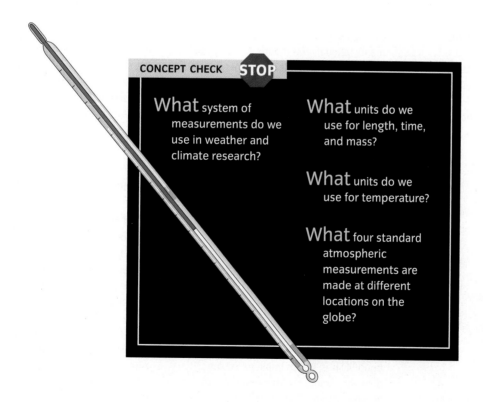

CONCEPT CHECK STOP

What system of measurements do we use in weather and climate research?

What units do we use for length, time, and mass?

What units do we use for temperature?

What four standard atmospheric measurements are made at different locations on the globe?

Globally, 2005 was considered one of the warmest years on record in the last 150. During that year, a record 27 tropical storms formed in the Atlantic, including Hurricane Katrina, which devastated New Orleans. In addition, severe droughts occurred in Australia, the central United States, and Brazil—which had its worst drought in 60 years.

This map shows temperature anomalies during 2005 compared with the 1951–1980 mean temperatures.

Which regions had the largest temperature anomalies?

Which regions had negative temperature anomalies?

Can you tell what type of map is presented here?

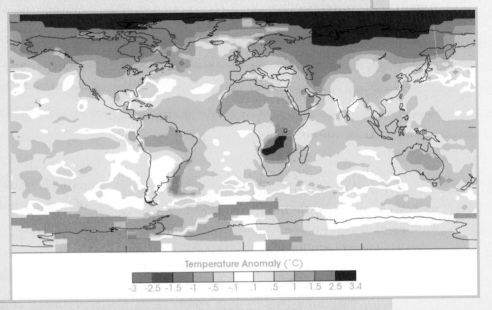

Temperature Anomaly (°C)

-3 -2.5 -1.5 -1 -.5 -.1 .1 .5 1 1.5 2.5 3.4

SUMMARY

1 Introducing Weather and Climate

1. **Weather** represents changes in the state of the **atmosphere** over the course of minutes to days. Weather systems are recurring patterns of atmospheric circulation that produce recurring changes in the weather, including changes in temperature, moisture, rainfall, or winds.

2. **Climate** describes the average state of the atmosphere in a given region. The average state can be based on measurements taken over a month, a year, even thousands of years. It represents the average of all the weather events that occurred during that time. Like weather, climate can change, again on time periods ranging from months to years to thousands of years.

3. Weather and climate are intimately related. Changes in the types of weather a given region experiences will produce changes in the climate of the region. At the same time, climate-change processes can produce changes in the type and severity of weather events that affect a given region.

2 Visualizing Weather and Climate

1. Scientific visualization tools include photos, drawings, maps, and graphs.

2. Graphs show the relation between one set of data and another. Often they present how one set of data changes with time, although any two sets of data can be plotted with respect to one another. Graphs typically are presented as continuous lines, which implies that the data vary smoothly from one point to another. At other times, data are presented as discrete values, such as with bar graphs.

3. Maps are used to present data at various locations on the Earth. Difficulties arise because representations of the curved Earth surface on a flat map, called **map projections**, can distort the size or shape of mapped objects or regions. Data can be presented by showing lines connecting regions with equal values, called **isolines**. Alternatively, each location can be color-coded based on the data value at that location.

3 Standard Measurements in Weather and Climate

1. In studies of weather and climate, the SI system of metric units is used. The basic units are meters (length), kilograms (mass), seconds (time), and Celsius or Kelvin (temperature).

2. From these basic units, we can derive other physical quantities we will encounter throughout the text, including velocity, acceleration, force, pressure, energy, and power.

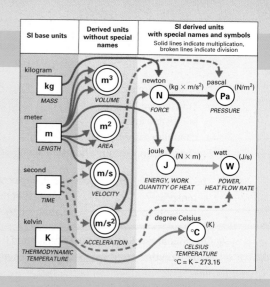

KEY TERMS

CRITICAL AND CREATIVE THINKING QUESTIONS

1. If you were to take a photo of the Earth from space, which of the weather systems might you be able to identify? Which ones could you probably not see? 1) Midlatitude Cyclones; 2) Fronts; 3) Tropical Cyclones; 4) Thunderstorms; 5) Tornadoes

2. Discuss the concept of time cycles, and give three examples that vary from short periods to very long periods of time.

3. If you measured the temperature outside, would you expect it to be the same as the average temperature for this time of year? If you took the same measurements over many consecutive days and found the average value, would you expect it to be closer to or farther away from the average temperature for this time of year?

4. What climate-change processes might you expect to witness during your lifetime? Which ones do you think take too long for you to experience?

5. If you wanted to plot the change in temperature over the course of a day, would you use a line graph or a bar graph? What type of graph would you use if you wanted to plot the instantaneous temperature at 20 different cities?

6. If an object moves 20 m in 4 s, what is its velocity? If its velocity then increases by 20 m/s over the next 5 s, what is its acceleration? If the object has a mass of 5 kg, what force is needed to produce this acceleration?

SELF-TEST

1. The Earth's atmosphere is _____.
   a. constantly changing
   b. stable and immutable
   c. changing extremely slowly
   d. changed only through human activity

2. The _____ encompasses all living organisms of the Earth.
   a. biosphere   c. hydrosphere
   b. lithosphere   d. atmosphere

3. The annual revolution of the Earth around the Sun generates a natural rhythm or _____ of incoming solar energy flow.
   a. time sequence   c. time cycle
   b. phase   d. repetition

4. On the diagram below, label the following: (1) the days of the year that had the coldest and warmest actual temperatures; (2) the days of the year that had the coldest and warmest record temperatures.

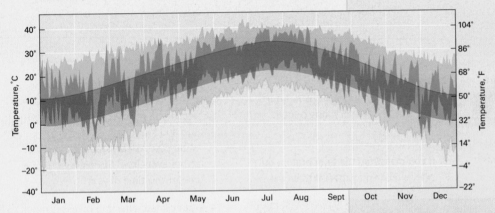

5. A _____ system generates similar changes in the atmospheric state of a given region.
   a. climate-change   c. polar
   b. topographic   d. weather

6. Climate refers to the state of the atmosphere measured over the course of _____.
   a. seconds   c. days
   b. hours   d. a month and longer

7. Which of these decisions would be based on what the weather in your location is like today?
   a. Whether to install an air conditioner
   b. Whether to plant wheat or a banana tree
   c. Whether to put on a raincoat or shorts
   d. Whether to buy flood insurance for your house

8. Which of these is *not* a climate-change process?
   a. Change of the seasons
   b. Onset of an ice age
   c. Formation of a tropical cyclone
   d. Formation of an El Niño

9. On the diagram below, label the following: (1) the 5-year period with the smallest number of extreme weather events affecting the United States; (2) the 5-year period with the largest number of extreme weather events affecting the United States; (3) the year with the smallest number of extreme weather events affecting the United States; (4) the year with the largest number of extreme weather events affecting the United States.

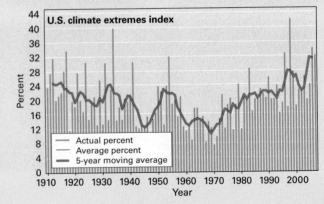

10. Which of the following would you use a bar graph to represent?
    a. Change in temperature over the course of one day
    b. Change in wind speed from one airport to the next
    c. Change in energy received from the Sun over the course of a year
    d. Change in the price of oil over a decade

11. On the diagram below, label the following: (1) the days of the month that had the coldest and warmest actual temperatures; (2) the days of the month that had the largest positive and negative departure from normal temperatures.

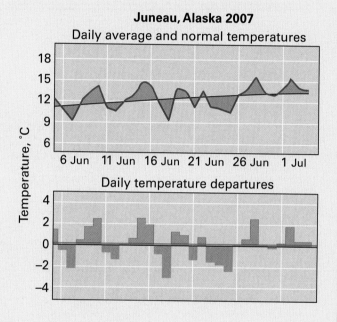

Juneau, Alaska 2007

12. The single problem all maps have in common is _____.
    a. that they are flat
    b. distortion
    c. shape projection
    d. that they cannot show the Atlantic and Pacific Oceans on the same map

13. A system for changing the geographic grid to a flat grid is a _____.
    a. map
    b. curve distortion
    c. map projection
    d. shape/area distortion

14. The _____ projection centers on the North or South Pole and displays the meridians as _____ lines.
    a. Mercator, curved
    b. Goode, straight
    c. conic, curved
    d. polar, straight

15. The _____ is the metric unit for _____.
    a. meter, temperature
    b. second, distance
    c. kilogram, mass
    d. watt, velocity

The Earth's Atmosphere

I n 1991, a volcano in the Philippines called Mount Pinatubo produced one of the most violent eruptions of the 20th century. The eruption blew off the top of the mountain, raining hundreds of cubic meters of sand, gravel, and ash down on the area surrounding the mountain. By the time it was over, several hundred people had died, and tens of thousands suffered the effects of mudslides, ash-clogged rivers, and crop devastation.

Reverberations from the eruption were also felt across the globe. The volcano blasted vast plumes of sulfur dioxide gas into the atmosphere. A haze of sulfuric acid droplets then spread throughout the stratosphere—a layer of the atmosphere 15 km above the Earth's surface— during the year following the eruption. These particles, although tiny, had a remarkable effect on the global climate. They reduced the sunlight reaching the Earth's surface by 2–3 percent, and average global temperatures fell by about 0.3°C (0.5°F) in a single year—equivalent to about half the temperature increase that has occurred through global warming over the last 50 years.

Our planet's air temperature is part of a delicate system involving interactions between the atmosphere, oceans, and the Earth. As the eruption of Mount Pinatubo showed us, this system can easily be thrown off balance. In this chapter, we discuss the chemical composition of the atmosphere, which was modified by this eruption, as well as the structure of the atmosphere, which caused the sulfuric acid to remain trapped in the stratosphere.

As Philippine farmers plow their fields, Mount Pinatubo erupts on July 8, 1991.

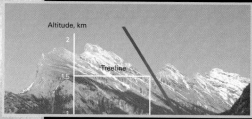

Composition of the Atmosphere

LEARNING OBJECTIVES

Describe the gases that make up the atmosphere.

Explain the vital role of ozone for life on the Earth.

Explain what aerosols are and why they are important.

The Earth is surrounded by an envelope of air—a mixture of various gases—that makes up our *atmosphere* (**FIGURE 2.1**). This envelope of air, held in place by the Earth's gravity, reaches a height of hundreds of kilometers. However, about 97 percent of the atmosphere lies within 30 km (19 mi) of the Earth's surface, which is the focus of most of our discussions in this text.

In these lower regions, pure, dry air is mostly made up of nitrogen (about 78 percent by volume) and oxygen (about 21 percent), as seen in **FIGURE 2.2**. The remaining 1 percent of dry air is mostly argon, with a very small amount of carbon dioxide (CO_2), amounting to about 0.035 percent. In the real atmosphere, we also have to account for gaseous water vapor, which can be between 0.5–2.0 percent of the atmosphere by volume. Here we examine the role some of these gases play in the atmosphere.

RADIATIVELY AND CHEMICALLY INACTIVE GASES

Nitrogen molecules in the atmosphere contain two nitrogen atoms (N_2). Nitrogen gas does not easily react with other substances, so we can think of it as a mainly neutral substance. Soil bacteria do extract very small amounts of nitrogen, which can be used by plants, but otherwise nitrogen is largely a "filler," adding inert bulk to the atmosphere. Argon is also a relatively inert chemical within the atmosphere.

The atmosphere

FIGURE 2.1

The Earth's atmosphere, shown here in profile, is a thin layer surrounding our planet. The lower layer of clouds and weather is called the *troposphere*.

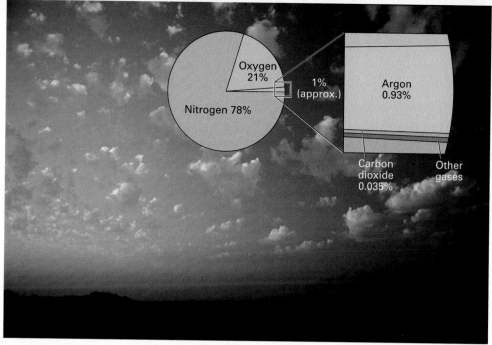

Values show percentage by volume for dry air near the surface. Nitrogen and oxygen form about 99 percent of our air. Other gases, principally argon and carbon dioxide, account for the final 1 percent. Depending upon the amount of water vapor at a given location, these percentages will be reduced slightly.

A Air is a mixture of gases dominated by nitrogen and oxygen. Water vapor, another important constituent, makes up 0.5–2.0 percent of the atmosphere. When water vapor condenses it can form clouds, such as these over Florence, Italy.

Composition of the atmosphere FIGURE 2.2

RADIATIVELY AND CHEMICALLY ACTIVE GASES

By contrast, oxygen gas (O_2) is highly chemically active, combining readily with other elements in the process of oxidation. Fuel combustion is a rapid form of oxidation, whereas certain types of rock decay (*weathering*) are very slow forms of oxidation. Living tissues require oxygen to convert food into energy. In addition, oxygen molecules can absorb very high intensity radiation coming from the Sun, thereby protecting living organisms from molecular damage that can be caused by this radiation.

Another important radiatively active component of the atmosphere is water vapor, the gaseous form of water. Individual water molecules in the form of vapor are mixed freely throughout the atmosphere, just like the other atmospheric component gases. Unlike the other component gases, however, water vapor can vary highly in concentration, both geographically and vertically. Usually, water vapor makes up less than 1 percent of the atmosphere. Under very warm, moist conditions, as much as 2 percent of the air can be water vapor. In addition, concentrations of water vapor tend to be much greater near the surface compared with higher up in the atmosphere.

Water vapor is extremely important to our understanding of weather and climate. It absorbs radiation emitted by the Earth before the radiation can escape to space; it also emits radiation back to the Earth's surface. This absorption and re-emission of radiation energy by water vapor in the atmosphere contributes to what we call the *greenhouse effect*. Because of this effect, water vapor is important for the heating of both the atmosphere and the Earth's surface.

Water in the atmosphere can also exist as liquid water and solid water. When the gaseous water vapor content is high, vapor can condense into liquid water droplets, forming low clouds and fog like those seen in FIGURE 2.2, or the vapor can be deposited as ice crystals, forming high clouds. Rain, snow, hail, and sleet—collectively termed *precipitation*—are produced when these condensation or deposition processes happen in significant amounts. The conversion from one form to another can also result in either the release or removal of immense amounts of energy from the atmosphere, which plays a major role in determining its temperature.

Although CO_2 is a very small part of our atmosphere, it is also an important contributor to the greenhouse effect and hence influences the temperature of the atmosphere and the Earth's surface. Carbon dioxide is also used by green plants, which convert it to chemical compounds used to build up their tissues, organs, and supporting structures, during photosynthesis. This process is described in FIGURE 2.3.

The carbon cycle FIGURE 2.3

Carbon moves through the cycle as a gas, as a liquid, and as a solid. In the gaseous portion of the cycle, carbon moves largely as carbon dioxide (CO_2), which is a free gas in the atmosphere and a dissolved gas in fresh and saltwater. In the sedimentary portion of its cycle, we find carbon in carbohydrate molecules in organic matter, as hydrocarbon compounds in rock (petroleum, coal), and as mineral carbonate compounds such as calcium carbonate ($CaCO_3$).

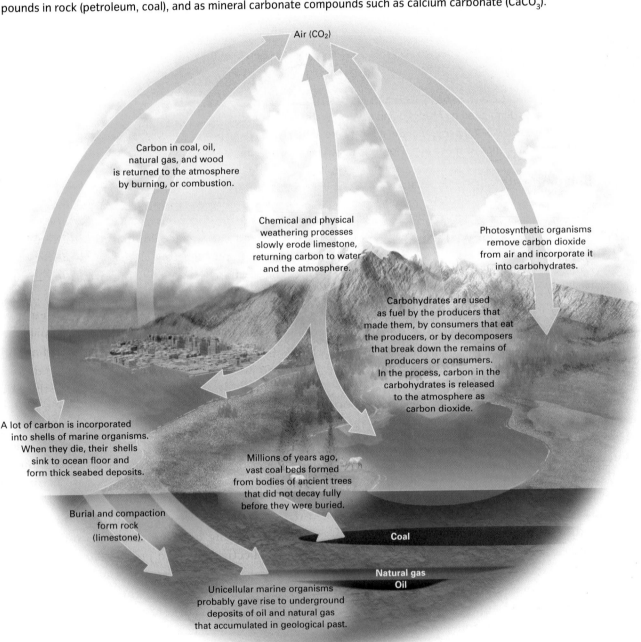

Air (CO_2)

Carbon in coal, oil, natural gas, and wood is returned to the atmosphere by burning, or combustion.

Chemical and physical weathering processes slowly erode limestone, returning carbon to water and the atmosphere.

Photosynthetic organisms remove carbon dioxide from air and incorporate it into carbohydrates.

Carbohydrates are used as fuel by the producers that made them, by consumers that eat the producers, or by decomposers that break down the remains of producers or consumers. In the process, carbon in the carbohydrates is released to the atmosphere as carbon dioxide.

A lot of carbon is incorporated into shells of marine organisms. When they die, their shells sink to ocean floor and form thick seabed deposits.

Millions of years ago, vast coal beds formed from bodies of ancient trees that did not decay fully before they were buried.

Burial and compaction form rock (limestone).

Coal

Natural gas
Oil

Unicellular marine organisms probably gave rise to underground deposits of oil and natural gas that accumulated in geological past.

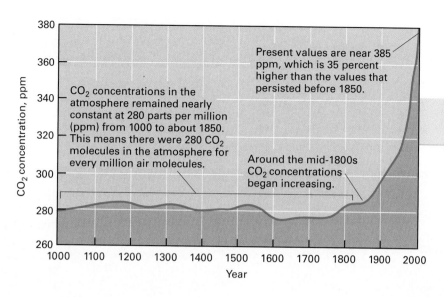

CO$_2$ concentrations in the atmosphere remained nearly constant at 280 parts per million (ppm) from 1000 to about 1850. This means there were 280 CO$_2$ molecules in the atmosphere for every million air molecules.

Present values are near 385 ppm, which is 35 percent higher than the values that persisted before 1850.

Around the mid-1800s CO$_2$ concentrations began increasing.

Changes in CO$_2$ concentrations over the last 1000 years FIGURE 2.4

Although CO$_2$ concentrations do not differ much from one location to another, they can change over time. In fact, there has been a recent increase in CO$_2$ concentrations over the last 100 years, as seen in FIGURE 2.4. This increase, which is predominantly driven by the burning of fossil fuels and deforestation, strengthens the greenhouse effect and hence increases the amount of radiation absorbed by the atmosphere and the Earth's surface.

Methane (CH$_4$) is another gas that can absorb and emit certain types of radiation and thus affects the greenhouse effect on the Earth. While its concentrations are much smaller than CO$_2$ or water vapor, it also is much more efficient at absorbing and emitting radiation, so it too can affect the temperature of the Earth's atmosphere and surface. Like CO$_2$, concentrations of methane in the atmosphere have been increasing over the last 100 years, again resulting in an increase in the greenhouse effect.

The atmosphere also contains countless tiny particles called **aerosols**. They differ from other components of the atmosphere in that they are solids or liquids, not gases.

As discussed in FIGURE 2.5 on pages 34–35, aerosols are swept into the air from dry desert plains, lakebeds, and beaches, or, as we saw in the chapter opener, released by

aerosols Tiny particles present in the atmosphere, so small and light that the slightest air movements keep them aloft.

exploding volcanoes. Strong winds blowing over the ocean lift droplets of spray into the air. These droplets of spray lose most of their moisture by evaporation, leaving tiny particles of watery salt that are carried high into the air. Forest fires and brushfires also throw up particles of soot as smoke. As meteors vaporize while traveling through the atmosphere, they leave behind dust particles in the upper layers of air. Closer to the ground, industrial processes that incompletely burn coal or fuel oil release aerosols into the air as well. Clouds, which are made up of either liquid water droplets or ice crystals, are also examples of aerosols. In fact, some aerosols are important because water vapor can condense on these specks of dust to form the tiny droplets that constitute clouds.

All of these aerosols can play a role in affecting the temperatures of the atmosphere when they occur in high concentrations. For instance, some aerosols scatter sunlight, brightening the whole sky while reducing the intensity of the solar radiation reaching the surface. Other aerosols, however, can absorb incoming solar radiation. While this decreases the amount of sunlight reaching the surface, it also increases the amount absorbed in the atmosphere, which increases the heating of the atmosphere.

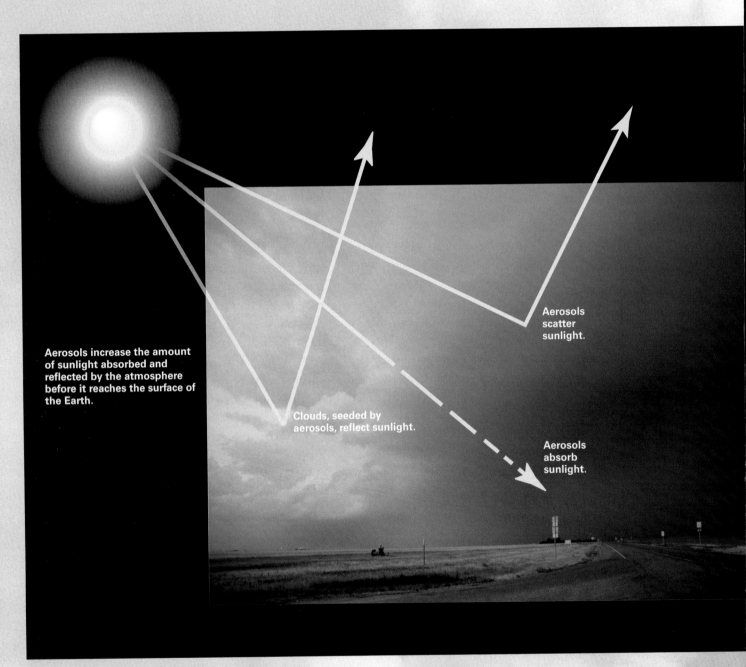

Aerosols increase the amount of sunlight absorbed and reflected by the atmosphere before it reaches the surface of the Earth.

Aerosols scatter sunlight.

Clouds, seeded by aerosols, reflect sunlight.

Aerosols absorb sunlight.

▲ **Impacts**

Aerosols decrease the amount of sunlight reaching the surface. In addition, they can impact clouds by making them more reflective, which further decreases the amount of sunlight reaching the surface.

Besides volcanoes, there are numerous sources of aerosols in the atmosphere. These aerosols can have a dramatic impact upon the amount of solar radiation that reaches us here at the surface of the Earth.

▲ **Dust**
The strong winds of this dust storm in the Painted Desert of Arizona sweep desert dust into the air.

▲ **Salt**
Breaking waves and bursting bubbles of foam add salt particles to the atmosphere.

▲ **Fuel burning**
Incomplete burning of fossil fuels, especially coal, releases smoke and other aerosols.

▲ **Meteors**
Meteors burn and vaporize, contributing molecules and particles to the upper atmosphere. This meteor shower was photographed at Knapps Castle in Santa Barbara, California.

▲ **Smoke**
This aerial photo of the North Fork Fire in Yellowstone National Park shows clouds of smoke rising to an altitude of 6100 m (20,000 ft).

How one CFC molecule destroys many ozone molecules FIGURE 2.6

In the atmosphere's stratospheric layer 15–35 km up, ozone (O_3) molecules absorb the Sun's high-energy ultraviolet (UV) radiation, reducing UV that reaches the Earth's surface, where it can harm us and other organisms. These protective ozone molecules get broken apart in two ways, shown below—directly **A** and indirectly **B–E**.

A How UV breaks apart ozone molecules

UV radiation strikes molecules of stratospheric ozone (O_3). As they absorb the energy, they break apart into O and O_2. These can recombine into ozone and absorb more UV. This continuing cycle maintains steady ozone levels, thus shielding Earth's surface from harmful UV radiation.

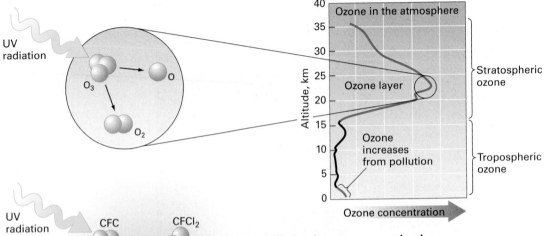

B How CFC's break apart ozone molecules

When UV bombards molecules of chlorofluorocarbon (CFC) in the stratosphere, a chain of reactions begins that can break apart thousands of ozone molecules. In the initial step at left, UV energy breaks apart a CFC molecule, releasing a chlorine atom (Cl). This chlorine atom proceeds to break apart many ozone molecules (steps **C**, **D**, **E**), allowing UV radiation to pass through the stratosphere unabsorbed and reach the surface of the Earth.

C The Cl atom interacts with an ozone molecule to form oxygen (O_2) and chlorine oxide (ClO).

D The ClO then encounters and interacts with a single-oxygen molecule (O). This forms O_2 and releases the Cl atom.

E The freed Cl atom reacts with another ozone molecule to form O_2 and ClO. These reactions (**C**, **D**, **E**) continue, allowing a single Cl atom from a decomposed CFC molecule to break apart thousands of ozone molecules over a month or two, typically during September–October over the Antarctic as the region emerges from its long, cold winter.

OZONE IN THE UPPER ATMOSPHERE

Another small but important constituent of the atmosphere is **ozone**—a form of oxygen in which three oxygen atoms are bonded together (O_3). Ozone is found mostly in the upper part of the atmosphere, about 14–50 km (9–31 mi) above the surface in a layer called the stratosphere.

ozone Form of oxygen with a molecule consisting of three atoms of oxygen, O_3.

The fate of ozone in the stratosphere is described in FIGURE 2.6. Ozone is produced in gaseous chemical reactions that require energy in the form of ultraviolet radiation. The reactions are quite complicated, but the net effect is that ozone (O_3), molecular oxygen (O_2), and atomic oxygen (O) are constantly formed, destroyed, and re-formed in the ozone layer, absorbing ultraviolet radiation with each transformation.

Because the ozone layer absorbs ultraviolet light from the Sun, it protects the Earth's surface from this damaging form of radiation. The presence of the ozone layer is thus essential for life on this planet to survive. If the full intensity of solar ultraviolet radiation ever hit the Earth's surface, it would destroy exposed bacteria and severely damage animal tissue.

Certain forms of air pollution, such as chlorofluorocarbons, or CFCs, reduce ozone concentrations substantially. CFCs are synthetic industrial chemical compounds containing chlorine, fluorine, and carbon atoms. Although CFCs were banned in aerosol sprays in the United States in 1978, they are still used as cooling fluids in some refrigeration systems. When appliances containing CFCs leak or are discarded, their CFCs are released into the air.

CFC molecules move up to the ozone layer where they break down, providing molecules of chlorine oxide (ClO). This compound attacks ozone, converting it to ordinary oxygen by a chain reaction. With less ozone, there is less absorption of ultraviolet radiation.

An unexpectedly large "hole" in the ozone layer was discovered over the continent of Antarctica in the mid-1980s. There, the ozone layer thins during the early spring of the southern hemisphere, reaching a minimum during the month of October (FIGURE 2.7). The hole then slowly shrinks and ultimately disappears in early December. Each year however, the ozone hole reappears and grows slightly larger than the year before. The largest ozone hole was recorded on October 8, 2006, and covered 29,500,00 km² (11,400,000 mi²).

Since 1978, surface-level ultraviolet radiation has been increasing—and not just over the Antarctic. Over most of North America, the increase has been about 4 percent per decade. Today, we are all aware of the dangers of harmful ultraviolet rays to the skin, and there is concern that this trend could lead to a rise in skin cancer cases. Phytoplankton, the base of the food chain in the ocean, may also suffer.

In response to the global threat of ozone depletion, 23 nations signed a treaty in 1987 to cut global CFC consumption by 50 percent by 1999. The treaty was effective: by late 1999, scientists confirmed that stratospheric chlorine concentrations had topped out

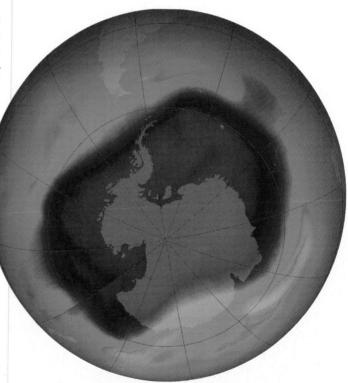

Ozone hole, September 24, 2006 FIGURE 2.7

The antarctic ozone hole of 2006 was the largest on record, covering about 29.5 million sq km (about 11.4 million sq mi). Low values of ozone are shown in purple, ranging through blue, green, and yellow. Ozone concentration is measured in Dobson units, and October 8, 2006, saw its lowest value—85 units. (NASA)

in 1997 and were continuing to fall. Although the ozone layer is not expected to be completely restored until the middle of the 21st century, this is a welcome observation.

OZONE IN THE LOWER ATMOSPHERE

Whereas ozone in the stratosphere acts as a protective shield, preventing harmful high-energy rays from the Sun from reaching the Earth's surface, ozone in the lower portion of the atmosphere—the *troposphere*—acts as a very strong pollutant. In addition, it is a principal component of *smog*, which greatly reduces visibility, as is evident in FIGURE 2.8. Because the ozone molecule is composed purely of oxygen, it can react readily with other chemicals, particularly organic chemicals in plants and animals. This process can corrode the cells that the ozone contacts, harming plant tissues and eventually killing sensitive plants. It can also be harmful to human lung tissue, aggravating bronchitis, emphysema, and asthma.

Ozone near the surface is formed through a complex interaction between naturally occurring and human-made organic chemicals, sunlight, and nitrogen/oxygen-based molecules produced mainly through burning of gasoline (cars) and other fossil fuels (energy production). Therefore, ozone is most prevalent in urban areas (where there is a great deal of automotive and industrial activity) and during the summer (when there is lots of sunlight).

In response to severe pollution events arising from increasing industrial activity, the United States adopted the Clean Air Act in 1963 and set up the Environmental Protection Agency (EPA) in 1970 to tackle the problem of monitoring and reducing pollution. Presently, ozone is monitored using the Air Quality Index, which rates air quality on a daily basis across the country. The index is used to notify people when air quality becomes unhealthy and potentially hazardous (particularly for the elderly and the very young). The index can also be used to monitor whether cities, towns, and states are in compliance with local and federal laws regarding levels of ozone and other pollutants.

Smog over Los Angeles FIGURE 2.8

This image shows smog over Los Angeles. A principal component of this smog is tropospheric ozone, which can corrode the cells in our lungs as well as those in plants. Continued exposure can lead to respiratory problems.

CONCEPT CHECK STOP

Which gases make up the air in our atmosphere?

What three forms can water take in the atmosphere?

How does ozone in the stratosphere and troposphere affect life on the Earth?

What are aerosols?

Why are aerosols important regulators of the temperature of the Earth's atmosphere and surface?

Temperature Structure of the Atmosphere

Define the different layers of the atmosphere.

Describe how temperature varies between these layers.

The temperature of the atmosphere is not constant but depends upon altitude. In the lower atmosphere, the air is generally cooler at higher altitudes. We call the decrease in air temperature with increasing altitude the **lapse rate**. We measure the temperature drop in °C per 1000 m (or °F per 1000 ft). For a typical summer day in the midlatitudes, the temperature drops at an average rate of 6.4°C/1000 m (3.5°F/1000 ft). The actual value, which can change with location and time, is known as the **environmental lapse rate** and is described in *What a Scientist Sees: Temperature Changes Along a Mountain Slope*. For example, when the air temperature near the surface is a pleasant 25°C (77°F), the air

lapse rate The rate at which temperature drops with increasing height.

environmental lapse rate The rate at which the actual temperature at a particular location and time drops with increasing height.

Temperature Changes Along a Mountain Slope

Even a casual observer would see a dramatic change in vegetation while moving up the side of this mountain. Near the base, the vegetation is lush and green. Going up the mountain, the vegetation becomes sparser and the plants tend to be smaller. Toward the crest, the vegetation becomes very sparse and consists mainly of small bushes and grasses. At the very top, vegetation disappears altogether. A scientist sees that this change in vegetation is partly related to the change in temperature along the mountain slope. At the base, temperatures tend to be warmer and can support more vegetation. Because the temperature decreases according to the environmental lapse rate, midway up the mountain, the temperatures are cooler than at the base, and hence vegetation is hardier and smaller. At the very top, where temperatures are coldest, trees can no longer survive, and the ground is largely barren. This region is said to be above the "tree line."

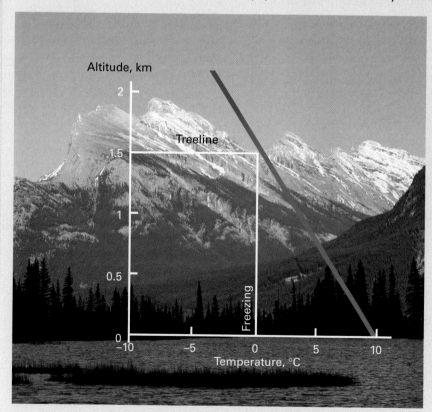

This image shows Mount Rundle in Banff National Park, Alberta, Canada. The graph shows the decreasing temperatures higher up the mountain, based on a fixed environmental lapse rate of 5°C/km.

What a Scientist Sees

at an altitude of 15 km (40,000 ft) might be a bone-chilling –71°C (–96°F). Keep in mind that this example of temperature change is an average value; on any given day or in a different location, the observed environmental lapse rate might be quite different.

Generally, for the first 12 km (7 mi) or so, temperature falls with increasing elevation. Between around 12–15 km (7–9 mi), however, the temperature stops decreasing. In fact, above that height, temperature slowly increases with elevation. Atmospheric scientists use this feature to define two important layers in the lower atmosphere—the troposphere and the stratosphere.

TROPOSPHERE

The **troposphere** is the lowest of the four atmospheric layers seen in FIGURE 2.9. All human activity takes place here. Everyday weather phenomena, such as clouds and storms, happen in the troposphere. Here, temperature generally decreases with increasing elevation. The troposphere is thickest in the equatorial and tropical regions, where it stretches from sea level to about 16 km (10 mi). It thins toward the poles, where it is only about 8 km (5 mi) thick.

> **troposphere** The lowest layer of the atmosphere, in which temperature falls steadily with increasing height.

Vertical structure of the atmosphere FIGURE 2.9

Temperature structure of the atmosphere.

Above the troposphere and stratosphere are the mesosphere and thermosphere. The homosphere, in which air's chemical components are well mixed, ranges from the surface to nearly 100 km altitude.

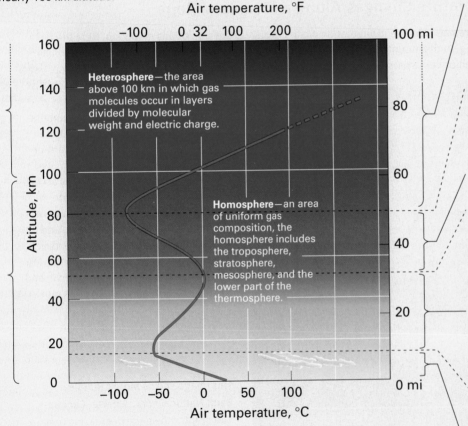

Thermosphere—air temperature goes up with altitude in the thermosphere, but the air density is thin and the air holds very little heat.

Mesopause—a boundary between the mesosphere and the thermosphere where air temperature stops falling with altitude.

Mesosphere—air temperatures in the mesosphere fall with altitude.

Stratopause—a boundary between the stratosphere and the mesosphere, the stratopause occurs where air temperatures stop increasing with altitude and start falling.

Stratosphere—air temperatures in the stratosphere become warmer as altitude increases. Much of the absorption of high-energy radiation from the Sun is absorbed by ozone in this layer.

Tropopause—a boundary between the troposphere and the stratosphere, the tropopause occurs where air temperatures stop decreasing with altitude and start increasing.

Troposphere—air in the troposphere becomes colder with increasing altitude. It is the lowest layer of the atmosphere, and the level where most weather activity takes place.

The troposphere gives way to a layer called the *stratosphere*. The boundary at which this occurs is called the **tropopause**, where temperatures stop decreasing with altitude and start to increase. The altitude of the tropopause varies somewhat with the season, so the troposphere is not uniformly thick at any location.

STRATOSPHERE

The **stratosphere** lies above the tropopause (Figure 2.9). Air in the stratosphere becomes slightly warmer as altitude increases. The stratosphere reaches up to roughly 50 km (about 30 mi) above the Earth's surface.

Air does not mix between the troposphere and stratosphere except under certain situations, so the stratosphere normally holds very little water vapor or dust. One exception is caused by volcanic activity. If a volcanic eruption is strong enough, it can deposit aerosols into the stratosphere. Once there, they can remain in the stratosphere for a year or two and spread around the globe. Because of this persistence, these aerosols can affect the temperature of the globe for more than a year after the actual eruption has occurred.

The stratosphere also contains the ozone layer, which shields earthly life from intense, harmful ultraviolet energy. As the ozone molecules absorb this ultraviolet energy, they warm the stratosphere, causing the temperature to increase with altitude in this region of the atmosphere.

Temperatures stop increasing with altitude at the *stratopause*. Above the stratopause we find the *mesosphere*. In the mesosphere, temperature falls with altitude. This layer ends at the *mesopause*, the level at which temperature stops falling with altitude. The next layer is the *thermosphere*. Here, temperature increases with altitude again, but because the density of air is very thin in this layer, the air holds little total heat energy.

The gas composition of the atmosphere is uniform for about the first 100 km (about 60 mi) of altitude, which includes the troposphere, stratosphere, mesosphere, and the lower portion of the thermosphere. We call this region the *homosphere*. Above 100 km, gas molecules tend to be sorted into layers by molecular weight and electric charge. This region is called the *heterosphere*.

THE IONOSPHERE

In the upper atmosphere we also find the **ionosphere**. The ionosphere—which extends from about 60 km (about 35 mi) above the Earth's surface up through 400 km (about 250 mi)—is characterized by the presence of electrically charged atoms and molecules called **ions** along with free electrons. These ions and electrons generate immense electric and magnetic fields spanning large portions of the globe. These electric fields control amazing displays of light as well as cause severe disruptions to our everyday life.

■ **tropopause** The level of the atmosphere between the troposphere and stratosphere, where temperatures stop decreasing with height and start increasing.

■ **stratosphere** The layer of atmosphere directly above the troposphere; here temperature increases with altitude.

■ **ionosphere** Layer of the upper atmosphere characterized by the presence of ions.

■ **ion** Electrically charged atom or molecule.

The northern lights—also called the *aurora borealis*—occur as particles from the Sun stream through the ionosphere and collide with charged atoms and molecules, energizing them. This energy is then released as light that we can see. The lights can take the form of sheets, streaks, and even spirals. Similar features are seen in the southern hemisphere—the southern lights or aurora australis.

One phenomenon common to the ionosphere is the occurrence of the aurora borealis, sometimes referred to as the *northern lights* (FIGURE 2.10). These multicolored sheets of undulating light brighten and fade as charged particles from the Sun move through the ionosphere, colliding with atoms and molecules. The atoms and molecules gain energy through these collisions, then emit this energy as visible light. Generally, the aurora borealis is most common at high latitudes, where the charged particles are drawn toward the magnetic polar regions of the Earth.

In addition, the ionosphere can produce disturbances in satellite communications. Heating of the lower portion of the ionosphere by the Sun during the day can generate bubbles of charged gases just after sunset that slowly move up through the ionosphere, much like the bubbles in a lava lamp. These bubbles interfere with electromagnetic radiation passing through them. All of our satellite communication and navigation systems, including those that involve cellular phones, Global Positioning System (GPS) navigation receivers, and satellite TV broadcasts, are transmitted via electromagnetic signals from satellites to the surface. As these signals pass through bubbles in the ionosphere, disruptions in communications across large portions of the globe can occur.

Similarly, in midlatitudes, large differences, or *gradients*, in the number of electrons between one location and another can be generated as charged particles from the Sun move through the ionosphere. As with the bubbles mentioned earlier, these gradients in electrons also disrupt satellite communication and navigation systems. In the midlatitudes, these disruptions are especially critical to the navigation system used to guide all commercial aircraft to safe landings at U.S. airports.

In the higher latitudes, the electric charge associated with the ionosphere can also directly affect our lives here at the Earth's surface. For instance, during auroral events, the interaction of the streaming particles from the Sun with the charged ions and electrons in the ionosphere can generate massive electric currents. These currents can induce similar currents in wires and power grids at the surface. If strong enough, these induced currents can overload and blow out transformers at power stations. One such event, in March 1989, overwhelmed the transformers of the Hydro Quebec power grid, knocking out power in most of eastern Canada for nine hours.

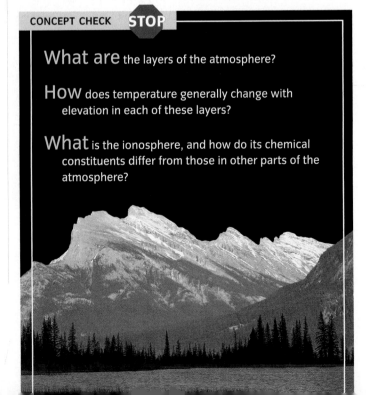

CONCEPT CHECK STOP

What are the layers of the atmosphere?

How does temperature generally change with elevation in each of these layers?

What is the ionosphere, and how do its chemical constituents differ from those in other parts of the atmosphere?

Atmospheric Pressure and Density

LEARNING OBJECTIVES

Define atmospheric pressure and density.

Explain how a barometer works.

Describe how pressure, density, and temperature are related.

We live at the bottom of a vast ocean of air—the Earth's atmosphere. Like the water in the ocean, the air in the atmosphere is constantly pressing on the Earth's surface beneath it and on everything that it surrounds.

The atmosphere exerts pressure because gravity pulls air molecules toward the Earth, creating the force that we call *weight*. Pressure is expressed as force per unit area—for example, the recommended air pressure in the tires of your car is probably around 32 pounds per square inch.

Atmospheric pressure is produced by the weight of a column of air above a unit area of the Earth's surface. At sea level, about 1 kilogram of air presses down on each square centimeter of surface (1 kg/cm²)—about 15 pounds on each square inch of surface (15 lb/in.²) (**FIGURE 2.11**).

The basic SI unit of pressure is the pascal (Pa). At sea level, the average pressure of air is 101,320 Pa. Many atmospheric pressure measurements are reported in bars and millibars (mb) (1 bar = 1000 mb = 10,000 Pa). A millibar is thus 100 pascals (100 Pa) or one hectopascal (1 hPa). In this book, we use the millibar as the SI unit of atmospheric pressure. The standard atmospheric pressure at sea-level is 1013.2 mb.

You probably know that a **barometer** measures atmospheric pressure, but do you know how it works? Certain types of barometers are based on the same principle as drinking soda through a straw. When using a straw, you create a partial vacuum in your mouth. In fact, though, it is the pressure of the atmosphere that actually forces soda up through the straw.

atmospheric pressure Pressure exerted by the atmosphere because of the force of gravity acting upon the overlying column of air.

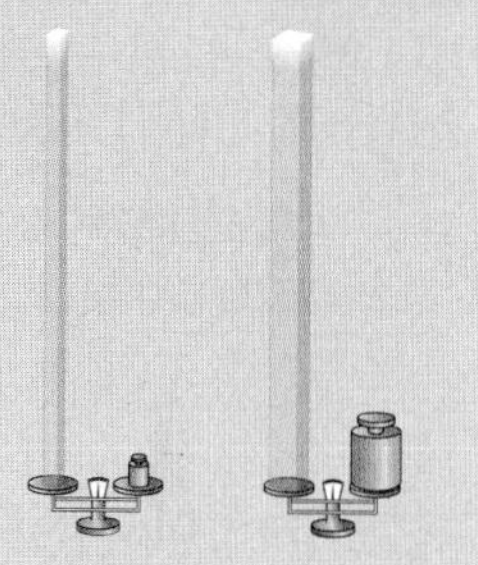

barometer Instrument that measures atmospheric pressure.

Column of atmosphere 1 cm in cross section

1kg

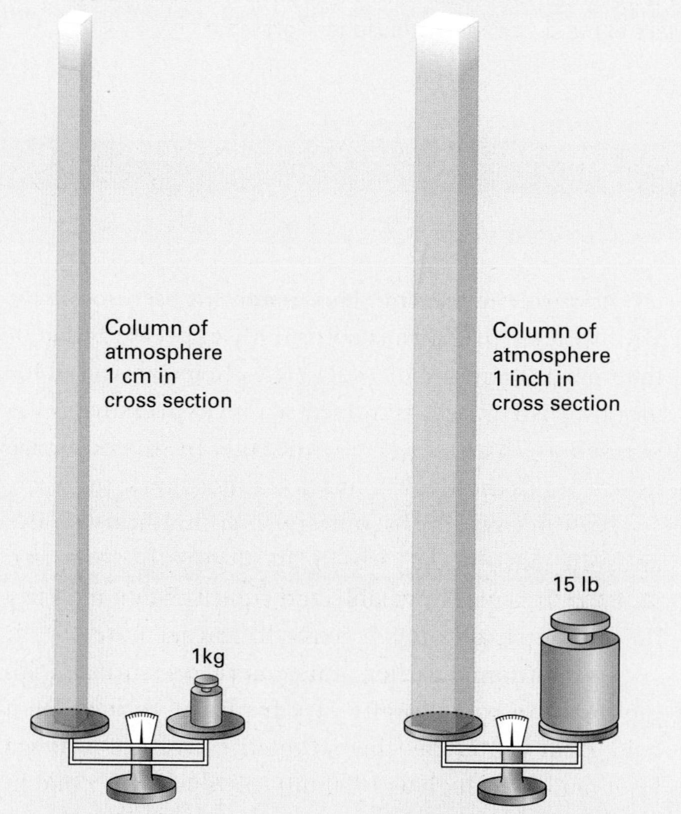

A Metric system—The weight of a column of air 1 cm on one side is balanced by the weight of a mass of about 1 kg.

Column of atmosphere 1 inch in cross section

15 lb

B English system—The weight of a column of air 1 in. on one side is balanced by a weight of about 15 lb.

Atmospheric pressure FIGURE 2.11

Here we can see atmospheric pressure depicted as the weight of a column of air.

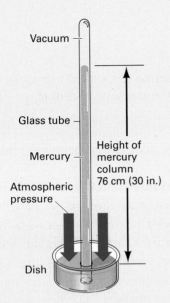

Vacuum

Glass tube

Mercury

Atmospheric pressure

Dish

Height of mercury column 76 cm (30 in.)

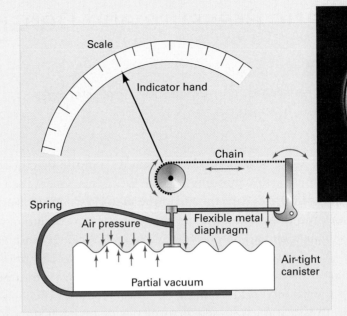

Scale

Indicator hand

Chain

Spring

Air pressure

Flexible metal diaphragm

Air-tight canister

Partial vacuum

A Mercury barometer In a mercury barometer, atmospheric pressure pushes the mercury upward into the tube, balancing the pressure exerted by the weight of the mercury column. As atmospheric pressure changes, the level of mercury in the tube rises and falls.

B Aneroid barometer An aneroid barometer uses a sealed canister of air that swells and contracts slightly as atmospheric pressure changes. The motion of the canister lid is transmitted mechanically to move an indicator hand along a scale, indicating rises and falls of the surrounding atmospheric pressure.

Types of barometers FIGURE 2.12

The oldest, simplest, and most accurate instrument for measuring atmospheric pressure, the mercury barometer, works the same way as a drinking straw (FIGURE 2.12A). A mercury barometer is created by taking a glass tube about 1 m (3 ft) long, sealed at one end, and filling it with mercury. The tube is inverted, and its open end is immersed in a dish of mercury. The mercury in the tube settles at a level about 76 cm (about 30 in.) above the surface of the mercury in the dish. Why? It is because atmospheric pressure on the surface of the mercury in the dish presses the mercury back upward into the tube. The pull of gravity on the mercury column exactly balances the pressure of the air pushing the mercury back into the tube, and the mercury level settles at a stable height. If the air pressure increases, it will the force the mercury in the tube to rise. Thus, this device measures atmospheric pressure directly as a height in centimeters or inches rather than in units of pressure.

Because the mercury barometer was used so widely, atmospheric pressure is commonly expressed in centimeters or inches of mercury. The chemical symbol for mercury is Hg, and standard sea-level pressure is expressed as 76 cm Hg (29.92 in. Hg). In this book, we use in. Hg as the English unit for atmospheric pressure.

Another type of barometer in common use is the aneroid barometer, which is shown in FIGURE 2.12B. It is more portable and robust than a mercury thermometer, although it is usually less accurate.

As mentioned earlier, atmospheric pressure at a single location can actually vary from day to day. On a cold, clear winter day, the barometric pressure near sea level may be as high as 1030 mb (30.4 in. Hg), while in the center of a storm system it may drop by about 5 percent to 980 mb (28.9 in. Hg). Indeed, changes in atmospheric pressure are typically associated with traveling weather systems and can be used to forecast coming weather.

HOW AIR PRESSURE AND DENSITY CHANGE WITH ALTITUDE

Atmospheric pressure decreases with altitude as well as with location (FIGURE 2.13). Why? We know that at the surface, the measure of atmospheric pressure is based on the weight of the overlying atmospheric molecules as they are pulled toward the surface by gravity. Moving up through the atmosphere, the number of atmospheric molecules decreases, and so does the pressure. The decrease of pressure with altitude is initially quite rapid. At higher altitudes, the decrease is much slower.

Because atmospheric pressure decreases rapidly with altitude near the surface, a small change in eleva-

Atmospheric pressure and density with altitude FIGURE 2.13

Atmospheric pressure and density decrease rapidly with altitude near the Earth's surface but much more slowly at higher altitudes. The bar on the right illustrates the decrease in the number of atmospheric molecules as altitude increases.

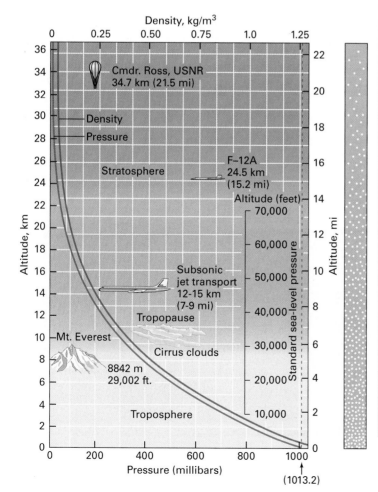

Climber on Mount Everest FIGURE 2.14

Most people who stay in a high-altitude environment for several days adjust to the reduced air pressure. Climbers who are about to ascend Mount Everest usually spend several weeks in a camp partway up the mountain before attempting to reach the summit.

tion will often produce a significant change in air pressure. You may have noticed that your ears pop from the pressure change during an elevator ride in a tall building. You may also have noticed your ears popping on an airplane flight. Aircraft cabins are pressurized to about 800 mb (24 in. Hg), which corresponds to an elevation of about 1800 m (5900 ft).

At the surface, the pressure of the atmosphere compresses the air molecules into a certain volume. The mass of air molecules within a given volume is called the **air density**. As the number of air molecules increases, so does the mass, and vice versa. With decreased air pressure, there is less force exerted on the air molecules, hence they decompress, or spread out, resulting in a decrease in air density.

> **air density** Mass of air molecules found within a fixed volume of air.

This decrease in air density means that fewer air molecules, including oxygen, surround you. Hence, when you are at higher altitudes, the decrease in the density of oxygen molecules produces a shortness of breath and fatigue. These symptoms, along with headache, nosebleed, or nausea, are referred to as *mountain sickness* and are likely to occur at altitudes of 3000 m (about 10,000 ft) or higher. Most people adjust to the reduced air pressure of a high-mountain environment after several days (FIGURE 2.14).

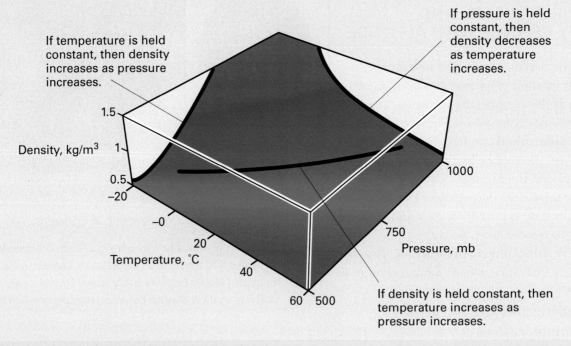

If temperature is held constant, then density increases as pressure increases.

If pressure is held constant, then density decreases as temperature increases.

If density is held constant, then temperature increases as pressure increases.

Density, kg/m³

1.5
1
0.5

Temperature, °C

−20
−0
20
40
60

500
750
1000

Pressure, mb

Temperature, pressure, and density FIGURE 2.15

Temperature, pressure, and density are related to one another by the ideal gas law. The graph shows this relationship for dry air.

▪ **ideal gas law** The law that describes the relationship between absolute temperature, pressure, and density of gases.

THE IDEAL GAS LAW

A basic relation between the pressure, density, and temperature of air (FIGURE 2.15) is given by the **ideal gas law**, which states:

$$\text{absolute temperature} \times \frac{\text{density}}{\text{pressure}} = \text{constant}.$$

If one of the three variable terms is held constant, then we can find the relation between the other two. For instance, if pressure is held constant, then:

$$\text{temperature} \times \text{density} = \text{constant}.$$

In this case, if temperature decreases then density must increase, which is why we say that cold air is more dense than warm air.

However, if we hold density constant, then:

$$\frac{\text{temperature}}{\text{pressure}} = \text{constant}$$

or

$$\text{temperature} = \text{constant} \times \text{pressure}.$$

In this case, if the pressure on an air mass increases, then its temperature will increase. Thus, temperature in the troposphere generally decreases with height because the surrounding pressures are also decreasing with height. It is important to realize, however, that only rarely is any one of these variables completely constant and that all three can change simultaneously.

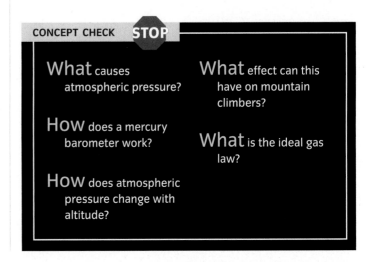

CONCEPT CHECK **STOP**

What causes atmospheric pressure?

How does a mercury barometer work?

How does atmospheric pressure change with altitude?

What effect can this have on mountain climbers?

What is the ideal gas law?

The force exerted by atmospheric pressure is a basic fact of life on the Earth. In this photo, a frog grips a smooth surface using suction cups.

▪ How do the suction cups keep the frog from falling?

▪ What would happen if the suction cups had small holes in them?

▪ Can you think of any other animals that use these principles to "defy gravity"?

SUMMARY

1 Composition of the Atmosphere

1. The Earth's atmosphere is dominated by nitrogen and oxygen. Although nitrogen is relatively inert, oxygen is very important for chemical reactions within the atmosphere and at the Earth's surface.

2. Water vapor, carbon dioxide, and methane can absorb and emit radiation, which makes them important contributors to the heating of the atmosphere and the Earth's surface.

3. **Ozone** (O_3) in the stratosphere helps absorb ultraviolet radiation, sheltering the Earth from ultraviolet rays. Industrial chemicals that reach the stratosphere can destroy ozone, allowing more high-energy solar radiation to reach the Earth's surface. Ozone in the troposphere is a pollutant that is hazardous to human, animal, and plant life.

4. **Aerosols** are suspended liquid and solid particles found in the atmosphere. Aerosols can be produced by the eruption of volcanoes, dust blowing off land surfaces, and the spray of sea salt into the air, as well as the condensation of water vapor into liquid and solid water. Aerosols can both absorb and reflect radiation, thereby affecting the temperatures of the atmosphere where they are found.

2 Temperature Structure of the Atmosphere

1. Air temperatures normally fall with altitude in the **troposphere**. The rate at which temperature decreases at a given location is called the **environmental lapse rate**.

2. In the **stratosphere**, temperatures increase slightly with altitude. The boundary between the regions where temperatures decrease and where they increase is called the **tropopause**.

3. In the **ionosphere**, electrons and electrically charged atoms and molecules—called **ions**—are found. The electric charge of these ions can interact with incoming charged particles from the Sun to produce the aurora borealis. In addition, they can affect electromagnetic communication from satellites to the Earth's surface.

3 Atmospheric Pressure and Density

1. The term **atmospheric pressure** describes the weight of air pressing on a unit of surface area. Atmospheric pressure is measured using a **barometer** and can vary across locations and time.

2. **Air density** measures the mass of gas molecules found in a fixed volume of air. Both atmospheric pressure and density decrease as altitude increases.

3. The pressure, density, and temperature of a parcel of air are all related through the **ideal gas law**. This law states that temperature × density / pressure remains constant. Hence, changes in one of these quantities will produce changes in one or both of the other two.

KEY TERMS

- aerosols p. 33
- ozone p. 37
- lapse rate p. 39
- environmental lapse rate p. 39
- troposphere p. 40

- tropopause p. 41
- stratosphere p. 41
- ionosphere p. 41
- ion p. 41
- atmospheric pressure p. 43

- barometer p. 43
- air density p. 45
- ideal gas law p. 46

CRITICAL AND CREATIVE THINKING QUESTIONS

1. Which component gas is most abundant in the atmosphere? Which is the most variable in concentration from place to place and from time to time?

2. Why are water vapor and carbon dioxide vapor levels in the atmosphere important?

3. How does the ozone layer help protect life on the Earth?

4. What are aerosols? How do they enter the atmosphere? Why are they important, in terms of climate?

5. Describe the changing atmospheric conditions that you might notice as you climb higher up a mountain. Explain what causes these changes.

6. What is atmospheric pressure? Why does it occur? How is it measured and in what units? What is the normal value of atmospheric pressure at sea level? How does it change with altitude?

1. _____ is the most abundant gas in the Earth's present atmosphere.
   a. Nitrogen
   c. Carbon dioxide
   b. Oxygen
   d. Water vapor

2. Of the following gases, which varies highly in concentration in the atmosphere?
   a. Nitrogen
   c. Argon
   b. Water vapor
   d. Oxygen

3. The ozone layer is located in the _____.
   a. lithosphere
   c. stratosphere
   b. hydrosphere
   d. thermosphere

4. The ozone layer is effective in shielding the Earth's surface from _____.
   a. ultraviolet radiation
   c. microwaves
   b. infrared radiation
   d. reflected rays

5. _____ cause(s) the most dramatic destruction of ozone.
   a. Nitrogen
   c. Chlorofluorocarbons (CFCs)
   b. Carbon dioxide
   d. Argon

6. Put the following labels on the figure: troposphere, stratosphere, mesosphere, thermosphere, tropopause, stratopause, homosphere, ozone layer.

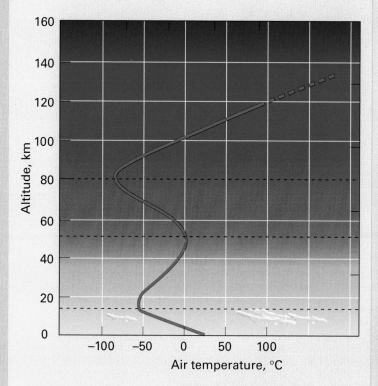

7. One-hundred percent of the Earth's atmosphere lies within 30 km of the Earth's surface.
   a. True
   b. False

8. The troposphere is noted for having _____.
   a. separation of atmospheric gases
   b. an increase in ozone concentration with height
   c. a decrease in temperature with height
   d. a lack of aerosols

9. Temperature increases with elevation in the _____.
   a. troposphere
   c. mesosphere
   b. stratosphere
   d. none of the above

10. The ozone layer is contained within the highest levels of the troposphere.
    a. True
    b. False

11. The _____ measures the decrease of air temperature with increasing altitude at a given location and time.
    a. environmental cooling lapse rate
    b. environmental temperature lapse rate
    c. environmental adiabatic lapse rate
    d. evapotranspiration lapse rate

12. Standard sea-level atmospheric pressure is _____.
    a. 1013.2 mb
    c. 998.3 mb
    b. 935.6 mb
    d. 1045.9 mb

13. Atmospheric pressure decreases with altitude.
    a. True
    b. False

14. A liquid thermometer is a commonly used instrument for determining barometric pressure.
    a. True
    b. False

15. The northern lights occur due to the excitation of charged particles in the _____.
    a. troposphere
    c. ionosphere
    b. stratosphere
    d. lithosphere

The Earth's Global Energy Balance

In 1989, our Sun released a massive burst of energy. At the Sun's surface, where the temperature is around 6000°C (about 11,000°F), swirling electrical currents focused this energy into a cloud of gas that erupted from the surface. Radiant energy and charged particles from this eruption traveled the 150 million km (93 million mi) from the Sun to the Earth in just over four days.

The solar surge knocked out electronics on four of the Navy's navigational satellites and destroyed half the solar panels on the $55 million GOES-7 weather satellite. The surge collided in the atmosphere with charged molecules and electrons, creating massive clouds of electrical gases that glowed and pulsed over the northern hemisphere. The currents created as the solar surge moved through these charged gases also induced a surge in the power wires at the Earth's surface, knocking out the power grid for most of eastern Canada for nine hours.

The Earth receives about 28,000 times more energy from the Sun than human society consumes each year. This chapter examines the flow of energy from the Sun to the Earth on a global scale, describing the process by which solar radiation is intercepted by our planet, flows through the Earth's atmosphere, and interacts with the Earth's land and ocean surfaces. It also describes how this radiation is converted into other forms of energy and how this energy eventually leaves the Earth system. This process is called the *global energy balance*.

Hot gases erupting from the Sun's surface form huge loops that radiate vast amounts of energy to space and toward the Earth.

Electromagnetic Radiation

s you lounge on an exotic tropical beach such as the one in FIGURE 3.1, you might be particularly aware of the Sun. Basking in its glow, you might be impressed by its ability to light your world and warm your body across millions of miles of black space. Actually, the Sun's power is even more impressive, considering that most natural phenomena that take place at the Earth's surface are directly or indirectly solar-powered. From the downhill flow of a river to the movement of a

A late afternoon at the beach FIGURE 3.1

Energy from the Sun provides both light and warmth here at the Earth's surface on this late afternoon in San Juan, Puerto Rico.

sand dune to the growth of a forest, solar radiation drives nearly all of the natural processes that shape the world around us (FIGURE 3.2). It is the power source for wind, waves, weather, rivers, and ocean currents.

The primary topic of this chapter is the energy balance of the Earth. Our study of the global energy system begins with the subject of radiation—that is, **electromagnetic radiation**. All surfaces—from the fiery Sun to the skin covering our bodies—constantly emit radiation. For instance, very hot objects, such as the Sun or a lightbulb filament, can give off, or emit, radiation in the form of light. Cooler objects than the Sun, such as Earth surfaces and even our own bodies, also emit radiation, which we refer to as *thermal radiation*. **Emission** is the term used to describe the process in which bodies radiate away energy.

Light is just one form of electromagnetic radiation. You can think of electromagnetic radiation as a collection of waves with a wide range of **wavelengths** that travel away from the surface of an object. Radiant energy can exist at any wavelength. In this book, we measure wavelength in micrometers. A micrometer is one millionth of a meter (10^{-6} m). The tip of your little finger is about 15,000 micrometers wide. We use the abbreviation μm for the micrometer. The first letter is the Greek letter μ, or mu.

electromagnetic radiation Wave-like form of energy radiated by any substance possessing heat; it travels through space at the speed of light.

emission Process in which bodies release radiation that subsequently travels through space.

wavelength The distance separating one wave crest from the next wave crest.

Solar-powered phenomena
FIGURE 3.2

MEXICO (YUCATAN)

JAMAICA

HONDURAS

Tropical cyclones Hurricane Mitch, with winds in excess of 195 mph, bears down on Central America. This hurricane and other cyclonic storms are driven indirectly by solar energy.

Solar boiler Movable mirrors reflect sunlight to a central receiver that vaporizes water into high-pressure steam. The steam is used to drive a generator.

Equatorial rainforest, Borneo Island, East Malaysia The Sun's energy drives plant growth through photosynthesis.

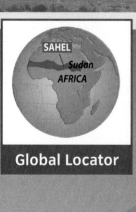

SAHEL

Sudan

AFRICA

Global Locator

Sand dune in the Sahara Desert, Africa Unequal heating of the Earth's surface, coupled with the Earth's rotation, produces global winds that create and move enormous sand dunes.

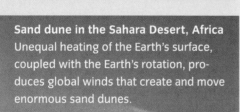

Wave erosion Ocean waves, powered by the Sun through the Earth's wind system, are actively eroding the eastern bluffs of Cape Cod.

Electromagnetic waves differ in wavelength throughout their entire range, or *spectrum,* shown in FIGURE 3.3. At the short-wavelength end of the spectrum are gamma rays and X-rays. Their wavelengths are normally expressed in nanometers. A nanometer is one one-thousandth of a micrometer, or 10^{-9} m, and is abbreviated nm. Gamma and X-rays have high energies and can be hazardous to health. Gamma- and X-ray radiation transition into ultraviolet radiation, which has wavelengths from about 10–400 nm or 0.4 μm. Like gamma and X-rays, ultraviolet radiation can be damaging to living tissues. Too much exposure to ultraviolet radiation, for instance, can cause sunburns.

The visible-light portion of the spectrum starts at about 0.4 μm with violet. Colors then transition through blue, green, yellow, orange, and red, reaching the end of the visible spectrum at about 0.7 μm. These are the colors we see in the rainbow in FIGURE 3.4. When white light from the Sun, which is composed of all these colors, passes through raindrops, these wavelengths are separated from one another, allowing us to see each individually. Next is near-infrared radiation, with wavelengths from 0.7 to 1.2 μm. This radiation is very similar to visible light in that most of it comes from the Sun. However, human eyes are not sensitive to radiation beyond about 0.7 μm, so we do not see near-infrared light.

The electromagnetic spectrum FIGURE 3.3

Electromagnetic radiation can exist at any wavelength. By convention, names are assigned to specific wavelength regions as shown in the figure.

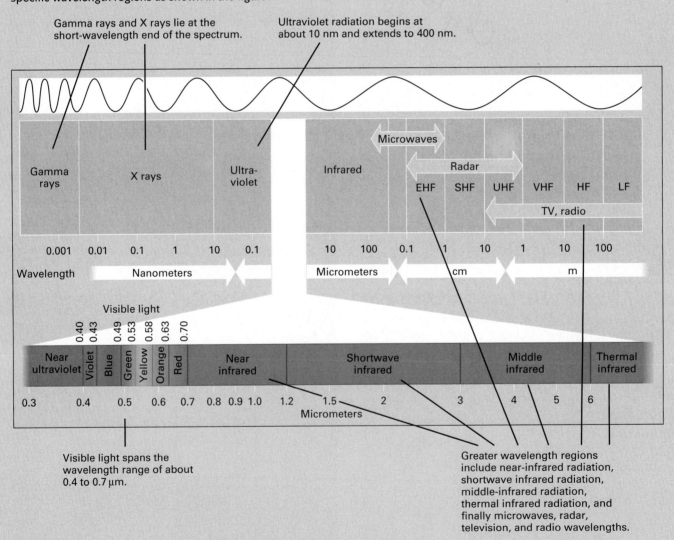

Gamma rays and X rays lie at the short-wavelength end of the spectrum.

Ultraviolet radiation begins at about 10 nm and extends to 400 nm.

Visible light spans the wavelength range of about 0.4 to 0.7 μm.

Greater wavelength regions include near-infrared radiation, shortwave infrared radiation, middle-infrared radiation, thermal infrared radiation, and finally microwaves, radar, television, and radio wavelengths.

Rainbow FIGURE 3.4

In the rainbow, tiny drops of water refract sunlight, splitting it into its component colors.

Next comes infrared radiation, which is generally divided into three separate categories, based on wavelength. Shortwave infrared wavelengths are in the range of 1.2–3.0 μm. Like near-infrared light, it comes mostly from the Sun. From 3 μm to about 6 μm is middle-infrared radiation. Again, much of this radiation comes from the Sun, but is also emitted by fires burning at the Earth's surface, such as forest fires or gas-well flames.

At wavelengths between 6 μm and about 300 μm, we have thermal infrared radiation. This radiation includes the radiation given off by bodies at temperatures normally found at the Earth's surface and in the atmosphere. Most thermal radiation from the Earth's surface and atmosphere lies in the range of 8–12 μm.

Radiation and temperature FIGURE 3.5

All objects emit radiation. Both the intensity and wavelength of emitted radiation depend upon the temperature of the object itself.

Beyond infrared wavelengths lie the domains of microwaves, radar, and communications transmissions, such as radio and television.

RADIATION AND TEMPERATURE

Two important physical principles govern the emission of electromagnetic radiation. The first is that hot objects radiate more energy than cooler objects—much more. In fact, the flow of radiant energy from the surface of an object is directly related to the absolute temperature of the surface, measured on the Kelvin temperature scale, raised to the fourth power. This equation, called the *Stefan-Boltzmann equation*, states that

$$\text{radiation intensity} = \text{constant} \times T^4.$$

Therefore, if you double the absolute temperature of an object, it will emit 16 times more energy from its surface. We can see in FIGURE 3.5 that even a small increase in temperature can mean a large increase in the radiation intensity given off by an object or surface. For example, water at room temperature emits about one-third more energy than when it is at the freezing point.

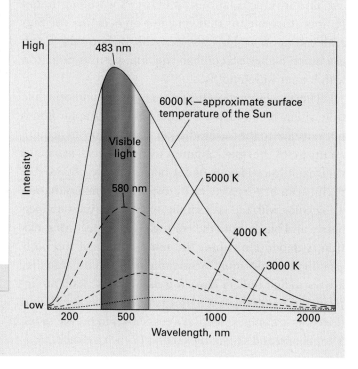

The second principle is that the hotter the object, the shorter the wavelengths of its emitted radiation. This equation, called *Wein's law*, states that

$$\frac{\text{wavelength of peak radiation}} {} = \text{constant} / \text{temperature.}$$

This inverse relationship between wavelength and temperature means that very hot objects like the Sun emit radiation at short wavelengths. Because the Earth is a much cooler object than the Sun, it emits radiation with longer wavelengths.

Thermal infrared radiation is no different in principle from visible light radiation. Although photographic film is not sensitive to thermal radiation, it is possible to acquire an image of thermal radiation using a special sensor. A thermal infrared image of a suburban home obtained at night, seen in **FIGURE 3.6**, displays thermal radiation using color, with red tones indicating the warmest temperatures and black tones the coldest. The intensity of radiation is dependent upon the temperature of the body emitting that radiation. Windows appear red because they are warm and radiate more intensely. House walls are intermediate in temperature and appear blue. Roads and driveways are cool, as are the trees, shown in purple tones. The ground and sky are the coldest (shown in black).

SOLAR RADIATION

The relation between the temperature of the emitting body and the intensity and wavelength of the emitted radiation allow us to differentiate between the fate of radiation emitted by the Sun and that emitted by bodies such as the Earth and its atmosphere.

The left side of **FIGURE 3.7** shows how the Sun's incoming electromagnetic radiation varies with wavelength. The uppermost line shows how a "perfect" Sun would supply solar energy at the top of the atmosphere. By "perfect," we mean a Sun radiating as a *blackbody*. A blackbody is a body that is perfectly efficient at emitting radiation and hence emits exactly the amount predicted by theoretical principles. Very few bodies are blackbodies—that is, perfect emitters. For instance, the solid line shows the actual output of the Sun as measured at the top of the atmosphere. It is quite close to the "perfect" Sun, except for ultraviolet wavelengths, where the real Sun emits less energy. Note that the Sun's output peaks in the visible part of the spectrum. Thus, our human vision system is adjusted to the wavelengths of solar light energy that are most abundant.

The line showing solar radiation reaching the Earth's surface is quite different from the others, however. As solar radiation passes through the atmosphere, it is affected by absorption and scattering. **Absorption** occurs when atoms, molecules, and particles in the atmosphere intercept and absorb radiation at particular wavelengths. This absorption is shown by the steep "valleys" in the graph—for example, at about 1.3 μm and 1.9 μm. At these wavelengths, molecules of water vapor and carbon dioxide absorb solar radiation very strongly and keep nearly all of it from reaching the Earth's surface. Note also that oxygen and ozone absorb almost all of the ultraviolet radiation at wavelengths shorter than about 0.3 μm, preventing it from reaching the surface and harming the cells of biological organisms.

Atmospheric absorption is important because it warms the atmosphere directly and constitutes one of the flows of energy in the global energy balance. In the case of ozone in the stratosphere, the absorption of shortwave radiation heats the atmosphere and contributes to the increase of temperature with height.

■ **absorption**

Process in which electromagnetic energy is transferred to heat energy when radiation strikes molecules or particles in a gas, liquid, or solid.

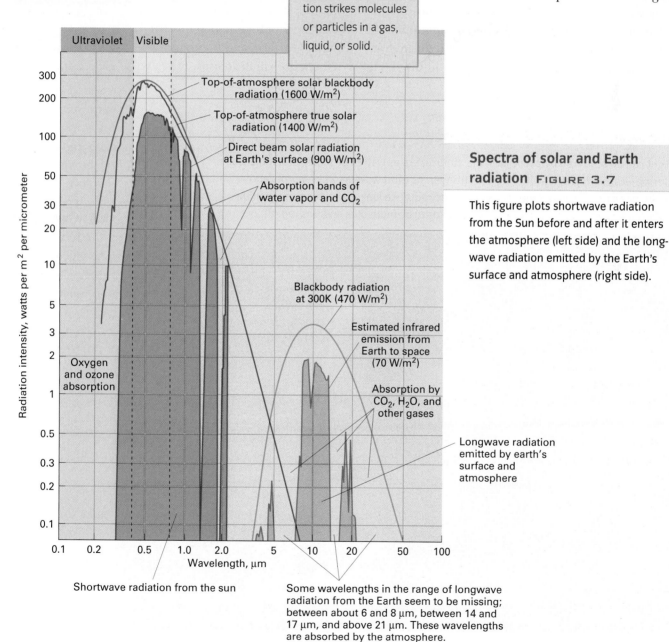

Ultraviolet   Visible

Top-of-atmosphere solar blackbody radiation (1600 W/m²)

Top-of-atmosphere true solar radiation (1400 W/m²)

Direct beam solar radiation at Earth's surface (900 W/m²)

Absorption bands of water vapor and CO_2

Blackbody radiation at 300K (470 W/m²)

Estimated infrared emission from Earth to space (70 W/m²)

Absorption by CO_2, H_2O, and other gases

Oxygen and ozone absorption

Longwave radiation emitted by earth's surface and atmosphere

Radiation intensity, watts per m² per micrometer

Wavelength, μm

Shortwave radiation from the sun

Some wavelengths in the range of longwave radiation from the Earth seem to be missing; between about 6 and 8 μm, between 14 and 17 μm, and above 21 μm. These wavelengths are absorbed by the atmosphere.

Spectra of solar and Earth radiation FIGURE 3.7

This figure plots shortwave radiation from the Sun before and after it enters the atmosphere (left side) and the longwave radiation emitted by the Earth's surface and atmosphere (right side).

Solar radiation passing through the atmosphere can also be **scattered**. By scattering, we refer to a turning aside of radiation by a molecule or particle so that the direction of a scattered ray is changed. Scattered rays may go upward toward space or downward toward the surface (FIGURE 3.8).

Solar radiation that is neither absorbed by the atmosphere nor scattered back into space is instead **transmitted** to the Earth's surface. Solar energy received at the surface ranges from about 0.3 to 3 μm and is known as **shortwave radiation**. These wavelengths are "short" compared with the longer wavelengths of energy that are emitted by the Earth and atmosphere.

The amount of shortwave energy transmitted to the Earth's surface can change from place to place, however. For instance, when the Sun is higher in the sky, its radiation travels a shorter distance through the atmosphere. The shorter this distance—its *path length*—the less likely it is that the radiation will be scattered or absorbed before it reaches the Earth's surface. Alternatively, if there is an abundance of gases, liquids, and particles in the intervening atmosphere—for instance cloud drops, dust, or soot—then it is more likely that the radiation will be scattered or absorbed before it reaches the Earth's surface, and the transmitted energy will decrease.

scattering Process in which particles and molecules deflect incoming solar radiation in different directions; atmospheric scattering can redirect solar radiation back to space.

transmission Process in which incoming solar radiation passes through the atmosphere without being absorbed or scattered.

shortwave radiation Electromagnetic energy in the range from 0.3 to 3 μm; shortwave radiation comes exclusively from the Sun.

Blue skies due to scattering FIGURE 3.8

Sunlight can be scattered by particles in the atmosphere as well as by molecules. Molecular scattering—called *Rayleigh scattering*—is stronger for short wavelengths—purple and blue—and less for long wavelengths—orange and red. If we look at the sky away from the Sun, the light we see is scattered light, which tends to be more blue and less red.

Molecular scattering affects all wavelengths but is strongest for short wavelengths—blue and purple.

Light coming from the sky is scattered light. It appears blue because blue is more strongly scattered than the other wavelengths.

Direct sunlight appears white, because it contains all wavelengths.

LONGWAVE RADIATION FROM THE EARTH

As discussed, an object's temperature controls both the range of wavelengths and the intensity of radiation the object emits. The Earth's surface and atmosphere are much colder than the Sun's surface, so we would deduce that our planet radiates less energy than the Sun and that this energy is emitted at longer wavelengths.

> **longwave radiation** Electromagnetic radiation in the range 3–30 μm; the type of radiation emitted by the Earth and atmosphere.

As with the Sun, the irregular orange line in Figure 3.7 on page 57, shows energy actually emitted by the Earth and atmosphere out to space, as measured at the top of the atmosphere. It ranges from about 3 to 30 μm and peaks at about 10 μm in the thermal infrared region. This thermal infrared radiation emitted by the Earth is also called **longwave radiation**.

Once again we can see that some wavelengths in this range seem to be missing, especially between about 6 and 8 μm, between 14 and 17 μm, and above 21 μm. These wavelengths are almost completely absorbed by the atmosphere before they can escape to space. Water vapor and carbon dioxide are the main absorbers. As with the absorption of solar radiation, the absorption of longwave radiation heats the atmosphere. The atmosphere can then subsequently reradiate this energy back to the Earth's surface, warming it as well. This absorption and reemission of longwave radiation by water vapor and carbon dioxide, as well as other gases, produces the *greenhouse effect.*

While much of the longwave energy emitted by the Earth and atmosphere is absorbed before it ever leaves the top of the atmosphere, there are still three regions where outgoing energy flow from the Earth to space is significant: 4–6 μm, 8–14 μm, and 17–21 μm. These wavelengths of longwave radiation are the primary way energy leaves the Earth and is transmitted back to space.

THE GLOBAL RADIATION BALANCE

We have seen that the Earth is constantly absorbing solar shortwave radiation and emitting longwave radiation. This process creates a global radiation balance, which is shown in more detail in **FIGURE 3.9** on pages 60–61.

In this balance, the Sun provides a nearly constant flow of shortwave radiation toward the Earth that is received at the top of the atmosphere. Part of this radiation is scattered away from the Earth and heads back into space without being absorbed. Clouds and dust particles in the atmosphere contribute to this scattering. Land and ocean surfaces also reflect some shortwave radiation back to space. The shortwave energy from the Sun that is not scattered or reflected is instead absorbed by either the atmosphere, the land, or the ocean. Once absorbed, solar energy raises the temperature of the atmosphere as well as the temperatures of the ocean and land surfaces.

The atmosphere, land, and ocean also emit energy in the form of longwave radiation. This radiation ultimately leaves the planet, headed for outer space. The longwave radiation outflow represents a loss of energy and hence tends to lower the temperature of the atmosphere, ocean, and land, thus cooling the planet. In the long run, these flows balance—incoming energy absorbed and outgoing radiation emitted are equal. Since the temperature of an object is determined by the overall (or net) amount of energy it absorbs and emits, the Earth's overall temperature tends to remain constant as long as these two are in balance.

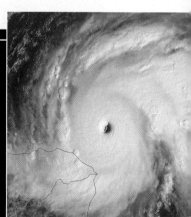

CONCEPT CHECK STOP

What is electromagnetic radiation?

What are the major regions of the electromagnetic spectrum?

How is an object's temperature related to the nature and amount of electromagnetic radiation it emits?

What wavelengths of radiation does the Sun emit?

What wavelengths does the Earth emit?

How does the global radiation balance result in the Earth's constant overall temperature?

The global radiation balance FIGURE 3.9

VIEW THIS IN ACTION
in your WileyPLUS course

The Earth continuously absorbs shortwave radiation from the Sun and emits long-wave radiation. In the long run, the gain and loss of radiant energy is balanced, and the Earth's average temperature remains constant.

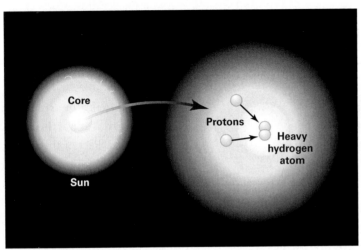

A Nuclear reactions in the Sun
Under intense heat and pressure, two hydrogen nuclei (protons) collide to form an atom of heavy hydrogen (deuterium), containing one proton and one neutron. This releases large amounts of energy. Many of these nuclear reactions continually occur, radiating bountiful energy into space.

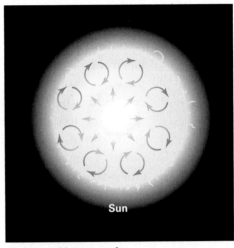

B Internal heat transfer
Heat is transferred from the Sun's core to its surface by slow convection of the dense, hot gases.

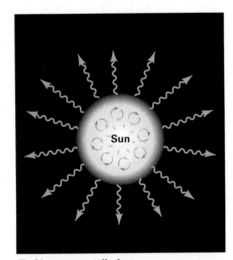

C Shortwave radiation
The Sun transmits this energy through space in all directions.

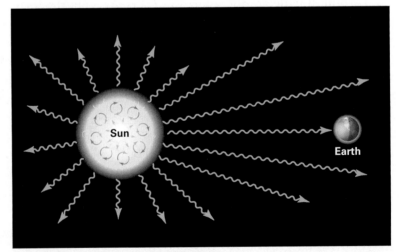

D Shortwave radiation reaches the Earth
The Sun provides a nearly constant flow of shortwave radiation to the Earth. The Earth is a tiny target in space, intercepting only about one-half of one-billionth of the Sun's total energy output.

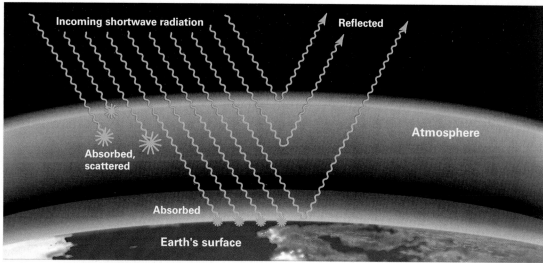

E Some shortwave radiation is reflected
Molecules and particles in the atmosphere absorb and scatter incoming shortwave radiation, as does the Earth's surface. So, some shortwave radiation is directly reflected back to space.

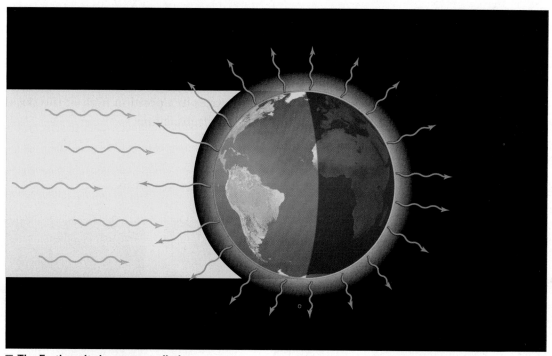

F The Earth emits longwave radiation
Shortwave radiation is absorbed by the Earth and the atmosphere, increasing earthly temperatures and generating longwave radiation that radiates back to space.

Geographic Variations in Energy Flow

insolation The flow of solar energy intercepted by an exposed surface, assuming a uniformly spherical Earth with no atmosphere.

Most natural phenomena of the Earth's surface are driven by the Sun, either directly or indirectly. Although the flow of solar radiation to the Earth as a whole remains constant, the amount received varies from place to place and from time to time.

INSOLATION OVER THE GLOBE

Incoming solar radiation is known as **insolation**. Insolation is a flow rate and has units of watts per square meter (W/m²). Daily insolation refers to the average of this flow rate over a 24-hour day. Annual insolation is the average flow rate over the entire year. Insolation is important because it measures the amount of solar power available to heat the land surface and atmosphere at a particular location.

Insolation depends on the angle of the Sun above the horizon, as seen in **FIGURE 3.10**. Insolation is greatest when the Sun is directly overhead and the Sun's rays are vertical. When the Sun is lower in the sky, the same amount of solar energy spreads over a greater area of ground surface, so insolation is lower.

Daily insolation at a location depends on two factors: (1) the angle at which the Sun's rays strike the Earth, and (2) the length of time of exposure to the rays. Both of these factors are controlled by the latitude of the location and the time of year. For example, in midlatitude locations in summer, days are long and the Sun rises to a position high in the sky, thus increasing the daily insolation.

Insolation and Sun angle FIGURE 3.10

The angle at which the Sun's rays strike the Earth varies from one geographic location to another because of the Earth's spherical shape.

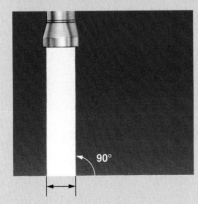

90°

1 unit of surface area
One unit of light is concentrated over one unit of surface area.

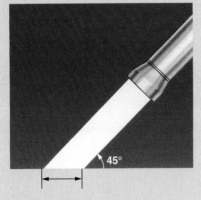

45°

1.4 units of surface area
One unit of light is dispersed over 1.4 units of surface area.

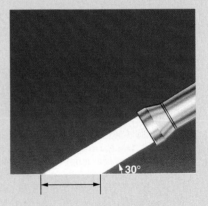

30°

2 units of surface area
One unit of light is dispersed over 2 units of surface area.

A Sunlight (represented by the flashlight) that shines vertically near the Equator is concentrated on the Earth's surface.

B and **C** Toward the poles, the light hits the surface more and more obliquely, spreading the same amount of radiation over a larger and larger area.

Annual insolation also depends upon the angle at which the Sun's rays strike the Earth and the length of time of exposure to the rays. For annual insolation, these factors are controlled only by the latitude of the location.

NET RADIATION, LATITUDE, AND THE ENERGY BALANCE

The heat level of our planet rises as it absorbs solar energy. At the same time, it radiates energy into outer space, cooling itself. Over time, these incoming and outgoing radiation flows balance for the Earth as a whole. However, these flows do not have to balance at each particular place on the Earth, nor do they have to balance at all times. At night, for example, there is no incoming radiation, yet the Earth's surface and atmosphere still emit outgoing radiation.

Net radiation is the difference between all incoming radiation and all outgoing radiation. In places where radiant energy flows in faster than it flows out, net radiation is positive, providing an energy surplus. In other places, net radiation can be negative. For the entire Earth and atmosphere, the net radiation is zero over the year.

> **net radiation** The difference between the amount of incoming radiation and the amount of outgoing radiation at a given location.

Solar energy input varies strongly with latitude. What is the effect of this variation on net radiation? FIGURE 3.11 shows the incoming shortwave radiation, outgoing longwave radiation, and net radiation profiles spanning the entire latitude range from 90° N to 90° S. Here we use yearly averages for each latitude, so that the effect of seasons is removed.

Between about 40° N and 40° S, there is a net radiant energy gain labeled "energy surplus." In other words, incoming solar radiation exceeds outgoing longwave radiation throughout the year. Poleward of 40° N and 40° S, the net radiation is negative and is labeled "energy deficit"—meaning that outgoing longwave radiation exceeds incoming shortwave radiation.

If you carefully examine the graph in Figure 3.11, you will find that the area labeled "surplus" is equal in size to the combined areas labeled "deficit." Thus the net radiation for the Earth as a whole is zero, as ex-

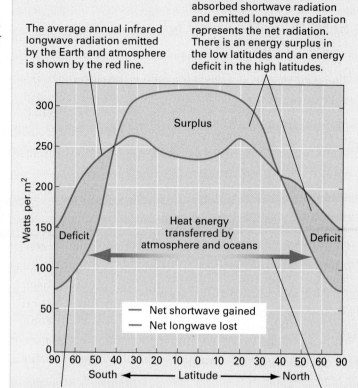

The average annual infrared longwave radiation emitted by the Earth and atmosphere is shown by the red line.

The difference between absorbed shortwave radiation and emitted longwave radiation represents the net radiation. There is an energy surplus in the low latitudes and an energy deficit in the high latitudes.

The average annual incoming solar shortwave radiation absorbed by the Earth and the atmosphere is shown by the blue line.

The oceans and atmosphere transport the surplus radiation energy from the low latitudes to the high latitudes.

Annual surface radiation from pole to pole

FIGURE 3.11

pected, with global incoming shortwave radiation exactly balancing global outgoing longwave radiation.

Because there is an energy surplus at low latitudes and an energy deficit at high latitudes, energy will flow from low latitudes to high. This energy is transferred poleward as warm ocean waters and warm, moist air move poleward, while cooler waters and cool drier air move toward the Equator.

This poleward heat transfer, driven by the imbalance in net radiation between low and high latitudes, is the power source for almost all broad-scale atmospheric circulation patterns, weather phenomena, and ocean currents. Without this circulation, low latitudes would heat up and high latitudes would cool down until a radiative balance was achieved, leaving the Earth with much more extreme temperature contrasts—and would make it very different from the planet that we are familiar with now.

SENSIBLE AND LATENT HEAT TRANSFER

The *net radiation* at a given location is the difference between the absorbed radiation (both shortwave and longwave) minus the emitted longwave radiation. When these two do not balance, the excess energy is converted into two other types of energy—sensible and latent heat.

Sensible heat is actually very familiar—it's what you feel when you touch a warm object. When we use a thermometer, we are measuring sensible heat. The transfer of sensible heat moves energy from warmer to colder objects when they are put into contact. When you touch a warm object, for example, sensible heat transfer

> **sensible heat** An indication of the intensity of kinetic energy of molecular motion within a substance; it is measured by a thermometer.

Melting of ice FIGURE 3.13

As these icicles melt, ice takes up heat from the air and changes state to become liquid water.

from the object to your hand occurs through *conduction*. In the atmosphere, sensible heat is conducted by random molecular motions directly from the surface to the adjacent air layer, as in FIGURE 3.12. This warm air is then mixed to higher levels and warms the air further away from the surface, which we refer to as *convection*. Some sensible heat can be conducted to lower levels in the soil as well.

By contrast, **latent heat**—or hidden heat—cannot be measured by a thermometer. It is heat that is taken up and stored as molecular motion when a substance changes state

> **latent heat** Heat absorbed or released as substances change from one phase to another.

> **phase** The state of a substance: either solid, liquid, or gas.

from a solid to a liquid, from a liquid to a gas, or from a solid directly to a gas. This process is called a change of **phase**. For example, to change the phase of water from solid to liquid as in FIGURE 3.13, or from liquid to gas, energy needs to be added to the water, which breaks the attractive bonds between the water molecules and allows them to move free of the surrounding water molecules, as seen in FIGURE 3.14. This latent heat is stored in the form of the fast random motion of the freely moving water vapor molecules in the gaseous state as compared with the relatively slower motions of the molecules in either the liquid or solid state.

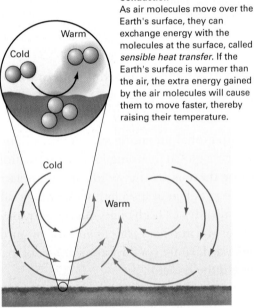

Conduction
As air molecules move over the Earth's surface, they can exchange energy with the molecules at the surface, called *sensible heat transfer*. If the Earth's surface is warmer than the air, the extra energy gained by the air molecules will cause them to move faster, thereby raising their temperature.

Convection
Vertical circulations carry the warmer surface air aloft, warming the air away from the surface. These circulations also bring cool air from aloft down to the surface, cooling the suface air and underlying ground.

Transfer of sensible heat from the Earth's surface to the atmosphere FIGURE 3.12

A similar process occurs during the cooling you feel in a breeze when your skin is wet. As the water evaporates from liquid to gas, it draws up heat and carries it away from your skin, keeping you cool.

The process can also work in reverse. When water vapor turns back to a liquid or solid, the fast random motion of the freely moving water vapor molecules is released and the molecules return to the relatively slower motions associated with either the liquid or solid phase. The latent heat that is released during this process can then go to warming the surrounding air molecules, increasing their temperatures.

In the Earth–atmosphere system, latent heat transfer to the atmosphere occurs when water from a moist

Ocean-atmosphere energy transfer FIGURE 3.15

Heat energy, in the form of latent heat, flows from the ocean to the atmosphere when ocean water evaporates. That energy is later released to the atmosphere when the water vapor condenses to form cloud droplets.

land surface or from open water evaporates (**FIGURE 3.15**). However, this latent heat is not released until the water recondenses, which can occur far from where the water originally evaporated. On a global scale, latent heat transfer by movement of moist air provides a very important mechanism for transporting large amounts of energy from one region of the Earth to another. In addition, it supplies energy for driving many different types of weather events.

Molecular motions associated with latent heating FIGURE 3.14

For liquid water to change to water vapor, it must acquire latent heat. This heat allows a water molecule to move fast enough to break the attractive bonds with other water molecules and move freely as a gas.

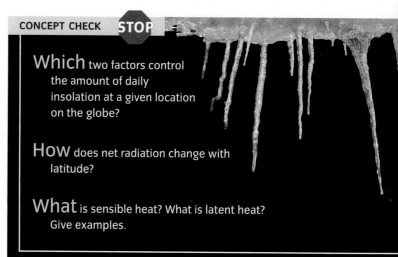

CONCEPT CHECK **STOP**

Which two factors control the amount of daily insolation at a given location on the globe?

How does net radiation change with latitude?

What is sensible heat? What is latent heat? Give examples.

The Global Energy System

The term *energy system* refers to the flows of energy reaching the Earth, which includes land and ocean surfaces and the atmosphere, and the flows of energy leaving it. The Earth's energy system controls the seasonal and daily changes in the Earth's surface temperature. Differences in energy flow rates from place to place also drive currents of air and ocean water. All these, in turn, produce the changing weather and rich diversity of climates we experience on the Earth's surface. For the globe as a whole, however, all of these flows have to balance. In this section, we examine this global energy balance in detail.

SOLAR ENERGY LOSSES IN THE ATMOSPHERE

Solar energy is the ultimate power source for the Earth's surface processes, so when we trace the energy flows between the Sun, Earth's surface, and atmosphere, we are really studying how these processes are driven.

Let's examine typical losses of incoming shortwave radiation as it penetrates the atmosphere on its way to the Earth's surface, shown in FIGURE 3.16. Gamma rays and X-rays from the Sun are almost completely absorbed by oxygen in the thin outer layers of the atmosphere. Lower down in the atmosphere, in the stratosphere, much of the ultraviolet radiation is absorbed, particularly by ozone.

As the radiation moves deeper through denser layers of the atmosphere, it can be scattered by gas molecules, dust, or other particles in the air, deflecting it in any direction. Apart from this change in direction, it is unchanged. Scattered radiation moving in all directions through the atmosphere is known as *diffuse radiation*. Some scattered radiation flows down to the Earth's surface, while some flows upward. This upward flow of diffuse radiation escaping back to space amounts to about 3 percent of incoming solar radiation if no clouds are present.

However, the presence of clouds can greatly increase the amount of incoming solar radiation scattered back to space. Scattering from the bright white surfaces of thick clouds—called *reflection*—deflects about 30–60 percent of incoming radiation back into space (FIGURE 3.17).

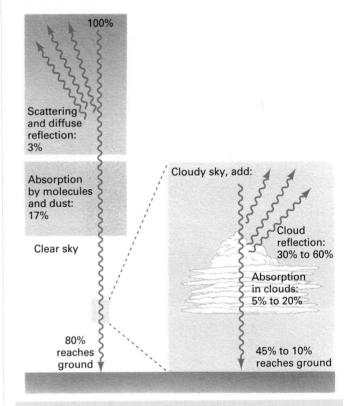

100%

Scattering and diffuse reflection: 3%

Absorption by molecules and dust: 17%

Clear sky

80% reaches ground

Cloudy sky, add:

Cloud reflection: 30% to 60%

Absorption in clouds: 5% to 20%

45% to 10% reaches ground

Fate of incoming solar radiation FIGURE 3.16

Losses of incoming solar energy are much lower with clear skies (left) than with cloud cover (right).

Cloud cover over the globe
FIGURE 3.17

Clouds are very effective reflectors of sunlight. Here, the western portion of North America is visible, while eastern North America, the Atlantic, and northern Europe are effectively shrouded in clouds.

What about absorption? Molecules and particles can absorb radiation as it passes through the atmosphere. Oxygen, ozone, and water vapor are the most important absorbers of incoming shortwave radiation. About 17 percent of incoming solar radiation is absorbed both in the upper levels and lower layers of the atmosphere, thereby raising the temperature of these layers. If clouds are present, the fraction of absorbed solar radiation can increase by 6–20 percent.

After taking into account absorption and scattering, we find that under clear-sky conditions about 80 percent of the radiation reaches the surface. Under cloudy conditions, this value can drop to 10–45 percent. Hence, the amount of cloud cover, including cloud type and altitude, is an important factor in determining changes in local temperatures as well as the overall climate of the planet.

When accounting for both cloudy skies and clear skies, on a global scale only about 55 percent of the total insolation at the top of the atmosphere makes it to the surface. Once this solar radiation reaches the surface, it can be absorbed, which heats the surface. However, it can also be reflected back to the atmosphere and out to space again.

ALBEDO

The proportion of shortwave radiant energy scattered upward by any surface is called its **albedo**. For example, a surface that reflects 40 percent of incoming

> **albedo** Proportion of solar radiation reflected upward from a surface.

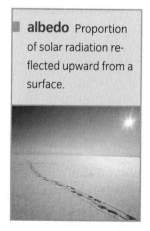

A Bright snow A layer of new, fresh snow reflects most of the sunlight it receives. Only a small portion is absorbed.

B Blacktop road Asphalt paving reflects little light and so appears dark or black. It absorbs nearly all of the solar radiation it receives.

Albedo contrasts FIGURE 3.18

shortwave radiation has an albedo of 0.40. Snowfields have a high albedo (0.45–0.85), reflecting most of the solar radiation that hits them, and absorbing only a small amount (FIGURE 3.18). By contrast, black pavement, which has a low albedo (0.03), absorbs nearly all the incoming solar energy. Because of the higher absorption of solar radiation, surface temperatures are warmer for low albedo than high albedo surfaces.

The albedo of a surface can depend both on the surface characteristics as well as on the angle at which the light hits the surface. For instance, the albedo of water depends strongly on the angle of incoming radiation. The albedo is very low (0.02) for nearly vertical rays hitting calm water. However, when the Sun shines on a water surface at a low angle, much of the radiation is directly reflected as *sun glint*, producing a higher albedo. Fields, forests, and bare ground have intermediate albedos, ranging from 0.03 to 0.25. As mentioned, we can also measure the albedo of clouds, which ranges from 0.30 to 0.60 (FIGURE 3.19).

Certain orbiting satellites carry instruments that can measure shortwave radiation at the top of the atmosphere, helping us estimate the Earth's average albedo. The albedo values obtained in this way vary between 0.29 and 0.34, which means that as a whole, the Earth–atmosphere system reflects slightly less than one-third of the solar radiation it receives back into space. *What a Scientist Sees: Surface Characteristics and Albedo* shows how the amount of reflected sunlight can change with time as the underlying surface characteristics change.

Albedos for different surfaces and Sun angles
FIGURE 3.19

This photograph shows the complexity of albedos at the Earth's surface. Clouds appear bright white because they reflect much of the incoming solar radiation. In addition, the low angle of the sunlight striking the water produces *sun glint*—the strong reflection of sunlight across the center of the ocean. Water surfaces away from this region are darker, indicating lower albedo. The lowest albedo of all is associated with the vegetation on the island to the left.

Surface Characteristics and Albedo

A view of the coast of Greenland provides a dramatic example of how changes in surface characteristics can alter the albedo of a location.

Blue regions show where the ice cap has been melting over the last 30 years. As it melts, the underneath soil and rock are exposed. In these regions, the albedo will change from very high values associated with snow cover to lower values associated with soils.

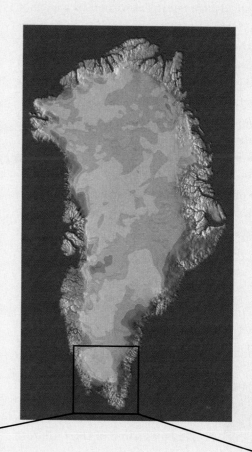

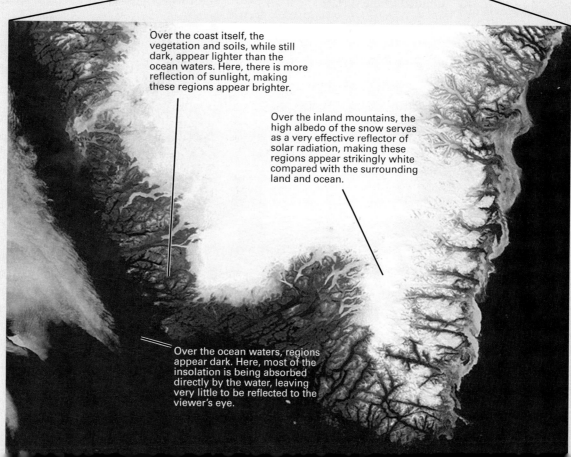

Over the coast itself, the vegetation and soils, while still dark, appear lighter than the ocean waters. Here, there is more reflection of sunlight, making these regions appear brighter.

Over the inland mountains, the high albedo of the snow serves as a very effective reflector of solar radiation, making these regions appear strikingly white compared with the surrounding land and ocean.

Over the ocean waters, regions appear dark. Here, most of the insolation is being absorbed directly by the water, leaving very little to be reflected to the viewer's eye.

COUNTERRADIATION AND THE GREENHOUSE EFFECT

The amount of shortwave energy absorbed by a surface is an important determinant of the surface's temperature. However, the surface is also warmed significantly by longwave radiation emitted by the atmosphere and absorbed by the ground. Let's look at the energy flows between the surface, atmosphere, and space in more detail (**FIGURE 3.20**). Here we can see the flow of shortwave radiation from the Sun to the surface. Some of this radiation is reflected back to space, but much is absorbed, warming the surface.

The surface emits radiation upward. A small fraction of this radiation escapes directly to space (A), but the largest part is absorbed by the atmosphere (B). What about longwave radiation emitted by the atmosphere? Although the atmosphere is colder than the surface, it also emits long-wave radiation, shown by flows C and D. This radiation is emitted in all directions, so some radiates upward to space (C), while the remainder radiates downward toward the surface (D). We call this downward flow **counterradiation**. It replaces some of the radiation emitted by the surface.

counterradiation Longwave atmospheric radiation moving downward toward the Earth's surface.

This mechanism, in which the atmosphere absorbs longwave radiation and reradiates it back to the surface, is known as the **greenhouse effect**. Unfortunately, the term greenhouse is not accurate. In a true greenhouse, glass windows permit entry of shortwave energy, as does our atmosphere. However, the primary reason that temperatures are warmer inside the greenhouse is because the warm inside air is not able to mix with the cooler air outside, and hence the inside temperatures continue to increase through solar heating.

In contrast, when applied to the atmosphere–Earth system, we use the term *greenhouse effect* to refer to any process in which longwave radiation that would have left the top of the atmosphere is absorbed and counterradiated back to the surface. This counterradiation from the atmosphere to the Earth's surface helps warm the surface through longwave heating, not shortwave heating as in a true greenhouse.

The counterradiation itself depends strongly on the presence of water vapor—which on a global scale absorbs more outgoing longwave radiation than any other greenhouse gas—and carbon dioxide in the atmo-

greenhouse effect Accumulation of heat in the lower atmosphere and at the Earth's surface; produced through the absorption of longwave radiation by the atmosphere and the re-emission back to the surface.

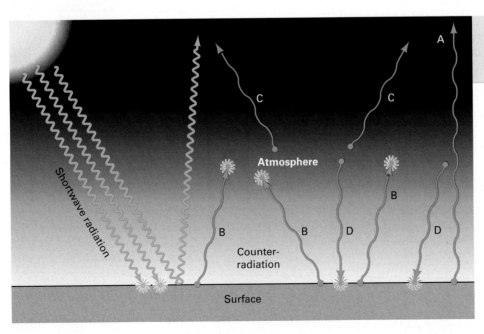

Counterradiation and the greenhouse effect

FIGURE 3.20

Shortwave radiation (left) passes through the atmosphere and is absorbed at the surface, warming the surface. The surface emits longwave radiation. Some of this flow passes directly to space (**A**), but most is absorbed by the atmosphere (**B**). In turn, the atmosphere radiates longwave energy back to the surface as counterradiation (**D**) and also to space (**C**). The return of outbound longwave radiation (**B**) by counterradiation (**D**) constitutes the greenhouse effect.

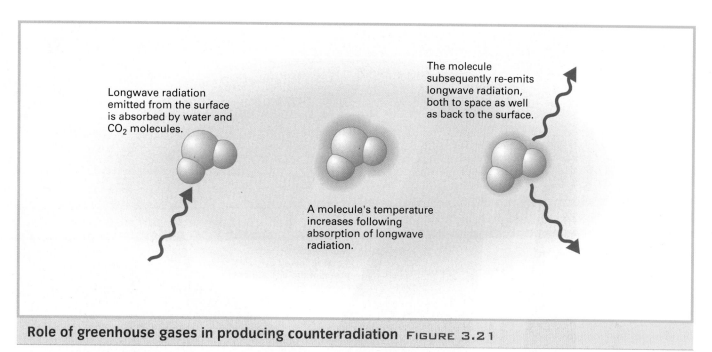

Longwave radiation emitted from the surface is absorbed by water and CO_2 molecules.

A molecule's temperature increases following absorption of longwave radiation.

The molecule subsequently re-emits longwave radiation, both to space as well as back to the surface.

Role of greenhouse gases in producing counterradiation FIGURE 3.21

Because they absorb longwave radiation, water vapor and carbon dioxide are effective greenhouse gases in the atmosphere.

sphere, as well as other gases such as methane and ozone. As shown in FIGURE 3.21, much of the longwave radiation emitted upward from the Earth's surface is absorbed by these gases. This absorbed energy raises the temperature of the atmosphere, causing it to emit more counterradiation towards the surface as well as radiation to space. Thus, the lower atmosphere, with its longwave-absorbing gases, acts like a blanket that traps heat underneath it (for this reason, greenhouse gases are sometimes referred to as *heat-trapping gases*). Cloud layers, which are composed of tiny water droplets, are also an important contributor in producing a blanketing effect, because liquid water is also a strong absorber of longwave radiation.

GLOBAL ENERGY BUDGETS OF THE ATMOSPHERE AND SURFACE

Thus far, we have described a number of important pathways by which incoming solar radiation is scattered, absorbed, and reradiated as longwave radiation. It is an important principle of physics that, except for nuclear reactions in which mass is converted to energy, energy is neither created nor destroyed. Just like a household-income budget, the energy flows we have been looking at between the Sun and the Earth's atmosphere and surface must balance over the long term. The global energy budget shown in FIGURE 3.22 on pages 72–73, takes into account all the important energy flows and helps us to understand how changes in these flows might affect the Earth's climate.

Incoming shortwave radiation
We begin with a discussion of solar shortwave radiation. Here we will trace the flow of solar radiation, which totals 100 units. Reflection by molecules and dust, clouds, and the surface (including the oceans) totals 31 percentage units. Expressed as a proportion (0.31), this is the albedo of the Earth and atmosphere as a combined system. In addition, the combined losses through absorption by molecules, dust, and clouds in the atmosphere average 20 percentage units. With 31 percentage units of the incoming solar energy flow reflected and 20 percentage units absorbed in the atmosphere, 49 units are left. These units are absorbed by the Earth's land and water surfaces. We have now accounted for all of the incoming flow of insolation.

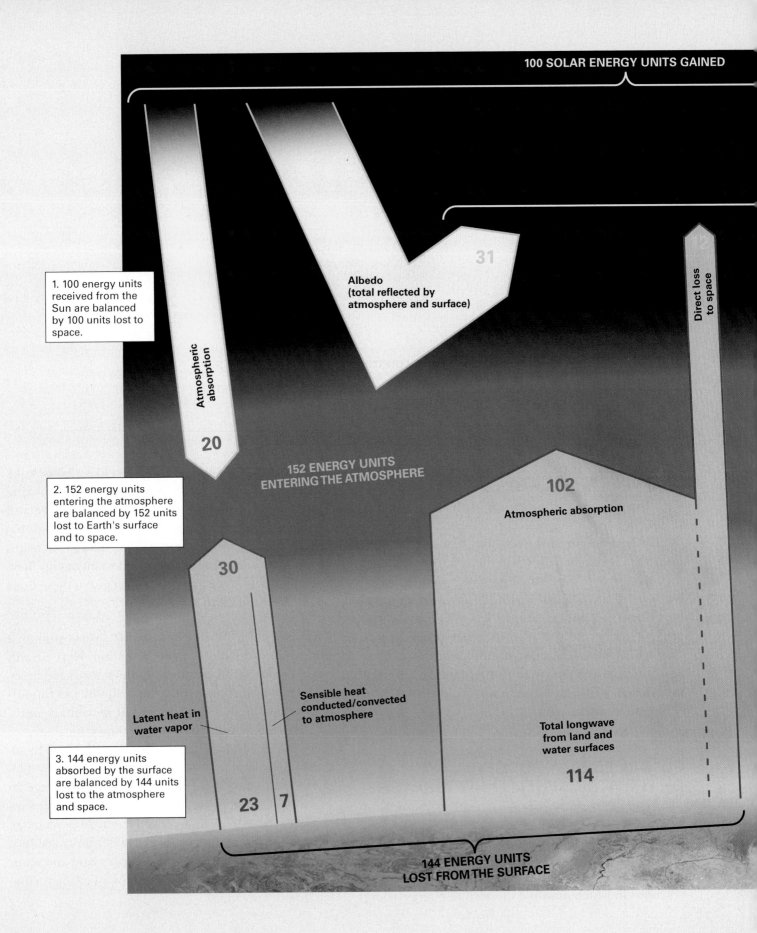

100 SOLAR ENERGY UNITS GAINED

Albedo
(total reflected by
atmosphere and surface)

31

Direct loss
to space

12

Atmospheric
absorption

20

1. 100 energy units
received from the
Sun are balanced
by 100 units lost to
space.

**152 ENERGY UNITS
ENTERING THE ATMOSPHERE**

102

Atmospheric absorption

2. 152 energy units
entering the atmosphere
are balanced by 152 units
lost to Earth's surface
and to space.

30

Sensible heat
conducted/convected
to atmosphere

Latent heat in
water vapor

Total longwave
from land and
water surfaces

114

3. 144 energy units
absorbed by the surface
are balanced by 144 units
lost to the atmosphere
and space.

23 **7**

**144 ENERGY UNITS
LOST FROM THE SURFACE**

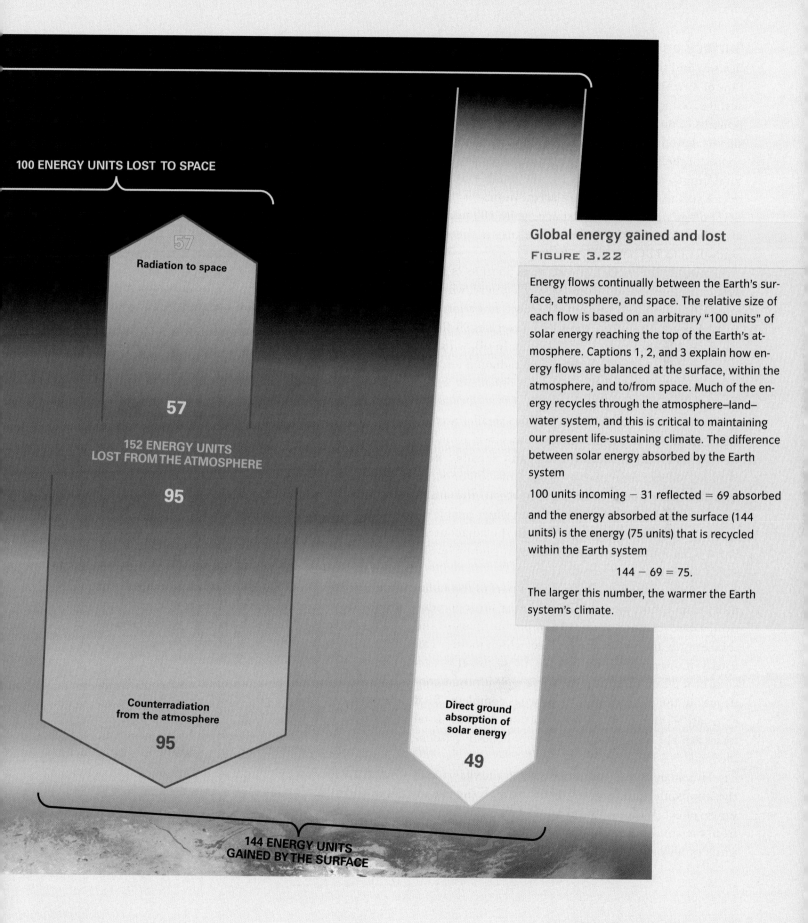

100 ENERGY UNITS LOST TO SPACE

57
Radiation to space

57

152 ENERGY UNITS LOST FROM THE ATMOSPHERE

95

Counterradiation from the atmosphere

95

Direct ground absorption of solar energy

49

144 ENERGY UNITS GAINED BY THE SURFACE

Global energy gained and lost

FIGURE 3.22

Energy flows continually between the Earth's surface, atmosphere, and space. The relative size of each flow is based on an arbitrary "100 units" of solar energy reaching the top of the Earth's atmosphere. Captions 1, 2, and 3 explain how energy flows are balanced at the surface, within the atmosphere, and to/from space. Much of the energy recycles through the atmosphere–land–water system, and this is critical to maintaining our present life-sustaining climate. The difference between solar energy absorbed by the Earth system

100 units incoming − 31 reflected = 69 absorbed

and the energy absorbed at the surface (144 units) is the energy (75 units) that is recycled within the Earth system

$$144 - 69 = 75.$$

The larger this number, the warmer the Earth system's climate.

Surface energy flows Let's now turn to energy flows to and from the Earth's surface. We've noted the flow of 49 units of shortwave radiation absorbed by the surface, but what about longwave radiation? The components of outgoing longwave radiation for the Earth's surface and the atmosphere are given by arrows in the center of the figure. The large arrow on the left shows total longwave radiation leaving the Earth's land and ocean surface. The long, thin arrow indicates that 12 percentage units are lost to space, while 102 units are absorbed by the atmosphere. The loss is therefore equivalent to 114 percentage units.

However, the surface also receives 95 units of longwave counterradiation from the atmosphere, which is in addition to the 49 units of shortwave radiation that are absorbed. Therefore, the surface receives 95 (longwave) + 49 (shortwave) = 144 units in all, which is more than the 114 units of longwave radiation emitted by the Earth's surface. The result is an imbalance in the net radiation at the surface.

On the far left of the figure are two smaller arrows, which show that this excess energy flows away from the surface as latent heat (23 units) and sensible heat (7 units). Evaporation of water from moist soil or ocean transfers latent heat from the surface to the atmosphere. Sensible heat transfer occurs when heat is directly conducted from the surface to the adjacent air layer. This warm air is then mixed to higher levels, which we refer to as convection.

These last two heat flows are not part of the radiation balance, because they are not in the form of radiation. However, they are a very important part of the total energy budget of the surface, which includes all forms of energy. Taken together, these two flows account for 30 units leaving the surface. With their contribution, the surface energy balance is complete. Total gains are 95 (longwave) + 49 (shortwave) = 144. Total losses are 114 (longwave) + 23 (latent heat) + 7 (sensible heat) = 144.

This analysis helps us to understand the vital role of the atmosphere in trapping heat through the greenhouse effect. Without the 95 units of counterradiation from the atmosphere, the surface would only receive 49 units of absorbed shortwave radiation, which would produce a temperature well below freezing (approximately −18°C or 0°F). Without the greenhouse effect, our planet would be a cold, forbidding place.

Energy flows to and from the atmosphere

What about the atmosphere? Let's first look at energy flowing into the atmosphere. Like the Earth's surface, the atmosphere gains energy by absorption of shortwave radiation, amounting to 20 units. The atmosphere also gains energy from latent heat transfer (23 units) and sensible heat transfer (7 units). With regard to longwave radiation, the atmosphere absorbs 102 units emitted by the surface. These units total 152.

Looking at energy leaving the atmosphere, we see a loss of 57 units of longwave radiation to space and a loss of 95 units in counterradiation to the surface. These total 152 units, so we see that the atmosphere's energy budget is also balanced.

Energy flows to and from space Finally, we can look at the energy balance for the entire Earth system. As mentioned, the Earth system receives 100 units of energy from the Sun in the form of shortwave radiation. Of this energy, 31 units of shortwave radiation are reflected back and do not interact with the Earth system. In addition, 12 units of longwave radiation from the surface are transmitted through the entire atmosphere and out to space. The main term that balances the incoming solar radiation is the outgoing longwave radiation emitted by the atmosphere (57 units). The 57 units emitted by the atmosphere, combined with the 12 units emitted by the Earth's surface, exactly balance the 69 units absorbed by the full Earth system.

CLIMATE AND GLOBAL CHANGE

Our analysis also helps us to understand how human activity might affect the Earth's climate (**FIGURE 3.23**). For instance, human activity around the globe has added extra carbon dioxide to the atmosphere. The extra carbon dioxide absorbs more longwave energy as the energy passes through the atmosphere. This absorption makes the atmosphere warmer, boosting counter-

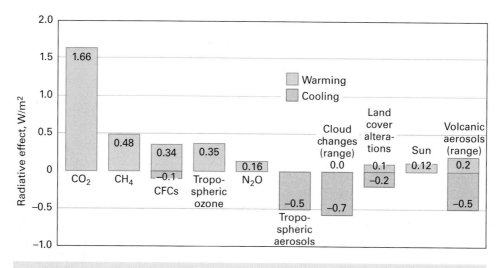

Factors affecting global warming and cooling FIGURE 3.23

Greenhouse gases such as CO_2, CH_4, CFCs, tropospheric ozone, and N_2O act primarily to enhance global warming, while aerosols, cloud changes, and land-cover alterations caused by human activity act to reduce global warming. Natural factors may be either positive or negative.

radiation and increasing the greenhouse effect. The increased counterradiation associated with the greenhouse effect can subsequently raise temperatures at the surface as well as in the atmosphere.

On the other hand, aerosols such as ash from forest fires and sulfates and nitrates from industrial activity tend to increase the albedo of the atmosphere and the Earth. In that case, less energy is absorbed by the Earth system, lowering its temperature. Similarly, clearing forests for agriculture also increases the surface albedo, which should also lower global temperatures if the change in the land cover is large enough.

Have these activities—such as fossil fuel burning and deforestation—irrevocably shifted the balance of energy flows? This is a very complex question. Overall, the energy flows between the Sun, surface, atmosphere, and space are all critical components of our climate system, as illustrated in FIGURE 3.24 on pages 76–77. If we want to understand the human impact on the Earth–atmosphere system, we need to examine how our activities change these different flows of energy in the global energy balance.

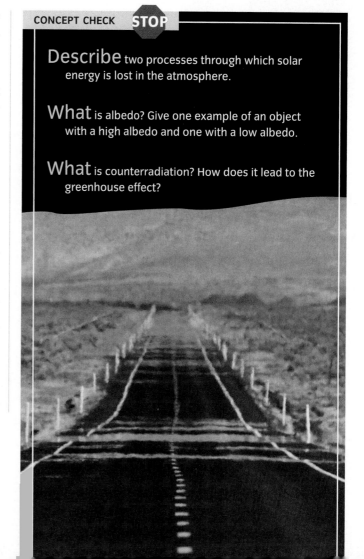

CONCEPT CHECK **STOP**

Describe two processes through which solar energy is lost in the atmosphere.

What is albedo? Give one example of an object with a high albedo and one with a low albedo.

What is counterradiation? How does it lead to the greenhouse effect?

THE GREENHOUSE EFFECT

The Earth's atmosphere acts like a blanket on a cold night, trapping heat to warm the Earth's surface.

Greenhouse heat trap ▶

The atmosphere acts like a greenhouse, allowing sunlight to filter through. Gases such as carbon dioxide, water vapor, methane, ozone, and nitrous oxide help the atmosphere hold heat. This heating is key in the Earth's ability to stay warm and sustain life.

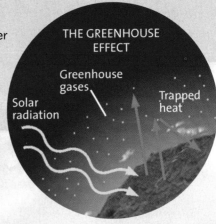

THE GREENHOUSE EFFECT

Greenhouse gases

Solar radiation

Trapped heat

◀ Fossil fuel burning

When fossil fuels are burned to produce power, CO_2 is released. As global CO_2 levels rise, the greenhouse effect is enhanced.

What shapes the Earth's climate ▶

Much of the Sun's heat (1) is held in the atmosphere (2) by greenhouse gases as well as in the top layer of oceans. Oceans (3) distribute heat; evaporation lifts moisture (4). Clouds (5) reflect sunlight and cool the Earth; they also warm it by trapping heat. Ice and snow (6) reflect sunlight, cooling the Earth. Land (7) can influence the formation of clouds and the albedo of the Earth's surface, and human use (8) can alter natural processes.

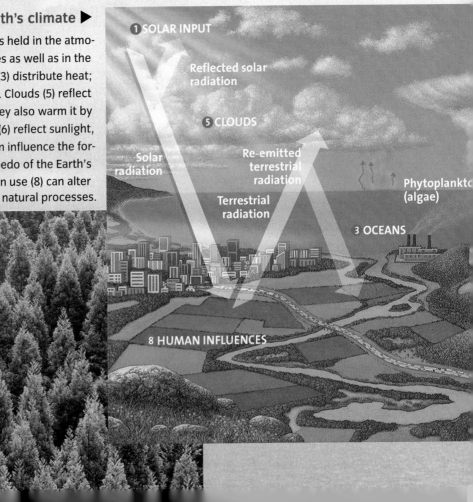

1 SOLAR INPUT

Reflected solar radiation

5 CLOUDS

Solar radiation

Re-emitted terrestrial radiation

Terrestrial radiation

Phytoplankton (algae)

3 OCEANS

8 HUMAN INFLUENCES

Global vegetation ▶

Plants take up CO_2 in photosynthesis, removing it from the atmosphere. But when the plants die, microorganisms digest the plant matter, returning the CO_2 to the atmosphere. CO_2 is also released by burning.

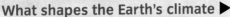

GLOBAL ENERGY FLOWS

Energy exchanges between oceans, lands, and atmosphere control the Earth's climate system.

◀ **Solar power**

Shortwave radiation from the Sun is the power source that heats the Earth.

▲ **Oceans**

Oceans play a key role in the climate system. Ocean currents carry warm water poleward and cool water equatorward, moving heat around the globe. Ocean waters evaporate, carrying latent heat upward into the atmosphere. When the water vapor condenses, the latent heat is released at a different location.

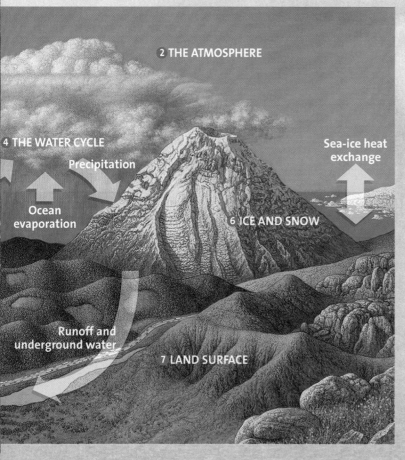

2 THE ATMOSPHERE

4 THE WATER CYCLE

Precipitation

Sea-ice heat exchange

Ocean evaporation

6 ICE AND SNOW

Runoff and underground water

7 LAND SURFACE

▲ **Clouds**

Clouds are also an important part of the climate system. Low, thick clouds reflect solar energy back to space, which tends to cool the Earth. High, thin clouds absorb upwelling longwave radiation, enhancing the greenhouse effect and warming the Earth.

What is happening in this picture ?

The Unaka Mountains in North Carolina are part of the Appalachian Mountain range, which also includes the Great Smoky Mountains and the Blue Ridge Mountains. In this range, evaporated water vapor from the surface and biogenic chemicals emitted by the trees can interact with incoming solar radiation in many ways.

- Where is reflection of sunlight greatest?

- How does this reflection compare with the reflection of sunlight by the trees at the surface?

- Why do the mountains in the background look more faint than those in the foreground?

SUMMARY

1 Electromagnetic Radiation

1. All objects give off **electromagnetic radiation**. Hotter objects emit greater amounts of electromagnetic radiation. They also emit radiation at shorter wavelengths than cooler objects.

2. The radiation from the Sun is called **shortwave radiation**, while the radiation from objects on the Earth and in the atmosphere is called **longwave radiation**.

3. On a global scale, the amount of shortwave radiation absorbed by the Earth system exactly balances the amount of longwave radiation the Earth emits back to space.

2 Geographic Variations in Energy Flow

1. **Insolation** is the rate of solar radiation flow at a location at a given time. It varies with the angle of the incoming sunlight as well as the length of time the sunlight hits a particular location.

2. At a given location and during a given time of year, there can be an imbalance between the amount of absorbed shortwave radiation and emitted longwave radiation. The difference between the two is called the **net radiation**. On average, at latitudes below 40 degrees, annual net radiation is positive, while it is negative at higher latitudes.

3. **Latent heat** is taken up or released when substances change between solid, liquid, and gas states. **Sensible heat** is a measure of the internal motions of the

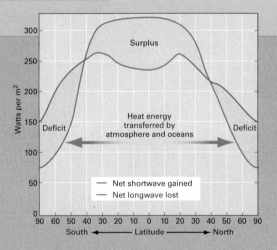

molecules within an object. Sensible heat is taken up or released to surrounding objects by conduction or convection.

4. The net radiation imbalance drives latent and sensible heat toward the poles, creating broad-scale atmospheric circulation patterns, weather events, and ocean currents.

3 The Global Energy System

1. Incoming solar radiation is partly absorbed or scattered by molecules in the atmosphere and at the surface. The **albedo** of an object gives the proportion of radiation that an object reflects.

2. The atmosphere absorbs longwave energy from the Earth. It also radiates longwave energy to space and counterradiates some back to the Earth. The **counterradiation** of longwave energy from the atmosphere back to the surface creates the **greenhouse effect**. The strength of the greenhouse effect is increased or decreased by the amount of greenhouse gases in the atmosphere.

3. On a global scale, the energy flows at the Earth's surface and atmosphere all balance. Changes in these flows, however, can produce changes in the temperature of the Earth's surface and hence its climate.

KEY TERMS

- **electromagnetic radiation** p. 52
- **emission** p. 52
- **wavelength** p. 52
- **absorption** p. 57
- **scattering** p. 58
- **transmission** p. 58

- **shortwave radiation** p. 58
- **longwave radiation** p. 59
- **insolation** p. 62
- **net radiation** p. 63
- **sensible heat** p. 64
- **latent heat** p. 64

- **phase** p. 64
- **albedo** p. 67
- **counterradiation** p. 70
- **greenhouse effect** p. 70

CRITICAL AND CREATIVE THINKING QUESTIONS

1. What is electromagnetic radiation? How is it characterized? Identify the major regions of the electromagnetic spectrum.

2. What is the Earth's global energy balance, and how are shortwave and longwave radiation involved?

3. How is sensible heat transferred from the Earth's surface to the atmosphere?

How is latent heat transferred? On a global scale, which of these two heat-transfer processes is larger?

4. Describe the counterradiation process and how it relates to the greenhouse effect.

5. What is net radiation? What is the role of poleward heat transport in balancing the net radiation budget by latitude?

6. Imagine that you are following a beam of either (a) shortwave solar radiation entering the Earth's atmosphere heading toward the surface, or (b) a beam of longwave radiation emitted from the surface heading toward space. How will the atmosphere influence the beam?

1. In the case of electromagnetic energy, an object that is hot
   _____.
   a. radiates much more energy than a cool object
   b. radiates much less energy than a cool object
   c. radiates the same amount of energy as a cool object
   d. absorbs a huge amount of energy

2. The highest energy, shortest wavelength form of electromagnetic radiation emitted by the Sun is _____.
   a. shortwave infrared
   b. visible light
   c. thermal infrared radiation
   d. ultraviolet radiation

3. On this figure, give the approximate wavelengths for each of the colors you see.

4. The amount of insolation falling on the surface of the Earth depends on _____.
   a. the angle the Sun's rays make with the Earth's surface
   b. the longitude of a location
   c. the amount of glacial coverage in a particular area
   d. the amount of ocean surface in a particular region

5. On the figure, give the approximate values for the amount of solar radiation involved in each process.

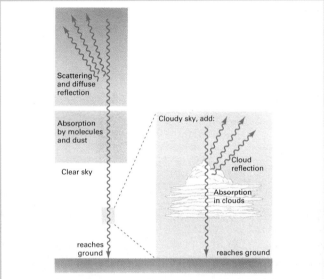

6. The transfer of sensible heat that occurs when two objects of unlike temperature are in contact is called _____.
   a. convection
   b. radiation
   c. latent transfer
   d. conduction

7. Latent heat is _____.
   a. similar to sensible heat in that it can be measured with a thermometer
   b. transferred through the mechanism of conduction
   c. found only in the various states of water
   d. stored heat and as such cannot be directly measured with a thermometer

8. What process is responsible for the turning aside of light rays in the atmosphere?
   a. Scattering
   b. Absorption
   c. Convection
   d. Transmission

9. The percentage of shortwave radiant energy scattered upward by a surface is termed its _____.
   a. outflow
   b. radiation
   c. albedo
   d. reflection

10. The ground can only radiate longwave radiation _____, but the atmosphere radiates longwave radiation _____.
   a. upward, downward
   b. upward, in all directions
   c. downward, upward
   d. upward, at right angles to the upward flow

11. On the figure, indicate which letters represent the following processes: transmitted shortwave radiation, reflected short-wave radiation, longwave radiation from the surface to the at-mosphere, longwave radiation from the surface to space, counterradiation from the atmosphere, longwave radiation from the atmosphere to space.

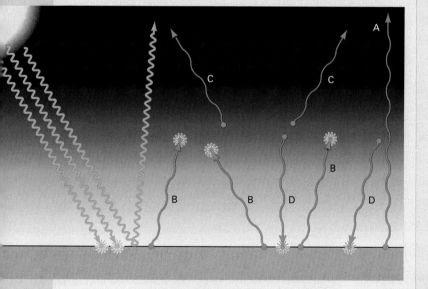

12. _____ produces the largest greenhouse effect in today's atmosphere:
   a. Carbon dioxide
   b. Methane
   c. Nitrous oxides
   d. Water vapor

13. Without the greenhouse effect, our planet _____.
   a. would be much warmer
   b. would be much colder
   c. would have a much lower surface albedo
   d. would not be any different than it is right now

14. _____ is the difference between all incoming radiation and all outgoing radiation.
   a. Net radiation
   b. Counterradiation
   c. The greenhouse effect
   d. Albedo

15. Poleward heat transfer involves the flow of heat by _____ and _____ from the lower latitudes to the polar regions.
   a. radiation, latent heat
   b. oceanic currents, atmospheric currents
   c. carbon dioxide, oceanic currents
   d. oceanic currents, radiation

Surface Temperature and Its Variation

Up near the Arctic Circle, winters in Yakutsk, Russia can be downright nasty. Temperatures dip below −40°C (−40°F). Daylight is scarce, and snow and ice cover the ground for more than six months.

The principal cause of these conditions is a lack of the Sun's warming rays. At these high latitudes, incoming solar radiation comes in at a glancing angle, spreading over a large area and thereby decreasing in intensity. This effect is stronger in winter, when the northern hemisphere is tilted away from the Sun so that the Sun is only above the horizon for a short period of time, and even then it sits very low in the sky. Latitude does not play the only role, however. Yakutsk sits in the middle of the great Eurasian continent. Here, decreases in solar heating cause much more dramatic temperature drops than at coastal locations, where the oceans moderate temperatures.

The people who live here have adapted to these conditions. Heavy winter clothing is a necessity. Buildings must be heated constantly and heavily insulated. Vehicles have studded tires. Planes have skis for landing gear. At times, horses and mules provide better transportation than any motorized vehicle.

In this chapter, we consider the factors that affect temperature at a given location as well as temperature variations over time. We also discuss how temperature is measured and what these measurements can tell us about temperature changes day-to-day as well as over many years.

Yakutsk, Russia is located not far from the Arctic Circle in Sakha, Russia. Its high latitude and its interior continental setting produce winters that are very cold.

YAKUTSK, RUSSIA

Global Locator

NATIONAL GEOGRAPHIC

The Earth's Rotation and Orbit

Since the time of Copernicus—the 16th century—we have known that the daily and seasonal motions of the Sun are a consequence of the Earth's motion rather than the Sun's. From day to day, the Earth's rotation on its axis produces the alternation of light and darkness that we experience in 24 hours. From month to month, the revolution of the Earth around the Sun produces slow changes in the length of daylight that create the rhythm of the seasons. These two great cycles continue, day after day, year after year, regulating the processes of life on the Earth.

EARTH'S ROTATION

Our planet spins slowly, making a full turn with respect to the Sun every day. We use the term *rotation* to describe this motion. One complete rotation with respect to the Sun defines the solar day. By convention, the solar day is divided into exactly 24 hours.

The Earth rotates on its axis, an imaginary straight line through its center. The two points where the axis of rotation intersects the Earth's surface are defined as the *poles*, one called the *North Pole* and the other the *South Pole*.

To determine the direction of Earth's rotation, you can use one of the following guidelines, as shown in **FIGURE 4.1**:

- Imagine you are looking down on the North Pole of the Earth. From this position, the Earth is turning in a counterclockwise direction.

- Imagine you are off in space, viewing the planet much as you would view a globe in a library, with the North Pole on top. The Earth is rotating from left to right.

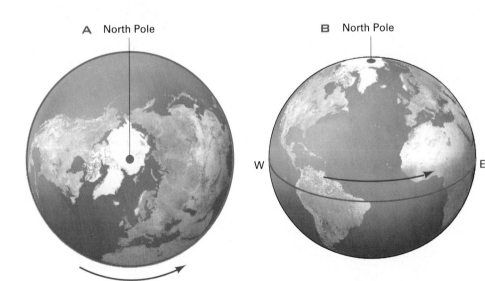

A North Pole

B North Pole

W   E

Direction of the Earth's rotation FIGURE 4.1

The direction of rotation of the Earth can be thought of as **A** counterclockwise at the North Pole, or **B** from left to right (eastward) at the Equator.

A **Daily solar cycle** Sunflowers turn their heads, following the Sun in the sky through the day.

B **Daily tidal cycle** Boats in the Bay of Fundy, stranded twice each day by low tide, also reveal the Earth's rotation.

Cycles driven by the Earth's rotation FIGURE 4.2

Environmental effects of the Earth's rotation

The Earth's rotation is important for three reasons. First, the axis of rotation serves as a reference in setting up the geographic grid of latitude and longitude. Second, it provides the day as a convenient measure of the passage of time. Third, the Earth's rotation affects the physical and life processes on Earth because it imposes a daily, or **diurnal**, rhythm to daylight, air temperature, and atmospheric and oceanic motions (**FIGURE 4.2**).

One effect of rotation is that all surface processes respond to the diurnal rhythm. Green plants receive and store solar energy during the day and consume some of it at night. Also, the amount of radiation absorbed and emitted at a given location changes with the course of the day, which in turn leads to changes in temperature.

A second environmental effect is that the flow paths of both air and water are in-

> **diurnal** A type of process that takes place or changes over the course of a day.

fluenced by the Earth's rotation. Indeed, the Earth's rotation actually introduces a force—termed the *Coriolis force*—that causes air and water in motion to veer from their direction of movement. This effect is of great importance in studying the Earth's systems of wind and ocean currents.

A third environmental effect of the Earth's rotation is upon the shape of the Earth. The familiar images of our round planet that we take for granted are a little misleading. As the Earth spins, the force needed to keep all objects on the Earth rotating—the *centripetal force*—lessens the local force of gravity. As the gravitational force weakens, it causes the Earth to bulge slightly at the Equator. The difference is very small—about three-tenths of 1 percent—but strictly speaking, the Earth's expanded shape is closer to what is known as an *oblate ellipsoid* than to a sphere.

THE EARTH'S REVOLUTION AROUND THE SUN

So far, we have discussed the importance of the Earth's rotation on its axis. Another important motion of the Earth is its *revolution*, or its movement in orbit around the Sun. The Earth completes a revolution around the Sun in 365.242 days—about one-fourth day more than the calendar year of 365 days. Every four years, the four extra one-fourth days add up to about one whole day, and we insert a 29th day in February in leap years to correct the calendar for this effect.

In which direction does the Earth revolve? If you were to look down on the North Pole of the Earth, you would see that the Earth travels counterclockwise around the Sun. This direction is the same as that of the Earth's rotation (FIGURE 4.3).

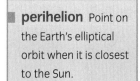

perihelion Point on the Earth's elliptical orbit when it is closest to the Sun.

aphelion Point on the Earth's elliptical orbit when it is farthest from the Sun.

In addition, the Earth's orbit around the Sun is shaped like an ellipse, or oval. This means that the distance between the Earth and Sun varies somewhat throughout the year. The Earth is nearest to the Sun at **perihelion**, which presently occurs on or about January 3. It is farthest away from the Sun at **aphelion**, on or about July 4. Because the distance between Sun and Earth varies only by about 3 percent during one revolution, we can regard the orbit as circular for most purposes, although on a time scale of tens of thousands of years, the change in the elliptical orbit of the Earth does become important.

Note that the Earth is closest to the Sun during the middle of winter in the northern hemisphere. Therefore, the change of the distance of the Earth from the Sun is not the cause of the seasons. Indeed, there is a much more important phenomenon that produces the seasons.

Revolution of the Earth around the Sun FIGURE 4.3

The Earth follows an elliptical path around the Sun.

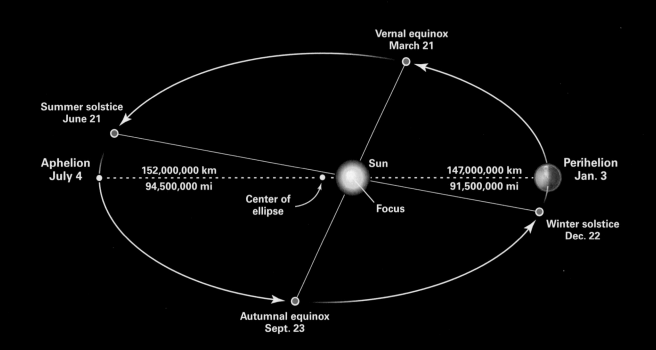

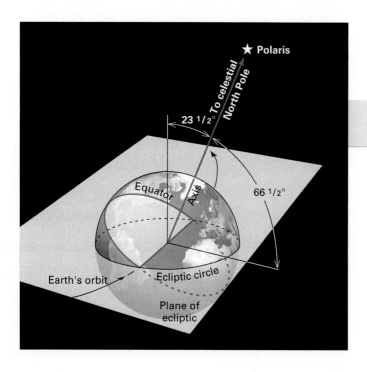

The tilt of the Earth's axis of rotation with respect to its orbital plane FIGURE 4.4

As the Earth moves in its orbit on the plane of the ecliptic around the Sun, its rotational axis remains pointed toward Polaris, the North Star, and makes an angle of 66 1/2° with the ecliptic plane. We can see that the axis of the Earth, which passes through its North and South Poles, is tilted at an angle of 23 1/2° away from a right angle to the plane of the ecliptic.

Tilt of the Earth's axis Depending on where you live in the world, the effects of the changing seasons can be large—but why do we experience seasons on Earth? Why do the hours of daylight change throughout the year—most extremely at the poles, and less so near the Equator?

Seasons arise because the Earth's rotational axis is not perpendicular to the plane containing the Earth's orbit around the Sun, which is known as the *plane of the ecliptic.* FIGURE 4.4 shows this plane as it intersects the Earth.

SOLSTICE AND EQUINOX

We know the Earth revolves in its orbit over the course of approximately one year. Because the direction of the Earth's axis of rotation is fixed, the North Pole is tilted away from the Sun during one part of the year and toward the Sun during the other part.

The point at which the North Pole is tilted at the maximum angle away from

winter (December) solstice Moment in time when the North Pole is directed 23.5° away from the Sun.

summer (June) solstice Moment in time when the North Pole is directed 23.5° toward the Sun.

equinox Time of year when the Earth's axis of rotation is neither pointed toward the Sun nor away from it.

the Sun is called the **winter solstice** in the northern hemisphere. While it is winter in the northern hemisphere, at this time the southern hemisphere is tilted toward the Sun and enjoys strong solar heating, producing summer in the southern hemisphere. You can therefore use the term **December solstice** to avoid any confusion.

When the Earth is on the opposite side of its orbit and the North Pole end of the axis is tilted at 23 1/2° toward the Sun, the event is called the **summer (or June) solstice.** The North Pole and northern hemisphere are tilted toward the Sun, while the South Pole and southern hemisphere are tilted away.

Midway between the solstice dates, the **equinoxes** occur. At an equinox, the Earth's axial tilt is neither toward nor away from the Sun. The vernal equinox occurs on March 21, and the autumnal equinox occurs on September 23. The Earth–Sun relationships are identical on the two equinoxes. Note also that the date of any

Earth–Sun relations through the year The four seasons occur because the Earth's tilted axis keeps a constant orientation in space as the Earth orbits about the Sun. This figure shows the Earth as it orbits around the Sun over the span of a year, passing through each of its four seasons.

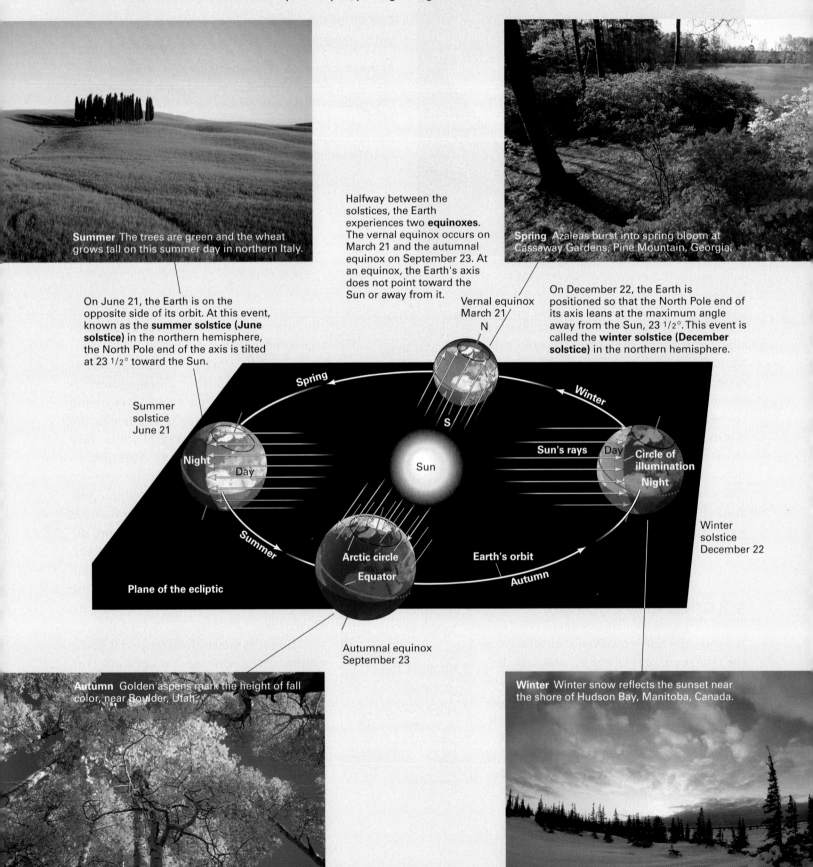

Summer The trees are green and the wheat grows tall on this summer day in northern Italy.

Spring Azaleas burst into spring bloom at Cassaway Gardens, Pine Mountain, Georgia.

Halfway between the solstices, the Earth experiences two **equinoxes**. The vernal equinox occurs on March 21 and the autumnal equinox on September 23. At an equinox, the Earth's axis does not point toward the Sun or away from it.

On June 21, the Earth is on the opposite side of its orbit. At this event, known as the **summer solstice (June solstice)** in the northern hemisphere, the North Pole end of the axis is tilted at 23 1/2° toward the Sun.

On December 22, the Earth is positioned so that the North Pole end of its axis leans at the maximum angle away from the Sun, 23 1/2°. This event is called the **winter solstice (December solstice)** in the northern hemisphere.

Vernal equinox
March 21
N

Summer
solstice
June 21

Spring

Winter

S

Night

Day

Sun's rays Day

Circle of
illumination

Night

Sun

Winter
solstice
December 22

Summer

Arctic circle

Earth's orbit

Equator

Autumn

Plane of the ecliptic

Autumnal equinox
September 23

Autumn Golden aspens mark the height of fall color, near Boulder, Utah.

Winter Winter snow reflects the sunset near the shore of Hudson Bay, Manitoba, Canada.

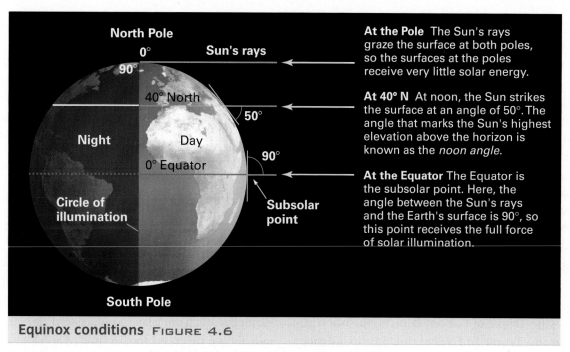

North Pole

0°

90°

Sun's rays

40° North

50°

Night　　**Day**

90°

0° Equator

Circle of illumination

Subsolar point

South Pole

At the Pole The Sun's rays graze the surface at both poles, so the surfaces at the poles receive very little solar energy.

At 40° N At noon, the Sun strikes the surface at an angle of 50°. The angle that marks the Sun's highest elevation above the horizon is known as the *noon angle.*

At the Equator The Equator is the subsolar point. Here, the angle between the Sun's rays and the Earth's surface is 90°, so this point receives the full force of solar illumination.

Equinox conditions FIGURE 4.6

At equinox, the Earth's axis of rotation is exactly at right angles to the direction of solar illumination. The circle of illumination passes through the North and South Poles. The subsolar point lies on the Equator. At both poles, the Sun is seen at the horizon.

solstice or equinox in a particular year may vary by a day or so, since the Earth's revolution period is not exactly 365 days (FIGURE 4.5).

Equinox conditions

The Sun's rays always divide the Earth into two hemispheres—one that is bathed in light and one that is shrouded in darkness. The **circle of illumination** is the circle that separates the day hemisphere from the night hemisphere. The **subsolar point** is the single point on the Earth's surface where the Sun is directly overhead at a particular moment.

At equinox, the circle of illumination passes through the North and South Poles (FIGURE 4.6). The subsolar point falls on the Equator. At other latitudes at noon, the Sun is less than 90° overhead. This angle is

circle of illumination The circle that separates the illuminated portion of the Earth from the portion that is not illuminated.

subsolar point The one point on Earth at any given time where the Sun is directly overhead.

termed the *noon angle*. During equinox conditions, it can be shown that for either hemisphere, noon angle = 90° − magnitude of latitude. At 40° N, the noon angle is therefore 50°.

Imagine you are standing at a point on the Earth at a latitude of 40° N. The Earth rotates from left to right, sweeping you along with it, so that you turn with the globe. You will complete a full circuit in 24 hours. At the equinox, you will spend 12 hours in twilight and darkness and 12 hours in sunlight. Now place yourself at a different latitude—say, at 65° S. The same will be true here too, because the circle of illumination passes through the poles, dividing every parallel exactly in two. Thus, one important feature of the equinox is that day and night are of equal length everywhere on the globe.

Solstice conditions   Now let's examine the solstice conditions in **FIGURE 4.7**. The summer (June) solstice is shown at the top. Only the Equator is divided exactly in two by the circle of illumination. On the Equator, daylight and nighttime hours will be equal during the solstice, as well as throughout the year. The farther north you go, the more the given latitude lies within the circle of illumination, indicating more daylight hours. North of 66 1/2° N—the Arctic Circle—all locations at a given latitude lie within the circle of illumination. Even though the Earth rotates through a full cycle during a 24-hour period, the area north of the Arctic Circle remains in continuous daylight.

> ■ **declination**   Latitude of the subsolar point at a moment in time.

We can also see that the subsolar point is at latitude 23 1/2° N—the Tropic of Cancer. Because the noon angle is 90° at the Tropic of Cancer during this solstice, solar energy is most intense here. Away from the subsolar point, during the summer solstice we find that in the northern hemisphere, noon angle = 90° − magnitude of latitude + 23 1/2°. In the southern hemisphere, noon angle = 90° − magnitude of latitude − 23 1/2°.

The conditions are reversed at the winter (December) solstice. While the Equator still experiences 12 hours of daylight and nighttime, locations to the north now lie more outside the circle of illumination, indicating fewer daylight hours. At the same time, locations to the south experience more daylight hours. All the area south of latitude 66 1/2° S—the Antarctic Circle—lies within the circle of illumination, indicating that these regions are inundated with 24 hours of daylight.

The subsolar point has shifted to a point on the parallel at latitude 23 1/2° S—the Tropic of Capricorn. Here the noon angle is 90°. Away from the subsolar point, we find in the northern hemisphere that noon angle = 90° − magnitude of latitude − 23 1/2°. In the southern hemisphere, noon angle = 90° − magnitude of latitude + 23 1/2°.

The solstices and equinoxes are four special events that occur only once during the year. Between these times, the latitude of the subsolar point—the Sun's declination—travels northward and southward in an annual cycle, looping between the Tropics of Cancer and Capricorn. As it does so, the length of daylight and the intensity of insolation at a given latitude change slightly from one day to the next. In this way, the locations on the Earth experience the rhythm of the seasons as the Earth continues its orbit around the Sun.

VARIATIONS IN INSOLATION OVER THE GLOBE

Although the flow of solar radiation to the Earth as a whole remains relatively constant, the amount received varies greatly from place to place and time to time. Previously, we referred to the flow of solar energy, intercepted by an exposed surface assuming a uniformly spherical Earth with no atmosphere, as the *insolation* (incoming solar radiation).

Daily insolation at a location depends on two factors: (1) the angle at which the Sun's rays strike the Earth, and (2) the length of time of exposure to the rays. Both of these factors are controlled by the Sun's daily path in the sky. When the Sun is high above the horizon—near noon, for example—the Sun's angle with regard to the horizon is greater, so insolation will be greater. In addition, if the Sun remains above the horizon for a longer period of time, the insolation will increase. However, the Sun's daily path in the sky at various latitudes changes from season to season. For example, in midlatitude locations in summer, days are long, and the Sun rises to a position higher in the sky compared with its position during winter, thus heating the surface more intensely.

Insolation and the path of the Sun in the sky

Insolation and the Sun's path in the sky vary considerably over the course of the seasons for latitude 40° N, which is typical of conditions found in midlatitudes in the northern hemisphere—for example, in New York

Solstice conditions FIGURE 4.7

At the solstice, the north end of the Earth's axis of rotation is fully tilted either toward or away from the Sun. Because of the tilt, polar regions experience either 24-hour day or 24-hour night. The subsolar point lies on one of the tropics at latitude 23 1/2° N or S.

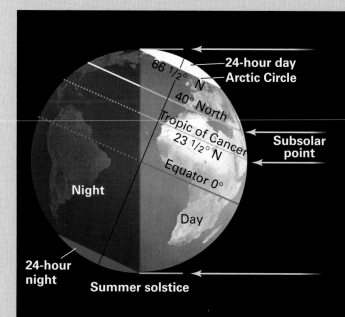

66 1/2° N (Arctic Circle) This latitude is positioned entirely within the daylight side of the circle of illumination. The day continues unbroken for 24 hours.

40° N The circle of illumination no longer divides this parallel into equal halves. Instead, daylight covers most of the parallel. Here, the day is about 15 hours and the night is about 9 hours.

23 1/2° N The subsolar point is at 23 1/2° N on the summer solstice. At noon, the angle between the Sun's rays and the Earth's surface is 90° here.

Equator The circle of illumination still divides this parallel into equal halves. Both day and night are 12 hours long here.

66 1/2° S (Antarctic Circle) This latitude is positioned entirely outside the daylight side of the circle of illumination. The night here continues unbroken for 24 hours.

40° N The circle of illumination no longer divides this parallel into equal halves. Instead, nighttime covers most of the parallel, so the day is about 9 hours and the night is about 15 hours.

23 1/2° S The subsolar point is at 23 1/2° S on the winter solstice. At noon, the angle between the Sun's rays and the Earth's surface is 90° here.

66 1/2° S (Antarctic Circle) This latitude is positioned entirely within the daylight side of the circle of illumination. The day here continues unbroken for 24 hours.

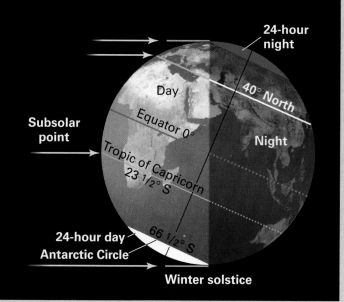

or Denver. The diagram in **FIGURE 4.8** shows a small area of the Earth's surface bounded by a circular horizon, which is the way things appear to an observer standing on a wide plain. The path of the Sun in the sky at equinox and solstices is shown.

By following the paths, we see that daily insolation will be greater at the June solstice than at the equinox, since the Sun is in the sky longer and reaches a higher angle at noon. At the December solstice, the Sun's path is low in the sky, reaching only 26 1/2° above the horizon, and the Sun is visible for only about nine hours. Thus, daily insolation reaching the surface at the December solstice will be less than at the equinox and much less than at the June solstice.

Now let's look at the Equator (**FIGURE 4.9**). During the equinox, the Sun rises directly to the east at 6 A.M. At noon it reaches a point directly overhead before setting at 6 P.M. directly to the west, 12 hours after it rose. At the December and June solstices, the Sun also rises at 6 A.M. and sets at 6 P.M., remaining above

the horizon for 12 hours. However, at noon, it is no longer directly overhead. Hence, daily insolation is smaller than at the equinoxes. In addition, for the Equator, there are two maximums of insolation—one at each equinox—and two minimums—one at each solstice. Thus, inhabitants at locations near the Equator may experience two hot seasons.

At the North Pole, shown in **FIGURE 4.10**, when the Sun is above the horizon it seems to trace a daily circle in the heavens that is parallel to the horizon. At the June solstice, the Sun reaches its highest point above the horizon. During the two equinoxes, the Sun's path follows the horizon. Between the September and March equinoxes, the Sun is below the horizon.

Hence, we see that daily insolation is greatest at the June solstice, when the Sun is at its highest position in the sky. Insolation then falls to zero at the September equinox, remains at zero between the September and March equinoxes, and then increases as the Sun rises above the horizon to reach its position at the June

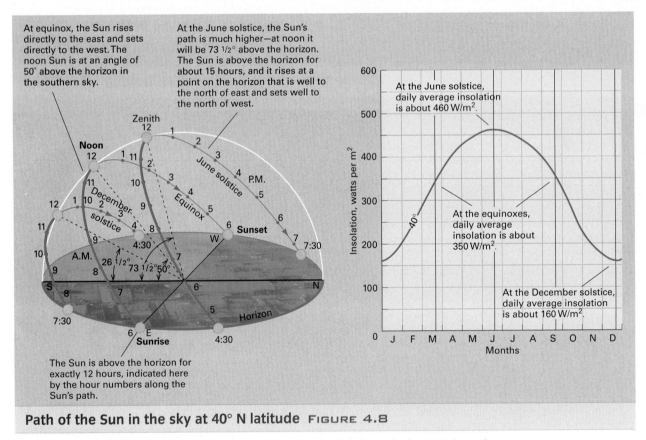

Path of the Sun in the sky at 40° N latitude FIGURE 4.8

The Sun's path in the sky can change greatly in position and height above the horizon through the seasons. The result is changes in daily insolation throughout the year, shown in the graph.

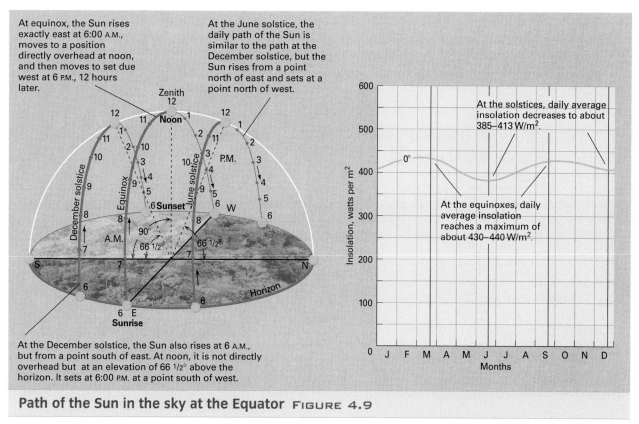

At equinox, the Sun rises exactly east at 6:00 A.M., moves to a position directly overhead at noon, and then moves to set due west at 6 P.M., 12 hours later.

At the June solstice, the daily path of the Sun is similar to the path at the December solstice, but the Sun rises from a point north of east and sets at a point north of west.

At the solstices, daily average insolation decreases to about 385–413 W/m².

At the equinoxes, daily average insolation reaches a maximum of about 430–440 W/m².

At the December solstice, the Sun also rises at 6 A.M., but from a point south of east. At noon, it is not directly overhead but at an elevation of 66 1/2° above the horizon. It sets at 6:00 P.M. at a point south of west.

Path of the Sun in the sky at the Equator FIGURE 4.9

The path of the Sun in the sky at the Equator does not produce as great a change in daily insolation as is found at 40° N. In addition, maximum values of daily insolation occur during the equinoxes, and minimums occur during the solstices.

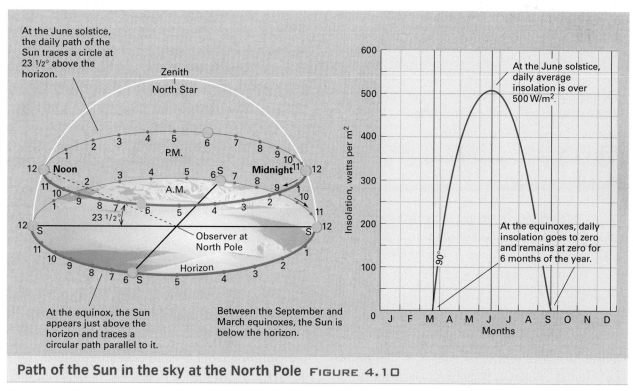

At the June solstice, the daily path of the Sun traces a circle at 23 1/2° above the horizon.

At the June solstice, daily average insolation is over 500 W/m².

At the equinoxes, daily insolation goes to zero and remains at zero for 6 months of the year.

At the equinox, the Sun appears just above the horizon and traces a circular path parallel to it.

Between the September and March equinoxes, the Sun is below the horizon.

Path of the Sun in the sky at the North Pole FIGURE 4.10

At the North Pole, the Sun remains above the horizon 24 hours a day for 6 months and below the horizon 24 hours a day for the other 6 months. The path of the Sun in the sky produces the most dramatic seasonal change in daily insolation of anywhere on the globe.

solstice again. Note that during the summer solstice, daily insolation—over 500 W/m²—is higher at the pole than at any other time of the year at either the Equator or 40° N, indicating the importance of day length in determining daily insolation.

Daily insolation through the year

What conclusions can we draw about daily insolation as it varies with latitude and season?

- Latitudes north of the Tropic of Cancer and south of the Tropic of Capricorn show greater daily insolation at the summer solstice and lower daily insolation at the winter solstice. The change in insolation over the course of the year gets larger the closer one moves toward the pole.

- Poleward of the Arctic and Antarctic Circles, the Sun is below the horizon for at least part of the year, and daily insolation drops to zero during that period.

- There are two maximums and two minimums in daily insolation at the Equator, occurring at the equinoxes and solstices, respectively.

- Poleward of the Equator but between the two tropics, there are also two maximum and minimum daily insolation values, but as the tropic is approached, the two maximum periods get closer and closer in time, and then merge into a single maximum at the tropic. The minimum in insolation occurs when the Sun is above the tropic of the opposite hemisphere.

These differences in daily insolation are important, because daily insolation measures the flow of solar power available to heat the Earth's surface and is one of the most important factors in determining air temperatures. The change in daily insolation with the seasons at a location is therefore a major determinant in producing changing weather in a given region as well as changes in climate.

Annual insolation by latitude

How does latitude affect annual insolation—the rate of insolation averaged over an entire year? FIGURE 4.11 shows two curves of annual insolation by latitude—one for the ac-

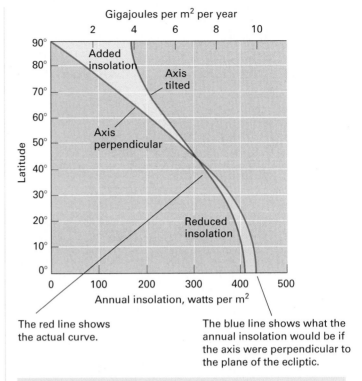

The red line shows the actual curve.

The blue line shows what the annual insolation would be if the axis were perpendicular to the plane of the ecliptic.

Annual insolation from Equator to pole for the Earth FIGURE 4.11

tual case of the Earth's axis tilted at 23 1/2°, and the other for an Earth with an untilted axis.

Let's look first at the real case of a tilted axis. We can see that annual insolation varies smoothly from the Equator to the pole and is greater at lower latitudes. However, high latitudes still receive a considerable flow of solar energy—the annual insolation value at the pole is about 40 percent of the value at the Equator.

Now let's look at what would happen if the Earth's axis were not tilted. With the axis perpendicular to the plane of the ecliptic, the situation is quite different. In this case, there are no seasons. Annual insolation is very high at the Equator, because the Sun passes directly overhead at noon every day throughout the year. However, annual insolation at the poles is zero, because the Sun's rays always skirt the horizon.

We can see that, without a tilted axis, our planet would be a very different place. The tilt redistributes a very significant portion of the Earth's insolation from the equatorial regions toward the poles, even without a contribution from oceanic and atmospheric transport. In fact, the difference in insolation between high and low latitudes determines how much excess energy needs to be transported by the ocean and atmosphere via weather and general-circulation processes. The decrease in the difference in insolation associated with the tilt of the Earth reduces the need for this energy transport.

World latitude zones   The annual insolation and seasonal changes of daily insolation can be used as a basis for dividing the globe into broad latitude zones, as shown in FIGURE 4.12. The zone limits shown in the figure and specified below should not be taken as

The Earth's diverse environments by latitude

FIGURE 4.12

Subarctic zone
Much of the subarctic zone is covered by evergreen forest, seen here with a ground cover of snow. Near Churchill, Hudson Bay region, Canada.

Midlatitude zone
A summer midlatitude landscape in the Tuscany region of Italy.

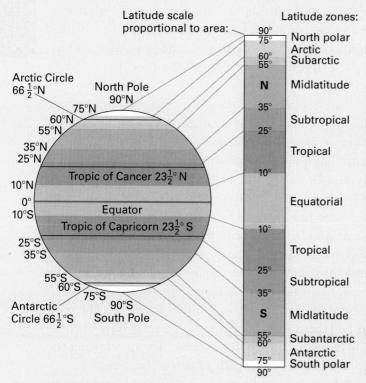

Latitude scale proportional to area:

Latitude zones:

Latitude	Zone
90°/75°	North polar
60°/55°	Arctic / Subarctic
	N Midlatitude
35°	Subtropical
25°	Tropical
10°	
	Equatorial
10°	
25°	Tropical
35°	Subtropical
	S Midlatitude
55°/60°	Subantarctic
75°	Antarctic / South polar
90°	

Arctic Circle 66½°N
North Pole 90°N
75°N
60°N
55°N
35°N
25°N
Tropic of Cancer 23½°N
10°N
0°
10°S
Equator
Tropic of Capricorn 23½°S
25°S
35°S
55°S
60°S
75°S
90°S
South Pole
Antarctic Circle 66½°S

World latitude zones A scientist's system of latitude zones, based on the seasonal patterns of daily insolation observed over the globe.

Tropical zone
The tropical zone is the home of the world's driest deserts. Pictured here is Rub' al Khali, Saudi Arabia.

Equatorial zone
An equatorial rainforest, as seen along a stream in the Gunung Palung National Park, Borneo, Indonesia.

absolute and binding, however. Rather, this system of names is a convenient way to identify general regional belts throughout this book.

The equatorial zone encompasses the Equator and covers the latitude belt roughly 10° north to 10° south. Here the Sun provides intense insolation throughout most of the year, and days and nights are of roughly equal length. Spanning the Tropics of Cancer and Capricorn are the tropical zones, ranging from latitudes 10° to 25° north and south. A seasonal cycle does exist in these zones, although high annual insolation produces warm temperatures throughout the year. In both of these regions, small differences in temperatures from one region to another prevent the formation of contrasting *air masses*—masses of air with markedly different temperature and moisture characteristics. Hence, weather phenomena in these regions differ from those found in the higher latitudes, which do involve the interaction of contrasting air masses.

Moving toward the poles from each of the tropical zones are transitional regions called the *subtropical zones*. For convenience, we assign these zones the latitude belts 25° to 35° north and south. At times, we may extend the label "subtropical" a few degrees farther toward the pole or Equator of these parallels. These zones have a strong seasonal cycle combined with an annual insolation nearly as large as the tropical zones.

The midlatitude zones are next, lying between 35° and 55° north and south latitude. In these belts, the Sun's height in the sky shifts through a wide range annually. Differences in day length from winter to summer are also large. Thus, seasonal contrasts in insolation are quite strong. In turn, these regions can experience a large range in annual surface temperature. In addition, large differences in temperature between different regions within the midlatitudes produce air masses with different temperature and moisture characteristics. When these differing air masses interact with one another, they produce *fronts*, which are a principal source of weather in these latitudes.

Bordering the midlatitude zones on the poleward side are the subarctic and subantarctic zones, 55° to 60° north and south latitudes. Astride the Arctic and Antarctic Circles from latitudes 60° to 75° N and S lie the arctic and antarctic zones. These zones have an extremely large yearly variation in day lengths, yielding enormous contrasts in insolation from solstice to solstice.

The polar zones, north and south, are circular areas between about 75° latitude and the poles. The polar regime of a six-month day and six-month night is predominant here. These zones experience the greatest seasonal contrasts of insolation of any area on Earth.

SOLAR HEATING OF LAND AND WATER

While insolation dictates the amount of solar energy intercepted by an exposed surface, other factors determine how this insolation affects local temperatures. The same insolation hitting two surfaces—one with a high albedo and one with a low albedo—will produce very different temperature changes in the two locations.

Another factor relates to the presence, or absence, of large bodies of water near the given location. If you have ever visited San Francisco, you probably noticed that it has quite a unique climate. Cool, damp weather prevails for most of the year. The cool climate is due to its location—on the tip of a peninsula, with the Pacific Ocean on one side and San Francisco Bay on the other. Ocean and bay water temperatures can be quite different from land temperatures, even if they receive the same insolation. In contrast, for a location far from the water, like Yuma, Arizona, air temperatures are much warmer on the average—about 28°C (82°F). The daily range is also much greater. Hot desert days become cool desert nights, with the clear, dry air allowing the ground to lose heat rapidly (**FIGURE 4.13**).

Why do these differences occur? The important principle is this: the surface layer of any extensive, deep body of water heats more slowly and cools more slowly than the surface layer of a large body of land when both are subjected to the same change in intensity of insolation. Because of this principle, daily and annual air temperature cycles will be quite different at coastal locations than at interior locations.

San Francisco's maritime location
San Francisco's climate is moderated by the surrounding waters of the Pacific Ocean and San Francisco Bay. You can observe the low clouds and fog sweeping in from the ocean on the right. The Golden Gate Bridge is in the foreground.

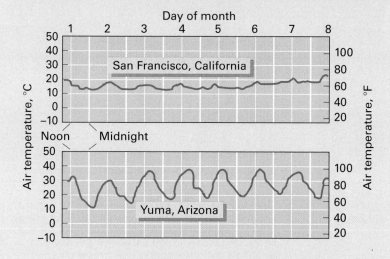

Daily temperatures at San Francisco, California and Yuma, Arizona
A recording thermometer made these continuous records of the rise and fall of air temperature for a week in summer at San Francisco, California, and at Yuma, Arizona.

Yuma's continental location
Yuma, Arizona is surrounded by dry desert. Its daily temperature cycle follows the desert, with hot days and cool nights.

Four important thermal differences between land and water surfaces account for the land–water contrast and explain why coastal areas enjoy more moderate temperatures than inland areas (FIGURE 4.14). The most important difference is that much of the downwelling solar radiation penetrates water, distributing the absorbed heat throughout a substantial water layer. In contrast, solar radiation does not penetrate soil or rock, so its heating effect is concentrated at the surface. The radiation therefore warms a thick water-surface layer only slightly, while a thin land-surface layer is warmed more intensely.

Process Diagram

How the sea moderates coastal land temperatures FIGURE 4.14

Compare land and water, and you will find four key thermal differences that make land heat faster than water. These differences explain why a coastal city has more moderate, uniform temperatures, but an inland city has more extremes. The seacoast city's summer heat is moderated by the presence of water, and so is cooler. In addition, the seacoast city's winter chill is moderated by the water, and so is warmer. Diagrams A, B, C, and D show why.

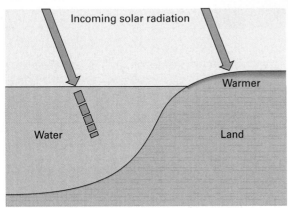

A Incoming solar radiation
Solar rays strike the land and water surfaces equally. On land, the radiation cannot deeply penetrate the soil or rock, so heating is concentrated just at the surface. On water, much of the radiation penetrates below the surface, distributing heat to a substantial depth.

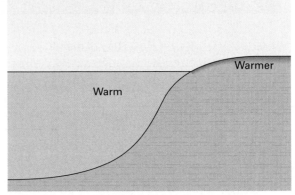

B Heat capacity of water and rock is very different
Heat capacity is the amount of heat that a substance can store. Rock (and soil) have a low heat capacity, requiring less energy to increase temperature. Water has a high heat capacity, requiring much more energy to raise its temperature. The same is true for cooling — after losing the same amount of energy, water temperature drops less than rock temperature. This means that water temperature remains more uniform than land temperature.

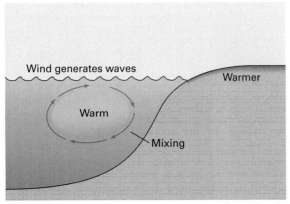

C Mixing
Water is a fluid, so it allows mixing, whereas rock is essentially immobile, preventing mixing. In water, the warmer surface water mixes with cooler water at depth to produce a more uniform temperature throughout. This mixing is driven by wind-generated waves. In rock and soil, no such mixing can occur.

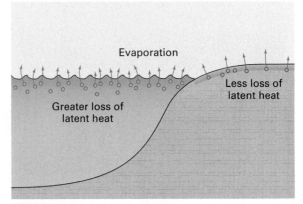

D Evaporation
Exposed water surfaces are easily cooled as water molecules evaporate, releasing latent heat to their surroundings. A water surface will always evaporate. Land surfaces also can be cooled by evaporation if water exists near the soil surface, but once the land surface dries, evaporation stops, and so does the loss of latent heat.

A second thermal factor is that water is slower to heat than dry soil or rock. The **heat capacity** of a substance describes how the temperature of a given mass of a substance changes with a given input of heat. The change in temperature of a given substance for a given input of heat is governed by the following equation:

temperature change = heat added / heat capacity.

heat capacity
Amount of heat needed to raise the temperature of one gram of a given substance by one degree Celsius.

Consider an experiment using a small amount (say, 1 kg) of water and the same amount of rock. Suppose we warm each by one degree. If we measure the amount of heat required, we will find that it takes about five times as much heat to raise the temperature of the water one degree as it does to raise the temperature of the rock one degree—that is, the heat capacity of water is about five times greater than that of rock. The same will be true for cooling: After losing the same amount of heat, the temperature of water falls less than the temperature of rock.

A third difference between land and water surfaces is related to mixing; that is, warm water in the surface layer can mix with cooler water below, producing a more uniform temperature throughout. For open water, the mixing is produced by wind-generated waves, while near the coast, mixing can be produced by breaking waves or currents. Clearly, no such mixing occurs on land surfaces.

A fourth thermal difference is that an open water surface can be cooled easily by evaporation. Land surfaces can also be cooled by evaporation, but only if water is present at or near the soil surface. When the surface dries, evaporation stops, and excess energy goes into sensible heat and begins to raise the temperature of the surface.

With these four important differences in mind, we can easily see that for the same change in heat input, air temperatures above water will change less than those over land, and hence be less variable.

CONCEPT CHECK STOP

How does the Earth's rotation on its axis affect insolation?

How does the tilt of the Earth's axis affect the intensity and duration of daily insolation at a given location?

When do the solstices and equinoxes occur?

What is the length of daylight at the North Pole on each of the equinoxes and solstices?

What about in the midlatitudes?

What about at the Equator?

How does the response of maritime and continental regions to heating differ?

Air Temperature

LEARNING OBJECTIVES

Explain what factors affect air temperature at a given location.

Describe how air temperature differs between urban and rural locations.

The familiar concept of temperature is a measure of the level of sensible heat of matter, whether it is gaseous (air), liquid (water), or solid (rock or dry soil). We know from experience that when a substance absorbs a flow of radiant energy, such as sunlight, its temperature rises. Similarly, if a substance loses energy, its temperature falls. This energy flow moves in and out of a substance at its surface. We can also measure **air temperature**—that is, the temperature of the air as observed at 1.2 m (4 ft) above the ground surface. Air temperature can be different from surface temperature. However, because conduction and convection transfer sensible heat to and from the surface to the air,

air temperature
Temperature of the air measured 1.2 m above the ground.

Surface type
The high albedo of the snow cover compared with the albedo of vegetation will produce lower overall temperatures.

Elevation
The high altitude of the mountain ranges keeps temperatures colder than at sea level, allowing snow to persist throughout the year.

Latitude
The Chugach Mountain Range near Anchorage, Alaska, is located at 61° N. The seasonal cycle in insolation is very large here, producing large seasonal variations in temperature.

Coastal versus interior location
The marshes seen here near the coast make the seasonal temperature changes more moderate than at interior land-based locations.

Environmental factors affecting temperature FIGURE 4.15

Five main factors affect temperature and its variability at a given location.

Atmospheric and oceanic circulations
During summer, winds off the Gulf of Alaska bring in warmer air from the south. During winter, however, winds bring cold arctic air from the north.

air temperatures tend to follow surface temperatures even if they are not equal.

Air temperature affects many aspects of our daily lives, from the clothing we wear to the amount of energy we use to heat or cool our homes. Air temperature and its cycles also act to select the plants and animals that make up the biological landscape of a region. Also, air temperature, along with precipitation, is a key determiner of climate.

Now we want to understand how and why air temperature at a given location changes from month to month and year to year.

The five fundamental factors that influence local air temperatures are shown in FIGURE 4.15.

1. *Latitude.* Daily and annual cycles of insolation vary systematically with latitude, causing air temperatures and air temperature cycles to vary as well. Yearly insolation decreases toward the poles, so less energy is available to heat the air. Thus, temperatures generally fall as we move poleward. Temperatures also become more variable over the year as latitude increases because of the greater range in insolation through the year.

2. *Surface type.* Urban surfaces of asphalt, roofing shingles, and rubber are dry compared with the moist soil surfaces of rural areas and forests. These urban surfaces heat more rapidly, because solar energy cannot be taken up in evaporation of water. They also have a lower albedo than vegetation-covered surfaces, so they absorb a greater portion of the Sun's energy. Because of these fac-

tors, urban air temperatures are generally higher than rural temperatures. Similarly, areas of barren or rocky soil surfaces, like those in deserts, heat more rapidly, because solar energy cannot be taken up in evaporation of water. However, they tend to have a higher albedo than vegetated surfaces, which means they absorb less insolation overall.

3. *Coastal versus interior location.* Locations near the ocean experience a narrower range of air temperatures than locations in continental interiors. The reason is that water heats and cools more slowly than land, so air temperatures over water bodies tend to be less extreme than temperatures over the land surface. Also, because winds can easily cause air to flow from water to land, a coastal location will more often feel the influence of the adjacent water.

4. *Elevation.* Temperatures tend to decrease with altitude in the atmosphere. By extension, at high elevations, average temperatures are cooler than at lower elevations, which allows snow to accumulate and remain longer on high peaks, for example. In addition, because there are fewer atmospheric gas molecules above high elevation locations, the greenhouse effect is reduced. Hence, during the night, high-altitude locations cool much more quickly through the emission of longwave radiation, while low-altitude locations cool less quickly. As a result, high-altitude locations tend to have large diurnal and seasonal variations in temperature.

5. *Atmospheric and oceanic circulations.* Local temperatures can be dramatically affected by the circulations of the overlying atmosphere and nearby oceans. During certain weather events, temperatures can drop over 20°C in a few hours as cold air from higher latitudes rushes into a region.

transpiration The process by which plants lose water to the atmosphere by evaporation through leaf pores.

evapotranspiration The combined water loss to the atmosphere by evaporation from soil and transpiration from plants.

Alternatively, coastal regions, particularly along the eastern portion of continents, can have average temperatures 5° to 10°C above coastal locations on the western portion of continents because of warm offshore currents.

These five factors act at different scales. Latitude effects vary at the global scale, whereas marine/continental locations and elevation affect temperature at the continental or regional scale. Surface type and ocean/atmosphere circulations affect both broad regions, like rainforests and deserts, and local areas, including cities and their surroundings.

URBAN HEAT ISLAND

We mentioned that temperatures can differ based on surface type. Human activity has altered much of the Earth's land surface. Vegetation has been removed to build cities, and soils have been covered with pavement.

Are these changes large enough to affect temperature patterns? Consider the example of walking across the parking lot on a hot day. The ground can become hot enough to burn your bare feet. You may have noticed that the situation is quite different when you step across a field—the ground does not feel as hot, even if the air temperature is similar. Clearly, surface material plays a large part in determining temperature.

Two of the key processes that help to keep rural surfaces cool are **transpiration** and *evaporation*, which together are known as **evapotranspiration**. Through evapotranspiration, excess radiational energy is removed as latent heat, thereby cooling the surface. With decreased or little evapotranspiration, excess radiational energy is converted into sensible heat, which in turn increases the temperature of the surface.

As a result of these effects, city centers tend to be several degrees warmer than the surrounding suburbs and countryside, as

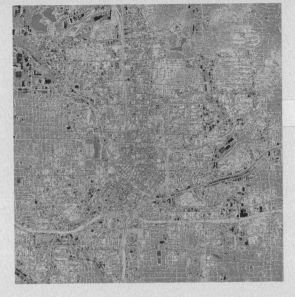

This image, taken at night in May, over downtown Atlanta, Georgia, shows the urban heat island. The main city area, in tones of red and yellow, is clearly warmer than the suburban area, in blue and green. The street pattern of asphalt pavement is shown very clearly as a red grid, with many of the downtown squares filled with red. (Courtesy NASA/EPA)

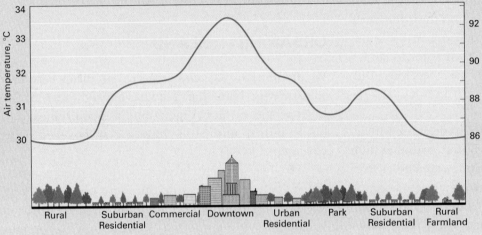

This diagram shows how air temperatures might vary across the urban and rural areas during the late afternoon. Downtown and commercial areas are warmest, while rural farmland is coolest.

seen in FIGURE 4.16. The sketch of a temperature profile across an urban area in the late afternoon shows this effect. We call the central area an **urban heat island**, because it has a significantly elevated temperature. Such a large quantity of heat is stored in the ground during the daytime hours that the heat island remains warmer than its surroundings during the night, too. The thermal infrared image of the Atlanta central business district at night demonstrates the heat island effect.

urban heat island
Area at the center of a city that has a higher temperature than surrounding regions.

The urban heat island effect has important economic consequences. We use more air conditioning and more electric power to combat higher temperatures in the city center. Also, smog is more likely to form in warm temperatures. Recognizing the important differences between urban and rural surfaces, many cities are now planting more vegetation in an at-tempt to counteract these problems. *What a Scientist Sees: Urban and Rural Surfaces* addresses this issue.

CONCEPT CHECK STOP

How does the latitude of a location affect the average temperature of the location? How does it affect the seasonal change in temperature?

What characteristics of a location's surface conditions can affect temperature?

When would temperatures near the ocean be warmer than at locations in the continental interior? When would they be colder?

What effect does elevation have on overall temperatures? What effect does it have on variations in temperature over the course of the day?

How does surface type affect air temperature in the city and in rural areas?

Urban and Rural Surfaces

The vegetation, while providing shade, also reflects away more incoming sunlight than paved surfaces and rooftops, thereby increasing the albedo and decreasing the amount of absorbed solar radiation.

The albedo of urban surfaces, particularly roadways and asphalt roofs, is very low, thereby increasing the amount of absorbed solar radiation.

The extensive use of asphalt and pavement, along with the lack of vegetation, reduces the amount of available moisture that can be turned into latent heat. Instead, excess energy goes into sensible heat, resulting in increased air temperatures.

The vegetation and moist soil support substantial evapotranspiration, which effectively removes excess energy in the form of latent heat, resulting in decreased temperatures.

NATIONAL GEOGRAPHIC

The Daily Cycle of Air Temperature

LEARNING OBJECTIVES

Describe the factors that influence the daily cycle of air temperature.

Explain why marine and continental locations have very different daily cycles of temperature.

 very day, the Earth's surface goes through a cycle of rising and falling temperatures caused by changes in insolation, radiation, and air temperature. These three factors are linked in important ways.

DAILY INSOLATION AND NET RADIATION

Because the Earth rotates on its axis, absorbed solar energy at a location can vary widely throughout the 24-hour period, while outgoing longwave energy remains more constant, as seen in **FIGURE 4.17**. During the day, net radiation is positive, and the surface gains heat. At night, net radiation is negative, and the surface loses heat by radiating it to the sky and space. Because the air next to the surface is warmed or cooled as well, air temperatures follow the same cycle. The result is the daily cycle of rising and falling air temperatures.

Let's look in more detail at how insolation, net radiation, and air temperature are linked in this daily cycle. The three graphs show average curves of daily absorbed shortwave and emitted longwave radiation, net radiation, and air temperature that we might expect for a typical observing station at latitude 40–45° N in the interior of North America. The time scale is set so that 12:00 noon occurs when the Sun is at its highest elevation in the sky.

Solar heating is greatest during noon, when the Sun reaches its highest point in the sky. The longwave radiation leaving the surface also changes with the time of day, although less dramatically. Generally, because temperatures are warmer in the afternoon, the longwave radiation emitted from the surface is greater than during night, when temperatures are colder.

Net radiation—the difference between the incoming shortwave radiation and outgoing longwave radiation—at noon has quite large positive values. These positive net radiation values begin just after sunrise and last through midafternoon. Once the Sun begins to descend in the sky, longwave emission from the surface becomes larger than incoming solar radiation and the net radiation becomes negative. Negative net radiation values then persist until the Sun rises and incoming solar radiation once again becomes larger than outgoing longwave radiation.

This daily cycle of surplus and deficit radiation drives the daily cycle of temperature. At night, temperatures continue to decrease as long as there is a net deficit of radiation. Hence, minimum temperatures tend to occur just after sunrise, when the net radiation starts to become positive. In addition, temperatures tend to increase as long as the net radiation is positive, which would suggest that maximum temperatures occur in late afternoon.

However, in the real world, temperatures tend to peak sooner in the day than suggested here (around 3:00 P.M. instead of 4:30–7:30 P.M.), particularly in summer. Why? On sunny days in the early afternoon, large convection currents can carry hot air near the surface upward and bring cooler air downward. As a result, the surface temperature is not determined simply by the net radiation but is also affected by changes in sensible and latent heat, complicating the simple pattern discussed here.

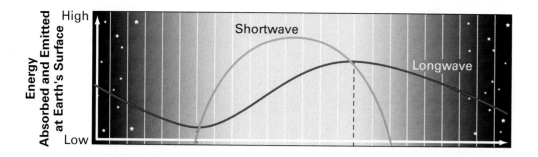

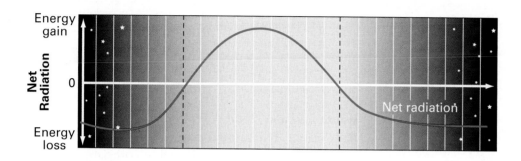

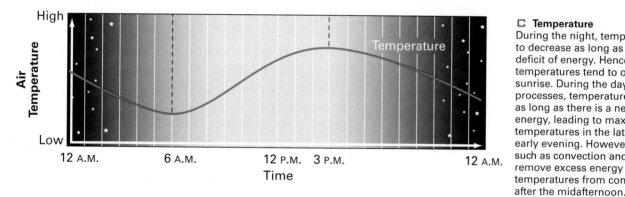

A Shortwave and longwave radiation
Incoming shortwave radiation changes with the daily path of the Sun across the sky. In addition, longwave radiation leaving the surface changes as the surface temperature changes. When incoming shortwave radiation is larger than outgoing longwave radiation, there is a net surplus of energy. When it is less, there is a net deficit of energy.

B Net radiation
During a period of surplus net radiation, temperatures tend to increase. In contrast, during a period of deficit, net radiation temperatures tend to decrease.

C Temperature
During the night, temperature continues to decrease as long as there is a net deficit of energy. Hence, minimum temperatures tend to occur shortly after sunrise. During the day, given no other processes, temperatures should increase as long as there is a net surplus of energy, leading to maximum temperatures in the late afternoon or early evening. However, other processes such as convection and evaporation remove excess energy and prevent temperatures from continuing to rise after the midafternoon.

Daily cycles of radiation and air temperature FIGURE 4.17

As we would expect, the height of the temperature curve varies with the seasons. In the summer, temperatures are warm, and the daily curve is high. In winter, the temperatures are colder, and the daily curve is low. In addition the September equinox conditions are considerably warmer than the March equinox conditions.

This discrepancy arises because the seasonal temperature curves lag behind the seasonal net radiation curves, just as the daily temperature curves lag behind the daily net radiation curves. Hence, temperatures in March are more like those in February, while temperatures in September are more like those in August.

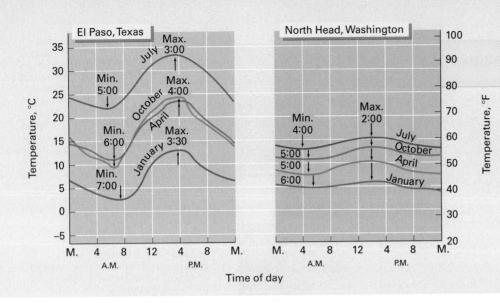

Maritime and continental temperatures FIGURE 4.18

The average cycle of air temperature throughout the day for four different months shows the effect of continental and maritime locations. Daily and seasonal ranges are great at El Paso, a station in the continental interior, but only weakly developed at North Head, Washington, a station on the Pacific coast. The seasonal effect on overall temperatures is also stronger at El Paso.

FACTORS THAT INFLUENCE DAILY TEMPERATURES

While we expect most locations to have a daily cycle in temperature driven by the daily cycle in absorbed shortwave radiation and longwave emissions, the variability of this daily cycle can change from one location to another. Many of the factors that affect the overall temperatures in a given location also affect the daily cycle of temperatures as well.

As an example, let's examine the daily cycle of air temperature for a continental and a marine location—El Paso, Texas, and North Head, Washington—in four months of the year, shown in FIGURE 4.18.

The El Paso curves show the temperature environment of an interior desert in the midlatitudes. Because soil moisture content is low and vegetation is sparse, evaporation and transpiration are not important cooling effects. Cloud cover is generally sparse. Under these circumstances, the ground surface heats intensely during the day and cools rapidly at night. Air temperatures show an average daily range of 11–14°C (20–25°F). This type of variation represents an interior temperature environment—typical of a station located in the interior of a continent, far from the ocean's influence.

North Head is located on the Washington coast. Here, prevailing westerly winds sweep cool, moist air off the adjacent Pacific Ocean. The average daily range at North Head is a mere 3°C (5.5°F) or less. Per-

sistent fogs and cloud cover contribute to the minimal daily range. The annual range of temperature also is greatly restricted, especially when compared with El Paso. North Head exemplifies a coastal temperature environment, typical of a station located in the path of oceanic air.

What factors affect the daily temperature cycle in other locations? We would expect urban surfaces—with their asphalt, roofing shingles, and rubber—and areas of barren or rocky soil surfaces to heat more rapidly than moist soil surfaces of rural areas and forests, because solar energy cannot be taken up in evaporation of water. Thus, a larger daily cycle of temperature is found over urban areas and regions with dry, rocky soils. High-altitude locations also tend to have larger daily cycles of temperature associated with reduced counterradiation, leading to more intense longwave cooling at night.

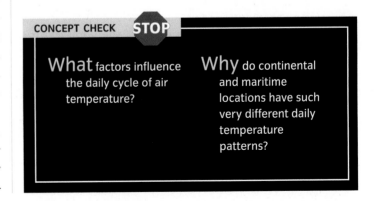

CONCEPT CHECK **STOP**

What factors influence the daily cycle of air temperature?

Why do continental and maritime locations have such very different daily temperature patterns?

The Annual Cycle of Air Temperature

Annual temperature cycles at a given location are driven mainly by the annual cycle of net radiation received through the year. However, other factors also contribute and can either moderate or amplify the annual cycle due to radiation changes.

NET RADIATION AND TEMPERATURE

Daily insolation varies over the seasons of the year because of the Earth's motion around the Sun and the tilt of the Earth's axis. That rhythm produces a net radiation cycle that, in turn, causes an annual cycle in mean monthly air temperatures (FIGURE 4.19). The difference between the maximum and minimum temperatures over the annual cycle is called the *annual temperature range.*

Where the net radiation cycle is large, the annual temperature range tends to be large, but other factors can affect it as well.

The relationship between net radiation and temperature FIGURE 4.19

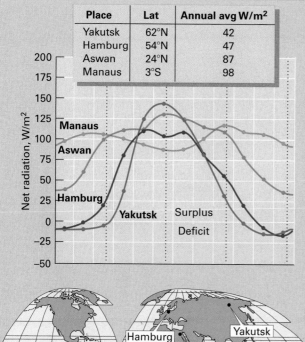

Place	Lat	Annual avg W/m²
Yakutsk	62°N	42
Hamburg	54°N	47
Aswan	24°N	87
Manaus	3°S	98

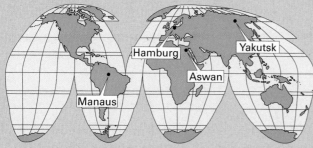

A Net radiation

At Manaus, Brazil, the average net radiation rate is strongly positive every month, with two minor peaks coinciding roughly with the equinoxes, when the Sun is nearly straight overhead at noon. The curve for Aswan, Egypt, shows a large surplus of positive net radiation every month, but values for June and July are triple those of December and January. The annual cycle of net radiation for Hamburg, Germany, is also strongly developed. During the long, dark winters in Yakutsk, Russia, the net radiation rate is negative, and the radiation deficit lasts about six months.

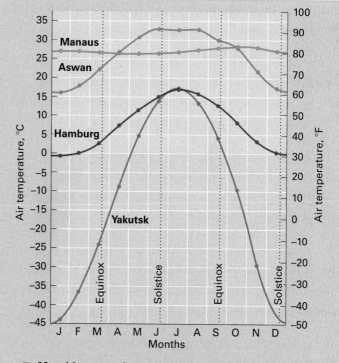

B Monthly mean air temperature

Manaus has uniform air temperatures, averaging about 27°C (81°F) for the year with only a small difference of 1.7°C (3°F) between the highest and lowest monthly temperature. Temperatures at Aswan follow the cycle of the net radiation rate curve, with an annual range of about 17°C (31°F). However, June, July, and August are very hot, averaging over 32°C (90°F). Hamburg has a similar annual cycle. However, summer months reach a maximum of just over 16°C (61°F), while winter months reach a minimum of just about freezing (0°C or 32°F). In Yakutsk, monthly mean temperatures in winter drop to –45°C (–50°F) but rise dramatically in the spring and the summer to 13°C (55°F), an annual temperature range of over 60°C (108°F).

107

LAND AND WATER CONTRASTS

While the largest effect upon the annual temperature cycle at a given location is caused by the annual cycle of net radiation, other characteristics of the location can also have an important influence. Let's first turn to the effects of land–water surface contrasts on the temperature cycle for the year by looking in detail at the annual cycle for another pair of stations—Winnipeg, Manitoba, located in the interior of the North American continent, and the Scilly Islands, off the southwestern tip of England, which are surrounded by the waters of the Atlantic Ocean (**FIGURE 4.20**). This time, the two stations chosen are at the same latitude, 50° N. As a result, they have the same insolation cycle and receive the same potential amount of solar energy for surface warming.

The temperature graphs for the Scilly Islands and Winnipeg confirm the effects we have already noted for North Head and El Paso—that the annual range in temperature is much larger for the interior station (39°C, 70°F) than for the coastal station (8°C, 14°F). Note that the nearby ocean waters keep the air temperature at the Scilly Islands well above freezing in the winter, while January temperatures at Winnipeg fall to near −20°C (−4°F).

Another important effect of land–water contrasts concerns the timing of maximum and minimum temperatures. Insolation reaches a maximum at summer solstice, but it is still strong for a long period afterward, so that net radiation is positive well after the solstice. Therefore, the hottest month of the year for interior regions is July, the month following the solstice. Similarly, the coldest month of the year for large land areas is January, the month after the winter solstice. The continued cooling after the winter solstice takes place because the net radiation is still negative even though the insolation has begun to increase.

Over the oceans and at coastal locations, maximum and minimum air temperatures are reached a month later than on land—in August and February, respectively. Because water bodies heat or cool more slowly than land areas, the air temperature changes more slowly. This effect is clearly seen in the Scilly Islands graph, which shows that February is slightly colder than January.

Maritime and continental annual air temperature cycles FIGURE 4.20

Annual cycles of insolation and monthly mean air temperature for two stations at latitude 50° N: Winnipeg, Canada and Scilly Islands, England.

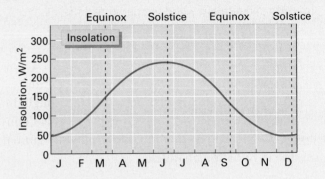

A Insolation is identical for the two stations.

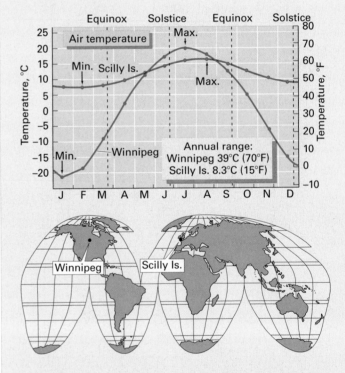

B Winnipeg temperatures clearly show the large annual range and earlier maximum and minimum that are characteristic of its continental location. Scilly Islands temperatures show a maritime location with a small annual range and delayed maximum and minimum.

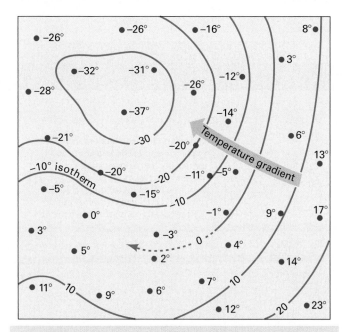

Isotherms FIGURE 4.21

Isotherms are used to make temperature maps. Each line connects points having the same temperature. Where isotherms close in a tight circle, a center exists. This example shows a center of low temperature.

WORLD PATTERNS OF AIR TEMPERATURE

In addition to a location's latitude and position within the continent or along the coast, we know that its surface type (urban or rural) and elevation can also influence air temperatures. Now we want to pull all these pieces of information together and see how they affect air temperature patterns around the world.

First, we need a quick explanation of air temperature maps. FIGURE 4.21 shows a map with temperature indicated by a set of **isotherms**—lines connecting locations that have the same temperature. Usually, we choose isotherms that are separated by 5° or 10°C, but they can be drawn at any convenient temperature interval.

> **isotherm** Line on a map drawn through all points with the same temperature.
>
> **temperature gradient** Rate of temperature change along a selected line or direction.

Isothermal maps clearly show centers of high or low temperatures. They also illustrate the directions along which temperature changes, which are known as **temperature gradients**.

FACTORS CONTROLLING AIR TEMPERATURE PATTERNS

We have already looked at the three main factors that explain world isotherm patterns. The first is latitude. As latitude increases, average annual insolation decreases, and so temperatures decrease as well, making the poles colder than the Equator. Latitude also affects seasonal temperature variation—for example, the poles receive more solar energy at the summer solstice than does the Equator. Therefore, we must remember to note the time of year and the latitude when looking at temperature maps.

The second factor is the maritime–continental contrast. As we have noted, coastal stations have more uniform temperatures and are cooler in summer and warmer in winter. Interior stations, on the other hand, have much larger annual temperature variations. Ocean currents can also have an effect, because they can keep coastal waters warmer or cooler than you might expect.

Elevation is the third important factor. At higher elevations, temperatures are cooler, so we expect to see lower temperatures near mountain ranges.

FIGURE 4.22 looks at world temperature maps for two months, January and July. It also shows the annual cycle in temperature, given by the difference between July temperatures and January temperatures. By comparing the maps, we can observe and explain many of the important features about the temperature patterns from January and July using the effects of latitude, interior or maritime location, surface type, and elevation. Overall, the global pattern of annual temperature range shows that tropical oceans have the smallest range, while northern hemisphere continental interiors show the largest range.

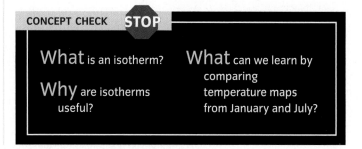

CONCEPT CHECK **STOP**

What is an isotherm?

Why are isotherms useful?

What can we learn by comparing temperature maps from January and July?

Mean monthly air temperatures for January and July

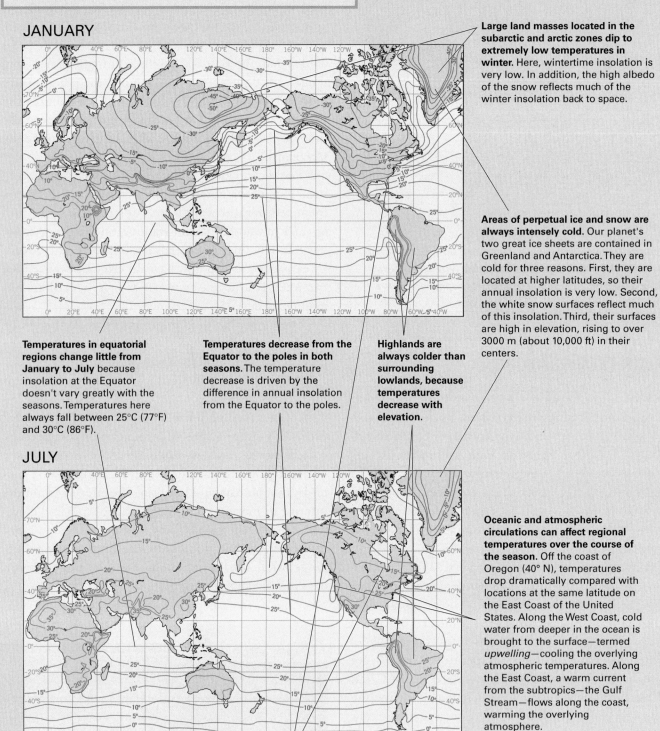

JANUARY

JULY

Large land masses located in the subarctic and arctic zones dip to extremely low temperatures in winter. Here, wintertime insolation is very low. In addition, the high albedo of the snow reflects much of the winter insolation back to space.

Areas of perpetual ice and snow are always intensely cold. Our planet's two great ice sheets are contained in Greenland and Antarctica. They are cold for three reasons. First, they are located at higher latitudes, so their annual insolation is very low. Second, the white snow surfaces reflect much of this insolation. Third, their surfaces are high in elevation, rising to over 3000 m (about 10,000 ft) in their centers.

Temperatures in equatorial regions change little from January to July because insolation at the Equator doesn't vary greatly with the seasons. Temperatures here always fall between 25°C (77°F) and 30°C (86°F).

Temperatures decrease from the Equator to the poles in both seasons. The temperature decrease is driven by the difference in annual insolation from the Equator to the poles.

Highlands are always colder than surrounding lowlands, because temperatures decrease with elevation.

Oceanic and atmospheric circulations can affect regional temperatures over the course of the season. Off the coast of Oregon (40° N), temperatures drop dramatically compared with locations at the same latitude on the East Coast of the United States. Along the West Coast, cold water from deeper in the ocean is brought to the surface—termed *upwelling*—cooling the overlying atmospheric temperatures. Along the East Coast, a warm current from the subtropics—the Gulf Stream—flows along the coast, warming the overlying atmosphere.

Isotherms make a large south-north shift from January to July over continents in the midlatitude and subarctic zones. The 15°C (59°F) isotherm lies over central Florida in January, but by July it has moved far north, cutting the southern shore of Hudson Bay and far up into northwestern Canada. This is because continents heat and cool more rapidly than oceans.

Annual range based upon difference between July and January mean temperatures

The annual range increases with latitude, especially over northern hemisphere continents. The increase is due to the contrast between summer and winter insolation, which increases with latitude.

The greatest ranges occur in the subarctic and arctic zones of Asia and North America. In these regions, summer insolation is nearly the same as at the Equator, while winter insolation is very low.

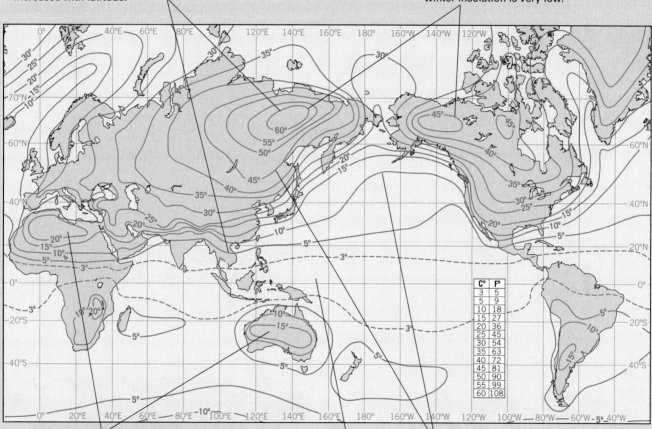

C°	F°
3	5
5	9
10	18
15	27
20	36
25	45
30	54
35	63
40	72
45	81
50	90
55	99
60	108

The annual range is moderately large on land areas in the tropical zone, near the Tropics of Cancer and Capricorn. Dry air and the absence of clouds and moisture allow these continental locations to become very cool in winter and very warm in summer, even though insolation contrasts with the season are not large.

The annual range is very small over oceans in the tropical zone. The range is less than 3°C (5°F), because insolation varies little with the seasons near the Equator, and water heats and cools slowly.

The annual range over oceans is less than that over land at the same latitude. This major difference is due to the fact that water heats and cools much more slowly than land, and hence a narrower range of temperatures is experienced.

Measurement of Air Temperature

LEARNING OBJECTIVES

Describe how air temperatures are measured.

Identify and **explain** different types of temperature indexes.

Air temperature is a piece of weather information that we encounter daily. Since air temperature can vary with height, it is measured at a standard level—1.2 m (4 ft) above the ground. FIGURE 4.23 shows a typical setup for measuring air temperature. A thermometer shelter is a louvered box that holds thermometers or other weather instruments at the proper height while sheltering them from the direct rays of the Sun. Air circulates freely through the louvers, ensuring that the temperature inside the shelter is the same as that of the outside air.

Many types of thermometers are put in these boxes. The traditional one is the familiar *liquid-in-glass thermometer*. It consists of a bulb and narrow tube with a liquid inside, typically mercury or colored alcohol. As the temperature of the surrounding atmosphere increases the temperature of the liquid in the bulb, the liquid expands and flows into the tube through a small opening. The height of the liquid indicates the temperature, which is marked on the thermometer tube. Another type of thermometer, more commonly used in weather and climate studies, is the *electrical thermometer*. This type can take the form of either a *thermistor* or an *electrical resistance thermometer*. Both types are based on the principle that the electrical resistance of certain materials changes with temperature. By measuring the resistance, it is possible to measure the temperature of the material and the atmosphere surrounding it.

A final type of thermometer measures temperature based on the radiation emitted by a given surface. We know that the intensity of the emitted radiation increases with temperature. By designing a sensor—termed a *radiometer*—to measure the intensity of the emitted radiation, we can estimate the temperature of the emitting surface, even from a great distance away.

Radiometers are used extensively in satellites because they allow the temperature to be sensed remotely.

Although temperature can be measured at regular intervals with some weather stations reporting temperatures hourly, most stations only report the highest and lowest temperatures recorded during a 24-hour period. These are the most important values in observing long-term trends in temperature. If a station employs an electrical thermometer, the digital data from the thermometer can be automatically fed into a *data logger* that records the temperature at specific time intervals. Using a computer, the maximum and minimum temperatures recorded during a given day are then retrieved.

However, a liquid-in-glass thermometer has to be modified to register the maximum and minimum temperatures. To measure maximum temperatures, a small constriction is placed in the tube just above the bulb. This constriction allows liquid to flow out as it is heating up. As the temperatures decrease and the liquid cools, the constriction prevents the liquid from flowing back into the bulb. As a result, the liquid height remains at its highest position, which can be recorded even after the temperature has cooled.

Weather recording instruments FIGURE 4.23

An instrument shelter houses a pair of thermometers. The shelter is constructed with louvered sides for ventilation and is painted white to reflect solar radiation.

To measure minimum temperature, a small dumbbell-shaped indicator is placed within the tube, and the thermometer is positioned horizontally. As the temperature of the liquid decreases, the indicator remains submerged in the fluid and moves down the tube, forced by the surface tension at the end of the fluid. When the temperature increases, the surface of the fluid moves back up the tube, and the fluid flows around the indicator. Hence, the location of the indicator in the tube marks the minimum temperature reached during that day.

APPLICATIONS OF TEMPERATURE DATA

Overall, temperature measurements are reported to governmental agencies charged with weather forecasting, such as the U.S. Weather Service or the Meteorological Service of Canada. These agencies typically make available daily, monthly, and yearly temperature statistics for each station using the daily maximum, minimum, and mean temperature. These statistics, along with others such as daily precipitation, are used to describe the weather and climate of the station and its surrounding area.

Temperature can also be used with other weather and climate data to produce *temperature indexes*—indicators of the temperature's impact upon environmental and human conditions.

Two of the more familiar indexes are the *wind chill index* and the *heat index*. The wind chill index is used to determine how cold temperatures feel to us, based on not only the actual temperature but also the wind speed. Air is actually a very good insulator, so when the air is still, our skin temperature can be very different from the temperature of the surrounding environment. However, as air moves across our skin, it removes sensible and latent heat and transports it away from our bodies. During the summer, this process keeps us cool as sweat is evaporated away, lowering our skin temperature. During the winter, it removes heat necessary to keep our bodies warm, thereby cooling our skin and making conditions feel much colder than the actual measured temperature.

The wind chill index, which is measured in °C (or °F), can be dramatically different from the actual temperature (FIGURE 4.24). For example, an actual temperature of 30°F (−1°C) and a wind-speed of 30 mph (13.45 m/s) produce a wind-chill of 15°F (−26°C).

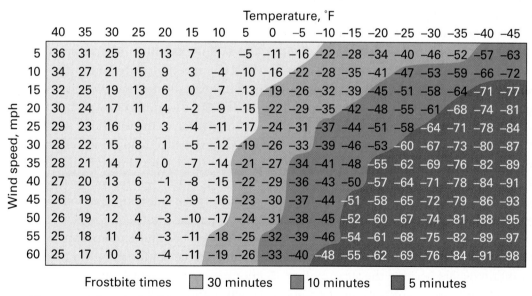

To convert the wind chill index from °F to °C, use the following equation: °C = (°F − 32) × 5/9

Wind chill conversion FIGURE 4.24

The wind chill index provides an indicator of how cold temperatures feel based on the actual temperature and the wind speed.

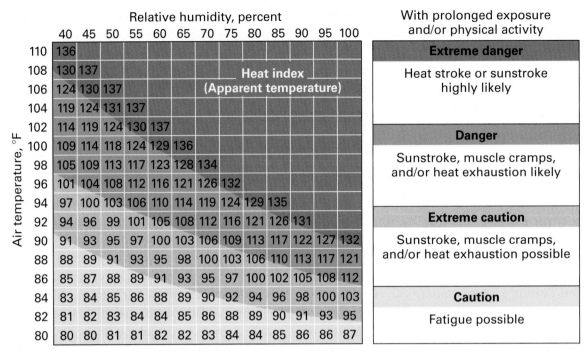

To convert the heat index from °F to °C, use the following equation: °C = (°F − 32) × 5/9

Heat index conversion FIGURE 4.25

The heat index provides an indicator of how hot temperatures feel based on the actual temperature and the relative humidity.

In comparison to the wind chill index, which gives an indication of how cold temperatures feel to us, the heat index gives an indication of how hot we feel based on the actual temperature and the *relative humidity*. Relative humidity is the humidity given in most weather reports and indicates how much water vapor is in the atmosphere as a percentage of the maximum amount possible: Low relative humidity indicates relatively dry atmospheric conditions, whereas high relative humidity indicates relatively humid atmospheric conditions. Why does relative humidity influence how hot the temperature feels? One of the ways our bodies remove excess heat is through the evaporation of sweat from our skin. This evaporation removes latent heat, thereby cooling our bodies. However, when the relative humidity is high, less evaporation occurs because the surrounding atmosphere is already relatively moist. Without evaporation, excess heat builds up in our bodies, and we feel hotter than we would if the heat was being removed via evaporation.

As with the wind chill index, the heat index is given in °F (or °C) and like the wind chill it can be very different from the actual temperature, as seen in **FIGURE 4.25**. For example, if the actual temperature is 90°F (32.2°C) and the relative humidity is 90 percent, the heat index indicates that the temperature will feel like 122°F (50°C)—a difference of 32°F (17.8°C)!

In addition to the wind chill and heat indexes, other indexes are used to help guide industrial and agricultural activities. Two indexes used by industries—particularly the energy industry—are *cooling degree days* and *heating degree days*. These indexes give an indication of the accumulated difference between the actual temperatures and our "preferred" living temperature. For example, the heating degree day for a single day represents the difference between the actual temperature and 65°F (18.3°C), which is the preferred "room" temperature for most buildings in the United States. If the actual temperature is 32°F (0°C) on a given day, the heating degree value for that day is 65 − 32 = 33°F (or 18.3 − 0 = 18.3°C).

These values can then be summed over the entire year to find the accumulated heating necessary to sustain the preferred temperature of 65°F (18°C) and hence how much energy is needed to keep our buildings at this temperature. Note, however, that the index is based on a preferred temperature of 65°F (18°C). If the preferred temperature is lower, then the heating degree days over the year—and the energy consumption needed to produce the preferred temperature—will be lower even though the actual temperatures remain the same.

Like heating degree days, cooling degree days are based on the actual temperature and the preferred room temperature of our buildings. In this case, however, it is the difference between the actual temperature and the temperature to which we need to *cool* our buildings to reach the preferred temperature of 65°F (18°C).

For agricultural purposes, one commonly used index is *growing degree days*. Like cooling degree days, it is a measure of how high the temperature is above a given threshold. For growing degree days, the threshold is set not based on human preference but rather on plant characteristics and the time at which a given type of crop begins to develop and mature. This temperature threshold—called the *base temperature*—is different for different crops. For corn, the value is 50°F (10°C), but for lettuce it is 40°F (4.4°C). These growing degree day values are then summed up over the course of the growing season until they reach a second threshold, at which point the crop is considered fully mature and is ready for harvesting. As with the base temperature, the growing degree days needed before a crop can be harvested change with each crop as well.

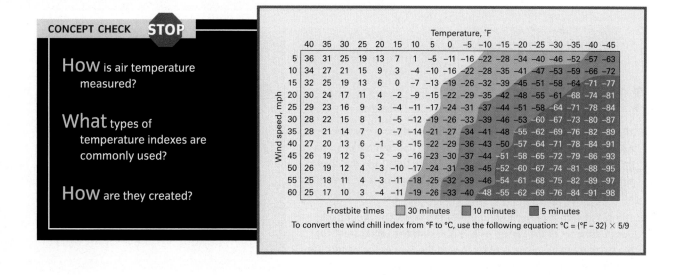

CONCEPT CHECK STOP

How is air temperature measured?

What types of temperature indexes are commonly used?

How are they created?

Temperature, °F

Wind speed, mph	40	35	30	25	20	15	10	5	0	−5	−10	−15	−20	−25	−30	−35	−40	−45
5	36	31	25	19	13	7	1	−5	−11	−16	−22	−28	−34	−40	−46	−52	−57	−63
10	34	27	21	15	9	3	−4	−10	−16	−22	−28	−35	−41	−47	−53	−59	−66	−72
15	32	25	19	13	6	0	−7	−13	−19	−26	−32	−39	−45	−51	−58	−64	−71	−77
20	30	24	17	11	4	−2	−9	−15	−22	−29	−35	−42	−48	−55	−61	−68	−74	−81
25	29	23	16	9	3	−4	−11	−17	−24	−31	−37	−44	−51	−58	−64	−71	−78	−84
30	28	22	15	8	1	−5	−12	−19	−26	−33	−39	−46	−53	−60	−67	−73	−80	−87
35	28	21	14	7	0	−7	−14	−21	−27	−34	−41	−48	−55	−62	−69	−76	−82	−89
40	27	20	13	6	−1	−8	−15	−22	−29	−36	−43	−50	−57	−64	−71	−78	−84	−91
45	26	19	12	5	−2	−9	−16	−23	−30	−37	−44	−51	−58	−65	−72	−79	−86	−93
50	26	19	12	4	−3	−10	−17	−24	−31	−38	−45	−52	−60	−67	−74	−81	−88	−95
55	25	18	11	4	−3	−11	−18	−25	−32	−39	−46	−54	−61	−68	−75	−82	−89	−97
60	25	17	10	3	−4	−11	−19	−26	−33	−40	−48	−55	−62	−69	−76	−84	−91	−98

Frostbite times ☐ 30 minutes ☐ 10 minutes ☐ 5 minutes

To convert the wind chill index from °F to °C, use the following equation: °C = (°F − 32) × 5/9

Extended heat waves, combined with lack of rainfall, can have devastating impacts, ruining crops and causing severe famine in many parts of the world. Here we see an image of a reservoir from southern Colorado during a prolonged heat wave.

- What do you notice about the soil?

- How will the soil affect subsequent temperature variations compared with the soil in a region where there is vegetation?

- Will the soil at this reservoir extend the heat wave or reduce it?

SUMMARY

1 The Earth's Rotation and Orbit

1. The Earth rotates on its axis once in 24 hours. The intersection of the axis with the Earth's surface marks the North and South Poles. The daily alternation of sunlight and darkness, the tides, and the diversion of the direction of motion of atmosphere and ocean currents arise from the rotation of the Earth.

2. The seasons change because of the Earth's revolution around the Sun and the tilt of the Earth's axis. At the **summer (June) solstice**, the northern hemisphere is tilted toward the Sun. At the **winter (December) solstice**, the southern hemisphere is tilted toward the Sun. When a given hemisphere is tilted toward the Sun, it experiences more direct insolation and longer days. At the **equinoxes**, neither hemisphere is pointed toward the Sun, so day and night are of equal length everywhere on the globe.

3. For the same amount of energy added, ocean surfaces tend to experience smaller increases in temperature than land surfaces because of four factors: (1) greater transparency, (2) higher **heat capacity**, (3) more evaporation, and (4) more mixing.

2 Air Temperature

1. Five factors affect the **air temperature** and its daily and seasonal variability at a given region: (1) latitude, (2) surface type, (3) coastal or interior location, (4) elevation, and (5) oceanic and atmospheric circulation.

2. Human modification of land surfaces, particularly with regard to the development of urban areas, can significantly increase the temperatures in certain regions. This effect is termed the **urban heat island effect**.

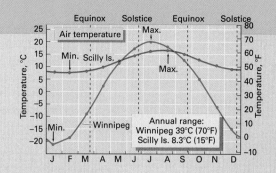

3 The Daily Cycle of Air Temperature

1. Daily and annual air temperature cycles are produced by the Earth's rotation and revolution, which create insolation and net radiation cycles. Temperatures continue to rise when net radiation is positive and continue to decrease when net radiation is negative. Therefore, maximum daily temperatures tend to occur during the afternoon, and minimum daily temperatures tend to occur just before or after sunrise.

2. Maritime locations show smaller ranges of both daily and annual temperature than do continental regions. The ranges are different because water heats more slowly, absorbs energy throughout a surface layer, and can mix and evaporate freely.

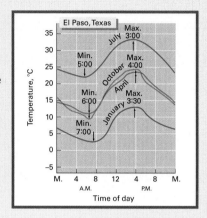

4 The Annual Cycle of Air Temperature

1. Annual air temperature cycles are influenced by annual net radiation patterns, which depend on latitude, as well as elevation, location relative to oceans and continental interiors, and surface type. Temperatures at the Equator vary little from season to season. Poleward, temperatures decrease with latitude, and continental surfaces at high latitudes and high elevations can become very cold in winter.

2. The annual range in temperature increases with latitude. It also increases over continental regions compared with marine locations. The annual range is greatest in northern hemisphere continental interiors.

5 Measurement of Air Temperature

1. Temperature can be measured in many different ways. Manually, it can be measured using a standard liquid-in-glass thermometer. It can also be measured electronically with a thermistor or electrical resistance thermometer, or it can be determined remotely using a radiometer.

2. Temperature measurements can be combined with other weather and climate data to form temperature indexes. These indexes are used to determine how hot or cold the temperature feels to us, how much energy is needed to heat or cool buildings to a desired temperature, or to determine when crops are ready for harvest.

KEY TERMS

1. Describe three environmental effects of the Earth's rotation on its axis.

2. What is meant by the "tilt of the Earth's axis"? How does this tilt cause the seasons?

3. Sketch a diagram of the Earth at equinox. Show the North and South Poles, the Equator, and the circle of illumination. Show the direction of the Sun's incoming rays on your sketch, and shade the night portion of the globe.

4. Why do large water bodies heat and cool more slowly than land masses? What effect does this have on daily and annual temperature cycles for coastal and interior stations?

5. Suppose that the Earth's axis were tilted at 40° to the plane of the ecliptic, instead of 23 1/2°. How would the seasons change at your location? What would be the global effects of the change?

6. Explain how latitude affects the annual cycle of air temperature through net radiation by comparing Manaus, Aswan, Hamburg, and Yakutsk.

7. On the figure below, describe and explain the following features on the map: (a) how annual range varies with latitude; (b) where the greatest ranges occur; (c) how the annual range differs over oceans and over land at the same latitude; (d) the size of the annual range in tropical regions.

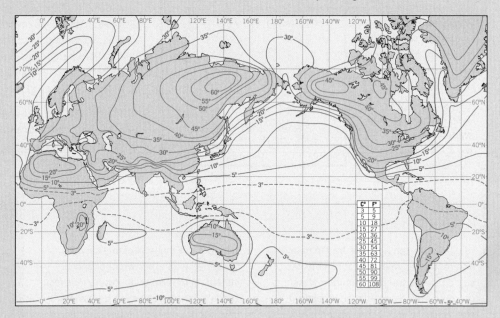

C°	F°
3	5
5	9
10	18
15	27
20	36
25	45
30	54
35	63
40	72
45	81
50	90
55	99
60	108

1. On each of the figures, draw the direction of the Earth's rotation.

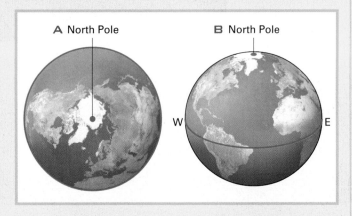

A North Pole

B North Pole

W   E

2. The Earth is nearest to the Sun at _____, which occurs on or about _____.
   a. aphelion, July 4
   b. perihelion, July 4
   c. perihelion, January 3
   d. aphelion, January 3

3. The tilt of the Earth's axis is _____.
   a. 23 1/2 degrees from the plane of the ecliptic
   b. 66 1/2 degrees from a perpendicular to the plane of the ecliptic
   c. 23 1/2 degrees from the Sun
   d. 23 1/2 degrees from a perpendicular to the plane of the ecliptic

4. The _____ divides the Earth into a sunlit side and a night side.
   a. international date line   c. circle of illumination
   b. prime meridian   d. Arctic Circle

5. During an equinox, the circle of illumination passes through the
   _____.
   a. North and South Poles   c. Arctic Circle
   b. Antarctic Circle   d. Equator

6. From the orientation of the Earth in this diagram, what day of the year is it? What is the specific name for this day of the year?

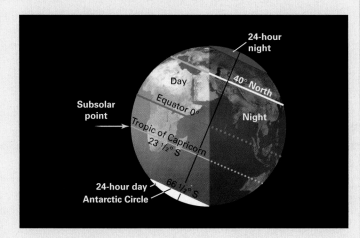

7. Temperature is _____.
   a. a measure of the level of sensible heat of matter
   b. only measured with a thermistor
   c. a measure of the level of latent heat of matter
   d. measured only through advection

8. Minimum daily temperatures usually occur _____.
   a. just after sunset
   b. at midnight
   c. one hour before sunrise
   d. about one-half hour after sunrise

9. The high temperature for the day is typically reached at midday.
   a. True   b. False

10. Urban surface temperatures tend to be warmer than rural temperatures during the day because _____.
   a. drier surfaces are cooler than wet soils
   b. drier surfaces have less water to evaporate than do moist soils
   c. paved surfaces reflect so much heat away into the air
   d. paved surfaces absorb little solar insolation

11. On the diagram provided, draw a line graph indicating temperatures associated with each location.

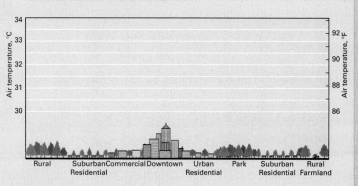

12. Since large bodies of water heat and cool more _____ compared to land surfaces, monthly temperature maximums and minimums tend to be delayed at coastal stations.
   a. slowly   c. constantly
   b. rapidly   d. randomly

13. _____ are lines of equal temperature drawn on a weather map.
   a. Isohyets   c. Isopachs
   b. Isobars   d. Isotherms

14. Relative to adjacent ocean surfaces, isotherms over a large continental land mass tend to shift _____ in latitude during the winter and _____ in latitude during the summer.
   a. north, south   c. west, east
   b. south, north   d. east, west

15. Using Figure 4.24, what would be the wind chill for a location with a temperature of 15°F and a wind speed of 20 mph?

Atmospheric Moisture

The rain began early on Wednesday, as the wind stirred the coastal palms of the northern coast of Honduras. Within a few hours, the wind was howling and the rain was falling in torrents. Hurricane Mitch had begun its week-long odyssey across the mountainous spine of Central America.

The fierce storm—a category 4 hurricane, with winds between 131 and 155 mph (59–70 m/s)—came ashore near the coastal city of Trujillo on Thursday, October 29, 1998. Laden with tropical moisture, Mitch then ambled inland across the coastal mountain ranges and the central Montañas de Comayagua, reaching Tegucigalpa, the capital city, by Saturday. As the saturated air moved up and over the mountains, strong cooling resulted, turning the warm, moist air into huge drops of liquid water that descended at a furious rate. A broad band along the coast received more than 20 in. (500 mm) of rain, with 10–15 in. (250–375 mm) falling near Tegucigalpa.

As the rain fell, the mountain slopes became saturated, turning to mud. Suddenly, whole mountainsides came sliding down, embedded with rocks, trees, and houses. More than 11,000 people across Central America died as Hurricane Mitch and its remnants churned across the region for eight days.

What causes precipitation? As we'll see in this chapter, rain and snow form when moist air is chilled and condensation or deposition takes place. The chilling is usually the result of uplift—for example, as air passes up and over a mountain barrier or is carried upward by convection.

TEGUCIGALPA, HONDURAS

Global Locator

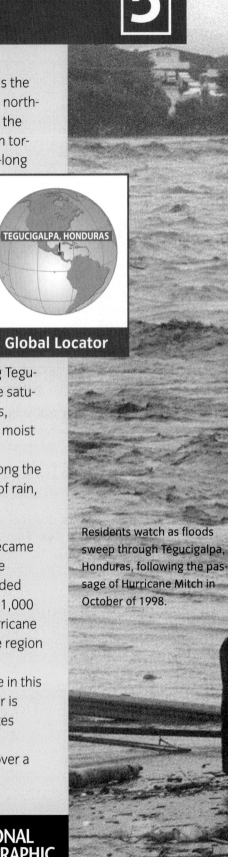

Residents watch as floods sweep through Tegucigalpa, Honduras, following the passage of Hurricane Mitch in October of 1998.

NATIONAL GEOGRAPHIC

Atmospheric Moisture and Precipitation

I n this chapter, we focus on water in the air, both as vapor and as liquid and solid water. Precipitation is the fall of liquid or solid water from the atmosphere that reaches the Earth's land or ocean surface. It forms when moist air is cooled, causing water vapor to form liquid droplets or solid ice particles. If cooling is sufficient, liquid and solid water particles will grow to a size too large to be held aloft by the motion of the atmosphere. They can then fall to the Earth. Before we begin our study of atmospheric moisture and precipitation, however, we will briefly review the three states of water and the conversion of one state to another.

THREE STATES OF WATER

Water can exist in three states—as a solid (ice), as a liquid (water), or as an invisible gas (water vapor), as shown in **FIGURE 5.1**. If we want to change the state of water from solid to liquid, liquid to gas, or solid to gas, we must put in heat energy. This energy, which is drawn in from the surroundings and stored within the water molecules, is called *latent heat*. When the change goes the other way, from liquid to solid, gas to liquid, or gas to solid, this latent heat is released to the surroundings.

We are all familiar with melting, freezing, evaporation, and condensation. *Sublimation* is the direct transition from solid to vapor. Perhaps you have noticed that old ice cubes left in the freezer shrink away from the sides of the ice cube tray and get smaller. They shrink through sublimation—never melting, but losing mass

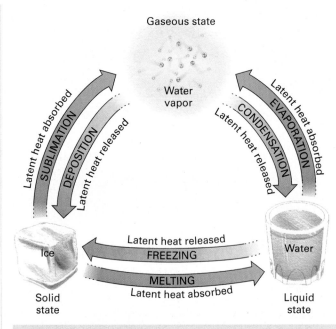

Three states of water FIGURE 5.1

Arrows show the ways that any one state of water can change into either of the other two states. Latent heat energy is absorbed or released, depending on the direction of change.

directly as vapor. In this book, we use the term *deposition* to describe the reverse process, when water vapor crystallizes directly as ice. Frost forming on a cold winter night is a common example of deposition.

THE HYDROSPHERE AND THE HYDROLOGIC CYCLE

Let's look at how water is distributed around the globe. The realm of water in all its forms, and the flows of water among ocean, land, and atmosphere, are known as the hydrosphere, shown in **FIGURE 5.2**. About 97.5 percent of the hydrosphere consists of ocean salt water. The remaining 2.5 percent is fresh water. The next largest reservoir is fresh water stored as ice in the world's ice sheets and mountain glaciers, which accounts for 1.7 percent of total global water.

A Distribution of water Nearly all the Earth's water is contained in the world's oceans. Fresh surface and soil water make up only a small fraction of the total volume of global water.

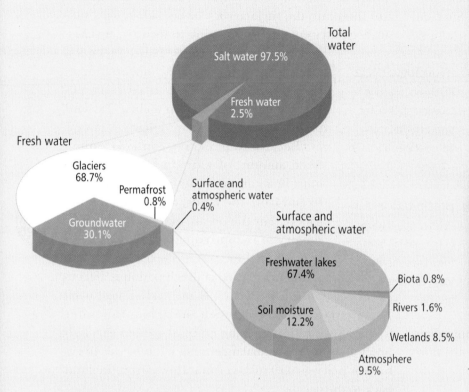

B Oceans Most of the Earth's water is held by its vast ocean. Here, a southern stingray swims along a shallow ocean bottom near Grand Cayman Island.

C Ice Ice sheets and glaciers are the second largest reservoir of water. Although glaciers are too cold and forbidding for most forms of animal life, sea ice provides a habitat for polar bears hunting seals and fish in arctic waters.

D Surface water Surface water, including lakes, is only a very tiny fraction of Earth's water volume. Here a bull moose wades along the margin of a lake in search of aquatic plants.

E Atmosphere Atmospheric water, although only 0.001 percent of total water, is a vital driver of weather and climate and sustains life on Earth.

Atmospheric Moisture and Precipitation   123

Fresh liquid water is found above and below the Earth's land surfaces. Subsurface water lurks in openings in soil and rock. Most of it is held in deep storage as ground water, where plant roots cannot reach. Ground water makes up 0.75 percent of the hydrosphere.

The small remaining proportion of the Earth's water includes the water available for plants, animals, and human use. Plant roots can access soil water. Surface water is held in streams, lakes, marshes, and swamps. Most of this surface water is about evenly divided between freshwater lakes and saline (salty) lakes. An extremely small proportion makes up the streams and rivers that flow toward the sea or inland lakes.

Only a very small quantity of water is held as vapor and cloud water droplets in the atmosphere—just 0.001 percent of the hydrosphere. However, this small reservoir of water is enormously important. Through precipitation, it supplies water and ice to replenish all freshwater stocks on land. In addition, this water, and its conversion from one form to another in the atmosphere, is an essential part of weather events across the globe. Finally, the flow of water vapor from warm tropical oceans to cooler regions provides a global flow of heat from low to high latitudes.

THE HYDROLOGIC CYCLE AND GLOBAL WATER BALANCE

The **hydrologic cycle** represents the flow of water among ocean, land, and atmosphere, shown in FIGURE 5.3. It moves water from land and ocean to the atmosphere. Water from the oceans and from land surfaces evaporates, changing state from liquid to vapor and entering the atmosphere. Total evaporation is about six times greater over oceans than land, because oceans cover most of the planet and because land surfaces are not always wet enough to yield much water.

> ■ **hydrologic cycle** Pathways of active movement of water between the ocean, atmosphere, and land surface.
>
> ■ **precipitation** Particles of liquid water or ice that fall from the atmosphere and may reach the ground.

Once in the atmosphere, water vapor can condense or deposit to form **precipitation**, which falls to the Earth as rain, snow, sleet, or hail. There is nearly four times as much precipitation over oceans than precipitation over land.

When precipitation hits land, it has one of three fates. First, it can evaporate and return to the atmosphere as water vapor. Second, it can sink into the soil and then into the surface rock layers below. This subsurface water emerges from below to feed rivers, lakes, and even ocean margins. Third, precipitation can run off the land, concentrating in streams and rivers that eventually carry it to the ocean or to a lake in a closed inland basin. This flow of water is known as *runoff*.

Because our planet contains only a fixed amount of water, a global balance must be maintained among flows of water to and from the lands, oceans, and atmosphere. For the ocean, evaporation leaving the ocean is approximately 420 cubic km per year (101 mi^3/yr), while the amount entering the ocean via precipitation is 380 cubic km per year (91 mi^3/yr). There is an imbalance between the amount of water lost to evaporation and the amount gained through precipitation. This imbalance is made up by the 40 cubic km per year (10 mi^3/yr) that flows from the land back to the ocean.

Similarly, for the land surfaces of the world, there is a balance. Of the 110 cubic km per year (27 mi^3/yr) of water that falls on the land surfaces, 70 cubic km per year (17 mi^3/yr) is re-evaporated back into the atmosphere. The remaining 40 cubic km per year (10 mi^3/yr) stays in the form of liquid water and eventually flows back into the ocean.

Of all these pathways, we will be most concerned with one aspect of the hydrologic cycle—the flow of water from the atmosphere to the surface in the form of precipitation. To understand this process, we first need to examine how water vapor in the atmosphere is converted into clouds and subsequently into precipitation.

The hydrologic cycle FIGURE 5.3

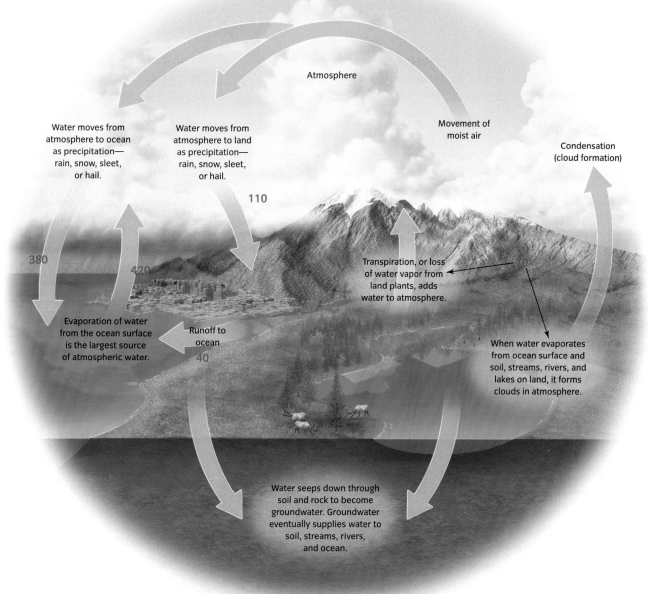

Atmosphere

Water moves from atmosphere to ocean as precipitation—rain, snow, sleet, or hail.

Water moves from atmosphere to land as precipitation—rain, snow, sleet, or hail.

Movement of moist air

Condensation (cloud formation)

110

380

420

Transpiration, or loss of water vapor from land plants, adds water to atmosphere.

Evaporation of water from the ocean surface is the largest source of atmospheric water.

Runoff to ocean

40

When water evaporates from ocean surface and soil, streams, rivers, and lakes on land, it forms clouds in atmosphere.

Water seeps down through soil and rock to become groundwater. Groundwater eventually supplies water to soil, streams, rivers, and ocean.

Calculating the global hydrologic balance

The natural flow of water between the oceans, the land and atmosphere, is a system in balance.
This table explains that balance as depicted in the figure.

Entering ocean	Leaving ocean	Entering land	Leaving land
380×1000 km^3/yr	420×1000 km^3/yr	110×1000 km^3/yr	70×1000 km^3/yr
40×1000 km^3/yr			40×1000 km^3/yr
Total: 420×1000 km^3/yr	Total: 420×1000 km^3/yr	Total: 110×1000 km^3/yr	Total: 110×1000 km^3/yr

CONCEPT CHECK **STOP**

What are the three states of water? Describe the six processes through which water changes state.

Which processes include taking in latent heat?

When is latent heat released?

What is the hydrologic cycle?

What is the hydrosphere?

What is precipitation?

Humidity

Blistering summer heat waves can be deadly, with the elderly and the ill at most risk. However, even healthy young people need to be careful, especially in hot, humid weather. High humidity slows the evaporation of perspiration from our bodies, reducing its cooling effect. Clearly, it is not only the temperature of the air that controls how hot weather affects us—the amount of water vapor in the air is important as well.

humidity General term for the amount of moisture in the air.

specific humidity Amount of water vapor (grams) contained within a kilogram of air.

The amount of water vapor present in the air, referred to as **humidity**, varies widely from place to place and time to time. In the cold, dry air of arctic regions in winter, the humidity is almost zero, while it can reach up to as much as 3 to 4 percent of a given volume of air in the warm wet regions near the Equator.

An important principle concerning humidity states that the maximum quantity of water vapor an air parcel can contain is dependent on the air temperature itself. Warm air can contain more water vapor than cold air—a lot more. Air at room temperature (20°C, 68°F) can contain about three times as much water vapor as freezing air (0°C, 32°F).

SPECIFIC HUMIDITY

The actual quantity of water vapor contained within a parcel of air is known as its **specific humidity** and is expressed as grams of water vapor per kilogram of air (g/kg). The equation for specific humidity is given as:

$$\text{specific humidity} = \frac{\text{mass of water vapor}}{\text{mass of total air}}.$$

Specific humidity is often used to describe the amount of water vapor in a large mass of air. Both humidity and temperature are measured at the same locations in standard thermometer shelters the world over and also on ships at sea. Specific humidity is largest at the warm, equatorial zones, and falls off rapidly toward the colder poles (**FIGURE 5.4**).

Global specific humidity and temperature
FIGURE 5.4

Pole-to-pole profiles of specific humidity (above) and temperature (below) show similar trends, because the amount of water vapor air can contain is limited by temperature.

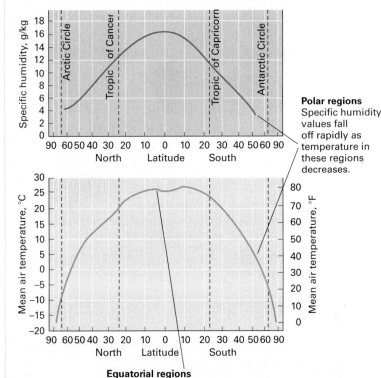

Polar regions
Specific humidity values fall off rapidly as temperature in these regions decreases.

Equatorial regions
More insolation is available at lower latitudes to evaporate water from oceans or moist land surfaces. Therefore, specific humidity and temperature values are high at low latitudes.

Extremely cold, dry air over arctic regions in winter may have a specific humidity as low as 0.2 g/kg, while the extremely warm, moist air of equatorial regions often contains as much as 18 g/kg. The total natural range on a worldwide basis is very wide. In fact, the largest values of specific humidity observed are from 100 to 200 times as great as the smallest values.

SATURATION SPECIFIC HUMIDITY

Although the actual amount of water in a given volume of air is called the specific humidity, this is not the same as the maximum quantity of moisture that a given volume of air can contain at any time. This maximum specific humidity, referred to as the **saturation specific humidity**, is dependent on the air's temperature.

■ **saturation specific humidity** The maximum amount of water vapor an air parcel can contain based on its temperature.

■ **saturation** The condition in which the specific humidity is equal to the saturation specific humidity.

■ **dew-point temperature** The temperature at which air with a given humidity reaches saturation when cooled without changing its pressure.

In FIGURE 5.5 we see, for example, that at 20°C (68°F), the maximum amount of water vapor that the air can contain—the saturation specific humidity—is about 15 g/kg. At 30°C (86°F), it is nearly doubled—about 26 g/kg. For cold air, the values are quite small. At –10°C (14°F), the maximum is only about 2 g/kg.

Another way of describing the water vapor content of air is by its dew-point temperature, also called simply the *dew point*. If air is slowly chilled, its saturation specific humidity decreases. This can continue until the saturation specific humidity is equal to the specific humidity. When this condition is reached, the air has reached **saturation**, because the air contains the maximum amount of water vapor possible. If further cooling continues, condensation begins. The temperature at which saturation occurs is therefore known as the **dew-point temperature**—that is, the temperature at which dew forms by condensation.

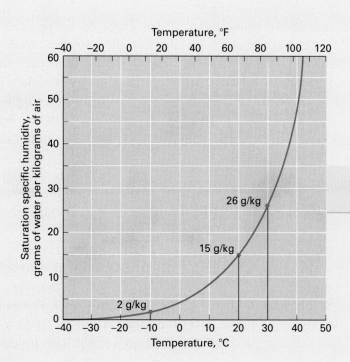

Saturation specific humidity and temperature FIGURE 5.5

The maximum specific humidity a mass of air can have—the saturation specific humidity—increases sharply with rising temperature.

Dew-point temperature
and actual temperature
FIGURE 5.6

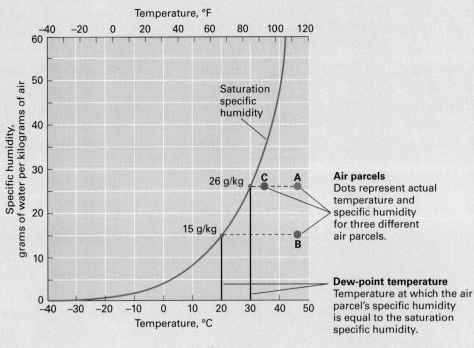

Saturation specific humidity

26 g/kg

15 g/kg

C   A

B

Air parcels
Dots represent actual temperature and specific humidity for three different air parcels.

Dew-point temperature
Temperature at which the air parcel's specific humidity is equal to the saturation specific humidity.

The dew-point temperature is the temperature to which you would have to cool an air parcel for it to reach saturation. It can be found by drawing a horizontal line from the present temperature and specific humidity to the line indicating the saturation specific humidity, then reading the dew-point temperature at that point.

A and B
Parcel A (moist air) has a higher specific humidity than parcel B (dry air). When A and B are cooled, A saturates first, so its dew point is higher.

A and C
Although air parcels A and C have different temperatures, they have the same specific humidity. When cooled, they saturate at the same dew point temperature.

As seen in FIGURE 5.6, the dew-point temperature provides information about the actual moisture in the air (i.e., the specific humidity). For this reason, many meteorologists and climatologists present maps of the dew-point temperature, because these provide a consistent way of characterizing the actual moisture content of the atmosphere.

RELATIVE HUMIDITY

When weather forecasters speak of humidity, they are usually referring to **relative humidity**. This measure compares the amount of water vapor present to the maximum amount that the air can contain at its given temperature. The relative humidity is expressed as a percentage given by:

$$\text{relative humidity} = 100 \times \frac{\text{specific humidity}}{\text{saturation specific humidity}}.$$

relative humidity
The amount of water vapor in an air parcel as a fraction of the maximum amount it can contain based on its temperature.

For example, if the air currently contains half the moisture possible at the present temperature, then the relative humidity is 50 percent. When the humidity is 100 percent, the air contains the maximum amount of moisture possible. The air is saturated, and its temperature is at the dew point. When the specific humidity and saturation specific humidity are not the same, the air is unsaturated. Generally, when the difference between the two is large, the relative humidity is small and vice versa.

The relative humidity of the atmosphere can change in one of two ways. First, the atmosphere can directly gain or lose water vapor, thereby changing the specific humidity of the air mass. For example, additional water vapor can enter the air from an exposed water surface or from wet soil. This process is slow, because the water vapor molecules must diffuse upward from the surface into the air layer above.

Relative humidity and air temperature FIGURE 5.7

Relative humidity changes with temperature because warm air can contain more water vapor than cold air. In this example, the amount of water vapor stays the same, and only the saturation specific humidity of the air mass changes.

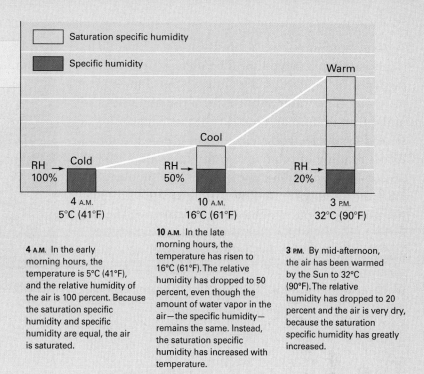

4 A.M. In the early morning hours, the temperature is 5°C (41°F), and the relative humidity of the air is 100 percent. Because the saturation specific humidity and specific humidity are equal, the air is saturated.

10 A.M. In the late morning hours, the temperature has risen to 16°C (61°F). The relative humidity has dropped to 50 percent, even though the amount of water vapor in the air—the specific humidity—remains the same. Instead, the saturation specific humidity has increased with temperature.

3 P.M. By mid-afternoon, the air has been warmed by the Sun to 32°C (90°F). The relative humidity has dropped to 20 percent and the air is very dry, because the saturation specific humidity has greatly increased.

The second way relative humidity changes is through a change of temperature. Even though no water vapor is added, an increase of temperature results in a decrease of relative humidity (FIGURE 5.7). Recall that the saturation specific humidity of air is dependent on temperature. When the air is warmed, the saturation specific humidity increases. The existing amount of water vapor, given by the specific humidity, then represents a smaller fraction of the saturation specific humidity.

Measuring relative humidity A simple method of measuring relative humidity uses two thermometers mounted together side by side in an instrument called a *sling psychrometer*, shown in FIGURE 5.8. After whirling the psychrometer in the air, the temperature difference between the wet-bulb thermometer and the dry-bulb thermometer can be used to derive the relative humidity.

There are also instruments that read relative humidity directly. One such instrument—called an

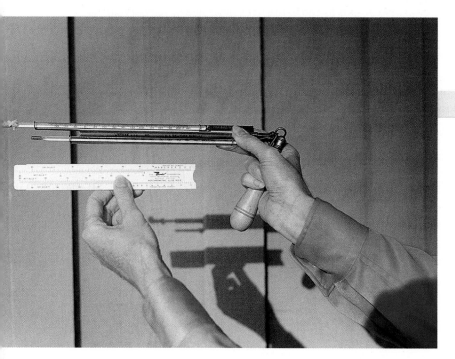

Sling psychrometer FIGURE 5.8

The sling psychrometer measures relative humidity using paired wet- and dry-bulb thermometers. The wet-bulb thermometer is wetted with water, and the thermometers are whirled overhead. As the water evaporates from the wet bulb, its temperature drops in proportion to the relative humidity of the air. The difference in temperature between the dry- and wet-bulb thermometers gives the relative humidity, which is read using the special scale held below the psychrometer.

Humidity 129

electrical hygrometer—uses a thin layer of a special material bonded to a metal film. The material absorbs water vapor in an amount dependent upon the relative humidity. The water vapor affects the ability of the metal film to hold an electric charge. This ability is sensed by an electronic circuit and is converted to a direct reading of relative humidity. Another type of sensor—an *infrared hygrometer*—uses a small device to emit and collect infrared radiation. Because water vapor absorbs infrared radiation, by measuring the difference between the amount emitted by the sensor and the amount received at the other end, we can determine the amount absorbed by water vapor in the intervening distance. The absorption amount can then be converted to specific humidity and, given the temperature, relative humidity.

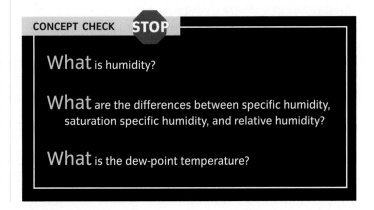

CONCEPT CHECK **STOP**

What is humidity?

What are the differences between specific humidity, saturation specific humidity, and relative humidity?

What is the dew-point temperature?

The Adiabatic Process

LEARNING OBJECTIVES

Describe the adiabatic principle.

Explain the role of the adiabatic process in cloud formation.

What makes the water vapor in the air turn into liquid or solid particles that can fall to the Earth? The answer is that the air is naturally cooled. When air cools to the dew point, the air is saturated with water. Think about extracting water from a moist sponge. To release the water, you have to squeeze the sponge—that is, reduce its ability to hold water. In the atmosphere, chilling the air beyond the dew point is like squeezing the sponge; it reduces the amount of water vapor the air can contain, forcing some water vapor molecules to change state to form water droplets or ice crystals.

One mechanism for chilling air is nighttime cooling. On a clear night, the ground surface can become quite cold as it loses longwave radiation. However, this is not enough to form precipitation. Precipitation only forms when a substantial mass of air experiences a steady drop in temperature below the dew point. This happens when an air parcel is lifted to higher and higher levels in the atmosphere.

DRY ADIABATIC RATE

If you have ever pumped up a bicycle tire using a hand pump, you might have noticed the pump getting hot. If

Adiabatic cooling and heating FIGURE 5.9

When air is forced to rise, it expands and its temperature decreases. When air is forced to descend, its temperature increases.

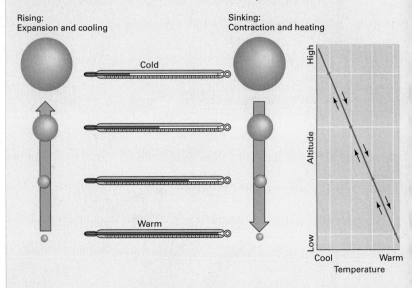

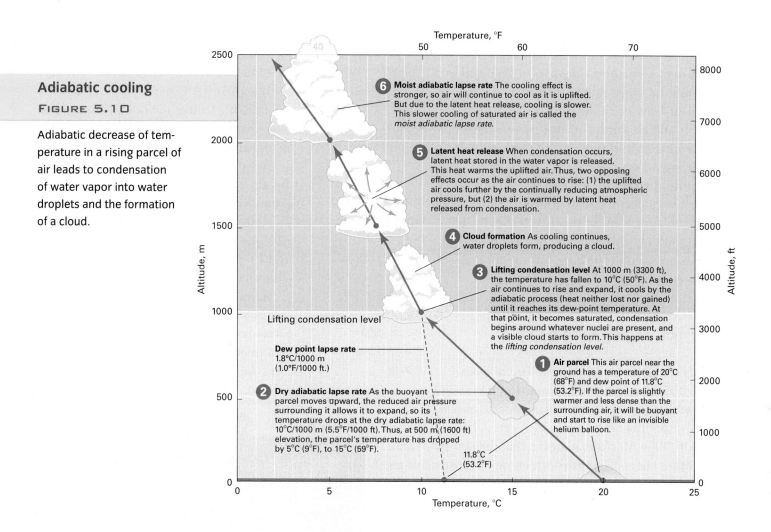

Adiabatic cooling

FIGURE 5.10

Adiabatic decrease of temperature in a rising parcel of air leads to condensation of water vapor into water droplets and the formation of a cloud.

6 **Moist adiabatic lapse rate** The cooling effect is stronger, so air will continue to cool as it is uplifted. But due to the latent heat release, cooling is slower. This slower cooling of saturated air is called the *moist adiabatic lapse rate.*

5 **Latent heat release** When condensation occurs, latent heat stored in the water vapor is released. This heat warms the uplifted air. Thus, two opposing effects occur as the air continues to rise: (1) the uplifted air cools further by the continually reducing atmospheric pressure, but (2) the air is warmed by latent heat released from condensation.

4 **Cloud formation** As cooling continues, water droplets form, producing a cloud.

3 **Lifting condensation level** At 1000 m (3300 ft), the temperature has fallen to 10°C (50°F). As the air continues to rise and expand, it cools by the adiabatic process (heat neither lost nor gained) until it reaches its dew-point temperature. At that point, it becomes saturated, condensation begins around whatever nuclei are present, and a visible cloud starts to form. This happens at the *lifting condensation level.*

Lifting condensation level

Dew point lapse rate
1.8°C/1000 m
(1.0°F/1000 ft.)

1 **Air parcel** This air parcel near the ground has a temperature of 20°C (68°F) and dew point of 11.8°C (53.2°F). If the parcel is slightly warmer and less dense than the surrounding air, it will be buoyant and start to rise like an invisible helium balloon.

2 **Dry adiabatic lapse rate** As the buoyant parcel moves upward, the reduced air pressure surrounding it allows it to expand, so its temperature drops at the dry adiabatic lapse rate: 10°C/1000 m (5.5°F/1000 ft). Thus, at 500 m (1600 ft) elevation, the parcel's temperature has dropped by 5°C (9°F), to 15°C (59°F).

11.8°C (53.2°F)

so, you have observed the *adiabatic principle.* This important law states that if no energy is added to a gas, its temperature will increase as it is compressed. Conversely, when a gas expands, its temperature drops. As you pump vigorously, compressing the air, the metal bicycle pump gets warm. In the same way, when a small jet of air escapes from a high-pressure canister, it feels cool.

Physicists use the term *adiabatic process* to refer to a heating or cooling process that occurs solely as a result of pressure change, not by heat flowing into or away from a volume of air.

How does the adiabatic principle relate to the uplift of air and to precipitation? The missing link is simply that atmospheric pressure decreases as altitude increases. As a parcel of air is uplifted, atmospheric

■ **dry adiabatic lapse rate** The rate at which rising air is cooled by expansion when no condensation is occurring; 10°C per 1000 m (5.5°F per 1000 ft).

pressure on the parcel becomes lower, and the air expands and cools, as shown in FIGURE 5.9. As a parcel of air descends, atmospheric pressure becomes higher, and the air is compressed and warmed.

We describe this behavior in the atmosphere using the **dry adiabatic lapse rate** for a rising air parcel that has not reached saturation, shown in the lower portion of FIGURE 5.10. This rate has a value of about 10°C per 1000 m (5.5°F per 1000 ft) of vertical rise. That is, if a parcel of air is raised 1 km, its temperature will drop by 10°C. In English units, if the parcel of air is raised 1000 ft, its temperature will drop by 5.5°F. This is the *dry* rate because no condensation occurs during this process. Conversely, an air parcel that descends will warm by 10°C per 1000 m.

There is an important difference to note between the dry adiabatic lapse rate and the *environmental lapse rate*. The environmental lapse rate is simply an expression of how the temperature of still air varies with altitude. This rate will vary from time to time and from place to place, depending on the state of the atmosphere. It is quite different from the dry adiabatic lapse rate. The dry adiabatic lapse rate applies to a mass of air moving vertically. It does not vary with time and place and is determined by physical laws, not the local atmospheric state.

MOIST ADIABATIC RATE

Let's continue examining the fate of a parcel of air that is moving upward in the atmosphere. As the parcel moves upward, its temperature drops at the dry adiabatic rate, 10°C/1000 m (5.5°F/1000 ft). Note, however, that the dew-point temperature changes slightly with elevation. Instead of remaining constant, it falls at the dew-point lapse rate of 1.8°C/1000 m (1.0°F/1000 ft). As the rising process continues, the air is eventually cooled to its dew-point temperature, and condensation starts to occur. This is shown in Figure 5.10 on page 131, as the **lifting condensation level**. The lifting condensation level is thus determined by the initial temperature of the air and its initial dew point and can differ from the example shown here.

If the parcel of saturated air continues to rise, a new principle comes into effect—latent heat release. That is, when condensation occurs, latent heat is released by the condensing water molecules and warms the surrounding air molecules. In other words, two effects are occurring at once. First, the uplifted air is being cooled by the reduction in atmospheric pressure. Second, it is being warmed by the release of latent heat from condensation.

Which effect is stronger? As it turns out, the cooling effect is stronger, so the air will continue to cool as it is uplifted. However, because of the release of latent

lifting condensation level Level of the atmosphere to which an air parcel must be lifted before condensation starts to occur.

moist adiabatic lapse rate Reduced rate at which rising air is cooled by expansion when condensation is occurring; ranges from 4 to 9°C per 1000 m (2.2–4.9°F per 1000 ft).

heat, the cooling will occur at a lesser rate. This cooling rate for saturated air is called the **moist adiabatic lapse rate** and ranges between 4 and 9°C per 1000 m (2.2–4.9°F per 1000 ft). Unlike the dry adiabatic lapse rate, which remains constant, the moist adiabatic lapse rate is variable, because it depends on the temperature and pressure of the air and its moisture content. For most situations, however, we can use a value of 5°C/1000 m (2.7°F/1000 ft). In Figure 5.10, the moist adiabatic rate is shown as a slightly curving line to indicate that its value changes with altitude.

Keep in mind that as the air parcel becomes saturated and continues to rise, condensation is occurring. This condensation produces liquid and solid ice particles that form clouds and eventually precipitation.

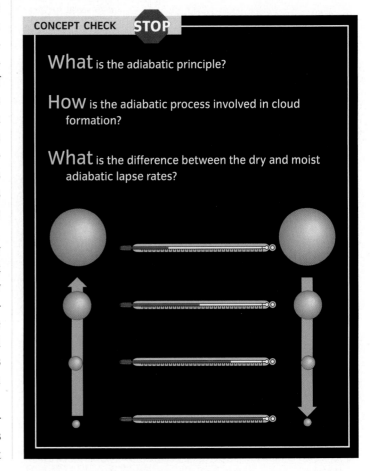

CONCEPT CHECK STOP

What is the adiabatic principle?

How is the adiabatic process involved in cloud formation?

What is the difference between the dry and moist adiabatic lapse rates?

Clouds

LEARNING OBJECTIVES

Explain how condensation nuclei help clouds to form.

Describe how clouds are classified.

Define fog.

Clouds are frequent features of the atmosphere. Views of the Earth from space show that clouds cover about half of the Earth at any given time. Low clouds reflect solar energy, thus cooling the Earth–atmosphere system, while high clouds absorb outgoing longwave radiation, thus warming the Earth–atmosphere system. One of the most familiar roles of clouds, however, is in producing precipitation.

Clouds are made up of water droplets, ice particles, or a mixture of both, suspended in air. These particles are between 20 and 50 µm (0.0008–0.002 in.) in diameter. Cloud particles cannot form in empty space. They need a tiny center of solid matter to grow around. This speck of matter is called a **condensation nucleus**, and typically has a diameter of 0.1–1 µm (0.000004–0.00004 in.).

> **condensation nucleus** A tiny bit of solid matter (aerosol) in the atmosphere on which water vapor condenses to form a tiny water droplet.

The surface of the sea is an important source of condensation nuclei. Droplets of spray from the crests of the waves are carried upward by turbulent air, shown in **FIGURE 5.11**. When these droplets evaporate, they leave behind a tiny residue of crystalline salt suspended in the air. This aerosol strongly attracts water molecules, helping begin cloud formation. Nuclei are also thrown into the atmosphere as dust in polluted air over cities, aiding condensation and the formation of clouds and fog.

If you ask, "What is the freezing point of water?" most people will reply that liquid water turns to ice at 0°C (32°F). This is true in everyday life, but when water is dispersed as tiny droplets in clouds, it behaves differently. Water in clouds can remain in the liquid state at temperatures far below freezing. In that case, we say the water is *supercooled*. In fact, clouds consist entirely of water droplets at temperatures down to about –12°C (10°F). As cloud temperatures drop, ice crystals begin to appear. The coldest clouds, with temperatures below –60°C (–76°F), occur at altitudes above 12 km (40,000 ft) and are made up entirely of ice particles.

Cloud condensation nuclei FIGURE 5.11

Cloud drops condense on small particulates called *cloud condensation nuclei*. Breaking or spilling waves in the open ocean, shown here from the deck of a ship, are an important source of condensation nuclei.

Cloud gallery FIGURE 5.12

A Cloud families and types Clouds are grouped into families on the basis of height. Individual cloud types are named according to their form.

Classification of clouds according to height and form

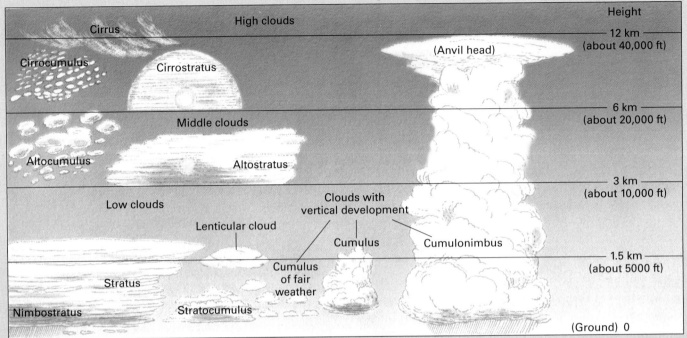

B Cirrus High, thin, wispy clouds drawn out into streaks are cirrus clouds. They are composed of ice crystals and form when moisture is present high in the air.

C Lenticular cloud A lenticular, or lens-shaped, cloud forms as moist air flows up and over a mountain peak or range.

D Cumulus Puffy, fair-weather cumulus clouds fill the sky above a prairie.

E Altocumulus High cumulus clouds, in a pattern sometimes called a *mackerel sky*, as photographed near sunset in Boston.

CLOUD FORMS

Anyone who has looked up at the sky knows that clouds come in many shapes and sizes, as seen in FIGURE 5.12. They range from the small, white, puffy clouds often seen in summer to the gray layers that produce a typical rainy day. Meteorologists name clouds by their vertical structure and the altitudes at which they occur. *Stratiform* clouds are blanket-like and cover large areas. A common type is *stratus*, a low cloud layer that covers the entire sky. Dense, thick stratus clouds can produce large amounts of rain or snow. Higher stratus clouds are referred to as *altostratus*. *Cirrus* clouds are high, thin clouds that often have a wispy or patchy appearance. When they cover the sky evenly, they form *cirrostratus*.

Cumuliform clouds are clouds with vertical development. The most common cloud of this type is the *cumulus* cloud, which is a globular cloud mass associated with small to large parcels of rising air starting near the surface. However, there are also *altocumulus*—individual, rounded clouds in the middle layers of the troposphere—and *cirrocumulus*—cloud rolls or ripples in the upper portions of the troposphere.

Nimbus clouds are clouds that produce rainfall. Thus, *nimbostratus* is a thick, flat, rain cloud, and *cumulonimbus* is a cumulus rain cloud.

Tall, dense cumulonimbus clouds with strong updrafts are often associated with intense rainfall, thunder, and lightning. Within these *thunderstorms*, a succession of warm, bubble-like air parcels rise up within a localized region. These bubbles are intensely cooled, according to the adiabatic process, producing precipitation. This precipitation can be water if the clouds are at the lower levels, mixed water and snow at intermediate levels, and snow at high levels where cloud temperatures are coldest. As the rising air parcels reach high levels, which may be 6–12 km (about 20,000–40,000 ft) or even higher, the rising rate slows. At such high altitudes, the winds are typically strong, dragging the cloud top downwind and giving the thunderstorm cloud its distinctive shape—resembling an old-fashioned blacksmith's anvil, shown in FIGURE 5.13.

Thunderstorm with anvil cloud FIGURE 5.13

This isolated thunderstorm has a characteristic horizontal cloud spreading downwind, referred to as an *anvil cloud*.

FOG

Fog is simply a cloud layer at or very close to the Earth's surface. For centuries, fog at sea has been a navigational hazard, increasing the danger of ship collisions and groundings. In our industrialized world, it can be a major environmental hazard. Dense fog on high-speed highways can cause chain-reaction accidents, sometimes involving dozens of vehicles. When flights are shut down or delayed by fog, it is inconvenient to passengers and costs airlines money. Polluted fogs, like London's "pea-soupers" in the early part of the 20th century, can injure urban dwellers' lungs and take a heavy toll in lives.

One type of fog, known as *radiation fog*, forms at night when the temperature of the air layer at the ground level falls below the dew point. This kind of fog forms in valleys and low-lying areas, particularly on clear winter nights when radiative cooling is very strong.

Another fog type—*advection fog*—results when a warm, moist air layer moves over a cold surface. As the warm air layer loses heat to the surface, its temperature drops below the dew point, and condensation sets in. Advection fog commonly occurs over oceans where warm and cold currents occur side by side. When warm, moist air above the warm current moves over the cold current, condensation occurs. Fogs form in this way off the Grand Banks of Newfoundland, because here the cold Labrador current comes in contact with the warmer waters of the Gulf Stream.

Advection fog is also frequently found along the California coast, as seen in FIGURE 5.14. It forms within a cool marine air layer in direct contact with the colder water of the California current. Similar fogs are also found on continental west coasts in the tropical latitude zones, where cool, equatorward currents lie parallel to the shoreline.

You can read about smog, the combination of smoke and fog, in *What a Scientist Sees: Smog.*

Fog FIGURE 5.14

A layer of advection fog entering San Francisco Bay obscures the Golden Gate Bridge.

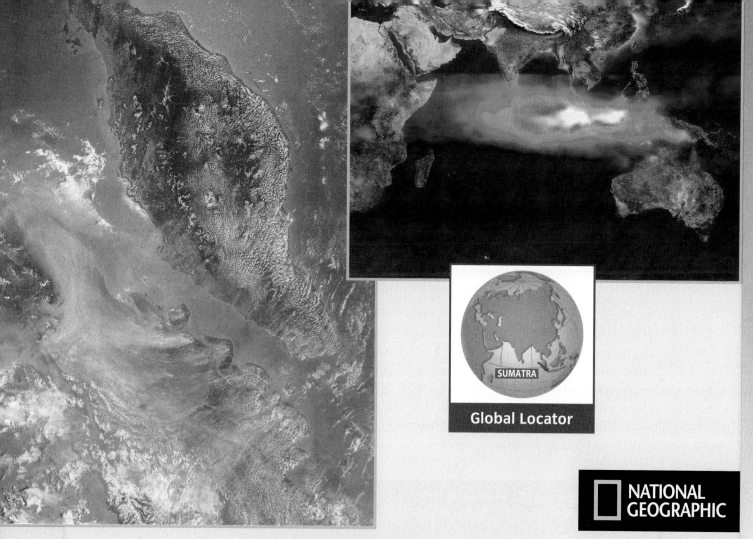

SUMATRA

Global Locator

NATIONAL GEOGRAPHIC

Smog

The term *smog* was coined by combining the words *smoke* and *fog*. In this photograph, which shows the island of Sumatra from overhead, burning fires produce smoke—seen as white streaks in the lower portion of the image. In this case, some hazy sunlight would still get through to the people on the ground. In other cases, however, smog can be dense enough to hide from view aircraft flying overhead.

A scientist, on the other hand, might look at the same scene using satellite data and "see" what appears in the inset. This image shows tropospheric ozone (in reds and greens) and smoke (in white and gray) emitted from Sumatra and other regions of Indonesia. While the smoke can irritate eyes and discolor structures, ozone, which is a principal constituent of smog, can harm plant tissues and cause respiratory problems in humans and other animals.

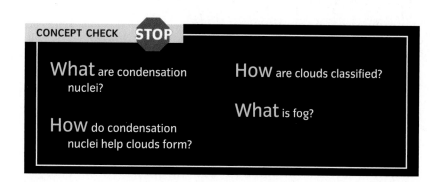

CONCEPT CHECK **STOP**

What are condensation nuclei?

How do condensation nuclei help clouds form?

How are clouds classified?

What is fog?

Precipitation

Clouds are the source of precipitation—the process that provides the fresh water essential for most forms of terrestrial life. Precipitation can form in two ways. In warm clouds, fine water droplets condense, collide, and coalesce into larger and larger droplets that can fall as rain. In colder clouds, ice crystals form and grow in a cloud that contains a mixture of both ice crystals and water droplets.

The first process occurs when saturated air rises rapidly and cooling forces additional condensation, discussed in **FIGURE 5.15**. For raindrops in a warm cloud, the updraft of rising air first lifts tiny suspended cloud droplets upward. By collisions with other droplets, some grow in volume. These larger droplets collide with other small droplets and continue to grow. Note that a droplet is kept aloft by the force of the updraft on its surface, and as the volume of each droplet increases, so does its weight. Eventually, the downward gravitational force on the drop exceeds the upward force, and the drop begins to fall. Now moving in the opposite direction to the fine cloud droplets, the drop sweeps them up and continues to grow. Collisions with smaller droplets can also split drops, creating more drops that can continue to grow in volume. Eventually, the drop falls out of the cloud, cutting off its source of growth. On its way to the Earth, it may suffer evaporation and decrease in size or even disappear.

VIEW THIS IN ACTION
in your WileyPLUS course

Process Diagram

Rain formation in warm clouds FIGURE 5.15

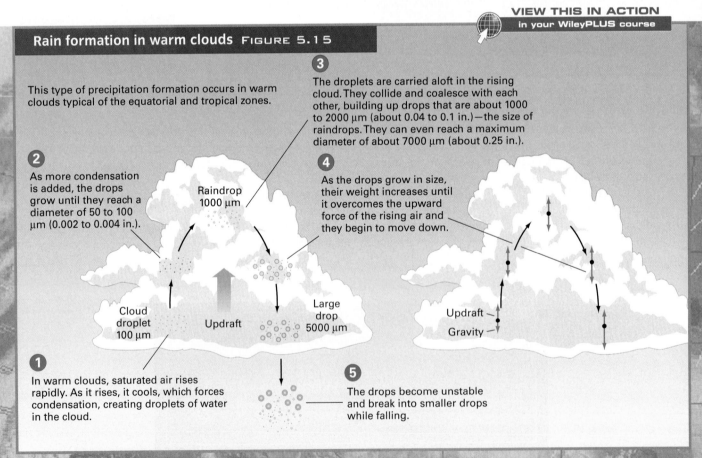

This type of precipitation formation occurs in warm clouds typical of the equatorial and tropical zones.

1 In warm clouds, saturated air rises rapidly. As it rises, it cools, which forces condensation, creating droplets of water in the cloud.

2 As more condensation is added, the drops grow until they reach a diameter of 50 to 100 μm (0.002 to 0.004 in.).

3 The droplets are carried aloft in the rising cloud. They collide and coalesce with each other, building up drops that are about 1000 to 2000 μm (about 0.04 to 0.1 in.)—the size of raindrops. They can even reach a maximum diameter of about 7000 μm (about 0.25 in.).

4 As the drops grow in size, their weight increases until it overcomes the upward force of the rising air and they begin to move down.

5 The drops become unstable and break into smaller drops while falling.

Raindrop 1000 μm

Cloud droplet 100 μm

Updraft

Large drop 5000 μm

Updraft

Gravity

The Bergeron process FIGURE 5.16

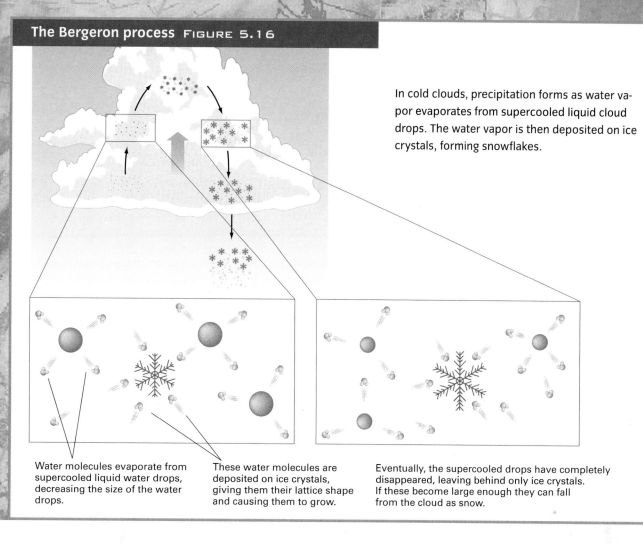

In cold clouds, precipitation forms as water vapor evaporates from supercooled liquid cloud drops. The water vapor is then deposited on ice crystals, forming snowflakes.

Water molecules evaporate from supercooled liquid water drops, decreasing the size of the water drops.

These water molecules are deposited on ice crystals, giving them their lattice shape and causing them to grow.

Eventually, the supercooled drops have completely disappeared, leaving behind only ice crystals. If these become large enough they can fall from the cloud as snow.

Within cool clouds, snow is formed in a different way, known as the *Bergeron process* (FIGURE 5.16). Cool clouds are a mixture of ice crystals and supercooled water droplets. The ice crystals take up water vapor and grow by deposition. At the same time, the supercooled water droplets lose water vapor by evaporation and shrink. In addition, when an ice crystal collides with a droplet of supercooled water, it freezes the droplet. The ice crystals then coalesce to form ice particles, which can become heavy enough to fall from the cloud.

orographic precipitation
Precipitation induced when moist air is forced to rise over a mountain barrier.

PRECIPITATION PROCESSES

Air that is moving upward is chilled by the adiabatic process, which leads, eventually, to precipitation. However, one key piece of the precipitation puzzle is still missing—what causes air to move upward in the first place?

Air can move upward in four ways. In this chapter, we discuss the first two: *orographic precipitation* and *convective precipitation*. A third way for air to be forced upward is through the movement of air masses and their interaction with one another. This type of process usually occurs at the boundaries—or the *fronts*—between air masses, so we call it *frontal precipitation*. The fourth way is by *convergence*, in which air currents converge together at a location from different directions, forcing air at the surface upward.

Orographic precipitation
Orographic precipitation occurs when a through-flowing current of moist air is forced to move upward over a mountainous barrier. The term orographic means "related to mountains." To understand the orographic precipitation

process, you can think of what happens to a mass of air moving up and over a mountain range (FIGURE 5.17). As the moist air is lifted, it is cooled, and condensation and rainfall occur. Passing over the mountain summit, the air descends the leeward slopes of the range, where it is compressed and warmed. Because the air is much warmer and much drier than when it started, little precipitation occurs in these regions, producing a **rain shadow** on the far side of the mountain.

California's rainfall patterns provide an excellent example of orographic precipitation and the rain shadow effect. FIGURE 5.18 contains maps of California mean annual precipitation that use lines of equal precipitation called **isohyets**. These lines clearly show the orographic effect on air moving across the mountains of California into America's great interior desert zone, which extends from eastern California and across Nevada.

■ **rain shadow** Belt of dry climate leeward of a mountain barrier, produced as a result of adiabatic warming of descending air.

■ **isohyets** Lines on a map connecting regions with equal amounts of rainfall.

Orographic precipitation FIGURE 5.17

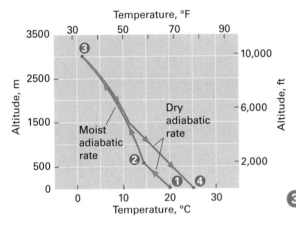

2 When the air has cooled sufficiently, water droplets begin to condense, and clouds will start to form. The cloud cools at the moist adiabatic rate until, eventually, precipitation begins. Precipitation continues to fall as air moves up the slope.

3 After passing over the mountain summit, the air begins to descend down the leeward slopes of the range. As it descends, it is compressed and so, according to the adiabatic principle, it gets warmer. This causes the water droplets and the ice crystals in the cloud to evaporate or sublimate. Eventually the air clears, and it continues to descend, warming at the dry adiabatic rate.

4 At the base of the mountain on the far side, the air is now warmer. It is also drier because much of its moisture has been removed by the precipitation. This creates a rain shadow on the far side of the mountain—a belt of dry climate extending down the leeward slope and beyond. Several of the Earth's great deserts are formed by rain shadows.

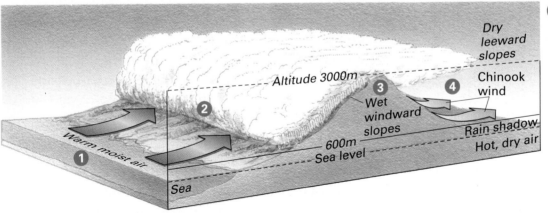

1 Air passing over a large ocean surface becomes warm and moist by the time it arrives at the coast. As the air rises on the windward side of the range, it is cooled by the adiabatic process, and its temperature drops according to the dry adiabatic rate.

California mountain ranges have a strong effect on precipitation in California because of the prevailing flow of moist oceanic air from west to east. The upper diagram shows lines of equal precipitation. Below it is a diagram showing the topography. We can see that centers of high precipitation coincide with the western slopes of mountain ranges, including the coastal ranges and Sierra Nevada. The desert regions lie to the east, in the mountains' rain shadows.

A. Coastal ranges
Prevailing westerly winds bring moist air in from the Pacific Ocean, first over the coast ranges of central and northern California, depositing rain on the ranges.

C. Sierra Nevada
The air continues up and over the great Sierra Nevada, whose summits rise to 4200 m (about 14,000 ft.) above sea level. Heavy precipitation, largely in the form of winter snow, falls on the western slopes of these ranges and nourishes rich forests.

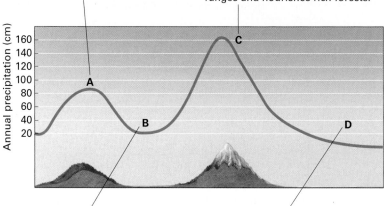

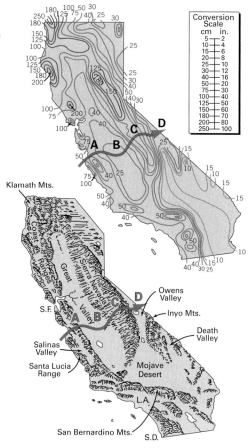

B. Central Valley
After rain is deposited on the coast ranges, the air descends into the broad Central Valley. Precipitation is low here, averaging less than 25 cm (10 in.) per year.

D. Interior desert
Passing down the steep eastern face of the Sierra Nevada, the air descends quickly into the Owens Valley. Adiabatic heating warms and dries the air, creating a rain shadow desert. In the Owens Valley, annual precipitation is less than 10 cm (4 in.).

Convective precipitation

Air can also be forced upward through convection, leading to **convective precipitation**. In this process, strong updrafts occur within convection cells—vertical columns of rising air that are often found above warm land surfaces. Air rises in a convection cell because it is warmer, and therefore less dense, than the surrounding air.

The convection process begins when a surface is heated unequally. Think of an agricultural field surrounded by a forest, for example. The field surface is largely made up of bare soil with only a low layer of

convective precipitation

Precipitation induced when warm, moist air is heated at the ground surface, rises, cools, and condenses to form water droplets, raindrops and, eventually, rainfall.

vegetation, so under steady sunshine the field will be warmer than the adjacent forest. This means that as the day progresses, the air above the field will grow warmer than the air above the forest.

The density of air depends on its temperature—warm air is less dense than cooler air. The hot-air balloon operates on this principle. The balloon is open at the bottom, and in the basket below a large gas burner forces heated air into the balloon. Because the heated air is less dense than the surrounding air, the balloon rises. The same principle will cause a bubble of air to

Formation of a cumulus cloud
FIGURE 5.19

A bubble of heated air rises above the lifting condensation level to form a cumulus cloud.

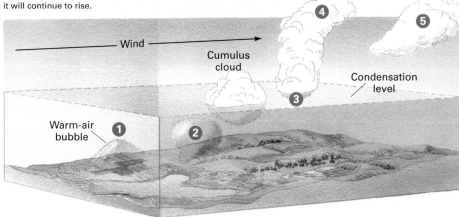

2 Adiabatic cooling As the bubble of air rises, it is cooled adiabatically. However, as long as the bubble is still warmer than the surrounding air, it will continue to rise.

3 Condensation If the bubble remains warmer than the surrounding air and uplift continues, adiabatic cooling chills the bubble to the dew point, and condensation sets in. The rising air column becomes a puffy cumulus cloud. The flat base of the cloud marks the lifting condensation level at which condensation begins.

4 Continued convection The bulging "cauliflower" top of the cloud is the top of the rising warm-air column pushing into higher levels of the atmosphere.

5 Dissipation A small cumulus cloud typically encounters winds aloft that mix it with the local air, reducing the temperature difference and slowing the uplift. After drifting some distance downwind, the cloud evaporates.

1 Surface heating Heated air is less dense than the surrounding air, causing a bubble of warm air to form over the field, rise, and break free from the surface.

form over the field, rise, and break free from the surface, as in **FIGURE 5.19**.

As the bubble of air rises, it is cooled adiabatically, and its temperature will decrease as it rises according to the dry or moist adiabatic lapse rate. The temperature of the surrounding air will normally decrease with altitude as well, but at the environmental lapse rate. For convection to occur, then, the temperature of the air bubble must always be warmer than the temperature of the surrounding atmosphere, even as it rises. The implication is that the environmental temperature must decrease more rapidly with altitude than the rising air parcel's temperature. Another way to state this relationship is that the environmental lapse rate must be greater than the dry or moist adiabatic lapse rates. **FIGURE 5.20** outlines how to determine whether the

Process Diagram

Determining stability FIGURE 5.20

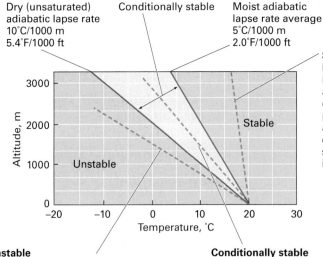

Dry (unsaturated) adiabatic lapse rate
10°C/1000 m
5.4°F/1000 ft

Conditionally stable

Moist adiabatic lapse rate average
5°C/1000 m
2.0°F/1000 ft

Stable
If the environmental lapse rate is less than both the dry and moist adiabatic lapse rates, the environment is *stable*. As an air parcel moves up through this atmosphere, it rapidly becomes colder and more dense than the surrounding environment, causing it to sink back down.

Unstable
If the environmental lapse rate is greater than both the dry and moist adiabatic lapse rates, the environment is *unstable*. As an air parcel moves up through the atmosphere, it is always warmer and less dense than the surrounding environment, causing it to continue to rise.

Conditionally stable
If the environmental lapse rate is less than the dry adiabatic lapse rate, but greater than the moist adiabatic lapse rate, the environment is *conditionally stable*. An unsaturated air parcel moving up in this environment becomes colder than the surrounding air and sinks back down. However, a saturated air parcel moving up in this same environment is warmer than the surrounding air and continues to rise.

atmosphere in a given region is conducive to producing convection. Environmental lapse rates in these regions have a flatter slope than the dry or moist adiabatic rates. Air with these characteristics is referred to as *unstable air*.

There are two adiabatic rates—dry and moist. The moist rate is about half the dry rate. While condensation is occurring, the lesser, moist rate applies. Thus, air in which condensation is occurring cools less rapidly with uplift. Because the temperature decrease of the air parcel is small, the rising air is more likely to stay warmer than the surrounding air, so uplift continues. A simple example, provided in **FIGURE 5.21**, makes the concept of convection in unstable air clearer.

Convection in unstable air FIGURE 5.21

When the air is unstable, a parcel of air that is heated enough to rise will continue to rise to great heights.

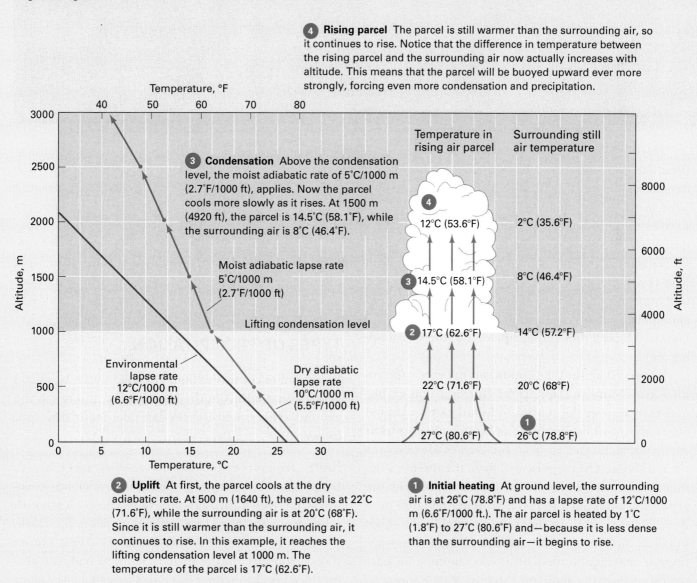

4 **Rising parcel** The parcel is still warmer than the surrounding air, so it continues to rise. Notice that the difference in temperature between the rising parcel and the surrounding air now actually increases with altitude. This means that the parcel will be buoyed upward ever more strongly, forcing even more condensation and precipitation.

3 **Condensation** Above the condensation level, the moist adiabatic rate of 5°C/1000 m (2.7°F/1000 ft), applies. Now the parcel cools more slowly as it rises. At 1500 m (4920 ft), the parcel is 14.5°C (58.1°F), while the surrounding air is 8°C (46.4°F).

2 **Uplift** At first, the parcel cools at the dry adiabatic rate. At 500 m (1640 ft), the parcel is at 22°C (71.6°F), while the surrounding air is at 20°C (68°F). Since it is still warmer than the surrounding air, it continues to rise. In this example, it reaches the lifting condensation level at 1000 m. The temperature of the parcel is 17°C (62.6°F).

1 **Initial heating** At ground level, the surrounding air is at 26°C (78.8°F) and has a lapse rate of 12°C/1000 m (6.6°F/1000 ft.). The air parcel is heated by 1°C (1.8°F) to 27°C (80.6°F) and—because it is less dense than the surrounding air—it begins to rise.

Cumulus clouds over South Dakota Figure 5.22

A gathering of storm clouds forms over the Badlands region of South Dakota. Convection within these clouds gives them their archetypical cumulus structure.

The key to the convective precipitation process, then, is latent heat release. When water vapor condenses into droplets or ice particles, it releases latent heat to the air parcel. This heat keeps the parcel warmer than the surrounding air, fueling the convection process and driving the parcel ever higher. When the parcel reaches a high altitude, most of its water will have condensed. As adiabatic cooling continues, less latent heat will be released, so the uplift weakens. Eventually, uplift stops, because the energy source, latent heat, is gone. The cell dies and dissipates into the surrounding air.

Where do we find the conditions necessary to produce convective precipitation? During summer, hot masses of air over the central and southeastern United States are typically unstable. Summer weather patterns sweep warm, humid air from the Gulf of Mexico over the continent. Over a period of days, the intense summer insolation strongly heats the air layer near the ground, producing a steep environmental lapse rate. Both of the conditions that create unstable air—high humidity and unstable environmental conditions—are present, making thunderstorms very common in these regions during the summer. FIGURE 5.22 shows an example of one of these storms.

Unstable air is also commonly found in the vast, warm, and humid regions of the equatorial and tropical zones. In these regions, convective showers and thundershowers are frequent. At low latitudes, much of the orographic rainfall is actually in the form of heavy showers and thundershowers produced by convection. In that case, the forced ascent of unstable air up a mountain slope easily produces rapid condensation, which then triggers the convection process.

TYPES OF PRECIPITATION

Precipitation consists of liquid or solid water drops and crystals that fall from the atmosphere and reach the ground. This precipitation can take many different forms.

Rain   Rain is precipitation that reaches the ground as liquid water. Raindrops can form in warm clouds as liquid water, which through collisions can coalesce with other drops and grow large enough to fall to the Earth. They can form through other processes as well. For instance, solid ice, in the form of snow or hail, can also produce rain by melting if it falls through a layer of air that is warm enough.

To fall from the sky and reach the ground, raindrops usually have to grow larger than 0.2 mm (0.008 in.). At these sizes, we refer to the drops as *mist* or *drizzle*. Once the drops reach 0.5 mm (0.02 in.), they are termed *raindrops*. Typically, raindrops can only have a maximum size of 5–8 mm (0.2–0.3 in.)—drops any larger than that become unstable and break into smaller drops as they fall through the atmosphere.

Snow

Snow forms as individual water vapor molecules are deposited upon existing ice crystals. If these ice crystals are formed entirely by deposition, they take the shape of snowflakes with their characteristic intricate crystal structure. However, most particles of snow have endured collisions and coalesce with each other and with supercooled water drops. As they do so, they lose their shape and can become simple lumps of ice.

Eventually, whether they are intricate snowflakes or accumulations of ice and supercooled water drops, these ice crystals become heavy enough to fall from the cloud. By the time this precipitation reaches the ground, it may have changed form. Snow produced in cold clouds reaches the ground as a solid form of precipitation if the underlying air layer is below freezing. Otherwise, the snow melts and arrives as rain.

Sleet and freezing rain

Perhaps you have experienced an ice storm. Ice storms occur when the ground is frozen and the lowest air layer is also below freezing. Actually, ice storms are more accurately named "icing" storms, because it is not ice that is falling but supercooled rain. Rain falling through the cold air layer is chilled and freezes onto ground surfaces as a clear, slippery glaze, making roads and sidewalks extremely hazardous. Ice storms cause great damage, especially to telephone and power lines and to tree limbs pulled down by the weight of the ice.

Hail

Hailstones are formed by the accumulation of ice layers on ice pellets that are suspended in the strong updrafts of thunderstorms. As these ice pellets—called *graupel*—move through subfreezing regions of the atmosphere, they come into contact with supercooled liquid water droplets, which subsequently freeze to the pellets in a thin sheet (**FIGURE 5.23**). This process, called *accretion*, results in a buildup of concentric layers of ice around each pellet, giving it its typical ball-like shape.

With each new layer, the ball of ice—now called *hail*—gets larger and larger. In addition, it gets heavier and heavier. When it becomes too heavy for the updraft to support, it falls to the Earth. When the updrafts are extremely strong, the hail remains aloft, slowly accumulating more mass and getting larger. In that case, hailstones can reach diameters of 3–5 cm (1.2–2.0 in.). **FIGURE 5.24** on pages 146–147, discusses hail and other forms of precipitation.

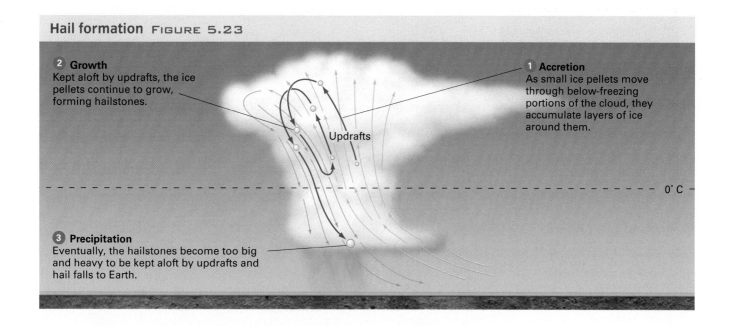

Hail formation FIGURE 5.23

2 Growth
Kept aloft by updrafts, the ice pellets continue to grow, forming hailstones.

1 Accretion
As small ice pellets move through below-freezing portions of the cloud, they accumulate layers of ice around them.

Updrafts

3 Precipitation
Eventually, the hailstones become too big and heavy to be kept aloft by updrafts and hail falls to Earth.

0° C

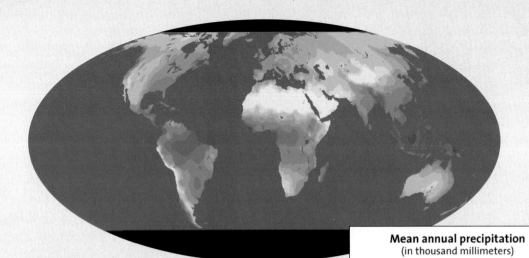

▲ Mean annual precipitation

On a global scale, rainfall is greatest in the equatorial regions, where abundant solar energy drives evaporation. The world's deserts, most visible here in a crescent from the Sahara to central Asia, occur where moisture sources are far away or where circulation patterns retard precipitation formation. Mountain chains generate orographic precipitation. At higher latitudes, precipitation is abundant where circulation patterns spawn storms.

Mean annual precipitation
(in thousand millimeters)

More than 3.0 | 2.0 - 3.0 | 1.5 - 2.0 | 1.0 - 1.5 | .60 - 1.0 | .40 - .60 | .20 - .40 | .10 - .20 | Less than .10 | Non-land area | No data available

▲ Monsoon rainfall

In tropical and equatorial regions, rainfall can be abundant and in some regions has a strong seasonal cycle. Here rice farmers plant a new crop while sheltered by woven rain shields called *patlas*.

▲ Snow

Where air temperatures are low, precipitation falls as snow. This bison in Yellowstone National Park is well prepared for the region's cold weather.

▲ Freezing rain

When rain falls into a surface layer of below-freezing air, clear ice coats the ground. The weight of the ice brings down power lines and tree limbs. In January of 1998, heavy rain fell into a layer of colder air, causing ice accumulations of four inches or more over a large area of northern New England and Quebec.

▲ Hail

Hail consists of lumps of ice ranging from pea- to grapefruit-sized—that is, with a diameter of 5 mm (0.2 in.) or larger. Most hail particles are roughly spherical in shape. In Montana, a horse receives a light coating of hail during an autumn storm.

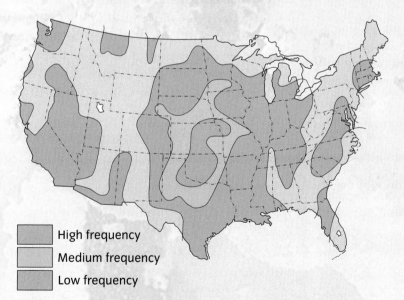

High frequency

Medium frequency

Low frequency

◀ Frequency of severe hailstorms

Severe hailstorms—defined as a local convective storm producing hailstones equal to or greater than 1.9 cm (0.75 in.) in diameter—can do severe damage to buildings and cars. In addition, crop destruction caused by hailstorms can add up to losses of several hundred million dollars. Damage to wheat and corn crops is particularly severe in the Great Plains, running through Nebraska, Kansas, Missouri, Oklahoma, and northern Texas.

MEASURING PRECIPITATION

We talk about precipitation in terms of the depth that falls during a certain time. For example, we use millimeters or inches per hour or per day. A millimeter (inch) of rainfall would cover the ground to a depth of 1 mm (1 in.) if the water did not run off or sink into the soil. Rainfall is measured with a rain gauge (sometimes written "raingage"). The simplest rain gauge is a straight-sided, flat-bottomed pan, which is set outside before a rainfall event. (An empty coffee can does nicely.) After the rain, the depth of water in the pan is measured.

A very small amount of rainfall, such as a millimeter or two, makes too thin a layer to be accurately measured in a flat pan. To avoid this difficulty, a typical meteorological rain gauge is constructed from a narrow cylinder with a funnel at the top (FIGURE 5.25). The funnel gathers rain from a wider area than the mouth of the cylinder, so the cylinder fills more quickly. The water level gives the amount of precipitation, which is read on a graduated scale.

Snowfall, in contrast, is measured by estimating the amount of liquid water it would produce if it were melted, thereby making the measurement equivalent to a rainfall amount. In this way, we can combine rainfall and snowfall into a single record of precipitation. Ordi-narily, a 10 mm (or 10 in.) layer of snow is assumed to be equivalent to 1 mm (or 1 in.) of rainfall, but this ratio may range from 30 to 1 in very loose snow to 2 to 1 in old, partly melted snow.

To make these measurements more accurately, we sometimes measure the weight of the snow. This measurement can be done manually by inserting into the snow a tube that extends to the ground and then removing a core of snow. The core can be weighed, and its mass will give us the mass of the liquid water equivalent. Alternatively, in certain locations, it is possible to put a scale—termed a *snow pillow*—on the ground before the first snowfall. As snow accumulates on the scale and melts away, the scale measures the change in weight of the overlying snowpack and records it digitally. After the last of the snow has melted away, the snow pillow can be retrieved and the data downloaded to a computer.

A final way to measure precipitation in both liquid and solid form is via *radar*, which is an acronym for **RA**dio **D**etection **A**nd **R**anging. Across the United States, there is a system of radar stations, termed *Doppler radar*, that can be used to determine the intensity of rain, snow, sleet, and hail. In this system, the radar sends out radio waves. Based on the amount of liquid and solid water in the atmosphere, a fraction of the signal reflects back to the radar. Using the intensity of the reflected signal received by the radar station, it is possible to estimate the intensity of the precipitation in the region. In addition, by determining the difference between the time the signal is emitted and the time the receiver acquires the return signal, it is possible to determine how far from the radar station the precipitation is occurring.

Different types of rain gauges FIGURE 5.25

This photo shows five different types of rain gauges installed in Arvada, Colorado. By placing the gauges next to one another, it is possible to compare how their precipitation measurements differ.

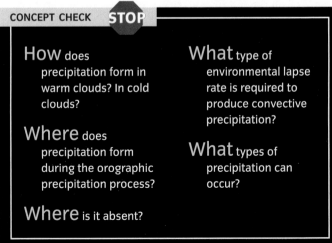

CONCEPT CHECK STOP

How does precipitation form in warm clouds? In cold clouds?

Where does precipitation form during the orographic precipitation process?

Where is it absent?

What type of environmental lapse rate is required to produce convective precipitation?

What types of precipitation can occur?

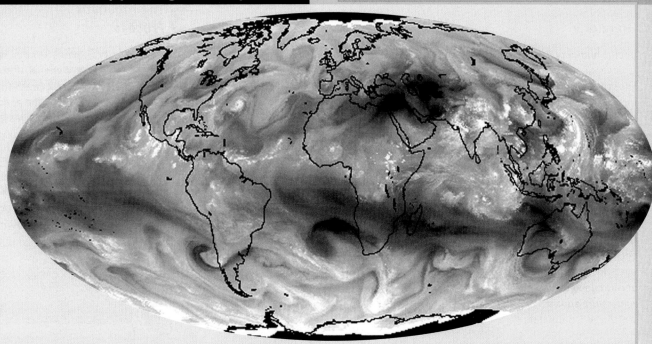

The Geostationary Operational Environmental Satellite (GOES) system is an important tool for weather forecasting. Shown here is a water vapor image from April 11, 2000. The brightest areas show regions of precipitation. Dark areas show low water vapor content. Storm systems, called *cyclonic storms*, are marked by dry air spiraling inward with moist air.

- Which areas have the most and least water vapor?

- Can you spot any cyclonic systems building up?

- At what latitudes are these cyclonic storms found?

SUMMARY

1 Atmospheric Moisture and Precipitation

1. Water can change state by evaporating, condensing, melting, freezing, and through sublimation and deposition.

2. The **hydrosphere** is the realm of water in all its forms. The fresh water in the atmosphere and on land in lakes, streams, rivers, and ground water is only a very small portion of the total water in the hydrosphere. Movement of water between these reservoirs is called the **hydrologic cycle**.

3. **Precipitation** is the fall of liquid or solid water from the atmosphere to reach the Earth's land or ocean surface.

2 Humidity

1. **Specific humidity** measures the mass of water vapor in a mass of air, in grams of water vapor per kilogram of air. Warm air can contain much more water vapor than cold air. **Saturation specific humidity** is a measure of how much water the air could contain based on its temperature.

2. The **dew-point temperature** is the temperature to which air would have to be cooled in order to reach 100 percent relative humidity. It is a measure of the specific humidity of the air—higher dew-point temperatures indicate higher specific humidities. If the dew-point temperature equals the actual temperature, the air is saturated.

3. **Relative humidity** measures water vapor in the air as the percentage of the maximum amount of water vapor that the air mass can contain at the given air temperature. A relative humidity of 100 percent indicates that the air is saturated.

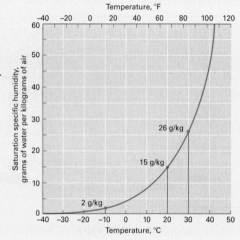

3 The Adiabatic Process

1. The adiabatic principle states that when a gas is compressed, it warms, and when a gas expands, it cools. When an air parcel moves upward in the atmosphere, it encounters a lower pressure and so expands and cools according to the adiabatic process.

2. The **dry adiabatic lapse rate** describes the rate of cooling with altitude. When condensation or deposition is occurring, the cooling rate is given by the **moist adiabatic lapse rate**.

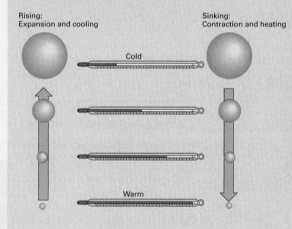

4 Clouds

1. Clouds are composed of droplets of water or crystals of ice that form on condensation nuclei.

2. Clouds typically occur in layers, as stratiform clouds, or in globular masses, as cumuliform clouds. **Fog** occurs when a cloud forms at ground level.

5 Precipitation

1. Precipitation from clouds occurs as rain, hail, snow, and sleet. There are four types of processes—orographic, convective, frontal, and convergence—that can produce precipitation.

2. In **orographic precipitation**, air moves up and over a mountain barrier. As it moves up, it is cooled adiabatically and rain forms. As the air descends the far side of the mountain, it is warmed, producing a **rain shadow**.

3. When a surface is heated unequally, air parcels can become warmer and less dense than the surrounding air. Because these air parcels are less dense, they rise. As they move upward, they cool, and condensation with precipitation may occur, leading to **convective precipitation**.

KEY TERMS

- **hydrologic cycle** p. 124
- **precipitation** p. 124
- **humidity** p. 126
- **specific humidity** p. 126
- **saturation specific humidity** p. 127
- **saturation** p. 127

- **dew-point temperature** p. 127
- **relative humidity** p. 128
- **dry adiabatic lapse rate** p. 131
- **lifting condensation level** p. 132
- **moist adiabatic lapse rate** p. 132
- **condensation nucleus** p. 133

- **orographic precipitation** p. 139
- **rain shadow** p. 140
- **isohyets** p. 140
- **convective precipitation** p. 141

CRITICAL AND CREATIVE THINKING QUESTIONS

1. What is the hydrosphere? Where, and in what amounts, is water found on our planet? How does water move in the hydrologic cycle?

2. How is the moisture content of air influenced by air temperature?

3. What happens when a parcel of moist air is chilled? Use the terms saturation, dew point, and condensation in your answer.

4. What is the adiabatic process? Why is it important?

5. Distinguish between dry and moist adiabatic lapse rates. When do they apply for a parcel of air moving upward in the atmosphere? Why is the moist adiabatic lapse rate lower than the dry adiabatic rate? Why is the moist adiabatic rate variable in amount?

6. How is precipitation formed? Describe the process for warm and cool clouds.

7. Compare and contrast orographic and convective precipitation. Begin with a discussion of the adiabatic process and the generation of precipitation within clouds. Then compare the two processes, paying special attention to the conditions that create uplift. Can convective precipitation occur in an orographic situation? Under what conditions?

SELF-TEST

1. _____ energy is released or absorbed as water changes from one state to another.
   a. Latent heat
   b. Sensible heat
   c. Conductive heat
   d. Convective heat

2. On the following diagram, label the processes in which water releases latent heat to the environment and those in which it absorbs latent heat from the environment.

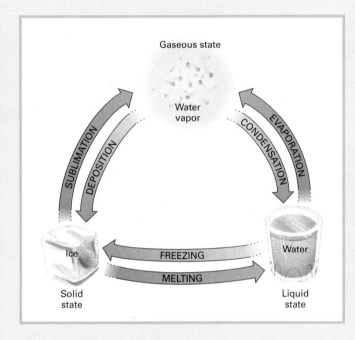

3. More water is stored in ice sheets and glaciers than in freshwater bodies.
   a. True
   b. False

4. Relative humidity _____.
   a. is the total amount of water vapor present in the air
   b. is responsible for life on the Earth
   c. depends on the volume of water present in the air unrelated to temperature
   d. is the amount of water vapor in the air compared to the amount it could contain

5. Relative humidity is usually lowest during the _____.
   a. early morning
   b. early afternoon
   c. evening
   d. night

6. The _____ of the air represents the actual quantity of water vapor within an air parcel.
   a. relative humidity
   b. saturation level
   c. specific humidity
   d. absolute saturation level

7. On the following diagram, determine the saturation specific humidity for: (a) 0°C, (b) 25°C, (c) 35°C.

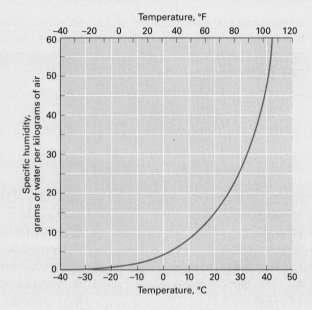

8. _____ temperature changes occur in parcels of air solely as a result of air expansion or compression.
   a. Adiabatic
   b. Compressive
   c. Latent
   d. Specific

9. The dry adiabatic lapse rate does not vary from place to place and time to time.
   a. True
   b. False

10. Because rising air cools less rapidly when condensation is occurring as a result of the release of latent heat, the _____ has a lesser value than the _____.
    a. dry adiabatic lapse rate; moist adiabatic lapse rate
    b. dry adiabatic lapse rate; environmental adiabatic lapse rate
    c. environmental adiabatic lapse rate; moist adiabatic lapse rate
    d. moist adiabatic lapse rate; dry adiabatic lapse rate

11. _____ fog forms when warm, moist air moves over a colder surface.
   a. Radiation
   b. Advection
   c. Sublimation
   d. Compression

12. _____ precipitation is a result of air being lifted over a highland area.
   a. Convective
   b. Orographic
   c. Convergence
   d. Frontal

13. The two conditions that promote thunderstorm development are _____.
   a. warm, moist air and an unstable environmental lapse rate
   b. cold, dry air and a neutral environmental lapse rate
   c. warm, moist air and a stable environmental lapse rate
   d. cold, dry air and a stable environmental lapse rate

14. The _____ lapse rate is in effect when a parcel of air rises above the condensation level.
   a. environmental
   b. adiabatic
   c. dry adiabatic
   d. moist adiabatic

15. The following diagram depicts a type of precipitation formation process called the _____.
   a. coalescence process
   b. Bergeron process
   c. collision process
   d. Brownian process

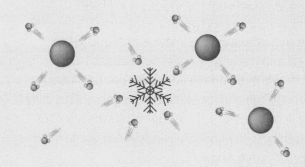

Winds

Here in the Columbia River Gorge, between Washington and Oregon, the winds pick up around ten in the morning. As the day progresses, strong temperature differences develop between the warmer, lower elevations to the west and the cooler, higher elevations to the east. Warmer surface air moves up the mountain through the gap provided by the gorge. This daily cycle of heating and cooling results in very consistent and dependable winds in the gorge from April through August.

The wind sets up perfect conditions for windsurfing and kiteboarding in the gorge. The steep walls of the gorge prevent winds from swirling in different directions. Instead, they blow straight up the canyon throughout the day. The winds also blow against the downstream flow of the river, producing steep waves that provide excellent takeoff ramps for aerial maneuvers.

In this chapter, we examine the forces responsible for producing and affecting the winds, both on a local scale as in the Columbia River Gorge and on a larger scale where wind systems can span continents.

A windsurfer catches hold of the prevailing winds along the Columbia River Gorge between Washington and Oregon.

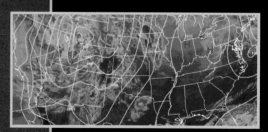

Winds

The air around us is always in motion. From a gentle, cooling breeze on a warm summer afternoon to the howling gusts of a major dust storm, seen in FIGURE 6.1, wind is a familiar phenomenon. Why does the air move? What are the forces that cause winds to blow? Why do winds blow more often in some directions than in others? What is the pattern of wind flow in the upper atmosphere? These are some of the questions we answer in this chapter.

Wind is defined as air moving relative to the Earth's surface. Meteorologists generally use the term to describe only the horizontal motion of the air, because vertical atmospheric motions are relatively small. When

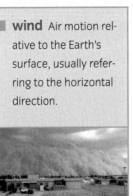

wind Air motion relative to the Earth's surface, usually referring to the horizontal direction.

referring to vertical movements, meteorologists either explicitly refer to *vertical winds* or use other terms such as updrafts or downdrafts.

Winds are caused by unequal heating of the Earth's atmosphere. When the atmosphere in one location is heated to a temperature that is higher than that in another location, a difference in pressure, or *pressure gradient*, results. This pressure gradient can cause air at the surface to move toward the warmer location and air at upper levels to move away from the warmer location. This process directly explains many types of local winds, such as the sea breeze you might experience on an afternoon at the beach.

Dust storm FIGURE 6.1

In a front reaching up to 1000 m (3300 ft) in altitude, a dust storm approaches an American facility in Iraq. The storm passed over in about 45 minutes, leaving a thick layer of dust in its wake.

Global Locator

IRAQ

Devices for wind measurement FIGURE 6.2

A Combination cup anemometer and wind vane The anemometer and wind vane observe wind speed and direction, which are displayed on the meter below. The wind vane and anemometer are mounted outside, with a cable from the instrument leading to the meter, which is located indoors.

B Radiosonde This buoyant weather balloon, known as a *radiosonde*, will carry instruments upward that radio back temperature and pressure. They also allow scientists to measure wind speed and direction at levels aloft. This balloon is at Sable Island, Nova Scotia.

Pressure gradients are also the cause of global wind motions. The equatorial and tropical atmosphere is heated more intensely by the Sun than is the atmosphere at mid- and higher latitudes. In response to this difference in heating, global-scale pressure gradients move vast bodies of warm air poleward and huge pools of cool air shift equatorward, strongly influencing the day-to-day weather in various regions, as well as the climate.

MEASUREMENT OF WIND

Like all motion, the movement of air—called its *velocity*—is defined by the wind's direction and speed. A number of devices are used to measure wind (FIGURE 6.2). One instrument for tracking wind direction is a simple vane with a tail fin that keeps it always pointing into the wind. To measure wind speed near the surface, we use anemometers (FIGURE 6.2A). The most common type consists of three funnel-shaped cups on the ends of the spokes of a horizontal wheel that rotates as the wind strikes the cups. Some anemometers use a small electric generator that produces more current when the wheel rotates more rapidly. This type is connected to a meter calibrated in meters per second or miles per hour.

To measure winds aloft, where there is no fixed station location, we use *radiosondes*. These are balloons launched concurrently at regular intervals (typically twice daily) all over the world (FIGURE 6.2B). Radiosondes have a package of small weather-recording instruments that measure temperature, humidity, and other atmospheric variables as they ascend. By tracking the balloon's vertical and horizontal distance from the initial launch location, it is possible to determine the wind speed and direction at various levels of the atmosphere. Typically, these balloons reach altitudes of 25–30 km before they rupture, giving us information about winds up through the troposphere and lower portions of the stratosphere.

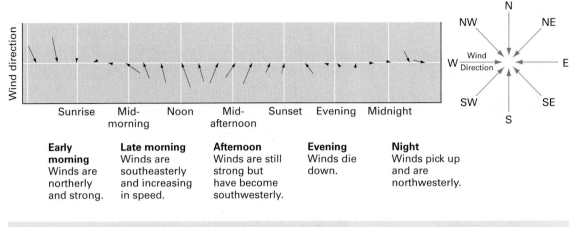

Wind direction	Sunrise	Mid-morning	Noon	Mid-afternoon	Sunset	Evening	Midnight

Early morning
Winds are northerly and strong.

Late morning
Winds are southeasterly and increasing in speed.

Afternoon
Winds are still strong but have become southwesterly.

Evening
Winds die down.

Night
Winds pick up and are northwesterly.

Change in wind speed and direction with time FIGURE 6.3

This plot shows how wind direction and speed at a given station can change during the day. The direction of the arrow shows the direction the winds are blowing, and the length gives the relative speed.

In addition, ground-based radar stations, termed *Doppler radar*, can be used to determine the intensity of winds blowing toward or away from the radar site up to 15 km above the surface. These stations emit radio waves of a particular wavelength. As the waves bounce off solid or liquid particles in the atmosphere, the wavelength is shifted slightly toward higher or lower wavelengths—called a *Doppler shift*. The principle is the same one you experience as a car or train approaches and then drives by. As the vehicle approaches you, the sound waves it emits are compressed, making them shorter and higher-pitched. As the vehicle drives away from you, the sound waves are expanded, making them longer and lower-pitched.

Similarly, objects moving toward the radar site cause the radio waves reflected from them to become compressed, while objects moving away from the site cause the reflected waves to expand. The reflected waves then travel back to the radar site. By measuring the change in the wavelength of the emitted radiation compared with the wavelength of the received radiation, it is possible to determine the velocity of the object from which they were reflected, with faster-moving objects causing a greater change in wavelength.

It is important to note here that in meteorology and climatology, wind direction is always given as the direction from which the wind is coming, as seen in FIGURE 6.3. For example, a west (or westerly) wind is one that comes from the west and moves to the east. Any direction on the compass can be used to provide the direction of the wind. For instance, northwesterly winds come from north of west. The metric unit for wind speed is meters per second (m/s), and the English unit is typically the knot. (A knot is one nautical mile—1.15 statute miles—per hour.) Throughout this book we give wind speed in m/s.

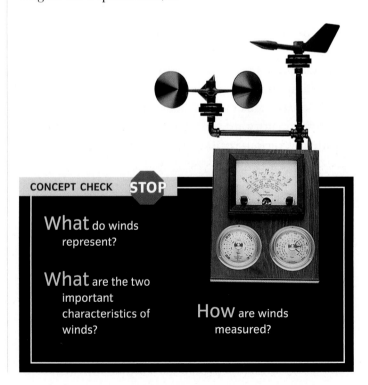

CONCEPT CHECK STOP

What do winds represent?

What are the two important characteristics of winds?

How are winds measured?

Winds and Pressure Gradients

The air in the atmosphere is constantly pressing on the Earth's surface beneath it and on everything that it surrounds. At sea level, the average pressure of air is 1013.2 mb. However, this does not mean that the pressure at the surface everywhere is equal to 1013.2 mb. Atmospheric pressures at a location can vary from day to day as well as from location to location. On a cold, clear winter day, the barometric pressures at sea level might be as high as 1030 mb (30.4 in. Hg), while in the center of a storm system, pressure might drop to 980 mb (28.9 in. Hg). These variations in pressure from one location to another can produce movement of atmospheric molecules and hence generate winds.

isobars Lines on a map drawn through all points having the same atmospheric pressure.

WINDS AND PRESSURE GRADIENTS

Wind is caused by differences in atmospheric pressure from place to place. In the absence of other forces, air tends to move from high to low pressure regions until the air pressures are equal. This tendency is similar to the simple physical principle that any flowing fluid (such as air) subjected to gravity will move until the surface level is uniform. For winds, the force applied to the air is related to the difference, or gradient, in pressure from one region to the next, so it is called the *pressure gradient force.*

On maps, lines called **isobars** connect locations of equal pressure, as illustrated in FIGURE 6.4. This simple map shows the pressure between two locations—Wichita and Columbus. At Wichita, the pressure is high (H), and the barometer (adjusted to sea-level pressure) reads 1028 mb (30.4 in. Hg). At Columbus, the barometer (again, adjusted to sea-level pressure) is low (L) and reads 997 mb (29.4 in. Hg).

At both locations, we first adjust the surface pressures at Wichita and Columbus to *sea-level pressure* values—the values Wichita and Columbus would have if they were at sea level. Obviously, neither Wichita nor Columbus is at sea level, so why do we adjust their surface pressures to sea-level pressures? If we used the *surface pressure* values—the values measured directly at the surface—we would most likely find that the pressure at Columbus is higher than at Wichita, because Columbus is 249 m above sea level and Wichita is 406 m above sea level. Because pressure decreases with altitude, we would expect the pressures at Columbus to be higher than those at Wichita.

The pressure difference between locations that arises solely from differences in altitude is called the *static pressure gradient* and is the result of differences in pressure in the vertical direction, not the horizontal direction. To remove this effect, we need to adjust all of

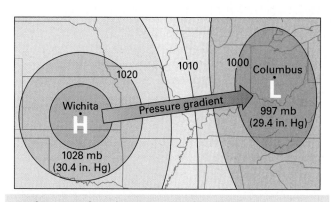

Isobars and a pressure gradient FIGURE 6.4

At Wichita, the pressure is high (H), and the barometer (adjusted to sea-level pressure) reads 1028 mb (30.4 in. Hg). At Columbus, the barometer is low (L) and reads 997 mb (29.4 in. Hg). The difference means there is a pressure gradient between Wichita and Columbus.

the surface pressure measurements to the same level—in this case sea level—and then find the difference between them. This will give us the pressure gradient responsible for producing horizontal movements of air—termed the *dynamic pressure gradient.*

Once we have done this, we find that there is a **pressure gradient** between Wichita and Columbus. Isobars on the map around Wichita and Columbus connect locations of equal pressure. As you move between Wichita and Columbus, you cross these isobars on the map, encountering changing pressures as a result of the drop from 1028 to 997 mb (30.4 to 29.4 in. Hg).

Because atmospheric pressure is unequal at Wichita and Columbus, a pressure gradient force will push air from Wichita toward Columbus. This pressure gradient force will produce a wind (although this wind does not necessarily blow from Wichita toward Columbus!). The greater the pressure difference between the two locations, the greater this force will be and the stronger the wind.

Although we have been discussing the pressure gradient between Wichita and Columbus, there are also pressure differences between one region and the next at all locations on the map. These pressure differences give rise to a pressure gradient force at all locations.

To determine the direction of the overall pressure gradient force at any one location, we can look at the isobars. The pressure gradient force is always perpendicular to the isobar running through a given region and always points from high to low pressures.

In addition, the magnitude of the pressure gradient force can be determined by the spacing of the isobars. The closer together the isobars are, the larger the pressure gradient force, because the same change in pressure occurs over a smaller distance. To find the pressure gradient between two points, we can use the equation:

$$\text{pressure gradient} = \frac{\text{difference in pressure}}{\text{distance between points}}.$$

PRESSURE AND TEMPERATURE GRADIENTS

How do pressure gradients come about? One way pressure gradients can develop is through unequal heating of the atmosphere. To understand this process, consider a region that has warmer temperatures to the south and cooler temperatures to the north (as you might expect in the northern hemisphere, for example), shown here in FIGURE 6.5. In the region of warmer temperatures, the air expands. As it does so, the top of the air column rises. In contrast, to the north the cold air contracts and the top of the air column descends. Now let's look at pressures *at a constant height*—say, 5000 m above the surface. For the

> **pressure gradient**
> Change of atmospheric pressure measured along a line at right angles to the isobars.

Relation between temperature and pressure FIGURE 6.5

Differences in temperature between two regions can produce a change in pressure both aloft and at the surface. As air expands in warmer regions, pressures aloft (at 5000 m, for instance) will increase, while pressures near the surface (at 50 m, for instance) will decrease. Conversely, in cooler regions, air contracts, and pressures aloft decrease while pressures near the surface increase.

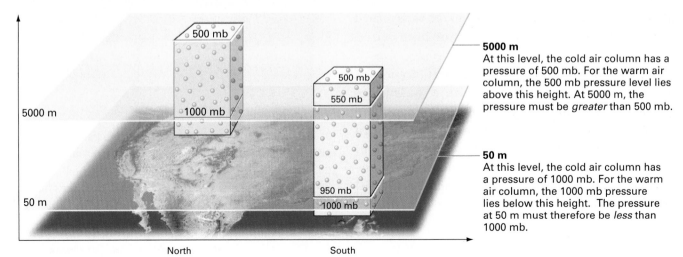

5000 m
At this level, the cold air column has a pressure of 500 mb. For the warm air column, the 500 mb pressure level lies above this height. At 5000 m, the pressure must be *greater* than 500 mb.

50 m
At this level, the cold air column has a pressure of 1000 mb. For the warm air column, the 1000 mb pressure lies below this height. The pressure at 50 m must therefore be *less* than 1000 mb.

Thermal circulations FIGURE 6.6

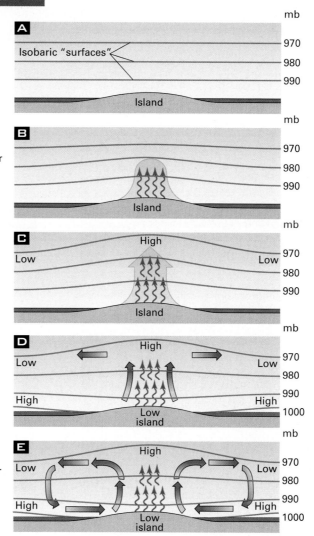

Uniform atmosphere (heated equally) Imagine a uniform atmosphere above a ground surface. The isobaric levels (or "surfaces") are parallel with the ground surface (*Isobar* = equal pressure).

A — Isobaric "surfaces" — 970 mb, 980, 990 — Island

Uneven heating Imagine now that the underlying ground surface, an island, is warmed by the Sun, with cool ocean water surrounding it. The warm surface air rises and mixes with the air above, warming the column of air above the island. Since the warmer air occupies a large volume, the isobaric levels are pushed upward.

B — 970 mb, 980, 990 — Island

Pressure gradient The result is that a pressure gradient is created and air at higher pressure (above the island) flows toward lower pressure (above the ocean).

C — High, Low, Low — 970 mb, 980, 990 — Island

Surface pressure Because air is moving away from the island and over the ocean surfaces, the surface pressure changes. There is less air above the island, so the ground pressure there drops. Since more air is now over the ocean surfaces, the pressure there rises.

D — High, Low, Low, High, High — 970 mb, 980, 990, 1000 — Low island

Thermal circulations The new pressure gradient at the surface moves air from the ocean surfaces toward the island, while air moving in the opposite direction at upper levels completes the two loops.

E — High, Low, Low, High, High — 970 mb, 980, 990, 1000 — Low island

warm air column, there are more air molecules above the 5000 m level. Because pressure is a measure of the mass of air molecules weighing down on a given region, the pressure at 5000 m in the warm air column is greater than the pressure in the cold air column. As a visual reference, high pressures at 5000 m correspond to regions with higher column heights, while regions with lower pressures correspond to regions with lower column heights.

What about near the surface? As the warm air column expands, the bottom of the column *sinks*. For the cold air column, the bottom of the air column rises as the air contracts. Now let's look at a constant height near the surface—say, 50 m. The pressure at this height in the cold-air region is 1000 mb. In the warm air region, however, some of the air molecules are below this height,

meaning there are fewer air molecules above it. Hence, the warm-air region has lower pressures at this level.

Again, as a visual reference, high pressures near the surface correspond to regions with higher column bases, while regions with lower pressures correspond to regions with lower column bases.

To understand how the change in temperature and pressure between two locations can then affect winds, we can look at the development of a simple thermal circulation, depicted in FIGURE 6.6.

LOCAL WINDS

There are many types of local wind systems, as shown in FIGURE 6.7 on pages 162–163.

▲ Sea breezes
Kiteboarders harness the onshore sea breeze during the late afternoon.

Sea and land breezes

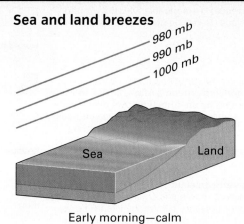

Early morning—calm

A Early in the day, winds are often calm.

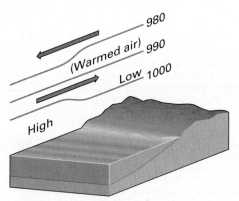

Afternoon—sea breeze

B During the day, the land warms the air above it. This warm air moves oceanward aloft, while surface winds bring cool marine air landward at the surface, creating a thermal circulation.

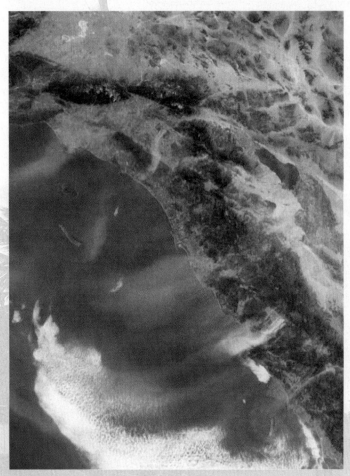

▲ Santa Ana winds
The gray streaks extending from the coast of California out over the Pacific Ocean represent smoke from fires being fanned by Santa Ana winds. These fires can spread very quickly under the influence of the hot, dry air as it is forced to descend from inland desert plateaus by high pressures situated to the east.

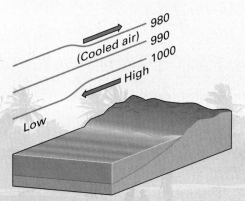

Night—land breeze

C At night, radiation cooling over land creates a reversal of the thermally-driven winds, developing a land breeze.

Examples of local wind systems

▼ Valley and mountain breezes

During the day, mountain hill slopes are heated intensely by the Sun, making the air expand and rise, creating a valley breeze. An air current moves up valleys from the plains below—upward over rising mountain slopes, toward the summits. At night, the hill slopes are chilled by radiation, setting up a reversal in the thermal circulation. The cooler, denser hill slope air moves valleyward, down the hill slopes, creating a mountain breeze.

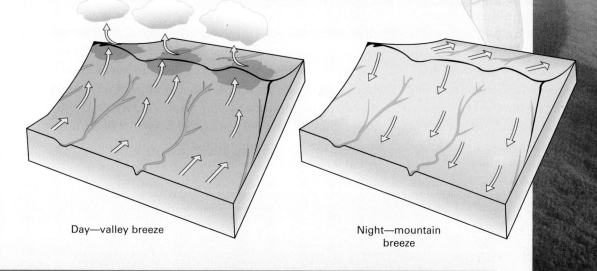

Day—valley breeze

Night—mountain breeze

▼ Foehn winds

These skiers struggle against foehn winds in Antarctica. The cloud cover seen over the mountains forms when these winds blow air from the high mountain peaks out to the plains. As the air descends, it warms causing sublimation of the surface ice into water vapor, which then condenses into these cloud banks seen streaming down the side.

▲ Valley breezes

A hang glider soars over Massanutten Mountain in Virginia, catching updrafts associated with the daytime valley breeze.

Sea and land breezes are examples of local wind systems generated by heating and cooling of different surface types. If you've ever vacationed at the beach in the summer, you may have noticed that in the afternoons a sea breeze often sets in. This wind brings cool air off the water, dropping temperatures and refreshing beachgoers and residents living close to the beach. During this time, the air over the land is warmed by the land surface, and thermal circulations form. Late at night, a land breeze may develop. This wind moves cooler air, which is chilled over land by nighttime radiant cooling, toward the water.

The sea breeze–land breeze provides an example of surface thermal high- and low-pressure zones; that is, low surface pressure is associated with warm air, and high surface pressure with cool air. At the same time, aloft are high pressures situated over warm regions and low pressures situated over cold regions.

Mountain and valley breezes are local winds that alternate in direction in a manner similar to the land and sea breezes. During the day, mountain slopes are heated by the Sun, causing the air in contact with the mountain surface to expand and rise. In comparison, the air at the same level situated over the valley is not in contact with the land surface and hence is not warmed as much, so it begins to sink toward the valley floor. This sinking creates an air current that moves from the plain up and along the mountain slope—the *upvalley breeze*. At night, with clear skies and light winds, the temperature of the slope decreases and cools the overlying atmosphere, creating a surface layer of cooler, denser air. This air then flows downslope, often concentrating in stream valleys, to form the mountain (or *downvalley*) breeze.

Other types of local winds include fall (or drainage) winds, foehn winds, and Santa Ana winds. Fall winds consist of cold, dense air that flows under the influence of gravity from higher to lower regions. In a typical situation, cold, dense air accumulates in winter over a high plateau or high interior valley. Under favorable conditions, some of this cold air spills over low divides or through passes, flowing out on adjacent lowlands as a strong, cold wind.

Winds and Fire

Fall winds occur in many mountainous regions of the world and go by various local names. The mistral of the Rhône Valley in southern France is a well-known example—it is a cold, dry local wind. On the ice sheets of Greenland and Antarctica, powerful fall winds move down the gradient of the ice surface and are funneled through coastal valleys. Both downvalley breezes and fall winds are types of *katabatic* winds because they are related to the sinking of cold, dense air from higher elevations to lower ones.

Another type of local wind—the foehn wind—results when strong regional winds pass over a mountain range and descend on the lee side. The descending air is warmed and dried. A foehn wind can sublimate snow or dry out soils very rapidly. Its name comes from winds that blow over the Alps during winter. Similar winds are found at other locations around the globe. For instance, those that blow over the Rocky Mountains are called *chinook winds*, and New Zealand has its *Canterbury northwester*.

Still another type of local wind occurs when the outward flow of dry air from a large high-pressure center is combined with the local effects of mountainous terrain. An example is the Santa Ana—a hot, dry easterly wind that sometimes blows from the interior elevated desert region of southern California across coastal mountain ranges to reach the Pacific coast. As this wind descends from the higher elevations, it warms adiabatically. In addition, its relative humidity drops, making it very dry. Because this wind is dry, hot, and strong, it can easily fan wildfires in brush or forest out of control, as discussed in *What a Scientist Sees: Winds and Fire.*

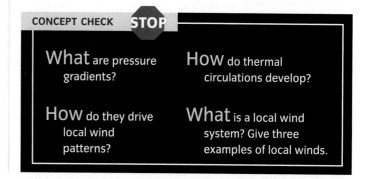

CONCEPT CHECK STOP

What are pressure gradients?

HOW do thermal circulations develop?

HOW do they drive local wind patterns?

What is a local wind system? Give three examples of local winds.

The photo shows homes in Simi Valley, California, threatened by the fires of October 2003. Fires like this are often fanned by strong downslope winds of hot, dry air associated with the Santa Anas.

Fire can also create its own local wind systems. The intense heat of a fire can produce strong convection at the head of the fire. The result is local low pressures, producing a pressure gradient that can create surface winds that feed the flames with oxygen. Adding oxygen can lead to fire blowups with rapid intensification of the fire.

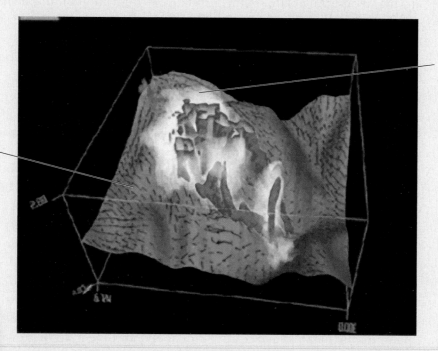

Arrows show winds blowing into the superheated region at the head of the fire.

White regions indicate smoke and convection while red indicates flames.

NATIONAL GEOGRAPHIC

The Coriolis Effect and Winds Aloft

LEARNING OBJECTIVES

Explain the Coriolis effect.

Describe how the Coriolis effect influences the motions of objects in the northern and southern hemispheres.

Discuss the geostrophic wind.

Define cyclones and anticyclones.

> ■ **Coriolis effect**
>
> Force produced by the Earth's rotation that appears to deflect a moving object on the Earth's surface to the right in the northern hemisphere and to the left in the southern hemisphere.

We have seen that the pressure gradient force moves air from high pressure to low pressure. For sea and land breezes, which are local in nature, this push produces a wind motion in about the same direction as the pressure gradient. For wind systems on a larger scale, however, the direction of air motion is somewhat different. The difference is due to the Earth's rotation, applied through the **Coriolis effect**.

The Coriolis effect was first identified by the French scientist Gaspard-Gustave de Coriolis in 1835. Because of the Coriolis effect, an object in the northern hemisphere moves as if a force were pulling it to the right (**FIGURE 6.8A**). In the southern hemisphere, objects move as if pulled to the left. This apparent deflection does not depend on direction of motion—it occurs whether the object is moving toward the north, south, east, or west. In each case, the object moves as if a force were pulling it sideward. The effect is strongest near the poles and decreases to zero at the Equator.

To visualize the Coriolis effect, imagine that a rocket is launched from the North Pole toward New York (**FIGURE 6.8B**). As the rocket travels toward New York, the Earth rotates from west to east beneath its straight flight path. If you were standing at the launch point on the rotating Earth below, you would see the rocket's trajectory curve to the right, away from New York and toward Chicago, despite the fact that the rocket has been flying in a straight line from the viewpoint of space. To reach New York, the rocket's flight path would have to be adjusted to allow for the Earth's rotation.

What would happen if a rocket were launched from New York heading north toward the pole? Because of the Earth's rotation, a point on the Earth's surface at the latitude of New York moves eastward as well. Although the rocket is aimed properly along the meridian, at its launch its motion will have an eastward velocity imparted by the rotation of the Earth along with a northward velocity needed to get it to the North Pole.

While the rocket retains the eastward velocity imparted to it when it was launched, as it moves northward, the velocity of the Earth underneath it is not moving eastward as rapidly. Why is this? It is because locations at higher latitudes rotate through a shorter distance over 24 hours than locations at lower latitudes, so their velocity does not have to be as great. Therefore, the rocket, which has an eastward velocity imparted at its launch location, will move eastward more quickly than the surface it passes over, and its path will be deflected to the right relative to a location on the surface that has a smaller eastward velocity.

From this analysis, we can see that the Coriolis effect is actually produced because observers on the ground are not stationary but are in motion along with the Earth. In most real situations, we are concerned with analyzing the motions of ocean currents or air masses relative to a fixed point rotating with the Earth. In this case, we need to account for the Coriolis effect by treating it as an additional force that can act upon winds and ocean currents, called the *Coriolis force*.

There are some important characteristics of the Coriolis force to remember. First, the Coriolis force always acts perpendicular to the direction of motion (perpendicular and to the right in the northern hemisphere and perpendicular and to the left in the southern hemisphere). Because the Coriolis force is directed perpendicular to the direction of motion, it cannot speed an object up or slow it down—it can only change the direction in which the object moves. In addition, the strength of the Coriolis force increases with the speed of motion but decreases with latitude. Hence, an object that is stationary with respect to the Earth will not feel the effects of a Coriolis force. To calculate the strength of the Coriolis force, we use the equation:

$$\text{Coriolis force} = 2 \times \frac{\text{rotation rate}}{\text{of the Earth}} \times \sin(\text{latitude}) \times \frac{\text{object}}{\text{speed}}$$

One final point: The Coriolis force tends to become important for winds or currents that travel over long distances (100 km or greater); given the typical wind speeds in the atmosphere (about 10 m/s), it takes air masses approximately a day or two to travel this far. Hence, the Coriolis force becomes an important force when considering circulations that are larger than about 100 km and that last for more than a day. If we treat the Coriolis effect this way, it will properly describe motion on the Earth.

The Coriolis effect FIGURE 6.8

B We can visualize the Coriolis effect by considering the flight of two rockets. As the rocket launched from the North Pole towards New York travels in a straight line south, the Earth rotates beneath it so that by the time it reaches the United States, it is situated over Chicago, not New York. As the rocket is launched from New York towards the North Pole, the rotation of the earth imparts an eastward velocity to it that is greater than the eastward velocity of regions in the high latitudes. Therefore, as it moves north, it also moves east compared to these locations. In both cases, the two rockets have moved to the right of their intended destinations.

A The Coriolis effect appears to deflect winds and ocean currents to the right in the northern hemisphere and to the left in the southern hemisphere. Blue arrows show the direction of initial motion, and red arrows show the direction of motion apparent to the Earth observer. The Coriolis effect is strongest near the poles and decreases to zero at the Equator.

THE GEOSTROPHIC WIND

How does the Coriolis force affect the movement of air? For the simplest case, we can look at higher levels of the troposphere. At these levels, air parcels move without feeling any frictional force from the surface. There are only two forces on the parcel—the pressure gradient force and the Coriolis force. Imagine a still parcel of air that begins to move in response to the pressure gradient force, as seen in FIGURE 6.9. At first, the parcel of air travels in the direction of the pressure gradient, but as it accelerates, the Coriolis force pulls it perpendicular and toward the right (in the northern hemisphere). As its velocity increases, the parcel turns increasingly rightward until the Coriolis force just balances the gradient force. At that point, the sum of

> **geostrophic wind**
> Wind at high levels above the Earth's surface blowing parallel to the isobars.

forces on the parcel is zero. With no net force acting on it, the air parcel's speed and direction remain constant.

We call this type of airflow in which the pressure gradient force exactly balances the Coriolis force the **geostrophic wind**. It occurs at upper levels in the atmosphere. Two important characteristics of geostrophic flow are that (1) it blows parallel to the isobars, not across them; and (2) it occurs when the

pressure gradient force is equal to and opposite the Coriolis force.

How does the geostrophic wind differ in the southern hemisphere? In that case, given the same pressure gradient, the air parcel would start to move across the pressure gradient toward low pressure. However, the Coriolis force deflects it to the *left*. This deflection continues until the Coriolis force again exactly balances the pressure gradient force, at which point the parcel feels zero net force and its speed and direction remain constant. Because the direction of the Coriolis force causes winds to be deflected to the left, the eventual direction of the winds would be exactly opposite to the winds in the northern hemisphere.

Geostrophic wind FIGURE 6.9

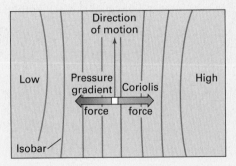

A At upper levels in the atmosphere, a parcel of air is subjected to a pressure gradient force and a Coriolis force.

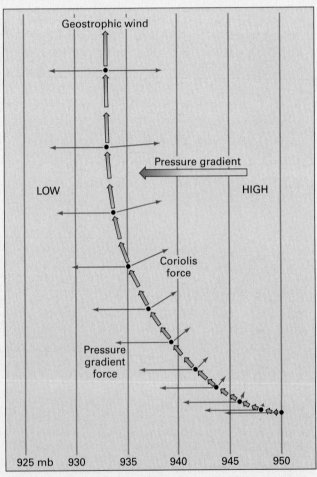

B The parcel of air moves in response to a pressure gradient. At the same time, it is turned progressively sideways until the pressure gradient force and Coriolis force balance, producing the geostrophic wind.

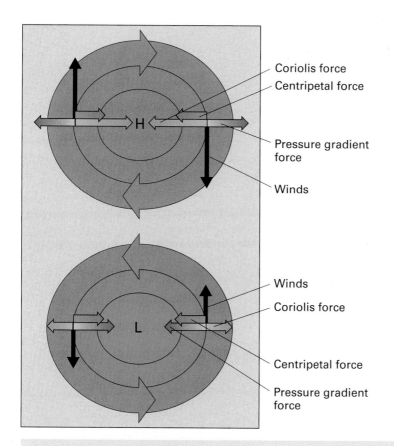

Anticyclone

In the northern hemisphere, circulation is clockwise around a high-pressure center. Because the Coriolis force has to balance the pressure gradient force and also supply a centripetal force to keep the air turning, it will be larger than it would be if it only balanced the pressure gradient force. Hence the winds that produce this Coriolis force need to be faster than the geostrophic winds.

Cyclone

In the northern hemisphere, circulation is counterclockwise around a low-pressure center. Because the pressure gradient force has to supply a centripetal force to keep the air turning, the balancing Coriolis force does not need to be as large as it would be if it had to balance the full pressure gradient force. Hence the winds that produce this Coriolis force will be slower than the geostrophic winds.

Cyclones and anticyclones aloft FIGURE 6.10

Upper-air winds circulate around the centers of cyclones and anticyclones. Because the Coriolis effect deflects moving air to the right in the northern hemisphere and to the left in the southern hemisphere, the direction of circulation reverses from one hemisphere to the other.

CYCLONES AND ANTICYCLONES IN UPPER AIR

We are used to seeing low- and high-pressure centers on daily weather maps. You can think of them as marking vast whirls of air in a spiraling motion. Low-pressure centers are known as **cyclones**, whereas high-pressure centers are known as **anticyclones**. To understand winds associated with cyclones and anticyclones, it is easiest to consider an idealized case as in FIGURE 6.10.

cyclone Center of low atmospheric pressure.

anticyclone Center of high atmospheric pressure.

We know that locally, air aloft blows parallel to isobars, so that the corresponding Coriolis force exactly balances the pressure gradient force. However, what happens when this air starts circulating around high and low pressure centers—that is, anticyclones and cyclones? In that case, there is another small but important force that needs to be considered—*centripetal force.*

Centripetal force is the force needed to keep the air parcel rotating; if it were not supplied, the air parcel would simply move in a straight line and would not follow a curved trajectory along the isobars.

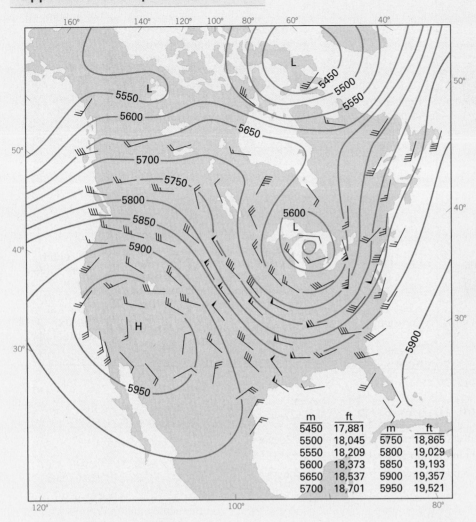

A An upper-air map for a day in late June. Lines are height contours for the 500-mb surface. High pressures—designated with an H—correspond with higher heights. Low pressures—designated with an L—correspond with lower heights.

m	ft	m	ft
5450	17,881		
5500	18,045	5750	18,865
5550	18,209	5800	19,029
5600	18,373	5850	19,193
5650	18,537	5900	19,357
5700	18,701	5950	19,521

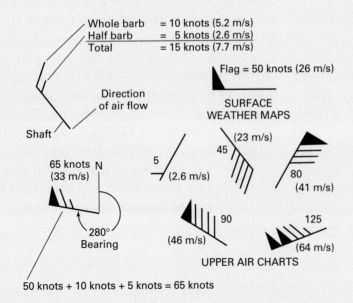

Whole barb = 10 knots (5.2 m/s)
Half barb = 5 knots (2.6 m/s)
Total = 15 knots (7.7 m/s)

Flag = 50 knots (26 m/s)

Direction of air flow

Shaft

SURFACE WEATHER MAPS

65 knots N (33 m/s)

280° Bearing

5 (2.6 m/s)

45 (23 m/s)

80 (41 m/s)

90 (46 m/s)

125 (64 m/s)

UPPER AIR CHARTS

50 knots + 10 knots + 5 knots = 65 knots

B Explanation of wind barbs. 1 knot (nautical mile per hour) = 0.514 meters per second. To see the wind direction, think of the wind arrows as moving along with the wind, with the barbs at the rear of the arrow and the point at the front.

In the case of anticyclones, in the northern hemisphere the winds blow from north to south—that is, northerly—along the eastern edge of the anticyclone and from south to north along the western edge. If we look closer at other locations, we see that the winds blow parallel to the isobars with the high-pressure center to the right of the winds. We find, then, that in the northern hemisphere, the winds will move clockwise around a high-pressure center.

To do so, however, a centripetal force needs to be applied to the winds, which in this case is supplied by the Coriolis force. Because the Coriolis force needs to not only balance the pressure gradient force but also supply the centripetal force, it needs to be larger than if it were simply balancing the pressure gradient force alone. In that case, the winds that produce this Coriolis force need to be stronger than the geostrophic winds—they must be *supergeostrophic.*

For cyclones in the northern hemisphere, the winds blow from south to north—that is, southerly—along the eastern edge of the cyclone and from north to south along the western edge. Again, if we look at any given location, we will find that the winds blow parallel to the isobars with the high pressure to the right of the winds (and the low pressure center to the left of the winds). Hence, in the northern hemisphere, the winds will move counterclockwise around a low-pressure center.

Again, to maintain this circulation, a centripetal force needs to be applied to the winds, which in this case is supplied by the pressure gradient force. Now, because the pressure gradient force is partly offset by the centripetal force, the Coriolis force does not need to be as large as it would be if it had to balance the full pressure gradient force. Hence, the winds that produce this Coriolis force do not need to be as strong—they are *subgeostrophic.*

In the southern hemisphere, we have to account for the fact that the Coriolis force is directed to the left of the wind fields. Hence, while anticyclones and cyclones are associated with high and low pressures respectively, and the direction of the pressure gradient forces and Coriolis forces are the same as in the northern hemisphere, the winds that generate these Coriolis forces are in the opposite direction. Therefore, air circulates clockwise around a cyclone in the southern hemisphere and counterclockwise around an anticyclone.

In the real world, cyclones and anticyclones are not perfectly circular. To see how winds behave in a more practical setting, let us look at an upper-air map for North America on a late June day (FIGURE 6.11). The contours show the height of the 500-mb pressure level. Low heights correspond to regions of low pressure, whereas high heights correspond to regions of high pressure. Also, we know that the pressure patterns aloft are related to the temperature of the intervening air. Because low pressures aloft are related to cold air, this map indicates that a mass of cold Canadian air has moved southward to dominate the eastern part of the continent. A large, but weaker, upper-air high is centered over the southwestern desert. This high is a mass of warm air, heated by intense surface insolation on the desert below.

The figure also shows upper-air wind strengths and directions using the wind barb symbol employed on weather maps. The winds generally follow the contours well, as expected for geostrophic flow. They are strongest where the contours are closest together, because there the pressure gradient force is the strongest. To balance this strong pressure gradient force, there is a correspondingly strong Coriolis force generated by the intense winds in these locations. In contrast, where we see weak pressure gradients—for example, around the high over the southwest—we see weak winds.

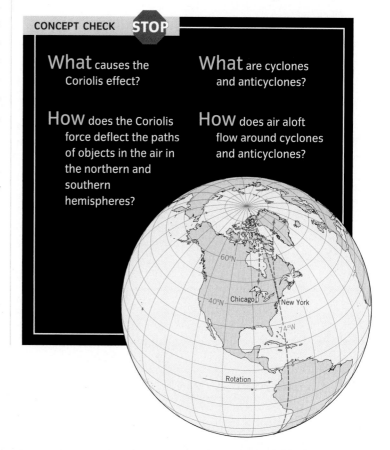

CONCEPT CHECK STOP

What causes the Coriolis effect?

What are cyclones and anticyclones?

How does the Coriolis force deflect the paths of objects in the air in the northern and southern hemispheres?

How does air aloft flow around cyclones and anticyclones?

Winds at the Surface

Explain the effect friction has on winds at the surface.

Describe how surface winds in the northern and southern hemispheres move relative to pressure centers.

Discuss the role surface pressure patterns play in influencing large-scale weather.

T he direction and strength of winds aloft are determined by the balance between the Coriolis force and the pressure gradient force, with the centripetal force acting to keep air circulating around cyclones and anticyclones. How do winds behave near the surface? To fully understand this motion, we have to introduce another force—the **frictional force**. The frictional force is related to the tendency for air to slow down as it blows over a fixed surface (either land or ocean). The frictional force acts in the opposite direction to the winds. In addition, the frictional force in-

> **frictional force**
> Force applied to atmospheric motions due to differences between the wind velocity and the velocity of the surface over which the wind moves.

creases with increasing wind speeds. Because of turbulence in the atmosphere, it is not only winds right at the surface that are affected by friction, but also all winds through the bottom kilometer or so.

Let's now consider winds created by a low-pressure and high-pressure center at the surface, as in **FIGURE 6.12**. In this case, as with winds aloft, the air tends to move down the pressure gradient under the influence of the pressure gradient force. As it does so, it comes under the influence of the Coriolis force. In the northern hemisphere, these winds are deflected to the right.

However, because of the frictional force, these winds are slowed and the Coriolis force is smaller. In that case, unlike for winds aloft, the Coriolis does not balance the pressure gradient force, and the winds continue to blow across isobars but at an angle.

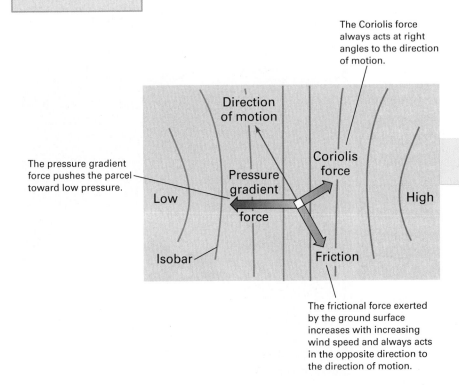

The Coriolis force always acts at right angles to the direction of motion.

The pressure gradient force pushes the parcel toward low pressure.

The frictional force exerted by the ground surface increases with increasing wind speed and always acts in the opposite direction to the direction of motion.

Balance of forces on a parcel of surface air FIGURE 6.12

A parcel of air in motion near the surface is subjected to three forces. The sum of these three forces produces motion toward low pressure but at an angle to the pressure gradient. The example shown here is for the northern hemisphere.

CYCLONES AND ANTICYCLONES AT THE SURFACE

When considering centers of low-pressure—*cyclones*—and centers of high pressure—*anticyclones*—at the surface, we need to keep in mind that winds do not blow along isobars but at an angle across them. When low pressure is at the center, the pressure gradient is straight inward, as seen in FIGURE 6.13. However, because of the Coriolis force and friction with the surface, locally the surface air moves at an angle to the gradient. When we consider the winds at all locations, we find that these local winds produce an inspiraling motion. In the northern hemisphere, the cyclonic inspiral is counterclockwise because the Coriolis force acts to the right. In the southern hemisphere the cyclonic inspiral is clockwise, because the Coriolis force acts to the left.

Cyclones and anticyclones at the surface FIGURE 6.13

Surface winds spiral inward toward the center of a cyclone but outward and away from the center of an anticyclone. Because the Coriolis effect deflects moving air to the right in the northern hemisphere and to the left in the southern hemisphere, the direction of inspiraling and outspiraling reverses from one hemisphere to the other.

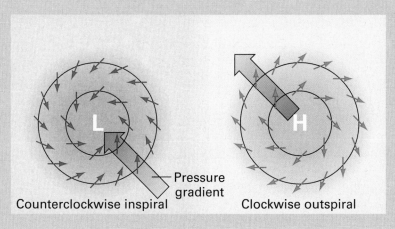

Counterclockwise inspiral   Pressure gradient   Clockwise outspiral

Northern hemisphere
With the frictional force now added, the direction of motion is not parallel to isobars but is angled toward low pressure and away from high pressure. This creates a counterclockwise inspiral around cyclones and a clockwise outspiral around anticyclones.

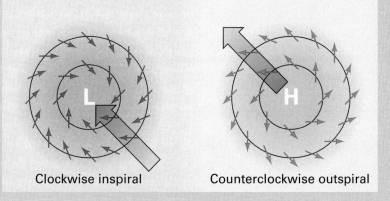

Clockwise inspiral   Counterclockwise outspiral

Southern hemisphere
As in the northern hemisphere, the frictional force angles the wind motion toward low pressure and away from high pressure. However, the outspiraling and inspiraling motions are reversed in direction because the Coriolis force deflects air to the left in the southern hemisphere.

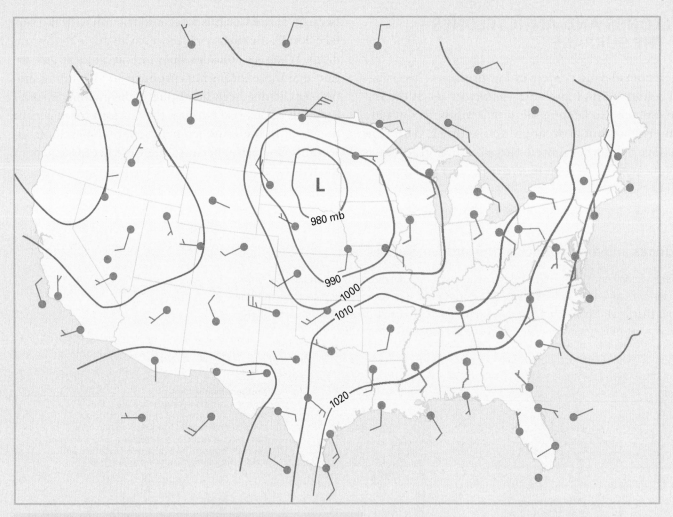

Surface pressures and wind fields FIGURE 6.14

Wind barbs indicate surface winds blowing counterclockwise and into this low-pressure center found over the central United States.

When high pressure is at the surface, the pressure gradient is straight outward. Again, because of the Coriolis force and friction with the surface, locally the surface winds move at an angle to the gradient. Now, however, the winds produce an outspiraling motion. In the northern hemisphere, the anticyclonic outspiral is clockwise, and in the southern hemisphere the anticyclonic outspiral is counterclockwise.

As with cyclones and anticyclones aloft, a centripetal force is needed to keep air circulating around the centers of high and low pressure. For air flow around cyclones, the centripetal force is supplied by the pressure gradient force, weakening it and thereby weakening the winds. For air flow around anticyclones,

the centripetal force is supplied by a combination of the Coriolis force and frictional force, meaning they have to be larger than if they were only balancing the pressure gradient force. Because both of these forces increase with wind speed, we expect wind speeds to increase in order to support the addition of the centripetal force.

As an example of how the system behaves in the real world, FIGURE 6.14 shows the wind fields and surface pressures associated with a cyclone situated over the central United States. From these measurements, we can see that the surface winds are circulating counterclockwise around the low and spiraling inward, crossing the isobars as they do so.

DIVERGENCE AND CONVERGENCE

You may know from experience that low-pressure centers (cyclones) are often associated with cloudy or rainy weather and that high-pressure centers (anticyclones) are often associated with fair weather. Why? The inward spiraling motion around surface cyclones produces **convergence** of air (or *mass convergence*) in the center and tends to result in rising air near the center, as seen in FIGURE 6.15. When surface air is forced aloft by this process, it is cooled according to the adiabatic principle, allowing condensation and precipitation to begin. As a result, cloudy and rainy weather often accompanies the inward and upward air motion of cyclones.

In contrast, the outward spiraling motion from an anticyclone produces **divergence** of air away from the pressure center (also called *mass divergence*). This divergence of air is replaced by air descending from above the pressure center. When air descends, it is warmed by the adiabatic process, so condensation cannot occur. Thus, anticyclones are often associated with fair weather.

These surface cyclones and anticyclones can be a thousand kilometers (about 600 mi) across, or more. Cyclones and anticyclones

convergence The movement of atmospheric mass into a region.

divergence The movement of atmospheric mass out of a region.

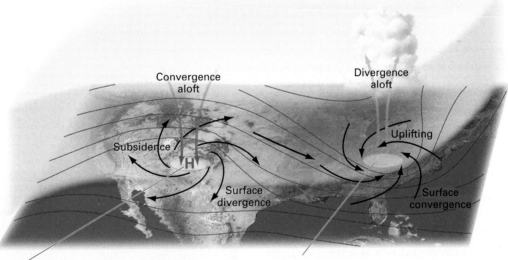

Anticyclone
As air spirals around and out of surface high-pressure centers, it produces divergence of air from the region. Air from above descends to replace the air that is removed. This air warms as it descends, inhibiting condensation and precipitation.

Cyclone
As air spirals around and into a surface low-pressure center, it produces convergence of air into the region. This convergence forces air at the surface to rise and cool adiabatically, leading to condensation and precipitation.

Convergence and divergence around surface pressure centers
FIGURE 6.15

The convergence and divergence of air into and out of cyclones and anticyclones at the surface can produce the rising motions necessary for condensation and precipitation.

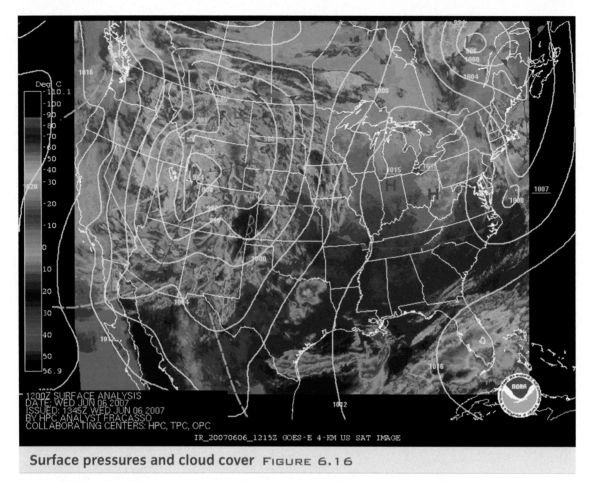

Surface pressures and cloud cover FIGURE 6.16

Low and high surface pressure patterns—adjusted to sea level and indicated by yellow isobars—are accompanied by regions of cloudy skies—designated by yellow and red colors—and clear skies—designated by green and blue colors.

can remain more or less stationary, or they can move, sometimes rapidly, to create weather disturbances in the regions they cross. FIGURE 6.16 shows two anticyclones—one over eastern Canada and another off the west coast of the United States—both of which are associated with clear skies. In contrast, the low-pressure system centered near the Nebraska-Iowa border lies under heavy cloud cover, as indicated by the yellow and red shading. As these cyclones and anticyclones move, the weather systems move with them.

Although we have discussed how surface cyclones and anticyclones generate convergence and divergence, and the resulting vertical motions, the reverse process can happen as well. That is, vertical motions can generate cyclones and anticyclones. For example, if air above a given region begins to rise through convection, surface pressure will decrease, producing a gradi-

ent force that induces a cyclonic circulation. This is the same process that occurs in thermal circulations, but on a much larger scale.

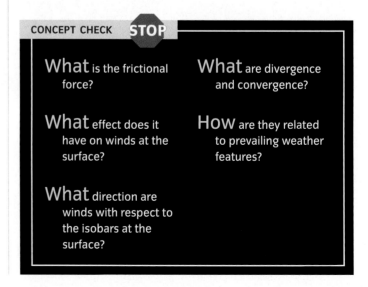

CONCEPT CHECK **STOP**

What is the frictional force?

What effect does it have on winds at the surface?

What direction are winds with respect to the isobars at the surface?

What are divergence and convergence?

How are they related to prevailing weather features?

What is happening in this picture ?

This image shows a brown plume of dust being transported from the African continent out over the Atlantic ocean on February 26, 2000. The dust in this image can be transported all the way to the Caribbean if the winds are strong enough.

- During this event, which way are the winds blowing, based upon this image?

- Do you think these winds are associated with local wind systems or larger cyclones and anticyclones?

SUMMARY

1 Winds

1. **Winds** are the horizontal movement of air molecules. The direction of the wind is given as the direction from which the wind is coming.

2. Wind direction is measured by a wind vane. Wind speed is measured by an anemometer. Wind speeds can also be measured using Doppler radar.

3 The Coriolis Effect and Winds Aloft

1. The Earth's rotation creates the **Coriolis effect**. The Coriolis effect acts as an additional force that deflects wind motion to the right in the northern hemisphere and to the left in the southern hemisphere.

2. The strength of the Coriolis force increases with increasing wind speed as well as with latitude. It is always directed perpendicular to the object's motion.

3. When only the pressure gradient force and Coriolis force affect the winds, they are called **geostrophic winds**. Geostrophic winds blow parallel to isobars. For geostrophic winds, the pressure gradient force is equal to and opposite the Coriolis force.

4. Because of the Coriolis force, air spirals around **cyclones** (centers of low pressure) and **anticyclones** (centers of high pressure). The direction of the spiral depends upon the hemisphere. For winds blowing around cyclones and anticyclones aloft, we also have to account for the centripetal force, which keeps the air circulating around pressure centers. The centripetal force affects the strength of the Coriolis force and hence the wind fields.

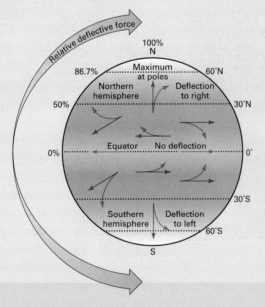

2 Winds and Pressure Gradients

1. Air motion is produced by horizontal **pressure gradients**. Horizontal pressure gradients represent differences in pressure between one location and another, after accounting for differences associated with changes in elevation. The direction of pressure gradients is from high to low pressures, perpendicular to the isobars. The strength increases with decreasing distance between the isobars.

2. Pressure gradients at the surface and aloft form when air is unevenly heated, creating thermal circulations. Sea and land breezes are examples of thermal circulations formed from unequal heating and cooling of land and water surfaces. Mountain and valley winds, Santa Ana, and foehn winds are other examples of local winds.

4 Winds at the Surface

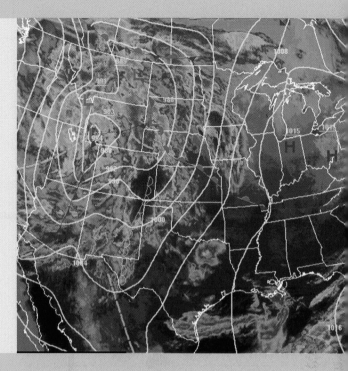

1. Winds at the surface are also affected by **friction**. The three important forces affecting these winds are the pressure gradient force, the Coriolis force, and friction.

2. Surface winds do not blow parallel to isobars, but at an angle across them. This angle causes winds to spiral into low pressures at the surface and out of high pressures at the surface. The direction of circulation again depends upon the hemisphere, with winds blowing into and counterclockwise around low pressures in the northern hemisphere and out of and clockwise around high pressures. The sense of rotation is reversed in the southern hemisphere.

3. Air parcels spiraling around cyclones at the surface generate a **convergence** of air at the surface. This convergence causes air at the surface to lift and cool adiabatically, producing condensation and cloud formation. Air parcels spiraling around anticyclones at the surface produce a **divergence** of air, causing air from above to descend to the surface.

KEY TERMS

- **wind** p. 156
- **isobars** p. 159
- **pressure gradient** p. 160
- **Coriolis effect** p. 166
- **geostrophic wind** p. 168
- **cyclone** p. 169
- **anticyclone** p. 169
- **frictional force** p. 172
- **convergence** p. 175
- **divergence** p. 175

CRITICAL AND CREATIVE THINKING QUESTIONS

1. Given a map with isobars, how can you find the direction of the pressure gradient force? How can you determine where the pressure gradient force is strongest and where it is weakest?

2. Describe a simple thermal circulation wind system. In your answer, explain why air motion occurs.

3. How do land and sea breezes form? How do they illustrate the concepts of pressure gradient and thermal circulation?

4. What is the Coriolis effect, and why is it important? What produces it? How does it influence the motion of wind and ocean currents in the northern hemisphere? In the southern hemisphere?

5. What is the geostrophic wind, and what is its direction with respect to the pressure gradient force?

6. Define cyclone and anticyclone. What type of weather is associated with surface cyclones and surface anticyclones? Why?

7. Draw four spiral patterns showing outward and inward flow in clockwise and counterclockwise directions. Identify which diagrams are associated with surface cyclones and anticyclones in the northern hemisphere and which are associated with surface cyclones and anticyclones in the southern hemisphere. Explain why different diagrams are needed for each hemisphere.

SELF-TEST

1. On weather maps, _____ connect lines of equal atmospheric pressure.
   a. isobars
   b. isotherms
   c. pressure contours
   d. isohyets

2. The closer together isobars are, the higher is the pressure gradient.
   a. True
   b. False

3. Wind speed is measured using an instrument called a(n) _____.
   a. barometer
   b. wind vane
   c. speedometer
   d. anemometer

4. A land breeze generally occurs _____.
   a. at nightfall, when the land cools below the surface temperature of the sea
   b. when strong winds blow in from the sea over the land
   c. only during certain restricted seasons
   d. during the day, when the land heats above the surface temperature of the sea

5. A sea breeze occurs when the sea surface heats and draws air from the shore.
   a. True
   b. False

6. Fall (or drainage) winds are present where winds drain from shallow (flat) plains out to the ocean.
   a. True
   b. False

7. The Coriolis effect is _____.
   a. a result of the Earth's counterclockwise orbit around the Sun
   b. a result of the Earth's counterclockwise rotation around its axis
   c. a result of the Earth's axis of rotation with respect to its axis of orbit
   d. unrelated to other physical phenomena on the Earth

8. In the southern hemisphere, the winds are deflected to the left by the Coriolis force.
   a. True
   b. False

9. At upper levels in the atmosphere, as a parcel of air moves in response to a pressure gradient, it is turned progressively sideward until the gradient and Coriolis forces balance to produce the _____.
   a. geostrophic wind
   b. tropospheric wind
   c. upper-air westerlies
   d. equatorial easterlies

10. On the following diagram, at points A and B, draw in arrows representing the direction of (1) the pressure gradient force; (2) the Coriolis force; (3) the centripetal force; and (4) the winds.

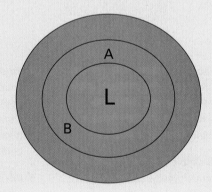

11. A cyclone is a spiraling center of high pressure that turns clockwise in the northern hemisphere.
a. True
b. False

12. On the following diagram, circle the wind barbs indicating the highest velocities. How fast are these winds blowing? In what direction? Use the wind barb legend from Figure 6.11.

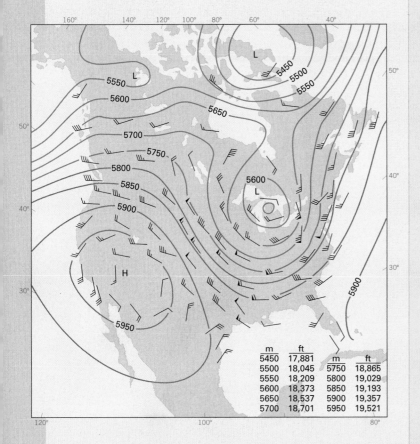

m	ft	m	ft
5450	17,881		
5500	18,045	5750	18,865
5550	18,209	5800	19,029
5600	18,373	5850	19,193
5650	18,537	5900	19,357
5700	18,701	5950	19,521

13. A parcel of air at the surface is subjected to three forces, and the balance among the _____, Coriolis, and frictional forces determines the direction of motion of the parcel of air.
a. gravitational
b. pressure gradient
c. centrifugal
d. divergent

14. A northern hemisphere anticyclone is a _____.
a. high-pressure system that rotates counterclockwise
b. low-pressure system that rotates clockwise
c. high-pressure system that rotates clockwise
d. low-pressure system that rotates counterclockwise

15. Cloudy, rainy weather is often associated with the inward and upward convergence of air within _____.
a. surface anticyclones
b. surface cold fronts
c. surface warm fronts
d. surface cyclones

Global Atmospheric and Oceanic Circulation 7

At the top of Mount Washington, high above the New Hampshire countryside, weather conditions can get downright nasty. Said by some to have the worst weather in the United States, Mount Washington has winds that can be greater than those found in hurricanes and tornadoes. In fact, the strongest recorded surface wind speed in history—373 km/hr (231 mph)—was measured here during a storm on April 12, 1934. During that same storm, wind speeds in the valley below Mount Washington were in the range of 48–80 km/hr (30–50 mph). Why are the winds on the mountain so strong?

The high-velocity winds found on Mount Washington are related to atmospheric processes 10 km up in the atmosphere. At these levels, a very fast moving, narrow current of air, called the *jet stream*, circles the globe. The jet stream is present in the atmosphere year round but shifts with the seasons and even day to day.

The jet stream is influenced by the unequal solar heating of the Earth, with the lower latitudes receiving more solar energy than the higher latitudes. This uneven heating sets up a global pressure gradient that gives rise to the jet stream. Without the retarding effects of friction with the Earth's surface, the jet stream can attain speeds of 300–400 km/h (180–250 mph). Although the jet stream is normally found near the tropopause, occasionally it dips down, affecting the wind speeds of elevated mountain ranges such as Mount Washington and producing the record-breaking winds of April 12, 1934.

The weather observatory atop Mount Washington in New Hampshire sits above the cloud tops in this picture taken during the middle of winter.

NATIONAL GEOGRAPHIC

Surface Winds

Hadley cell Low-latitude atmospheric circulation cell with rising air over the equatorial regions and sinking air over the subtropical belts.

Near the surface of the Earth, large-scale wind fields are subject to three important forces—the pressure gradient force, the Coriolis force, and frictional force. Here we consider the largest-scale wind systems—those found on an ideal Earth, without a complicated pattern of land and water and no seasonal changes.

The most important features for understanding the global wind patterns—called the *general circulation* of the atmosphere—are **Hadley cells**, shown on the cross section of **FIGURE 7.1**. Insolation is strongest when the Sun is directly overhead, so the surface and atmosphere at the Equator are heated more strongly than other places on this featureless Earth. When this air is heated, two thermal circulations form—the Hadley cells—one in the northern hemisphere and one in the southern hemisphere. In each Hadley cell, air rises over the Equator and generates high pressures aloft. Similar to the upper-air pressure gradient that is formed during the sea breeze, the air aloft moves poleward, away from high pressures over low latitudes and toward low pressures associated with cooler air found over the high latitudes. Eventually, the air aloft cools and descends at about 30° latitude, completing the thermal circulation. The Hadley cell is named for George Hadley, who first proposed its existence in 1735.

How does the thermal circulation associated with the Hadley cell affect the surface pressure patterns? Vertical motions at the Equator remove air from the

Hadley cells FIGURE 7.1

Warm, moist, unstable air over the equatorial region rises along the intertropical convergence zone (ITCZ). Flowing north and south aloft, the air eventually descends around 30° N and 30° S, creating two high-pressure centers—the subtropical high-pressure cells. As the return flow moves back toward the Equator, it comes under the influence of the Coriolis force, producing the northeast and southeast trade winds.

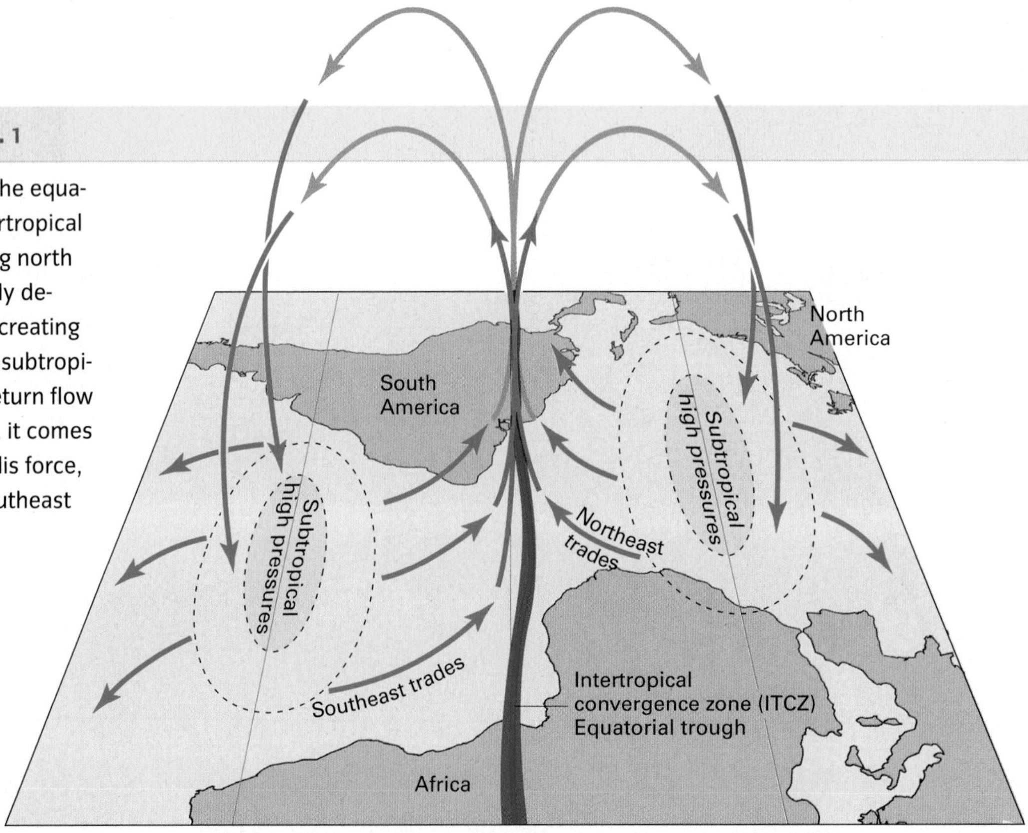

region, creating a zone of surface low pressure known as the *equatorial trough*. Air in both hemispheres moves toward the equatorial trough, where it converges and then rises as part of the Hadley cell circulation. Convergence occurs in a narrow zone, named the **intertropical convergence zone (ITCZ)**.

On the poleward side of the Hadley cell circulation, air descends, increasing the surface pressures underneath the descending air. This produces two **subtropical high-pressure belts**, each centered at about 30° latitude.

How do these pressure patterns affect the surface winds? Within the ITCZ, the air is generally moving vertically, not horizontally, so winds in the ITCZ are light and variable. In earlier centuries, mariners on sailing vessels referred to this region as the doldrums, where they were sometimes becalmed for days at a time. Similarly, in the centers of the subtropical high-pressure belts, air is descending, producing vertical motions as opposed to horizontal ones; thus winds also tend to be weak.

intertropical convergence zone (ITCZ) Zone of convergence of air masses along the equatorial trough.

subtropical high-pressure belt Belt of persistent high atmospheric pressure centered approximately on latitudes 30° N and 30° S.

trade winds Easterly winds found in the tropical regions north and south of the Equator.

In the subtropical high-pressure centers, winds spiral outward and move toward the Equator as well as to middle latitudes. The winds moving equatorward are the strong and dependable **trade winds**. North of the Equator, these winds are influenced by the Coriolis force, and hence veer to the right. Because they come from the northeast, they are referred to as the *northeast trades*. To the south of the Equator, the Coriolis force moves the northward-moving winds to the left, so they are called the *southeast trades*. The trade winds converge in the equatorial trough of the ITCZ, completing the thermal circulation associated with the Hadley cells on either side of the Equator.

Poleward of the subtropical highs, air spiraling outward again comes under the influence of the Coriolis force, seen in FIG-URE 7.2. In the northern hemisphere, the rightward deflection produces south-westerly winds, whereas in the southern hemisphere, the leftward deflection produces northwesterly winds. In both hemispheres, these winds are called the *midlatitude westerlies*.

Circulations in the midlatitudes and high latitudes FIGURE 7.2

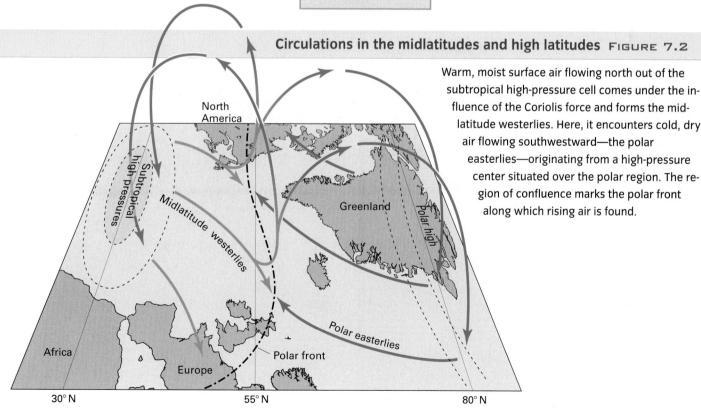

Warm, moist surface air flowing north out of the subtropical high-pressure cell comes under the influence of the Coriolis force and forms the midlatitude westerlies. Here, it encounters cold, dry air flowing southwestward—the polar easterlies—originating from a high-pressure center situated over the polar region. The region of confluence marks the polar front along which rising air is found.

Between about the 30° and 60° latitudes, the pressure and wind pattern becomes more complex. This latitudinal belt is a zone of conflict between air bodies with different characteristics, called *air masses*. From the low latitudes, air moves into the region from the subtropical high-pressure centers. In addition, at the poles, the air is intensely cold. As a result, surface high pressures form. Outspiraling winds around the polar anticyclone move air from this region into the lower latitudes, heading westward under the influence of the Corio-

lis force. The region where the warm, low-latitude air meets the cold, high-latitude air is known as the **polar front**. Pressures and winds can be quite variable in the midlatitudes from day to day and week to week as disturbances form along the polar front, producing midlatitude storms. On average, however, winds are more often from the west, so the region is said to have prevailing westerlies.

Although the cold anticyclones centered on the poles should be surrounded by easterly surface winds, polar easterlies

> **polar front** Boundary between cold polar air masses and warm tropical air masses.

Global surface winds on an ideal Earth FIGURE 7.3

Here we see the global surface winds and pressures on an ideal Earth with and without rotation. Surface winds are shown on the disk of the Earth, while the cross section in Part B shows winds aloft.

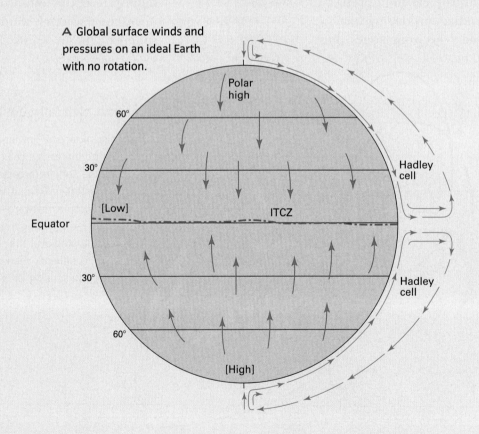

A Global surface winds and pressures on an ideal Earth with no rotation.

only exist in the south polar region. In the north polar region, winds tend to have an easterly component, but there is too much variation in direction to consider polar easterlies the dominant winds there.

As the midlatitude westerlies meet northerly and easterly polar winds, the convergence causes air at the surface to rise around 50° N (and 50° S). As with the surface winds in this region, this rising air represents the average conditions and can vary from day to day and location to location. As the air rises, it eventually moves poleward, where it feeds into the descending air over the polar regions, as well as equatorward, where it joins the descending air over the subtropics. This upper-air flow completes two more *circulation cells*, the Ferrel cell in the midlatitudes and the polar cell in the high latitudes. Unlike the Hadley cell, however, these circulation cells are only seen in the average wind patterns, not in the day-to-day circulations.

FIGURE 7.3 shows how heating/cooling and the rotation of the Earth combine to produce the winds and pressure patterns around the globe.

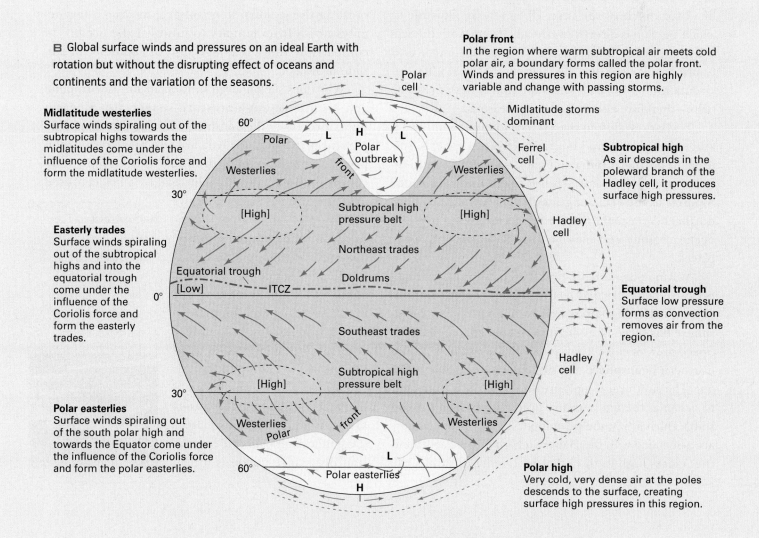

B Global surface winds and pressures on an ideal Earth with rotation but without the disrupting effect of oceans and continents and the variation of the seasons.

Midlatitude westerlies
Surface winds spiraling out of the subtropical highs towards the midlatitudes come under the influence of the Coriolis force and form the midlatitude westerlies.

Easterly trades
Surface winds spiraling out of the subtropical highs and into the equatorial trough come under the influence of the Coriolis force and form the easterly trades.

Polar easterlies
Surface winds spiraling out of the south polar high and towards the Equator come under the influence of the Coriolis force and form the polar easterlies.

Polar front
In the region where warm subtropical air meets cold polar air, a boundary forms called the polar front. Winds and pressures in this region are highly variable and change with passing storms.

Subtropical high
As air descends in the poleward branch of the Hadley cell, it produces surface high pressures.

Equatorial trough
Surface low pressure forms as convection removes air from the region.

Polar high
Very cold, very dense air at the poles descends to the surface, creating surface high pressures in this region.

GLOBAL WIND AND PRESSURE PATTERNS

So far, we have discussed the wind pattern for a seasonless, featureless Earth. Let's turn now to actual global surface wind and pressure patterns. We will use the global maps of wind and pressure for January and July seen in FIGURE 7.4. The pressures and winds shown are averages over many years for all daily observations in either January or July. They are corrected for the elevation of the recording station so that pressures are shown for sea level.

The engine for the general circulation of the atmosphere is the Hadley cell circulation. The rising branch of the Hadley cell circulation forms along the ITCZ. Here, convection is driven by intense solar heating, because the noontime Sun is almost directly overhead for much of the year. However, the latitude at which the Sun is directly overhead changes with the seasons, migrating between the Tropics of Cancer and Capricorn. We see that the elements of the Hadley cell circulation—the ITCZ and subtropical high-pressure belts—therefore also shift with the seasons.

The largest shift in the ITCZ is found around 90° E and entails a seasonal migration from northern Australia to northern India and back—a shift of about 40° of latitude. A significant shift is also found over South America as the ITCZ migrates from central Brazil into northern Venezuela. Smallest shifts are found over the ocean regions and may be only 5° latitude in some places.

Moving poleward, we find that the subtropical high-pressure belts, created by the Hadley cell circulation, also shift. The high-pressure belt in the southern hemisphere has three large high-pressure cells, each developed over oceans, that persist year round. In the northern hemisphere, the situation is somewhat different. The subtropical high-pressure belt shifts from land to ocean as the temperature gradient between the two shifts. During July, the prominent high-pressure centers are associated with the Hawaiian high in the Pacific and the Azores high in the Atlantic. During this time, the high pressures are centered over the ocean, because the temperatures here are generally cooler than over the surrounding land surfaces.

During January, however, when the land surfaces cool substantially, the high-pressure centers move from the ocean to the land. The most prominent signature of this is the extension of the Siberian high into the subtropical regions. Note that the Hawaiian high and Azores high are still present in January; however, they have weakened significantly and have moved south.

Before moving to circulation features in the higher latitudes, let's examine these low-latitude features in more detail.

The ITCZ and the monsoon circulation

In general, the shift of the ITCZ with the seasons is moderate in the western hemisphere, moving a few degrees north from January to July over the oceans. In comparison, there is a huge shift of about 40° of latitude in Asia.

Why does such a large shift occur in Asia? In January, an intense high-pressure system, the Siberian high, forms over the continent. During this time, temperatures in northern Asia are very low, so this high pressure is to be expected. In July, this high-pressure center is absent, replaced instead by a low centered over the Middle Eastern desert region. The Asiatic low is produced by the intense summer heating of the landscape.

The subsequent movement of the ITCZ and the change in the pressure pattern with the seasons create an almost complete reversal of the winds in Asia known as the **monsoon**. Although similar to smaller-scale thermal circulations, the extent of the Asian monsoon can affect the general atmospheric circulation over large portions of the

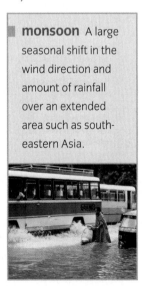

monsoon A large seasonal shift in the wind direction and amount of rainfall over an extended area such as southeastern Asia.

Atmospheric pressure maps—Mercator FIGURE 7.4

These global maps show average sea-level pressures and winds in January and July.
The average barometric pressure is 1013 mb (29.2 in. Hg). Values greater than this are
"high" and are shown in red, while lower values are "low" and are shown in green.
Pressure units are millibars adjusted to sea level.

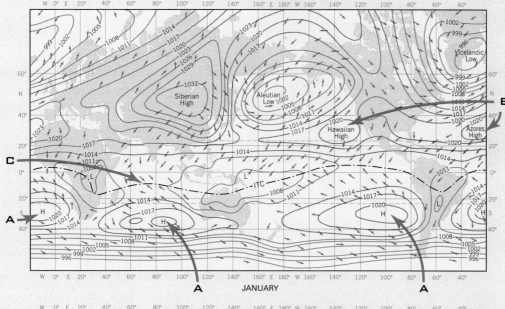

JANUARY

A Southern hemisphere subtropical high-pressure belt
The most prominent features of the maps are the subtropical high-pressure belts, created by the Hadley cell circulation. The southern hemisphere belt has three large high-pressure cells, each developed over oceans, that persist year round. A fourth, weaker high-pressure cell forms over Australia in July, as the continent cools during the southern hemisphere winter.

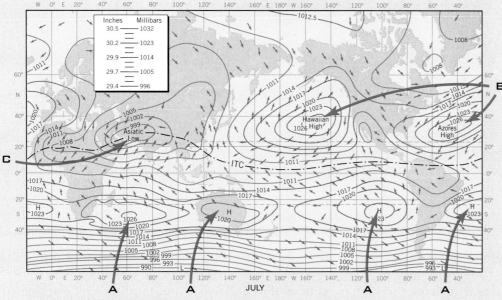

JULY

B Northern hemisphere subtropical high-pressure belt
The situation is different in the northern hemisphere. The subtropical high-pressure belt shows two large anticyclones centered over oceans—the Hawaiian high in the Pacific and the Azores high in the Atlantic. From January to July, these highs intensify and move northward.

C Seasonal shift in the ITCZ
You can see a huge shift in Africa and Asia by comparing the two maps. In January, the ITCZ runs south across eastern Africa and crosses the Indian Ocean to northern Australia at a latitude of about 15° S. In July, it swings north across Africa along the south rim of the Himalayas, in India, at a latitude of about 25° N—a shift of about 40 degrees of latitude!

Surface Winds   189

A **Monsoon rains** Heavy monsoon rains turn streets into canals in Delhi, India.

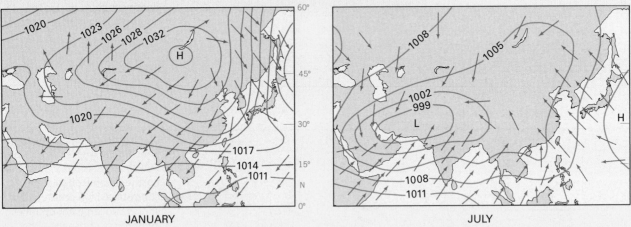

JANUARY

JULY

B **Monsoon wind patterns** The Asiatic monsoon winds alternate in direction from January to July, responding to reversals of barometric pressure over the large continent.

Monsoon wind patterns FIGURE 7.5

globe, as seen in FIGURE 7.5. During the winter monsoon, dry conditions prevail as cold, descending air is situated over the continent. In addition, there is a strong outflow of dry, continental air from the high pressures situated over China and India, leading to northeasterly flow over India and the Middle East and a northwesterly flow over China and Southeast Asia.

In the summer, the inspiraling of warm, humid air from the Indian Ocean and the southwestern Pacific moves into the continental low-pressure system, pro-

ducing southwesterlies over India and southeasterlies over Indochina and China. In addition, the inspiraling produces convergence over the continent. This convergent airflow and subsequent uplift of the surface air produces strong adiabatic cooling and heavy rainfall in southeastern Asia during the summer.

North America does not have the remarkable extremes of monsoon winds experienced in Asia. Even so, in summer there is a prevailing tendency for warm, moist air originating in the Gulf of Mexico to move northward across the central and eastern part of the United States. At times, moist air from the Gulf of California also invades northwestern Mexico and the desert Southwest of the United States. Together, these circulations comprise the North American monsoon system. In winter, the airflow pattern across North America changes, and dry, continental air from Canada moves south and east, reducing precipitation. Similarly, over South America, a surface low-pressure system forms over the Amazon during the southern hemisphere summer, accompanied by low-level convergence and increased precipitation. During the southern hemisphere winter, cooler, descending air is found over Brazil, where a surface high pressure forms, resulting in drier conditions in Brazil, Paraguay, Bolivia, and Argentina.

Subtropical high-pressure belts

The next prominent features we want to examine are the subtropical high-pressure belts created by the Hadley cell circulation. A schematic map of two large high-pressure cells bounded by continents (**FIGURE 7.6**) shows the features of circulation around the subtropical high-pressure cells. The outspiraling circulation produces the easterly trade winds in the tropical and equatorial zones and the westerlies in the subtropical zones and poleward.

In these high-pressure cells, air on the east side subsides more intensely because of cold-water currents typically found off the west coast of continents. This means that winds spiraling outward on the eastern side are drier and colder (since they are coming from the higher latitudes). On the west side, subsidence is less strong. In addition, these winds travel long distances across warm, tropical ocean surfaces before reaching land, and so they pick up moisture and heat on their journey.

When the Hawaiian and Azores highs intensify and move northward in July, the east and west coasts of North America feel the effects of the subtropical highs most prominently. On the west coast, dry subsiding air from the Hawaiian high dominates, so fair weather and rainless conditions prevail. On the east coast, warm, moist air from the Azores high flows across the continent from the southeast, producing generally hot, humid weather for the central and eastern United States.

In January, these two anticyclones weaken and move to the south—leaving the east coast of North America to the mercy of colder winds and air masses from the north and west.

Those regions located predominantly under the subtropical high-pressure centers can be extremely dry. Here, the descending air inhibits the vertical motions necessary to produce condensation and precipitation. In addition, this descending air warms adiabatically, resulting in an increase in the saturation specific humidity and hence a reduction in the relative humidity. There is a subsequent re-evaporation of any remaining liquid water, producing clear skies and blistering solar heating. Most of the world's great deserts—including the Sahara Desert in Africa, the Gobi Desert in China, the Arabian Desert in the Middle East, the Mojave Desert (including Death Valley) in the United States, and the Great Western Desert (or outback) in Australia—are found beneath subtropical high-pressure centers.

Subtropical high-pressure cells FIGURE 7.6

Over the oceans, surface winds spiral outward from the subtropical high-pressure cells, feeding the trades and the westerlies. On the eastern side of the cells, air subsides more strongly, producing dry winds. On the western sides, subsidence is not as strong, and a long passage over oceans brings warm, moist air to the continents.

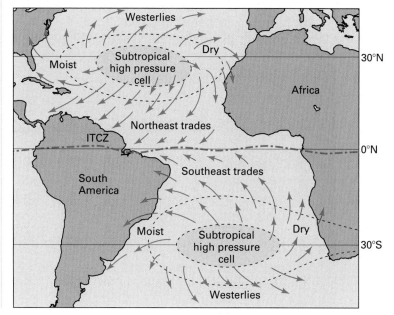

North America

BRAZIL

South America

Global Locator

Deserts and rainforests FIGURE 7.7

Some of the driest and wettest regions of the world result from the positioning of the intertropical convergence zone (ITCZ) and subtropical high-pressure centers. On the left, the Sahara desert of Algeria, situated at 30° N, stretches out in all directions. On the right, a waterfall in the equatorial rainforests of Brazil, located on the Equator, returns heavy rainfall produced by convection in the ITCZ back to the ocean.

The circulations out of the subtropical high-pressure centers and into the equatorial trough produce both the wettest and driest regions of the globe (FIGURE 7.7). Within the ITCZ, the strong convergence of warm, moist air produces strong convection and persistent rainfall. Many of the world's tropical rainforests, including the Amazon in South America and the Congo Basin in Africa, are situated along the Equator in the vicinity of the ITCZ.

Wind and pressure features of higher latitudes
If we turn to the higher latitudes, we find that the northern and southern hemispheres are quite different in their geography. The northern hemisphere has two large continental masses, separated by oceans. An ocean is also at the pole. In the southern hemisphere, we find a large ocean with a cold, glacier-covered continent at the center. These differing land–water patterns strongly influence the development of high- and low-pressure centers with the seasons.

Let's examine the northern hemisphere first. Continents are cold in winter and warm in summer, as compared to oceans at the same latitude. Cold air is associated with surface high pressure and warm air with surface low pressure. Thus, continents show high pressure in winter and low pressure in summer.

This pattern is very clear in the northern hemisphere polar maps of FIGURE 7.8 and is represented by the strong Siberian high in Asia and the Canadian high in North America. From these high-pressure centers, air spirals outward, bringing cold air to the south. Over the oceans, two large centers of low pressure are striking—the Icelandic low and the Aleutian low. These two low-pressure centers are not actually large stable features that we would expect to find on every daily world weather map in January. Rather, they are regions of average low pressure and rising air associated with the poleward branch of the Ferrel cell.

Atmospheric pressure maps—polar FIGURE 7.8

These global maps show the average sea-level pressure and wind for daily observations in January and July. Values greater than average barometric pressure (1013 mb; 29.2 in. Hg) are shown in red, and lower values are in green. Pressure units are millibars adjusted to sea level.

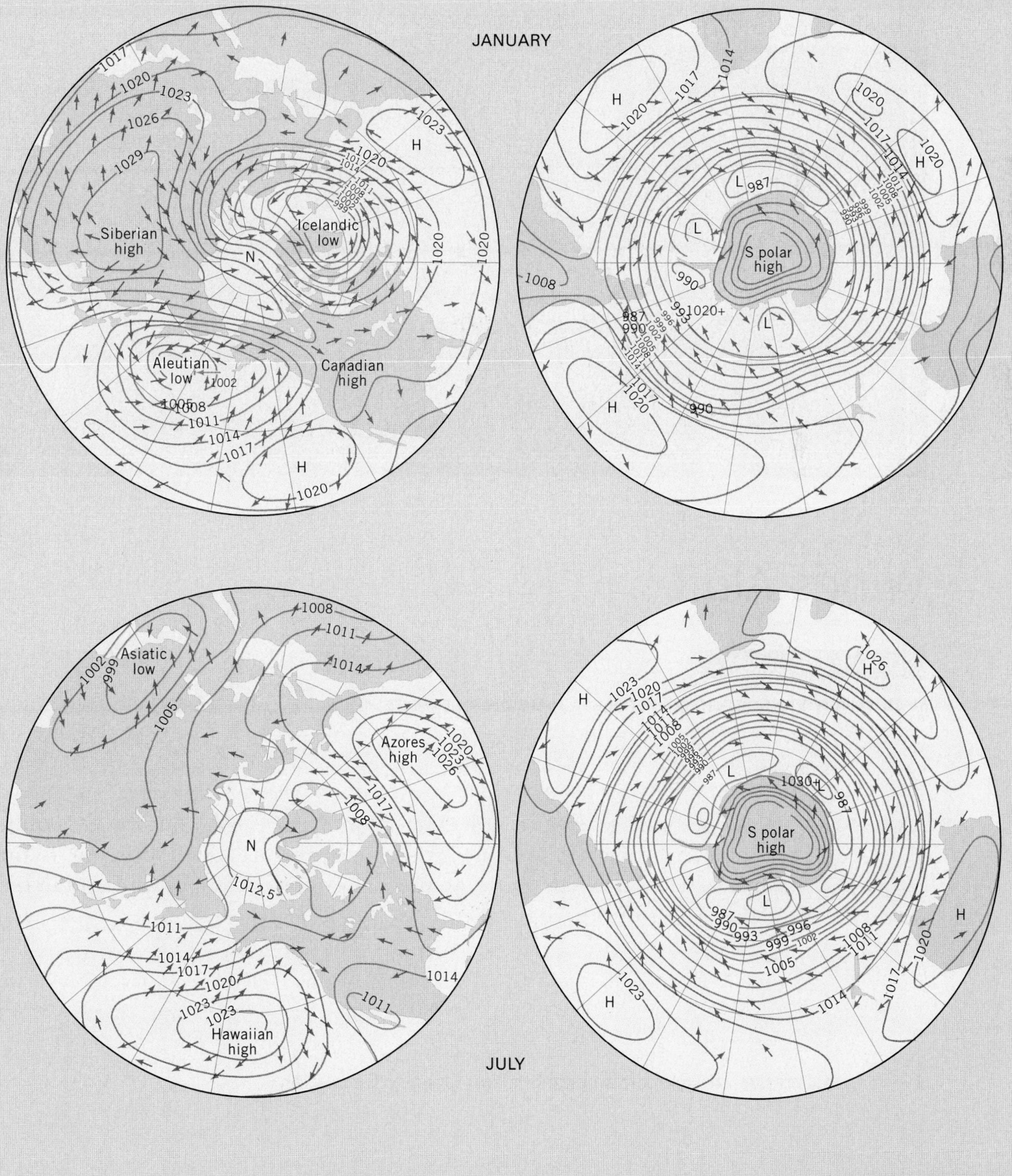

A January
In winter, air spirals outward from the strong Siberian high and its weaker cousin, the Canadian high. Circulations around the Icelandic low and Aleutian low bring moisture-laden winter storm systems eastward and northward onto the continents.

B July
In summer, the continents show generally low surface pressure, while high pressure builds over the oceans. Inspiraling winds of the strong Asiatic low bring warm, moist Indian Ocean air over India and Southeast Asia. A lesser low forms over the deserts of the southwestern United States and northwestern Mexico. The proximity of the Hawaiian high and Azores high to the west coasts of North America and Europe keeps these regions warm and dry, while outspiraling air from these highs brings warm moist air onto the east coasts of North America and Asia.

JANUARY

JULY

In summer, the pattern reverses. The continents show generally low surface pressure, while high pressure builds over the oceans. This pattern is easily seen on the northern hemisphere July polar map. The Asiatic low is strong and intense. The two subtropical highs, the Hawaiian high and the Azores high, strengthen and dominate the Atlantic and Pacific Ocean regions, where temperatures are cooler than the surrounding land surfaces.

In contrast, the higher latitudes of the southern hemisphere are characterized by a polar continent surrounded by a large ocean. Since Antarctica is covered by a glacial ice sheet and is cold at all times, a permanent anticyclone, the South Pole high, is centered there. Easterly winds spiral outward from the high-pressure center. Surrounding the high is a band of deep, low pressure, with strong, inward-spiraling westerly winds. As early mariners sailed southward, they encountered this band, in which wind strength intensifies toward the pole. Because of the strong prevailing westerlies, they named these southern latitudes the "roaring forties," "flying fifties," and "screaming sixties."

CONCEPT CHECK **STOP**

How do Hadley cells develop?

How do they influence wind and pressure patterns?

What is the ITCZ?

How do the subtropical high-pressure belts influence surface wind patterns?

How would you describe the general circulation of the Earth's atmosphere?

Winds Aloft

LEARNING OBJECTIVES

Describe how pressure gradients intensify at upper levels.

Define jet streams.

Explain how disturbances in the jet stream develop and grow.

he surface wind systems we have examined describe airflows at or near the surface. How does air move at the higher levels of the troposphere? As with air near the surface, air aloft moves in response to pressure gradients and is influenced by the Coriolis effect, but without the retarding influence of friction.

GLOBAL CIRCULATION AT UPPER LEVELS

How do pressure gradients arise at upper levels? Aloft, regions of warm air produce higher pressures (as compared to regions of cold air) as the air column expands and lifts the upper portions of the air column. In addition, the pressure difference between the warm-air regions and cold-air regions increases as one moves higher in the atmosphere, as seen in FIGURE 7.9. For each incremental increase in altitude, the distance between the pressure levels—called *isobaric surfaces*—in the warm-air region is greater than in the cold-air region, due to expansion of the intervening air. By the time one reaches the upper levels of the atmosphere, the cumulative effect of these incremental increases produces a very large pressure difference between the warm-air region and the cold-air region. This in turn produces a very large pressure gradient force that drives high-velocity winds in these levels of the atmosphere.

While this principle holds for any large-scale temperature difference between two regions, we are interested in how it affects upper-air winds associated with the general circulation of the atmosphere.

Upper-air pressure gradient FIGURE 7.9

The isobaric surfaces slope downward from the low latitudes to the pole, creating a pressure gradient force. The height between one level and the next is smaller in the cold-air region than in the warm-air region. The cumulative effect is to increase the pressure gradient force with altitude, producing strong winds at high altitudes.

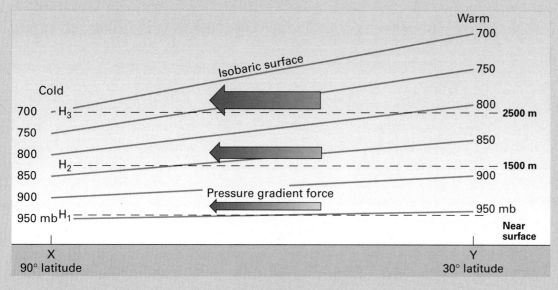

At 2500 m
In the warm-air region, the pressure at H_3 is 810 mb, while in the cold-air region it is 700 mb, a difference of 90 mb.

At 1500 m
In the warm-air region, the pressure at H_2 is 890 mb, while in the cold-air region it is 825 mb, a difference of 65 mb.

Near the surface
In the warm-air region, the pressure at H_1 is 955 mb, while in the cold-air region it is 940 mb, a difference of 15 mb.

In FIGURE 7.10, we see the general pattern of airflows at higher levels in the troposphere. Because the Earth's insolation is greatest near the Equator and least near the poles, there will be a general temperature gradient from the Equator to the poles. This gradient creates a pressure gradient force aloft. Because there is no friction in these upper levels, this pressure gradient force is balanced by the Coriolis force, resulting in geostrophic winds in the upper atmosphere.

These upper-air winds comprise westerlies that blow in a complete circuit about the Earth, from about 25° latitude almost to the poles. In the midlatitudes, they often exhibit large undulations from their westerly track, which are associated with the passage of midlatitude storms. At high latitudes, the westerlies form a huge circumpolar spiral, circling a great polar low-pressure center.

Toward lower latitudes, atmospheric pressure rises steadily, forming a tropical high-pressure belt at 15° to 20° N and S latitude. This high-pressure belt shifts north and south with the seasons and is associated with the latitude of the ITCZ. When the high-pressure ridge is centered off the Equator, there can be a zone of lower pressure near the Equator, setting up a pressure gradient that can generate easterly geostrophic winds aloft. These winds are called the *equatorial easterlies*, although winds in these regions can change direction and magnitude depending upon the season and location.

Thus, the overall picture of upper-air wind patterns is really quite simple—a band of weak easterly winds in the equatorial zone, belts of high pressure near the Tropics of Cancer and Capricorn, and westerly winds, with some variation in direction and magnitude, spiraling around polar lows.

Global upper-level winds FIGURE 7.10

In this generalized plan of global winds high in the troposphere, strong west winds dominate the mid- and high-latitude circulation.

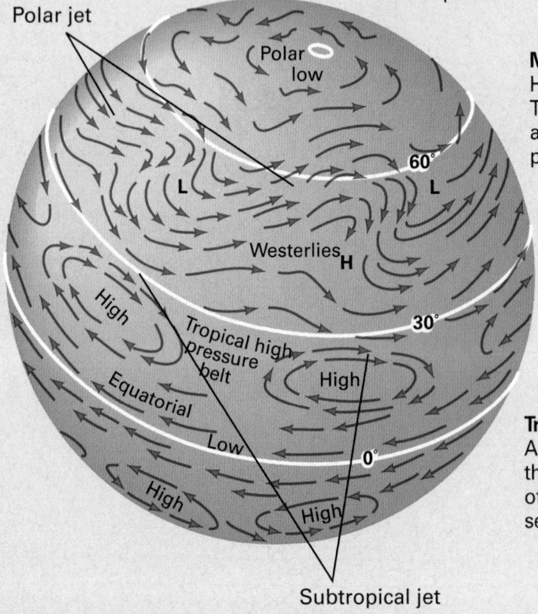

Polar regions
Westerlies form a circumpolar spiral, circling a low-pressure center situated over the poles.

Midlatitudes
Here the westerlies become very strong. They often sweep to the north or south around varying centers of high and low pressure aloft.

Subtropics
Atmospheric pressure continues to rise from the poles toward the Equator, hence westerlies extend equatorward to about 25° N and S latitude, with maximum values near 30° N and 30° S.

Tropics
A weak easterly wind pattern prevails, called the *equatorial easterlies*. However, the direction of winds at these latitudes can shift with the season and location on the globe.

JET STREAMS AND THE POLAR FRONT

The smooth westward flow of the upper-air westerlies is punctuated by regions of very intense, narrow bands of westerly winds called **jet streams**. Jet streams are important features of upper-air circulation. They are narrow zones at a high altitude in which wind streams reach great speeds, averaging around 175–200 km/hr (100–125 mph), as shown in FIGURE 7.11. They occur where atmospheric pressure gradients are strong. Along jet streams, pulse-like movements of air follow broadly curving tracks. The greatest wind speeds—called *jet streaks*—occur in the center of a jet stream, with velocities decreasing away from the center.

jet stream High-speed air flow in narrow bands within the upper-air westerlies and along certain other global latitude zones at high levels.

There are three kinds of jet streams. Two are westerly wind streams, and the third is a weaker jet with easterly winds that develop in Asia as part of the summer monsoon circulation.

The most poleward type of jet stream is located along the polar front and is designated the *polar-front jet*

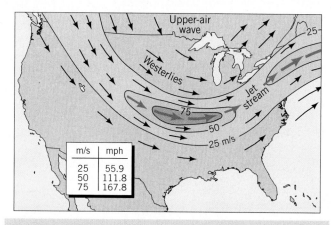

m/s	mph
25	55.9
50	111.8
75	167.8

Jet streams FIGURE 7.11

The jet stream shown on this map by lines of equal wind speed is characterized by a narrow, fast-moving current of air located near the tropopause. (National Weather Service)

stream (or simply, the "polar jet"). Located generally between 35° and 65° latitude, it is present in both hemispheres. The polar jet follows the boundary between cold polar air and warm subtropical air, as seen in FIGURE 7.12. In the region of the polar front, there is a very strong temperature gradient. This large north–south temperature gradient in turn creates a strong pressure gradient aloft that powers the jet stream.

Polar jet stream and the polar front
FIGURE 7.12

The polar jet stream is normally located over the polar front. Here, the strong temperature gradient between warm air to the south and cold air to the north produces a strong pressure gradient that can support the high-velocity jet stream.

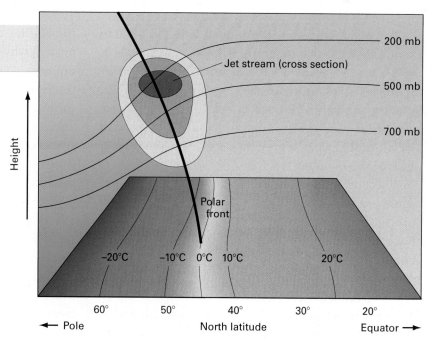

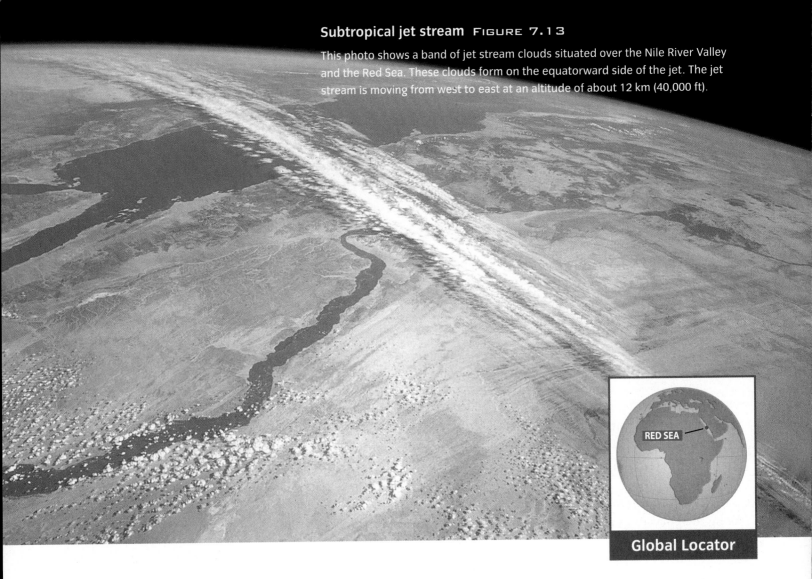

This photo shows a band of jet stream clouds situated over the Nile River Valley and the Red Sea. These clouds form on the equatorward side of the jet. The jet stream is moving from west to east at an altitude of about 12 km (40,000 ft).

RED SEA

Global Locator

The polar jet is typically found at altitudes of 10 to 12 km (about 33,000 to 40,000 ft), and wind speeds in the jet range from 175 to as much as 400 km/hr (about 100–250 mph). In the northern hemisphere, traveling aircraft often use the polar jet to increase ground speed when flying eastward. In the westward direction, flight paths are chosen to avoid the strong winds of the polar jet.

A second type of jet stream forms in the subtropical latitude zone—the *subtropical jet stream* (FIGURE 7.13). It occupies a position at the tropopause just above the subtropical high-pressure cells in the northern and southern hemispheres. Here, westerly wind speeds average 250 km/hr in winter (about 150 mph) and can reach speeds of 370 km/hr (about 230 mph). This jet is associated with the Coriolis force applied to the poleward moving air found in the upper branch of

the Hadley cell. In fact, the influence of the Coriolis force is why the upper-branch of the Hadley cell does not simply flow all the way to the poles—it is diverted to the right by the Coriolis force in the northern hemisphere (and to the left in the southern hemisphere), and becomes the eastward-flowing subtropical jet stream. The result is a river of swiftly moving air that flows west to east, cooling as it goes via longwave emission. Eventually, it cools enough that the air sinks to the surface as the descending branch of the Hadley cell.

A third type of jet stream is found at even lower latitudes. Known as the *tropical easterly jet stream*, it runs from east to west—opposite in direction to the polar-front and subtropical jet streams. The tropical easterly jet occurs only in the summer season and is limited to a northern hemisphere location over Southeast Asia, India, and Africa, when the ITCZ is shifted north of the Equator.

DISTURBANCES IN THE JET STREAM

While the jet stream is typically a region of confined, high-velocity westerly winds found in the midlatitudes, the jet stream can actually contain broad wave-like undulations. Many processes can produce disturbances in the eastward flow of the jet stream. One of the most important is related to *baroclinic instability*, which makes the atmosphere unstable to small disturbances in the jet stream and allows these disturbances to grow over time.

The development of these disturbances is shown in **FIGURE 7.14**. For a period of several days or weeks, the jet stream flow may be fairly smooth. Then, an undulation develops. As the undulation grows, warm air pushes poleward, while a tongue of cold air is brought to the south. Eventually, the cold tongue is pinched off, leaving a pool of cold air at a latitude far south of its original location. This cold pool may persist for some days or weeks, slowly warming with time. Because of its cold center, it will contain low pressures aloft. In addition, the cold air in the core will descend and diverge at the surface, creating surface high pressure. Similarly, a warm air pool will be pinched off far to the north of its original location. Within the core of the warm pool will be rising air, with convergence and low pressure at the surface, as well as high pressure aloft.

Because the circulation associated with the disturbance brings warm air poleward and cold air equatorward, it is the primary mechanism of poleward heat transport in the midlatitudes. It is also the reason weather in the midlatitudes is often so variable, as pools of warm, moist air and cold, dry air alternately invade midlatitude land masses, producing changes in temperature, winds, and precipitation.

Disturbances in the jet stream FIGURE 7.14

Disturbances form in the upper-air westerlies of the northern hemisphere, marking the boundary between cold polar air and warm tropical air.

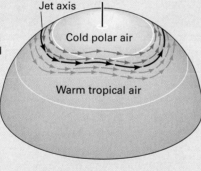

A The flow of air along the front will be fairly smooth for several days or weeks, but then it will begin to undulate.

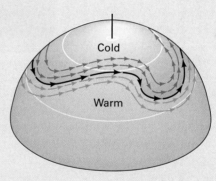

B As the undulation becomes stronger, disturbances in the jet stream begin to form. Warm air pushes poleward, while cold air is brought to the south.

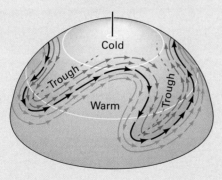

C The disturbances become stronger and more developed. A tongue of cold air is brought to the south, where it occupies low-pressure troughs.

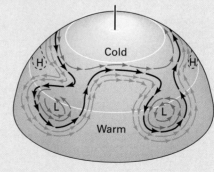

D Eventually, the tongue is pinched off leaving a pool of cold air at a latitude far south of its original location. These pools of cold air form cyclones, which can persist for some days or weeks.

Growth of disturbances in the jet stream FIGURE 7.15

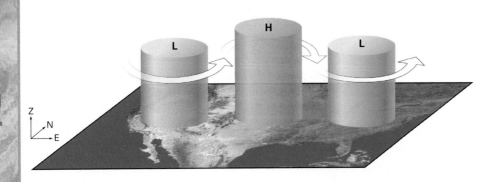

Step 1 Pressure disturbances in the upper atmosphere are associated with gradients in temperature. Here, the warmer air column produces high pressures aloft while the cooler air columns produce lower pressures. The pressure patterns produce disturbances in the jet stream, which circulates geostrophically around the pressure centers.

Step 2 Looking down from above, the change in temperature from one location to another results in a disturbance in the global-scale north–south temperature gradient between low and high latitudes, as shown by the isotherms. The counterclockwise flow around the low-pressure center aloft brings warm, tropical air from the south into the vicinity of the warm air column, which tends to heat this region even further. Conversely, the high-pressure center aloft brings cold air from the north into the vicinity of the cold air column, cooling it even further.

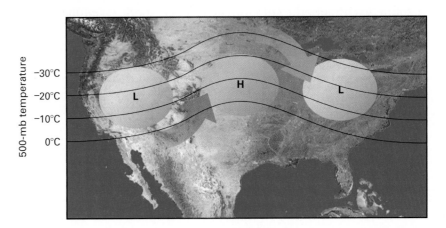

Step 3 As the warm air column continues to warm, it expands, producing even higher pressures aloft. In addition, the cold air column continues to cool, thereby decreasing pressures further. The pressure gradient between the cold-air region and the warm-air region intensifies, and so the wind speeds also intensify.

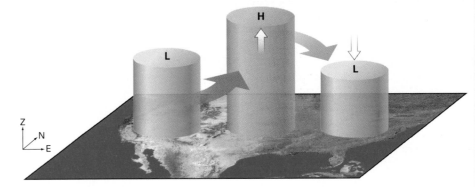

What causes disturbances in the jet stream to grow over time? This growth is actually a complicated process and involves the interaction of the upper-air circulations—formed by the high- and low-pressure centers—with the global temperature gradient between the warmer, low-latitude air and colder, high-latitude air. This growth process is described in FIGURE 7.15.

Consider first what happens to a slight disturbance in the jet stream. These disturbances arise from pressure differences between one location and another, resulting in geostrophic wind variations. In addition, the pressure differences are accompanied by temperature differences, with high pressures aloft found over regions of warm air and low pressures aloft over regions of cool air.

As air circulates around these disturbances, the counterclockwise circulations associated with the low-pressure system tend to bring warm, low-latitude air into a region that is already warm. In contrast, the clockwise circulation around the high-pressure region tends to bring cold, high-latitude air into a region that is already cold. As a result, the warm regions tend to get warmer and the cold regions tend to get colder. In turn, the high pressures associated with the warm-air region tend to get higher and the low pressures associated with the cold-air region tend to get lower. This process intensifies the pressure gradient between the two regions, resulting in stronger winds and an even larger disturbance.

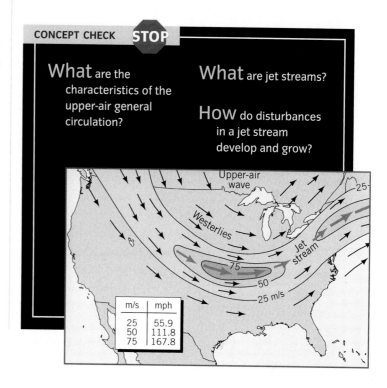

CONCEPT CHECK STOP

What are the characteristics of the upper-air general circulation?

What are jet streams?

How do disturbances in a jet stream develop and grow?

m/s	mph
25	55.9
50	111.8
75	167.8

Oceanic Circulation

LEARNING OBJECTIVES

Explain how gyre circulations form.

Describe how ocean circulations differ on the western and eastern sides of ocean basins.

Define the thermohaline circulation and describe its characteristics.

Just as there is a circulation pattern to the atmosphere, so too there is a circulation pattern to the oceans. Like the atmospheric circulation pattern, oceanic circulation is driven by differences in density and pressure acting along with the Coriolis force and frictional force. Like air, warm water is less dense than cold water, so unequal heating of ocean water can create pressure differences that induce water flow. Differences of salinity can also affect density, because saltier water is more dense than fresh water. The force of wind on surface water is another important factor in creating oceanic circulation. Before we examine ocean currents, however, let's take a broad look at the thermal structure of the ocean.

TEMPERATURE LAYERS OF THE OCEAN

As with the atmosphere, the ocean has a layered structure. Ocean layers are recognized in terms of temperature, with temperatures generally highest at the sea surface and decreasing with depth. This trend is not surprising, since the sources of heat that warm the ocean are solar insolation and heat supplied by the overlying atmosphere, both of which act to warm the water at or near the surface.

> **mixed layer** Upper portion of the ocean in which the water is well mixed by waves and wind, resulting in uniform temperatures and salinity.

> **thermocline** Region below the mixed layer in which there is a rapid decrease in ocean temperature over a relatively short vertical distance.

> **deep ocean** Region below the thermocline in which water is very cold. This region typically extends to the bottom of the ocean floor.

The ocean's layered temperature structure is shown in **FIGURE 7.16**. At low latitudes and in middle latitudes, there is a layer of warm water near the top. Here waves and winds mix heated surface water with the water below it to form a warm layer (called the **mixed layer**) whose thickness depends on the intensity of mixing. Below the warm layer, temperatures drop rapidly in a zone known as the **thermocline**. Below the thermocline is a layer of very cold water extending to the ocean floor, called the **deep ocean**. Temperatures near the base of the deep ocean range from 0°C to 5°C (32°–41°F). In arctic and antarctic regions, the warm layer and thermocline are absent. Note that because the warm layer is less dense than the cold deep-ocean water, the ocean is quite stable across the thermocline, which makes it difficult for water from the upper layers and deeper layers to mix.

SURFACE CURRENTS

An **ocean current** is any persistent, dominantly horizontal flow of ocean water. Current systems act to exchange heat between low and high latitudes and are essential in sustaining the global energy balance. We can identify both surface currents and deep currents. Surface currents are driven by prevailing winds. Deep currents are powered by changes in temperature and density occurring in surface waters that cause them to sink and flow along the ocean bottom.

The patterns of surface ocean currents are strongly related to prevailing surface winds. Energy is transferred from wind to water by the friction of the air blowing over the water surface. Because of the Coriolis effect, in the northern hemisphere the actual direction of water drift at the surface is deflected about 45° to the right from the direction of the driving wind.

The general features of the circulation of the ocean bounded by two continental masses (seen in **FIGURE 7.17**) include two large circular movements, called **gyres**, that are centered at latitudes of 20°–30°. These gyres track the movements of air around the subtropical high-pressure cells. An equatorial current with westward flow marks the belt of the trade winds. Although the trades blow to the southwest and northwest at an angle across the parallels of latitude, the surface water movement follows the parallels. The equatorial currents are separated

> **ocean current** Persistent, horizontal flow of ocean water.

> **gyre** Circulation of large-scale currents around ocean basins bounded by continents.

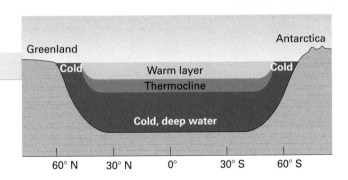

Ocean temperature structure FIGURE 7.16

A schematic north–south cross section of the world ocean shows that the warm surface-water layer disappears in arctic and antarctic latitudes, where very cold water lies at the surface. The thickness of the warm layer and thermocline is greatly exaggerated.

Ocean current gyres

FIGURE 7.17

Two great gyres, one in each hemisphere, dominate the circulation of ocean waters from the surface through 500–1000 m (1600–3300 ft).

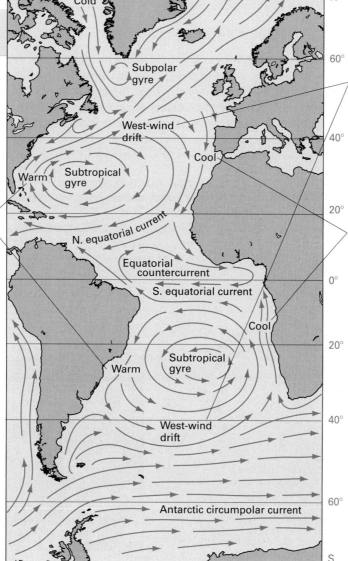

Warm poleward currents
In the tropical region, ocean currents flow westward, pushed by northeast and southeast trade winds. As they approach land, these currents turn poleward and become narrow, fast-moving currents called *western boundary currents*.

West-wind drift
The west-wind drift is a slow eastward motion of water in the zone of westerly winds. As west-wind drift waters approach the western sides of the continents, they are deflected equatorward along the coast.

Cold equatorward currents
The equatorward flows along the eastern portion of the ocean basins are cool currents, often accompanied by upwelling along the continental margins. In this process, colder water from greater depths rises to the surface.

western boundary currents Narrow, fast-moving currents found along the western edge of gyre circulations.

ocean upwelling Process in which warm surface waters are replaced by colder waters from below.

by an equatorial countercurrent. A slow, eastward movement of surface water over the zone of the westerlies is named the west-wind drift. It covers a broad belt between latitudes 35° and 45° in the northern hemisphere and between latitudes 30° and 60° in the southern hemisphere.

Along the western and eastern edges of the ocean basin, the currents are blocked by the presence of land masses. On the western edge of gyres, the equatorial currents are turned poleward and form narrow, fast-moving currents called **western boundary currents**. These currents are strong and well defined because they are compressed against the continental boundary by the Earth's rotation. Learn more about western boundary currents in *What a Scientist Sees: Three Images of the Gulf Stream.*

Over the eastern portion of the basin, the west-wind drift encounters the continents and is turned equatorward. Unlike the western boundary currents, these eastern boundary currents have the same velocity as the east–west currents in the center of the basin. These equatorward flows are cool currents. In addition, during the summer, they are often accompanied by **upwelling** along continental margins. The upwelling occurs because the subtropical high over the ocean intensifies, resulting in

Three Images of the Gulf Stream

Western boundary currents are narrow, fast-moving, warm-water currents found along the western edge of ocean gyres. A typical example of a western boundary current is the Gulf Stream. The Gulf Stream moves along the coast from Florida to North Carolina, then heads off the coast in a northeastward direction. However, unlike the image of a true stream, the path of the Gulf Stream changes with time and contains wave-like progressions in the flow, much like disturbances in the jet stream. As these wave-like features develop, warm and cold bodies of water are cut off to float freely, forming warm-core and cold-core rings. In fact, when floats are released into the current and tracked, they typically make tiny loops and veer in and out of the Gulf Stream, indicating that the water does not simply follow a uniform path like a stream.

Early navigation maps
A map drawn by Benjamin Franklin shows the approximate location of the Gulf Stream. Note the view of the Gulf Stream as a real stream with fixed boundaries.

Floating buoys
A "spaghetti diagram" tracks the movement of thousands of different floats as they move around the Atlantic. Off Florida, a very narrow concentration follows a single path. However, north of this location, the floats spread out into a large fan. The path of a single float often does not follow the Gulf Stream but instead meanders and loops around.

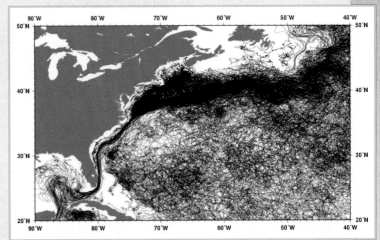

Biology and ecosystems
A satellite image shows marine plant life—phytoplankton—acquired by the NOAA-7 orbiting satellite. Rich plant life appears in red and yellow tones, and barren waters are in blue and purple tones. The waters along the coast of the Carolinas—to the west and north of the Gulf Stream—are rich in plant life and marine ecosystems, while the open ocean to the east of the Gulf Stream is relatively barren. However, loops in the Gulf Stream can capture islands of phytoplankton that survive for months to years out in the open ocean.

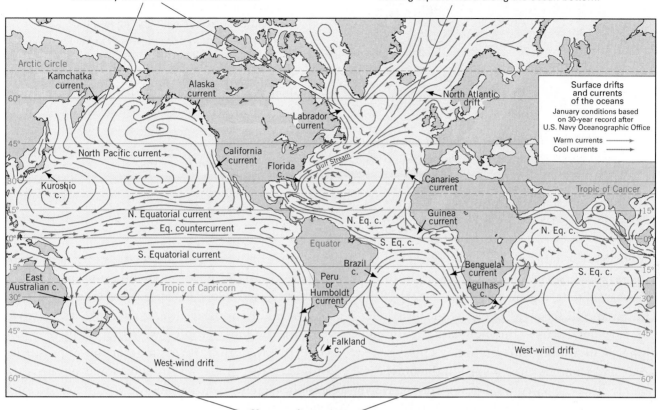

Cold western boundary currents
In the northern hemisphere, where there are significant continental boundaries in the high latitudes, cold equatorward currents flow along the east sides of continents and meet up with the poleward-flowing western boundary currents in the midlatitudes.

North Atlantic drift
In the northeastern Atlantic Ocean, the west-wind drift forms a relatively warm current, which spreads around the British Isles, into the North Sea, and along the Norwegian coast. Eventually, the current cools and sinks, flowing equatorward along the ocean bottom.

Surface drifts and currents of the oceans
January conditions based on 30-year record after U.S. Navy Oceanographic Office
Warm currents →
Cool currents →

Circumpolar current
In the southern hemisphere, the absence of continental boundaries allows the strong west winds around Antarctica to produce a continuously flowing current called the *Antarctic circumpolar current.*

January ocean currents FIGURE 7.18

Surface drifts and currents of the oceans in January.

equatorward winds along the coast. Moving along the north–south coastline, these winds produce surface currents at an angle of 45° away from the coast. As this surface water is removed from the coast, it is replaced by colder water from greater depths, further cooling these eastern-boundary current regions. The cold surface currents keep weather on the western coasts of continents cool, even in the height of summer.

While many of the general ocean circulation features hold for all ocean basins, a world map of the actual surface currents in January, seen in FIGURE 7.18, shows features unique to each. In the northern hemisphere, the western boundary currents in the Pacific and Atlantic Oceans are well developed and comprise the Gulf Stream of eastern North America and the Kuroshio current of Japan. In the southern hemisphere, the lack of significant continental barriers produces weaker western boundary currents, although they can be found in the Atlantic (Brazil current), the Pacific (the East Australian current), and the Indian Oceans (the Aguhlas current).

As with western boundary currents, equatorward-flowing eastern boundary currents are found in all the ocean basins. Examples of these cool, upwelling currents are the Humboldt (or Peru) current, off the coast of Chile and Peru, the Benguela current, off the coast of southern Africa, and the California current. Many of the world's most productive fisheries are also found along these coasts. The upwelling brings nutrient-rich waters from the ocean bottom to the surface. The nutrients,

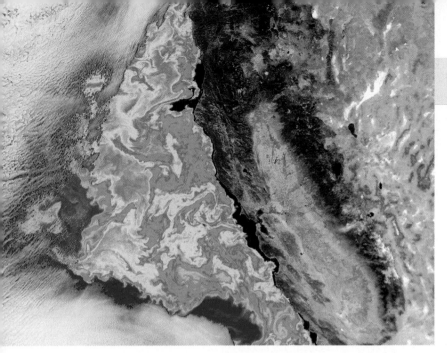

In upwelling regions, like these off the coast of California, nutrient-rich waters are brought to the surface. These waters support the growth of marine plant life—phytoplankton. Rich plant life appears here in red and yellow tones; barren waters are in blue and purple tones. Phytoplankton serve as the basis for the ocean food chain and support many of the world's largest fisheries along the west coasts of continents.

combined with sunlight, stimulate the growth of tiny ocean plants called *phytoplankton* that support the diverse ocean ecology of these regions (FIGURE 7.19).

The largest differences between the two hemispheres, however, occur in the high latitudes. In both hemispheres, west-wind drift water moves poleward to join arctic and antarctic circulations. In the northeastern Atlantic Ocean, the west-wind drift forms the North Atlantic drift, which extends past the British Isles, into the North Sea, and along the Norwegian and Greenland coasts. The warm surface currents of the North Atlantic drift, which is an extension of the Gulf Stream, keep winter temperatures in the British Isles from falling much below freezing in winter. In addition, the Russian port of Murmansk, on the Arctic Circle, remains ice-free year round because of this warm drift current.

Eventually, this water becomes very dense, due to cooling and an increase in salt concentration by evaporation and ice formation, and it sinks to the ocean floor, where it flows back toward the Equator. Some also returns equatorward along the surface as the Labrador current, which flows between Labrador and Greenland to reach the coasts of Newfoundland, Nova Scotia, and New England.

In the Pacific, because of the presence of the North American and Asian continents to the north, west-wind drift does not flow as far north, never cooling enough to sink. Instead, this water circles in a counterclockwise direction, eventually flowing equatorward as a cold western boundary current along the Asian coast across from Alaska, called the Kamchatka current.

In the southern hemisphere, the strong west winds around Antarctica produce an antarctic circumpolar current of cold water. Because there is an absence of continental boundaries, these currents can become very intense and, coupled with the strong westerly winds of the region, form some of the highest seas found on the globe. Eventually, some of this flow branches equatorward along the west coast of South America, adding to the Humboldt current, as well as reaching the western coast of Africa, where it flows equatorward as the Benguela current.

DEEP CURRENTS AND THERMOHALINE CIRCULATION

The ocean, on average, is 4 km (2.5 mi) deep. In the deep ocean—typically from about 1 km (0.62 mi) deep down to the ocean bottom—currents move water in a slow circuit across the floors of the world's oceans, shown in FIGURE 7.20. These currents are generated when surface waters become more dense and slowly sink downward. The deep currents are coupled with very broad and slow surface currents that link all of the world's oceans. The full, three-dimensional circulation is referred to as **thermohaline circulation**, be-

thermohaline circulation The global-scale, three-dimensional circulation of water through all of the ocean basins driven by the sinking of cold, dense water in the high latitudes of the Atlantic.

cause it depends on the temperature and salinity of North and South Atlantic Ocean waters, which in turn affect the density of seawater.

The thermohaline circulation starts in the high latitudes of the North and South Atlantic Ocean. In the North Atlantic—in the Greenland, Iceland, and Norwegian Seas—salty water from the low latitudes cools significantly. As it freezes, it exudes salt, increasing the salinity of the remaining ocean water. This cold, salty water is very dense and sinks to the ocean bottom. A similar process occurs in the South Atlantic, where sinking occurs in the Weddell Sea near the tip of South America.

The dense, bottom water from the North Atlantic flows south, where it joins the deep water formed in the South Atlantic. This water flows around Antarctica, similar to the circumpolar current near the surface. Eventually, some of the bottom water flows north into the Indian and Pacific Oceans, where it forces the less dense water in these ocean basins to lift toward the surface.

Some of this lifted water mixes with the overlying gyre circulations. However, some begins its return

Deep ocean circulation FIGURE 7.20

Deep ocean currents, generated by the sinking of cold, salty water in the northern Atlantic, circulate seawater in slowly moving coupled loops involving the Atlantic, Pacific, Indian, and Southern Oceans. Actual flows are much broader than the narrow flow paths shown here.

A Warm Atlantic surface water slowly moves northward through the equatorial and tropical zones. As this surface layer warms, evaporation occurs and it becomes saltier, which increases the density of the layer slightly.

B As this water moves into higher latitudes, it cools and becomes even more dense. Eventually, the surface layer becomes dense enough to sink. This sinking occurs in localized regions of the North and South Atlantic.

C Carried along the bottom, the cold, dense water flows southward, eventually reaching the Southern Ocean. Here, it flows around the Antarctic continent.

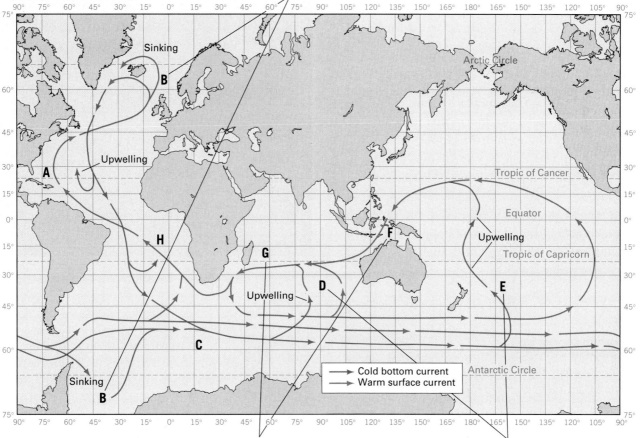

H Some of the upwelled water flows southward and rejoins the circumpolar current. However, some of the flow escapes to the west around the southern tip of Africa to enter the South Atlantic, where it joins the Gulf Stream and completes the entire circuit.

F, G The upwelled water from the Pacific moves through the Indonesian Seas (F) and into the Indian Ocean, where it joins with the upwelling water there.

D, E Eventually, the deep waters circulating the Antarctic are carried into the Indian and southern Pacific Oceans. As these waters flow into the other basins, they force less-dense water to upwell toward the surface.

circulation toward the Atlantic. The upwelled water in the Pacific flows past Indonesia, where it joins the upwelled waters in the Indian Ocean. From there, some of the upwelled water flows south again and joins the circumpolar current around Antarctica. Some, though, flows westward and northward around the tip of Africa and toward South America, where it crosses the Equator and joins the Gulf Stream. From there, it continues to flow north as the North Atlantic drift, replacing the sinking water in the high-latitude north Atlantic.

Although we have depicted the circulation as a continuous current, like the Gulf Stream this circulation represents the average movement of many different water parcels over a long period of time. In fact, if one hypothetical water parcel were to complete the entire journey, it would take 1000–1500 years!

The thermohaline circulation plays a vital role in the regional and global climate of the Earth. On a global scale, it affects the carbon cycle by moving CO_2-rich surface waters into the ocean depths. The deep ocean circulation provides a conveyor belt for storage and release of CO_2 in a cycle of about 1500 years' duration. This allows the ocean to moderate rapid changes in atmospheric CO_2 concentration, such as those produced by human activity through fossil fuel burning, and thus moderate the global climate changes that accompany rising CO_2 concentrations.

On a regional scale, the thermohaline circulation, through the North Atlantic drift, supplies warm waters to the high latitudes of the North Atlantic, significantly warming Europe compared to other locations at the same latitude.

Some scientists have observed that the thermohaline circulation could be slowed or stopped by inputs of fresh water into the North Atlantic. Such freshwater inputs could have come from the sudden drainage of large lakes formed by melting ice at the close of the last glacial period. The input of the fresh water would decrease the density of the ocean water, keeping the water from becoming dense enough to sink. Without this sinking, the thermohaline circulation would slow down. In turn, the warm waters of the Gulf Stream would not flow as far north, and the Gulf Stream would become more like the Kuroshio current in the Pacific. This shift would interrupt a major flow pathway for the transfer of heat from equatorial regions to the northern midlatitudes. This mechanism could result in relatively rapid climatic change and is one explanation for the periodic cycles of warm and cold temperatures experienced in these regions since the melting of continental ice sheets about 12,000 years ago.

CONCEPT CHECK STOP

What causes gyre circulations in the ocean to form?

What is a western boundary current?

Where do they form?

How do ocean currents in the northern hemisphere differ from those in the southern hemisphere?

How do the currents in the Atlantic differ from those in the Pacific?

What is the thermohaline circulation?

How does water in the thermohaline circulation traverse the globe?

Heat and Moisture Transport

LEARNING OBJECTIVES

Explain how the atmosphere redistributes heat and moisture from low latitudes to high latitudes.

Discuss the role the oceans play in redistributing heat to the high latitudes.

The general circulation of the atmosphere and oceans is driven by the uneven heating between high and low latitudes. At the same time, this general circulation, combined with jet stream disturbances in the midlatitudes, serves to redistribute excess heat—and moisture—from the Equator to the poles. FIGURE 7.21 on page 210, shows the various mechanisms by which this heat and moisture redistribution takes place.

An important feature of this redistribution is the Hadley cell circulation—a global convection loop in which moist air converges and rises in the intertropical convergence zone (ITCZ) while subsiding and diverging in the subtropical high-pressure belts.

The Hadley cell convection loop acts to pump heat from warm equatorial oceans poleward to the subtropical zone. Near the surface, air flowing toward the ITCZ picks up water vapor evaporated by sunlight from warm ocean surfaces, increasing the moisture and latent heat content of the air. With convergence and uplift of the air at the ITCZ, the latent heat is converted to sensible heat as condensation occurs. Air traveling poleward in the return circulation aloft retains much of this heat, although some is lost to space by radiant cooling. When the air descends in the subtropical high-pressure belts, the sensible heat becomes available at the surface. The net effect is to gather heat from tropical and equatorial zones and release the heat in the subtropical zone, where it can be conveyed farther poleward by jet stream disturbances into the midlatitudes.

In the mid- and high latitudes, poleward heat transport is produced almost exclusively by these disturbances in the jet stream. Because the jet streams flow west to east, there can be no direct poleward transport of heat as in the Hadley cells. Instead, lobes of cold, dry polar or arctic air associated with growing jet stream disturbances plunge toward the Equator, while tongues of warmer, moister air originating in the subtropics flow toward the poles. Within these disturbances, cyclones also develop along the polar front, producing convergence of air at the surface. The subsequent uplift releases latent heat by condensation. Both of these processes provide a heat flow that warms the mid- and higher latitudes well beyond the insolation they receive.

Just as atmospheric circulation plays a role in moving heat from one region of the globe to another, so do oceanic circulations. How is this heat transported? Near the surface, the water in the low latitudes is exposed to intense insolation in the equatorial and tropical zones and is subsequently warmed. It is then transported by the western boundary currents, which bring warm, tropical waters to higher latitudes. As the water travels

northward, it cools, losing heat and warming the atmosphere above it. The cooler waters then return south as part of the gyre circulation, where they cool the overlying atmosphere in the tropics.

Heat is also transported by the thermohaline circulation. By carrying warm surface water poleward, this loop acts like a heat pump in which sensible heat is acquired in tropical and equatorial regions and is moved northward into the North Atlantic, where it is transferred to the air. Since wind patterns move air eastward at higher latitudes, this heat ultimately warms Europe. The amount of heat released is quite large—a recent calculation shows that it is equal to about 35 percent of the total solar energy received by the Atlantic Ocean north of 40° latitude! This type of circulation does not occur in the Pacific or Indian Oceans. Hence, the high-latitude regions of the North Pacific and even eastern North America are significantly colder than Europe at the same latitude.

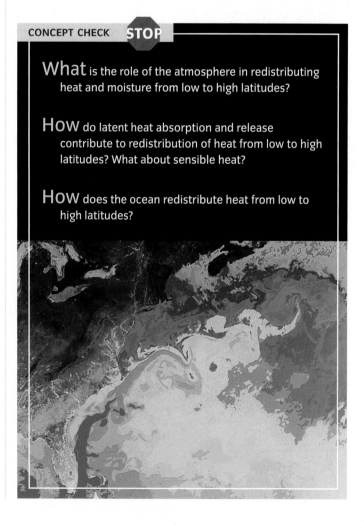

CONCEPT CHECK STOP

What is the role of the atmosphere in redistributing heat and moisture from low to high latitudes?

How do latent heat absorption and release contribute to redistribution of heat from low to high latitudes? What about sensible heat?

How does the ocean redistribute heat from low to high latitudes?

Global atmospheric transport of heat and moisture

Major mechanisms of heat and moisture transport in the atmosphere include intertropical convergence, the Hadley cell circulation, and jet stream disturbances in the midlatitudes.

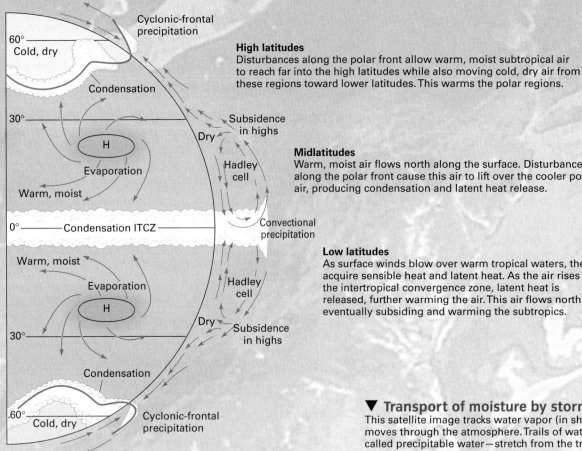

High latitudes
Disturbances along the polar front allow warm, moist subtropical air to reach far into the high latitudes while also moving cold, dry air from these regions toward lower latitudes. This warms the polar regions.

Midlatitudes
Warm, moist air flows north along the surface. Disturbances along the polar front cause this air to lift over the cooler polar air, producing condensation and latent heat release.

Low latitudes
As surface winds blow over warm tropical waters, they acquire sensible heat and latent heat. As the air rises in the intertropical convergence zone, latent heat is released, further warming the air. This air flows north, eventually subsiding and warming the subtropics.

▼ Global atmospheric transport of moisture
Water vapor is carried far from its source by atmospheric circulation patterns, including Hadley cell circulations, tropical cyclones, and midlatitude storms.

▼ Transport of moisture by storm systems
This satellite image tracks water vapor (in shades of white) as it moves through the atmosphere. Trails of water vapor—also called precipitable water—stretch from the tropical Pacific across the eastern United States and into the high latitudes of the Atlantic, marking the poleward transport of water and energy. The release of latent heat as this water vapor condenses drives tropical cyclones, like the one seen over the Gulf of Mexico, and midlatitude cyclones, pictured here as swirling bands of white over the North Atlantic and North Pacific.

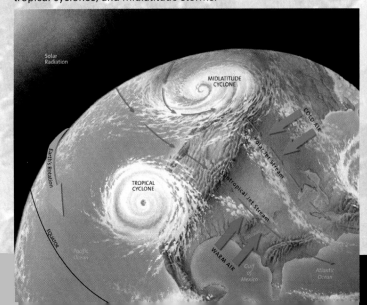

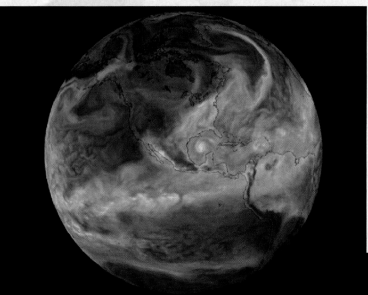

Excess heat is transported poleward by the general circulation of the atmosphere and ocean. The atmospheric circulations also transport water vapor from the moist low-latitudes to the drier high-latitudes.

Oceanic heat transport

The three ocean basins all contribute to the poleward transport of heat from the Equator. The total north–south heat transport by the oceans is nearly equivalent to the heat transport in the atmosphere.

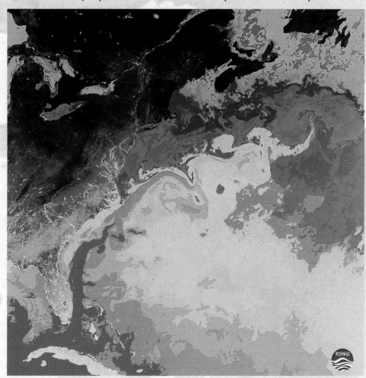

◄ Gulf Stream

A satellite image shows sea-surface temperature for a week in April from data acquired by the NOAA-7 orbiting satellite. Cold water appears in green and blue tones; warm water is in red and yellow tones. The narrow tongue of warm water—the Gulf Stream—transports large quantities of heat from the subtropics into the subarctic and arctic regions.

Gulf Stream

Warm less-salty shallow current

Cold salty deep current

◄ Thermohaline circulation

This three-dimensional figure shows the "conveyor belt" of deep ocean currents. Warm surface waters in the tropics move poleward, losing heat to the atmosphere on route. The cooler waters sink at higher latitudes, flow equatorward, and eventually upwell to the surface in the tropics, cooling these regions significantly.

What is happening in this picture ?

ANTOFAGASTA, CHILE

Global Locator

Considered one of the driest places on the Earth, some parts of the Atacama Desert in Chile have never had recorded rainfall. In this photo, we see a vast dry lake bed, the Salar de Atacama, encrusted with salt—a remnant of an earlier, wetter climate period. The lake bed is located at 23° S latitude between the high Andes Mountains to the east and Pacific coastal ranges to the west.

- Why do you think this region is so much drier than other regions?

- What effects do latitude, topography, and the nearby ocean have?

SUMMARY

1 Surface Winds

1. **Hadley cells** develop because the equatorial and tropical regions are heated more intensely than the higher latitudes, causing thermal circulations. These loops drive the northeast and southeast **trade winds**, the convergence and lifting of air at the **intertropical convergence zone (ITCZ)**, and the sinking and divergence of air in the **subtropical high-pressure belts**.

2. Surface air moving poleward from the subtropical highs forms the midlatitude westerlies. The **polar front** forms where this warm subtropical air meets the cold, dry air moving equatorward out of the polar highs.

3. The ideal circulation over the Earth is modified by the seasonal changes in land–sea temperature contrasts. These contrasts produce **monsoon** circulations over low-latitude continents. In addition, they produce seasonal changes in the midlatitude circulations over the northern hemisphere continents and oceans.

2 Winds Aloft

1. Winds in the atmosphere are dominated by a global pressure gradient force between the tropics and pole in each hemisphere.

2. The global pressure gradient force and the Coriolis force generate strong westerly geostrophic winds in the upper atmosphere called **jet streams**. These jet streams are concentrated westerly wind streams with high wind speeds located near the tropopause.

3. Disturbances in the jet stream develop in the upper-air westerlies, bringing cold, polar air equatorward and warmer, subtropical air poleward.

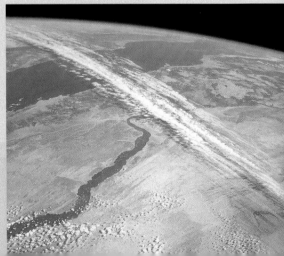

3 Oceanic Circulation

1. Oceans show a warm surface **mixed layer**, a **thermocline**, and a deep cold layer. Near the poles, the warm layer and thermocline are absent.

2. Ocean surface currents are dominated by huge **gyres** that are driven by the global surface wind pattern. Equatorial currents move warm water westward and then poleward along the east coasts of continents. Return flows bring cold water equatorward along the west coasts of continents.

3. Slow, **deep ocean currents** are driven by the sinking of cold, salty water in the northern and southern Atlantic. This **thermohaline circulation** pattern involves nearly all the Earth's ocean basins. It affects regional climates by warming nearby coastal locations. It also acts to moderate the buildup of atmospheric CO_2 by moving CO_2-rich surface waters to ocean depths.

4 Heat and Moisture Transport

1. The general circulation of the atmosphere transports both heat and moisture from low-latitude regions to high-latitude regions. In the low latitudes, this meridional transport is accomplished by the Hadley cell circulation. Across the midlatitudes, the jet stream prevents direct transport because it runs east–west, not north–south. Here, meridional transport is accomplished by the constant growth and decay of disturbances in the jet stream.

2. The ocean circulations also help transport heat from low latitudes to high latitudes. The warm, **western boundary currents** bring warm, tropical waters poleward, while circulation of the gyres brings colder water from the poles toward the equator. For all ocean basins combined, the oceans transport about as much heat poleward as the atmosphere.

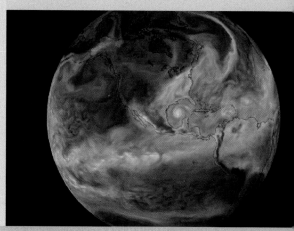

KEY TERMS

CRITICAL AND CREATIVE THINKING QUESTIONS

1. Sketch on an ideal Earth (without seasons or ocean and continent features) its global wind system and pressure patterns. Label the following on your sketch: doldrums, equatorial trough, Hadley cell, ITCZ, northeast trades, polar easterlies, polar front, polar outbreak, southeast trades, subtropical high-pressure belts, and westerlies.

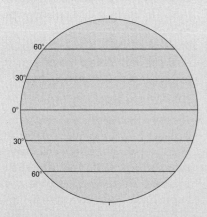

2. What is the Asian monsoon? What are its features in summer and winter? How is the ITCZ involved? How is the monsoon circulation related to seasonal high- and low-pressure centers in Asia?

3. An airline pilot is planning a nonstop flight from Los Angeles, California, to Sydney, Australia. What general wind conditions can the pilot expect to find in the upper atmosphere as the airplane travels? What jet streams will be encountered? Will they slow or speed the aircraft on its way?

4. For an ocean basin in the northern hemisphere, sketch the general pattern of ocean surface current circulation. Now sketch the overlying surface wind patterns. Describe how they are related.

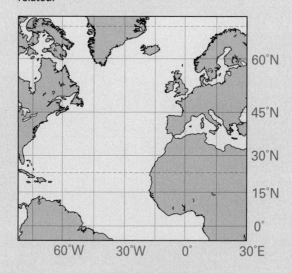

5. What process drives the thermohaline circulation? Where does this process occur? Describe the ocean bottom currents associated with this thermohaline circulation.

6. How does the atmosphere transport heat and moisture from low latitudes to high latitudes? How does the ocean also help transport heat poleward?

SELF-TEST

1. The standard sea-level atmospheric pressure in millibars (mb) is _____.

  a. 1000.0   c. 1130.0
  b. 1013.2   d. 2013.2

2. Upper-air disturbances and midlatitude storms form near the _____.

  a. polar fronts
  b. ITCZ
  c. subtropical high-pressure centers
  d. polar highs

3. Strong wind convergence in association with the ITCZ generally occurs at _____ latitude.

  a. 0°   c. 45°
  b. 30°   d. 60°

4. In the Hadley cell, air rises at the ITCZ and descends in the _____.

  a. polar high-pressure cells
  b. subpolar low-pressure cells
  c. subtropical high-pressure cells
  d. polar low-pressure cells

5. Hadley cells are driven by heat from the Sun.

  a. True   b. False

6. Most of the world's great deserts are located near the _____.

  a. polar fronts
  b. ITCZ
  c. subtropical high-pressure centers
  d. polar highs

7. In monsoon regions, average conditions tend to be
_____ during the summer and _____ during the
winter.
 a. dry and hot; wet and cool
 b. dry and cool; wet and hot
 c. wet and cool; dry and hot
 d. wet and hot; dry and cool

8. On the following diagram, draw the surface winds and isobars
associated with the monsoon during the winter and summer
seasons.

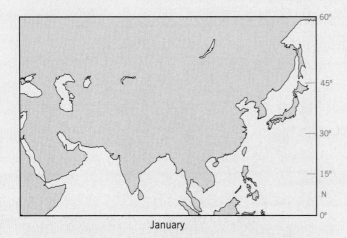

January

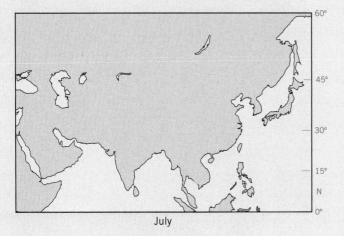

July

9. During the winter in the midlatitudes, we expect to find
_____ surface pressures over the land and _____
surface pressures over the ocean. During the summer, we expect this pressure pattern to _____.
 a. low; high; remain the same
 b. high; low; remain the same
 c. low; high; switch
 d. high; low; switch

10. Which of the following is *not* true about a jet stream?
 a. It is a narrow band of high velocity winds.
 b. It is due to the temperature difference between oceans and
 land.
 c. It is characterized by westerly winds in the subtropics and
 polar regions.
 d. It is located near the tropopause.

11. _____ mark the boundary between cold polar air
and warm tropical air.
 a. Hadley waves   c. Jet stream disturbances
 b. Ferrel waves   d. Polar fronts

12. The diagram shows the formation of jet stream disturbances.
Label (a) the jet axis, (b) the regions of high pressure, (c) the regions of low pressure, (d) the regions of warm air, and (e) the regions of cold air.

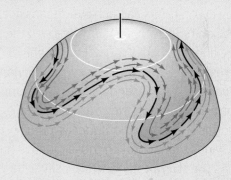

13. Disturbances in the polar jet over the midlatitudes help transport cold air to the _____ and warm air to the _____.
 a. high latitudes; low latitudes
 b. high longitudes; low longitudes
 c. low latitudes; high latitudes
 d. low longitudes; high longitudes

14. In the ocean, the boundary between warmer surface water and
colder subsurface water is marked by the _____.
 a. thermohaline   c. thermogyre
 b. thermocline   d. chemocline

15. Global surface ocean currents are dominated by huge, wind-
driven circular _____ centered near the subtropical high-
pressure cells.
 a. gyres   c. circulations
 b. currents   d. gyrocurrents

Midlatitude Weather Systems

They call it Jaws. This surfing location, situated along a reef barrier a half mile from the north shore of Maui, is renowned for having some of the largest surfable waves on the planet. Waves here can reach 15–23 m (50–75 ft) in height and travel at 48 km/hr (30 mph) over the sharp coral reef beneath the surface. Only recently, however, has anyone been able to ride them. Even then, given the waves' high speed and huge size, surfers can only catch them by getting towed behind a jet ski first. Once on a wave, however, surfers can achieve speeds in excess of 73 km/hr (45 mph) as they descend the wave face ahead of the plunging crest behind.

Where do these waves come from? They're actually generated thousands of miles away, by giant storms situated off the coast of Alaska. These storms—called *midlatitude cyclones*—can produce winds that approach those found in hurricanes. As the storms stall and intensify within the region of the Aleutian low, the persistent winds associated with the storms generate larger and larger waves. These waves then travel across the broad Pacific with little to inhibit their progress. Eventually, they come upon the abrupt rise of the volcanic islands of Hawaii. Here, they grow to become 15 m (50 ft) high monsters before plunging against the coral reef below.

Here, a half mile off the north shore of Maui, giant waves generated by storms near Alaska come ashore.

NATIONAL GEOGRAPHIC

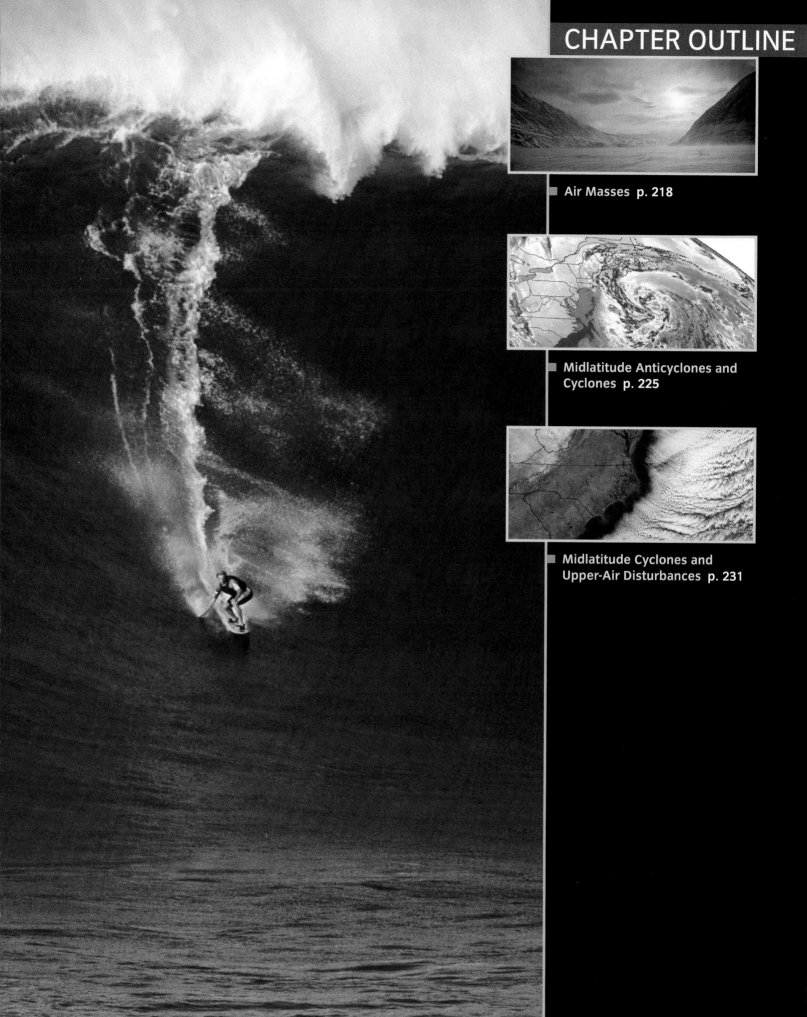

Air Masses

LEARNING OBJECTIVES

Define air mass.

Explain how air masses are classified.

Describe cold, warm, and occluded fronts.

We know that the Earth's atmosphere is in constant motion, driven by the planet's rotation and its uneven heating by the Sun. The horizontal motion of the wind moves air from one place to another, allowing air to acquire characteristics of temperature and humidity in one region and then carry those characteristics into another region. In addition, as winds at the surface converge and diverge, they produce vertical motions that affect clouds and precipitation. When air is lifted, it is cooled, enabling clouds and precipitation to form. When air descends, it is warmed, inhibiting the formation of clouds and precipitation. In this way, the Earth's wind systems influence the weather we experience from day to day—the temperature and humidity of the air, cloudiness, and the amount of precipitation.

Some patterns of wind circulation occur commonly and so present recurring patterns of weather. For example, traveling low-pressure centers (cyclones) of converging, inspiraling air often bring warm, moist air in contact with cooler, drier air, with clouds and precipitation as the result. We recognize these recurring circulation patterns and their associated weather as **weather systems**.

weather system
Recurring pattern of atmospheric circulation associated with characteristic weather, such as a cyclone or anticyclone.

Tropical oceans
An air mass with warm temperatures and high water vapor content develops over a warm equatorial ocean.

ALDABRA ISLANDS

Global Locator

Source regions FIGURE 8.1

Air masses form when large bodies of air acquire the temperature and moisture characteristics of the underlying surface conditions.

Subtropical deserts
Over a large subtropical desert, slowly subsiding air forms a hot air mass with low humidity.

SIMPSON DESERT

Global Locator

Arctic land masses
A very cold air mass with low water vapor content is generated over cold, snow-covered land surfaces in the arctic zone in winter.

POINT LAKE

Global Locator

Global air masses and source regions

FIGURE 8.2

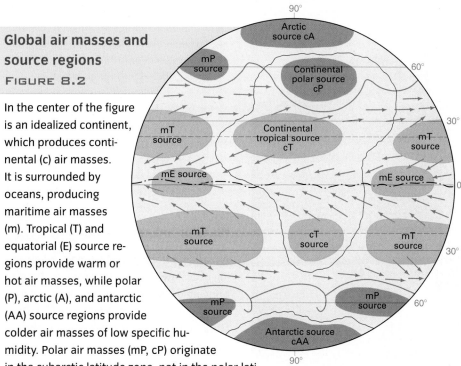

In the center of the figure is an idealized continent, which produces continental (c) air masses. It is surrounded by oceans, producing maritime air masses (m). Tropical (T) and equatorial (E) source regions provide warm or hot air masses, while polar (P), arctic (A), and antarctic (AA) source regions provide colder air masses of low specific humidity. Polar air masses (mP, cP) originate in the subarctic latitude zone, not in the polar latitude zone. Meteorologists use the word *polar* to describe air masses from the subarctic and subantarctic zones, and we will follow their usage when referring to air masses.

Source regions

Air mass	Symbol	Source region
Arctic	A	Arctic Ocean and fringing lands
Antarctic	AA	Antarctica
Polar	P	Continents and oceans, lat. 50–60° N and S
Tropical	T	Continents and oceans, lat. 20–35° N and S
Equatorial	E	Oceans close to equator

Surface types

Air mass	Symbol	Surface type
Maritime	m	Oceans
Continental	c	Continents

Weather systems range in size from a few kilometers, in the case of the tornado, to a thousand kilometers or more, in the case of a large traveling anticyclone. A system may last for hours or weeks, depending on its size and strength. Some forms of weather systems—tornadoes and hurricanes, for example—involve high winds and heavy rainfall and can be very destructive to life and property.

In the midlatitudes, weather systems are often associated with the motion of **air masses**—large bodies of air with fairly uniform temperature and moisture characteristics. An air mass can be several thousand kilometers or miles across and can extend upward to the top of the troposphere. We characterize each air mass by its surface temperature, environmental lapse rate, and surface specific humidity. Air masses can be searing hot, icy cold, or any temperature in between. Moisture content can also vary widely between different air masses.

Air masses acquire their characteristics in source regions. In a source region, air moves slowly or not at all, which allows the air to acquire temperature and mois-

air mass Extensive body of air in which temperature and moisture characteristics are fairly uniform over a large area.

ture characteristics from the region's surface (**FIGURE 8.1**). For example, a warm, moist air mass develops over warm equatorial oceans. In contrast, a hot, dry air mass forms over a large subtropical desert. Over cold, snow-covered land surfaces in the arctic zone in winter, a very cold air mass with very low water vapor content is found.

Pressure gradients and upper-level wind patterns drive air masses from one region to another. When an air mass moves to a new area, its properties change, because it is influenced by the new surface environment. For example, the air mass may lose heat or take up water vapor. However, these processes are fairly slow, and air masses can retain their initial temperature and moisture characteristics for weeks before they equilibrate to the new surrounding environment. This is one of the most important properties of air masses—they have the temperature and moisture characteristics of their original source regions, even as they move away from those regions.

Therefore, we classify air masses by the latitude and surface type of their source regions (**FIGURE 8.2**).

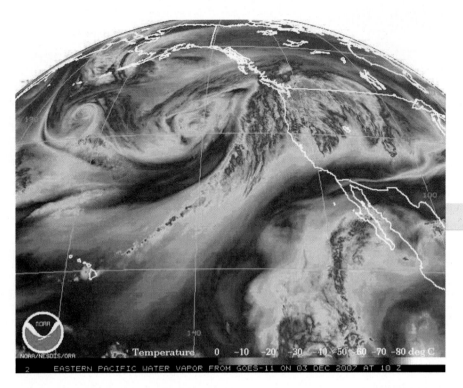

Temperature   0 -10 -20 -30 -40 -50 -60 -70 -80 deg C

2   EASTERN PACIFIC WATER VAPOR FROM GOES-11 ON 03 DEC 2007 AT 10 Z

Pineapple Express FIGURE 8.3

A maritime tropical (mT) air mass, originating over Hawaii, traveled across the Pacific, bringing moisture to feed precipitation over the western portion of North America. On Monday, December 3, 2007, this weather system caused heavy precipitation—indicated by yellow and red colors—and flooding in the Seattle area.

Their latitudinal position is important, because it determines the surface temperature and the environmental temperature lapse rate of the air mass. For example, air mass temperature can range from $-46°C$ $(-51°F)$ for arctic air masses to $27°C$ $(81°F)$ for equatorial air masses. The nature of the underlying surface—continent or ocean—usually determines the moisture content of an air mass given a latitudinal zone. Specific humidity of an air mass can range from 0.1 g/kg over the frozen ground of the arctic to as much as 19 g/kg over a warm ocean. In other words, maritime equatorial air can contain about 200 times as much moisture as continental arctic air.

Combining these two types of labels produces a list of six important types of air masses. The maritime tropical air mass (mT) and maritime equatorial air mass (mE) originate over warm oceans in the tropical and equatorial zones. They are quite similar in temperature and water vapor content. With their high values of specific humidity, both types can produce heavy precipitation. The continental tropical air mass (cT) has its

source region over subtropical deserts of the continents. Although this air mass may have a substantial water vapor content, it has very low relative humidity because of its high temperature and typically does not reach a point of saturation.

The maritime polar air mass (mP) originates over midlatitude oceans. It contains less water vapor than the maritime tropical air mass, so the mP air mass yields only moderate precipitation. The continental polar air mass (cP) originates over North America and Eurasia in the subarctic zone. It has low specific humidity and is very cold in winter. Last is the continental arctic (and continental antarctic) air mass type (cA, cAA), which is extremely cold and contains almost no water vapor.

The air masses that form near North America and their source regions have a strong influence on the weather. FIGURE 8.3 is an example of this—a "Pineapple Express" that came from Hawaii to cause floods in Washington and Oregon. FIGURE 8.4 shows the air masses and source regions that influence North American weather.

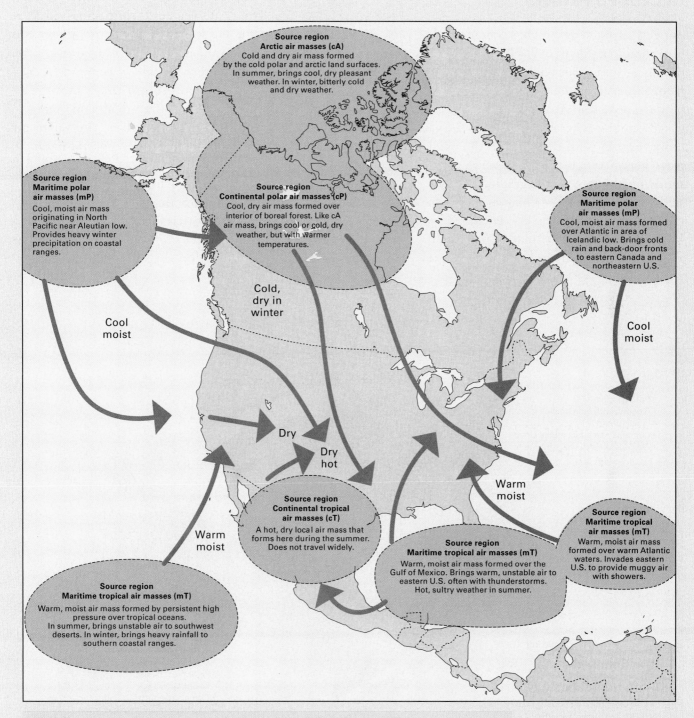

**Source region
Arctic air masses (cA)**
Cold and dry air mass formed
by the cold polar and arctic land surfaces.
In summer, brings cool, dry pleasant
weather. In winter, bitterly cold
and dry weather.

**Source region
Maritime polar
air masses (mP)**
Cool, moist air mass
originating in North
Pacific near Aleutian low.
Provides heavy winter
precipitation on coastal
ranges.

**Source region
Continental polar air masses (cP)**
Cool, dry air mass formed over
interior of boreal forest. Like cA
air mass, brings cool or cold, dry
weather, but with warmer
temperatures.

**Source region
Maritime polar
air masses (mP)**
Cool, moist air mass formed
over Atlantic in area of
Icelandic low. Brings cold
rain and back-door fronts
to eastern Canada and
northeastern U.S.

Cold,
dry in
winter

Cool
moist

Cool
moist

Dry

Dry
hot

Warm
moist

Warm
moist

**Source region
Continental tropical
air masses (cT)**
A hot, dry local air mass that
forms here during the summer.
Does not travel widely.

**Source region
Maritime tropical air masses (mT)**
Warm, moist air mass formed over the
Gulf of Mexico. Brings warm, unstable air to
eastern U.S. often with thunderstorms.
Hot, sultry weather in summer.

**Source region
Maritime tropical
air masses (mT)**
Warm, moist air mass
formed over warm Atlantic
waters. Invades eastern
U.S. to provide muggy air
with showers.

Warm
moist

**Source region
Maritime tropical air masses (mT)**
Warm, moist air mass formed by persistent high
pressure over tropical oceans.
In summer, brings unstable air to southwest
deserts. In winter, brings heavy rainfall to
southern coastal ranges.

North American air mass source regions and trajectories FIGURE 8.4

Air masses acquire temperature and moisture characteristics in their source regions, then
move across the North American continent.

COLD, WARM, AND OCCLUDED FRONTS

A given air mass usually has a sharply defined boundary between itself and a neighboring air mass. This boundary is termed a **front**. An example of a front is the contact between polar and tropical air masses associated with the general circulation of the atmosphere. We call this feature the *polar front*, across which we find a very strong south–north temperature gradient responsible for supporting the polar jet stream in the upper air.

In addition to this polar front, we can also have situations in which a cold air

front Surface of contact between two air masses with different temperature and moisture characteristics.

cold front Moving weather front along which a cold air mass moves underneath a warm air mass, lifting the warm air mass.

mass temporarily invades a zone occupied by a warm air mass during the passage of a weather system. The result is a **cold front**, shown in FIG-URE 8.5. Because the cold air mass is colder and therefore more dense than the warmer air mass, it remains in contact with the ground. As it moves forward, it forces the warmer air mass to rise above it. As it rises, the warm air mass cools adiabatically so that water vapor will start to condense and clouds will form. If the warm, moist air is unstable, severe convection may also develop. A cold front often forms a long line of massive cumulus clouds stretching for tens of kilometers.

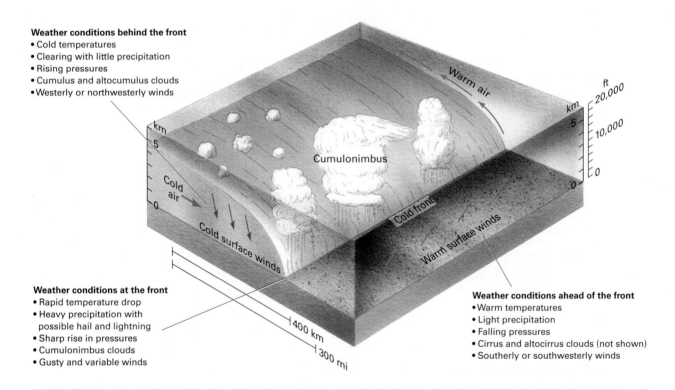

Weather conditions behind the front
- Cold temperatures
- Clearing with little precipitation
- Rising pressures
- Cumulus and altocumulus clouds
- Westerly or northwesterly winds

Weather conditions at the front
- Rapid temperature drop
- Heavy precipitation with possible hail and lightning
- Sharp rise in pressures
- Cumulonimbus clouds
- Gusty and variable winds

Weather conditions ahead of the front
- Warm temperatures
- Light precipitation
- Falling pressures
- Cirrus and altocirrus clouds (not shown)
- Southerly or southwesterly winds

Cold front FIGURE 8.5

In a cold front, a cold air mass lifts a warm air mass aloft. The upward motion can set off a line of showers or thunderstorms. The frontal boundary is actually much less steep than is shown in this schematic drawing.

In contrast to a cold front, a **warm front** is a front in which warm air moves into a region of colder air, shown in FIGURE 8.6. Again, the cold air mass remains in contact with the ground because it is more dense. As before, the warm air mass is forced aloft, but this time it rises up on a long ramp over the cold air below. This rising motion—called *overrunning*—creates stratus—large, dense, blanket-like clouds that often produce precipitation ahead of the warm front. If the warm air is stable, the precipitation will be steady. If the warm air is unstable, convection cells can develop, producing cumulonimbus clouds with heavy showers or thunderstorms.

Cold fronts normally move along the ground at a faster rate than warm fronts, because the cold dense air behind the cold front can more easily push through the warm, less dense air ahead of it. In contrast, the warm, less dense air behind a warm front simply rides up and over the cooler, more dense air ahead of the front—in fact, a warm front moves not because the warm air is pushing into the cooler air ahead of the warm front but because the cooler air ahead of the front is actually retreating. Thus, when a cold front and a warm front are in the same region, the cold front can eventually overtake the warm front. The result is an **occluded front**. ("Occluded" means closed or shut off.) The colder air of the fast-moving cold front remains next to the ground, forcing both the warm air and the less cold air ahead to rise over it,

> **warm front**
> Moving weather front along which a warm air mass slides over a cold air mass, leading to the production of stratiform clouds and precipitation.

> **occluded front**
> Weather front along which a moving cold front has overtaken a warm front, forcing the warm air mass aloft.

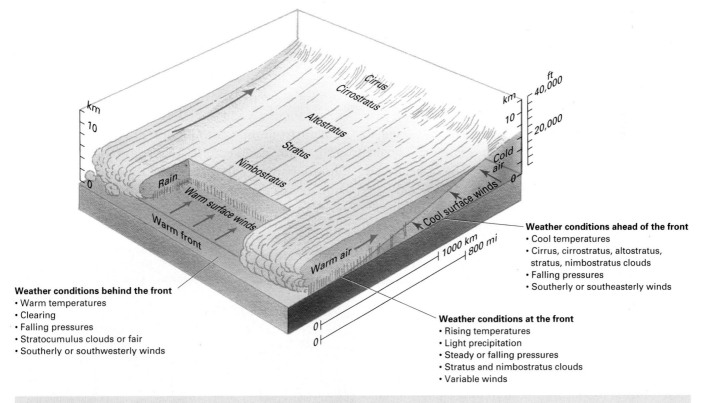

Weather conditions behind the front
• Warm temperatures
• Clearing
• Falling pressures
• Stratocumulus clouds or fair
• Southerly or southwesterly winds

Weather conditions at the front
• Rising temperatures
• Light precipitation
• Steady or falling pressures
• Stratus and nimbostratus clouds
• Variable winds

Weather conditions ahead of the front
• Cool temperatures
• Cirrus, cirrostratus, altostratus, stratus, nimbostratus clouds
• Falling pressures
• Southerly or southeasterly winds

Warm front FIGURE 8.6

In a warm front, warm air advances toward cold air and rides up and over the cold air, called *overrunning*. A notch of cloud is cut away to show rain falling from the dense stratus cloud layer.

Air Masses 223

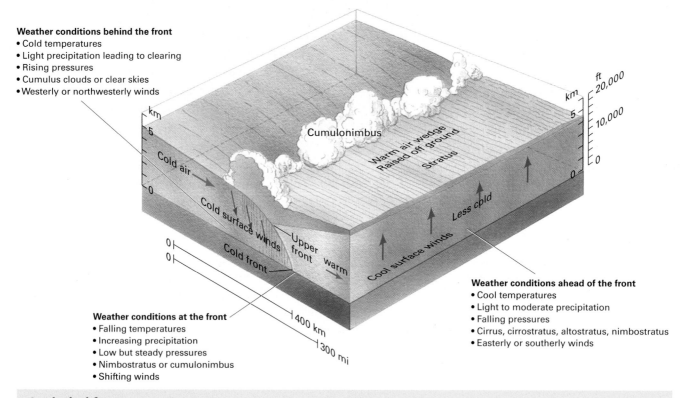

Weather conditions behind the front
- Cold temperatures
- Light precipitation leading to clearing
- Rising pressures
- Cumulus clouds or clear skies
- Westerly or northwesterly winds

Cumulonimbus

Warm air wedge
Raised off ground

Stratus

Cold air

Cold surface winds

Upper
front

Cold front

warm

Less cold

Cool surface winds

Weather conditions at the front
- Falling temperatures
- Increasing precipitation
- Low but steady pressures
- Nimbostratus or cumulonimbus
- Shifting winds

400 km

300 mi

Weather conditions ahead of the front
- Cool temperatures
- Light to moderate precipitation
- Falling pressures
- Cirrus, cirrostratus, altostratus, nimbostratus
- Easterly or southerly winds

Occluded front FIGURE 8.7

In an occluded front, a cold front overtakes a warm front. The warm air is pushed aloft, so that it no longer touches the ground. This abrupt lifting by the denser cold air produces precipitation.

as shown in FIGURE 8.7. The warm air mass is lifted completely free of the ground.

The example in Figure 8.7 is actually called a *cold occluded front* because the air mass behind the occluded front is colder than the air mass ahead of it and hence remains near the surface. Another type of occluded front—called a *warm occluded front*—occurs when the air mass behind the occluded front is warmer than the air mass ahead of it and hence rises up and over the air out ahead of the front. This type of occluded front is common along western portions of continents when cool, moist air from the ocean comes ashore and encounters very cold, dense air over the interior land masses.

A fourth type of front is known as the *stationary front*, in which two air masses are in contact but there is little or no relative motion between them. Stationary fronts often arise when a cold or warm front stalls and stops moving forward. Clouds and precipitation that were caused by earlier motion will often remain in the vicinity of the now-stationary front.

A final type of front—called a *dry line*—can form along the boundary between hot, dry, continental trop-

ical (cT) air and warm, moist, marine tropical (mT) air. Dry lines are usually found out ahead of cold fronts where southerly and southwesterly winds bring subtropical air from the continental regions and marine regions into contact with one another. Very strong thunderstorms can form along these dry lines as the hot, dry air mass mixes with the warm, moist air mass, increasing its temperature and making it very unstable.

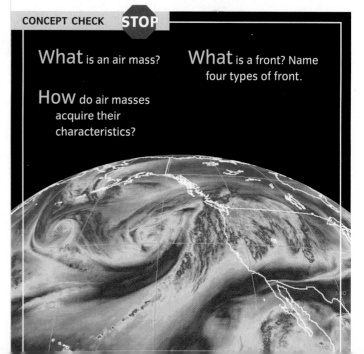

CONCEPT CHECK **STOP**

What is an air mass?

What is a front? Name four types of front.

How do air masses acquire their characteristics?

Midlatitude Anticyclones and Cyclones

A ir masses are set in motion by wind systems—typically, masses of air moving in a spiral. Air can spiral inward and converge in a cyclone, or spiral outward and diverge in an anticyclone. Most types of cyclones and anticyclones are large features spanning hundreds to thousands of kilometers that move slowly across the Earth's surface, bringing changes in the weather as they move. These are referred to as *traveling cyclones* and *anticyclones*.

ANTICYCLONES

Anticyclones are associated with fair skies, except for occasional puffy cumulus clouds that sometimes develop in a moist surface air layer. For this reason, we often call anticyclones *fair-weather systems*. Within a traveling anticyclone, the air is warmed adiabatically as it descends and diverges, so condensation does not occur. Toward the center of an anticyclone, the pressure gradient is weak, so winds are light and variable. We find traveling anticyclones in the midlatitudes, typically associated with ridges or domes of clear, dry air that move eastward and equatorward. *What a Scientist Sees: Surface Winds and Cloud Cover Associated with Midlatitude Cyclones and Anticyclones* shows an example of a large anticyclone centered over eastern North America, bringing fair weather and cloudless skies to the region.

Surface Winds and Cloud Cover Associated with Midlatitude Cyclones and Anticyclones

This geostationary satellite image shows a large clearing over the western portion of the United States. A scientist would identify the cloudless skies as a signature of an anticyclone in this region. The clockwise circulation of the surface wind barbs confirms that a large anticyclone is centered over eastern Colorado, with outspiraling winds stretching from Nevada to the Mississippi River and from Mexico to Canada. To the west, however, a strong cyclonic storm is raging off the coast of California. It can be identified by the inspiraling cloud formations and the strong, counterclockwise circulation of the surface wind barbs. Another cyclonic storm is situated over eastern Canada. Here, heavy cloud cover indicates convergence and ascent of air in the center of a low-pressure center. The counterclockwise circulation of the surface wind barbs confirms the location of a low-pressure center in this region.

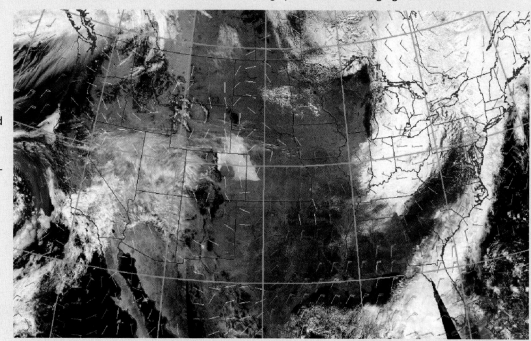

What a Scientist Sees

CYCLONES

In a cyclone, the air converges and rises, cooling adiabatically as it does so. If the cooling air reaches saturation, this can cause condensation leading to precipitation. Many cyclones are weak and pass overhead with little more than a period of cloud cover and light precipitation. However, some cyclones have very intense pressure gradients associated with them. In those cases, the large pressure gradients will generate strong, intense winds. In addition, the inspiraling motion associated with these winds can result in significant convergence, and heavy rain or snow can ac-

cyclonic storm Intense weather disturbance within a traveling cyclone, generating strong winds, cloudiness, and precipitation.

Visualizing

The Perfect Storm FIGURE 8.8

Cyclonic storms can be very dangerous. These satellite images show the development of the monster "perfect storm" of October 1991, which formed when a weakened tropical cyclone merged with a midlatitude cyclone. Generating 30-m (100-ft) waves and winds of 36 m/s (80 mph), it was perhaps the worst North Atlantic storm in a century.

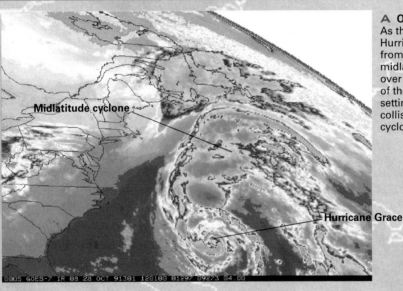

A October 28, 1991
As the remnants of Hurricane Grace approach from the south, a midlatitude cyclone moves over the eastern seaboard of the United States, setting the stage for a collision of the two cyclonic storms.

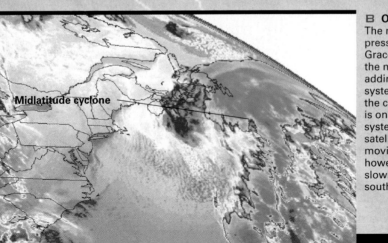

B October 29, 1991
The moisture and low pressures of Hurricane Grace are incorporated into the midlatitude cyclone, adding latent heat to the system and intensifying the circulation. Now there is only one large storm system visible in the satellite image. Instead of moving out to sea, however, it begins to slowly drift to the southwest.

company the cyclone. In that case, we call the disturbance a **cyclonic storm**, described in FIGURE 8.8.

There are three types of traveling cyclones. First is the *midlatitude cyclone* of the midlatitude, subarctic, and subantarctic zones, sometimes also called an *extratropical cyclone*. These cyclones range from weak disturbances to powerful storms. Second is the *tropical cyclone* found in the tropical and subtropical zones. Tropical cyclones range from mild disturbances to highly destructive hurricanes or typhoons. A third type is the *tornado*, a small, intense cyclone of enormously powerful winds. The tornado is much smaller in size than other cyclones and is related to strong, localized convective activity.

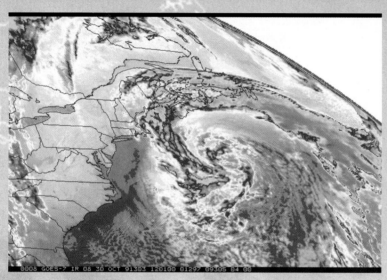

C October 30, 1991
The storm continues to intensify off the eastern seaboard of the United States. Winds generated by the storm reach hurricane strength. Devastating waves pound the shore from the Carolinas up to Nova Scotia. During this time, the storm continues to move southwest, back toward shore.

D November 1, 1991
As the storm moves back toward shore, it drifts over the warm, subtropical waters of the Gulf Stream. The added sensible and latent heat feed convection within the storm, lowering the central pressure even further. Eventually, a closed circulation forms around the central low pressure and an eye appears. A new, unnamed hurricane has formed. It eventually makes landfall in Nova Scotia on November 2.

MIDLATITUDE CYCLONES

The **midlatitude cyclone** is the dominant weather system in middle and high latitudes. It is a large

> **midlatitude cyclone** Traveling cyclone of the midlatitudes involving interaction of cold and warm air masses along sharply defined fronts.

inspiraling of air that repeatedly forms, intensifies, and dissolves along the polar front. The polar front sits between two large anticyclones—the subtropical high to the south and the polar high to the north, shown in the upper right of FIGURE 8.9. The air in the polar high comprises a cold, dry polar air mass, and the air in the subtropical high comprises a warm, moist maritime air mass. The airflow converges from opposite directions on the two sides of the front, with northeasterly winds to the north of the polar front and southwesterly winds to the south of the polar front.

These wind motions lead to a counterclockwise circulation, which we know is related to cyclonic circulations in the northern hemisphere. The result is a low-pressure trough created between the two high-pressure cells. The boundary between the two is unstable, and small disturbances in the boundary can grow over time. Midlatitude cyclones represent a local intensification of this surface low-pressure system caused by the growth of these disturbances. Figure 8.9 shows the life history of a midlatitude cyclone and associated warm and cold fronts and explains how a midlatitude cyclone forms, grows, and eventually dissolves.

How does weather change as a midlatitude cyclone passes through a region? As the midlatitude cyclone and its accompanying fronts move westward, a fixed location—like a point south of the Great Lakes—can experience weather ranging from warm, mild conditions with light winds to periods of heavy precipitation with gusty winds to cold, dry conditions with clear skies overhead. All of these can take place in a span of 24–36 hours as the fronts pass by. We can see these changes in FIGURE 8.10 on page 230.

Life history of a midlatitude cyclone FIGURE 8.9

Midlatitude cyclones are large features, spanning 1000 km (about 600 mi) or more. These are the "lows" that meteorologists show on weather maps. They typically last 3 to 6 days. In the midlatitudes, a cyclone normally moves eastward as it develops, propelled by prevailing westerly winds aloft. **A–D** show key stages in the life of a cyclone.

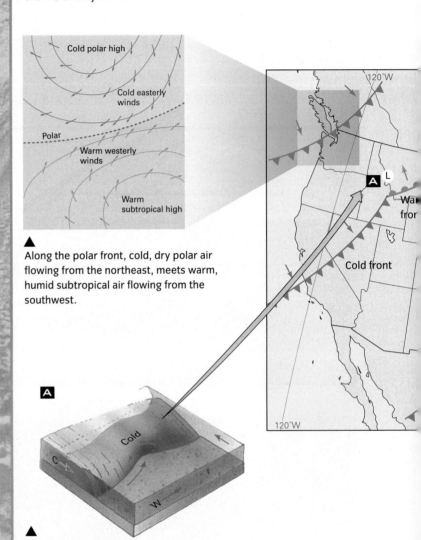

Along the polar front, cold, dry polar air flowing from the northeast, meets warm, humid subtropical air flowing from the southwest.

Early stage

An undulation or disturbance begins at a point along the polar front. Cold air is turned in a southerly direction and warm air in a northerly direction, so that each advances on the other. This creates two fronts—a cold front indicated by a line with blue triangles and a warm front indicated by a line with red half-circles. As the fronts begin to move, precipitation starts.

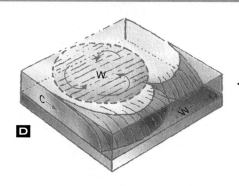

D

Dissolving stage
Eventually, the polar front is reestablished, but a pool of warm, moist air remains aloft and north of the polar front. As its moisture content reduces, precipitation dies out, and the clouds gradually dissolve. Soon, another midlatitude cyclone will form along the polar front and move across the continent.

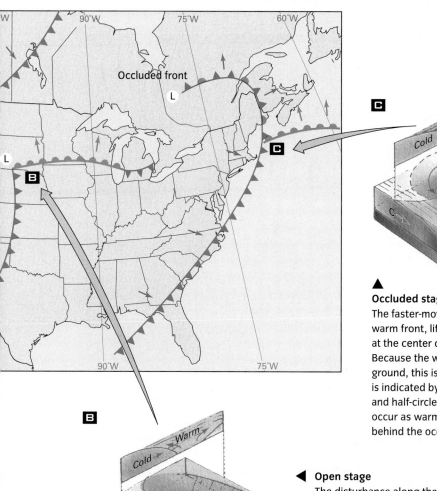

C

Occluded stage
The faster-moving cold front overtakes the warm front, lifting the warm, moist air mass at the center completely off the ground. Because the warm air is shut off from the ground, this is called an occluded front and is indicated by a line with alternating triangles and half-circles. Precipitation continues to occur as warm air is lifted ahead of and behind the occluded front.

B

Open stage
The disturbance along the cold and warm fronts deepens and intensifies. Cold air actively pushes southeastward along the cold front, and warm air actively moves northward along the warm front. Precipitation zones along the two fronts are now strongly developed. The precipitation zone along the warm front is wider than the zone along the cold front.

These maps show weather conditions on two successive days in the eastern United States. The three kinds of fronts are shown by special line symbols. Areas of precipitation are shown in gray. We can understand the movement of the respective air masses that generate the fronts by looking at the circulation around the surface low-pressure center, indicated by the isobars. The white line indicates the 0°C isotherm.

A Open stage

The isobars show a surface low-pressure center with inspiraling winds. The cold front is pushing south and east, supported by a flow of cold, dry continental polar air circulating around the low-pressure center. Note that the wind direction changes abruptly ahead of the cold front, shifting from southerly to northwesterly. The temperature behind the cold front drops sharply as cP air moves into the region. The warm front is moving north and somewhat east, with warm, moist maritime tropical air circulating around the low pressure as well. The precipitation pattern includes a broad zone near the warm front and the central area of the cyclone. A thin band of intense precipitation extends down the length of the cold front. Generally, there is cloudiness over much of the cyclone.

B Occluded stage

This map shows conditions 24 hours later. The cyclone track is shown by the red line. The center has moved about 1600 km (1000 mi) in 24 hours—a speed of just over 65 km (40 mi) per hour. In this time, the cold front has overtaken the warm front, forming an occluded front in the central part of the disturbance. A high-pressure area, or tongue of cold polar air, has moved into the area west and south of the cyclone, and the cold front has pushed far south and east. Within the cold-air tongue, the skies are clear. Winds shift from southeasterly to westerly as the occluded front passes. Now precipitation is found in a broad region across the occluded front and throughout the central area of the cyclone. Behind the occluded front, conditions are clear and cold.

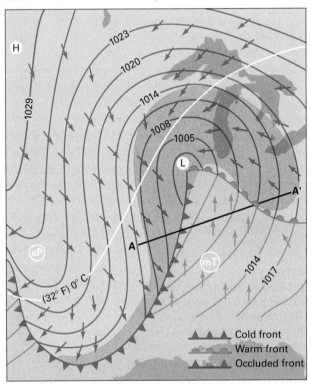

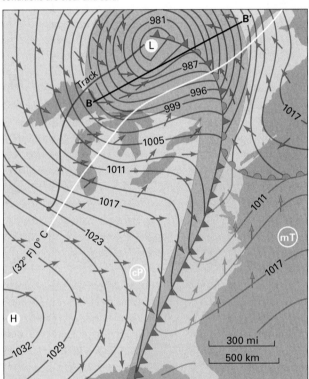

Cold front
Warm front
Occluded front

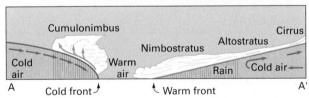

C Cross section of open stage

A cross section along the line A–A' shows how the fronts and clouds are related. A broad layer of stratus clouds ahead of the warm front takes the form of a wedge with a thin leading edge of cirrus. Westward, this wedge thickens to altostratus, then to stratus, and finally to nimbostratus with steady rain. Within the sector of warm air, the sky may partially clear with scattered cumulus. Along the cold front are cumulonimbus clouds associated with thunderstorms. These yield heavy rains but only along a narrow belt.

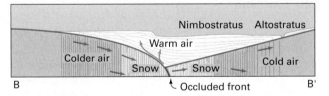

D Cross section of occluded stage

A cross section shows conditions along the line B–B', cutting through the occluded part of the storm. Note that the warm air mass is lifted well off the ground and yields heavy precipitation both ahead of and behind the occluded front.

CONCEPT CHECK STOP

What is a midlatitude cyclone? How does it develop?

What weather patterns accompany midlatitude cyclones and traveling anticyclones?

Midlatitude Cyclones and Upper-Air Disturbances

LEARNING OBJECTIVES

Explain how the development of midlatitude cyclones is linked to upper-air disturbances.

Describe processes that can generate cyclonic vorticity in the atmosphere leading to midlatitude cyclones.

Define storm tracks and families and their importance for the movement of midlatitude cyclones.

What causes midlatitude cyclones to grow over time? This growth is actually related to the growth of disturbances in the upper-air jet stream. In the upper atmosphere, the jet stream tends to develop ridges and troughs that intensify through the redistribution of warm and cold air. These disturbances produce variations in horizontal winds of the jet stream. In addition, there are also vertical circulations associated with them that affect the pressure patterns at the surface.

To understand this process, look at **FIGURE 8.11**. As the jet stream moves southward and eastward between the high-pressure disturbance—or *ridge*—and low-pressure disturbance—or *trough*—in the upper atmosphere, it tends to squeeze together or converge. As it does so, the winds around the low-pressure accelerate, forming a **jet streak**. On the eastward side of the low pressure, the winds slow down as they spread apart or diverge.

jet streak Localized regions of very high winds embedded within the overall jet stream.

Upper-air jet streak
FIGURE 8.11

This upper-air pressure and wind map shows a jet streak over the southwestern portion of the United States.

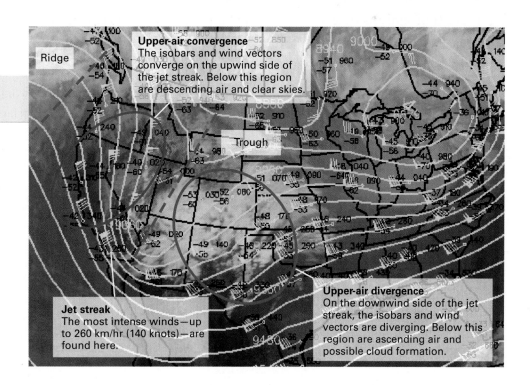

Upper-air convergence
The isobars and wind vectors converge on the upwind side of the jet streak. Below this region are descending air and clear skies.

Ridge

Trough

Jet streak
The most intense winds—up to 260 km/hr (140 knots)—are found here.

Upper-air divergence
On the downwind side of the jet streak, the isobars and wind vectors are diverging. Below this region are ascending air and possible cloud formation.

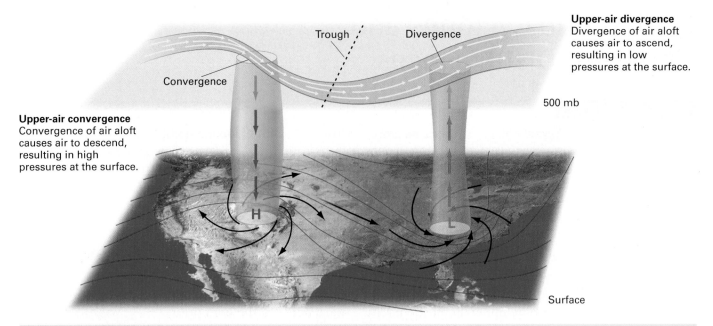

Upper-air divergence
Divergence of air aloft causes air to ascend, resulting in low pressures at the surface.

Trough

Divergence

Convergence

Upper-air convergence
Convergence of air aloft causes air to descend, resulting in high pressures at the surface.

500 mb

H

L

Surface

Upper-air and surface pressure patterns FIGURE 8.12

As air diverges and converges around an upper-air low-pressure trough, surface highs and lows are generated.

As the air aloft converges on the upwind side of the jet streak, it produces a descent of air toward the surface. The descending air subsequently produces an anticyclone (or high-pressure center) at the surface, as seen in FIGURE 8.12. Conversely, as the air diverges on the downwind side of the jet streak, air ascends from below, resulting in a cyclone (or low-pressure center) at the surface.

Thus, the life history of the low-level midlatitude cyclone and its accompanying fronts follows the life history of the upper-air disturbance, as shown in FIGURE 8.13. As the upper-air circulation disturbance intensifies and moves, the vertical circulations cause the midlatitude cyclones and anticyclones at the surface to intensify and move along with the disturbance. Eventually, the upper-air jet stream disturbance gets so large it causes the jet stream to circle back on itself, pinching off the upper-air low-pressure center and reestablishing the east–west flow of the jet stream. At the surface, we recognize this as the point at which the midlatitude cyclone occludes and re-establishes the polar front.

VORTICITY

Thus far, we have seen that the development of midlatitude cyclones at the surface is related to disturbances in the upper-air jet stream. As these disturbances grow over time, low-level pressure patterns and circulations intensify and develop into mature midlatitude cyclones and anticyclones. In fact, any atmospheric process that tends to enhance cyclonic circulations at the surface will tend to initiate the development of midlatitude cyclones, which can then intensify over time.

We measure the strength of cyclonic circulations based on their *vorticity*. Vorticity simply describes how fast something is spinning. If something is spinning counterclockwise, its vorticity is positive, and if it is spinning clockwise, its vorticity is negative.

Life history of an upper-air disturbance and accompanying midlatitude cyclone FIGURE 8.13

As an upper-air disturbance in the jet stream develops, vertical circulations in the regions of upper-air convergence and divergence produce changes in the surface pressures. As the upper-air disturbance moves and intensifies, so do the accompanying midlatitude cyclones and anticyclones.

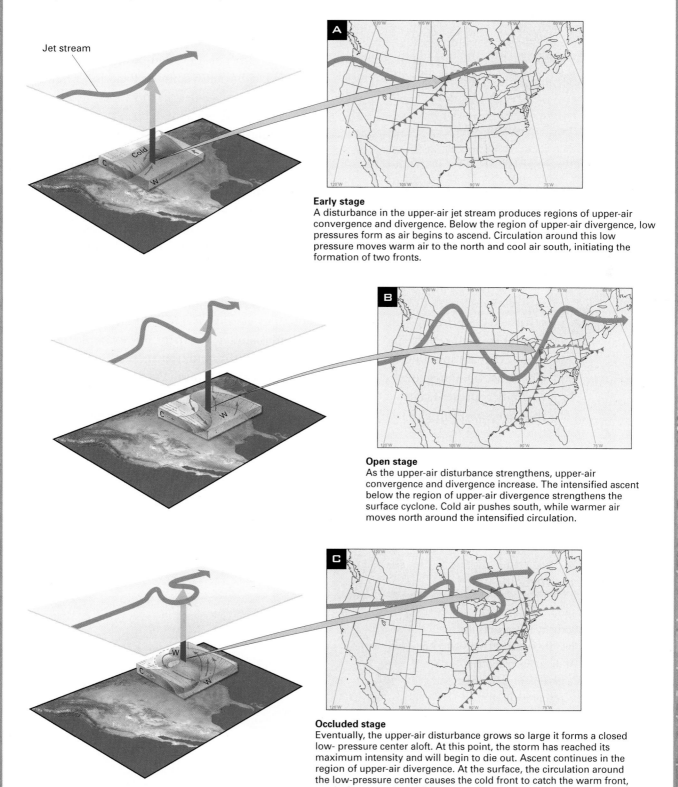

A

Early stage
A disturbance in the upper-air jet stream produces regions of upper-air convergence and divergence. Below the region of upper-air divergence, low pressures form as air begins to ascend. Circulation around this low pressure moves warm air to the north and cool air south, initiating the formation of two fronts.

B

Open stage
As the upper-air disturbance strengthens, upper-air convergence and divergence increase. The intensified ascent below the region of upper-air divergence strengthens the surface cyclone. Cold air pushes south, while warmer air moves north around the intensified circulation.

C

Occluded stage
Eventually, the upper-air disturbance grows so large it forms a closed low- pressure center aloft. At this point, the storm has reached its maximum intensity and will begin to die out. Ascent continues in the region of upper-air divergence. At the surface, the circulation around the low-pressure center causes the cold front to catch the warm front, producing an occluded front and a closed midlatitude cyclone.

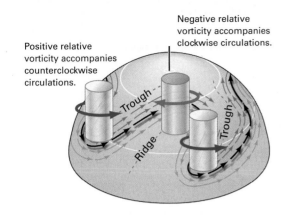

Positive relative vorticity accompanies counterclockwise circulations.

Negative relative vorticity accompanies clockwise circulations.

Relative vorticity FIGURE 8.14

Relative vorticity describes how fast and in what direction air is circulating.

There are three principal types of vorticity involved in atmospheric motions. The first is **relative vorticity**, which represents how fast and in what direction the at-mospheric circulations them-selves are spinning. Positive relative vorticity represents circu-lations that are spinning counter-clockwise, while negative relative vorticity represents circulations that are spinning clockwise, as

relative vorticity Measure of local rota-tion with respect to an observer on the Earth's surface.

seen in FIGURE 8.14. This relation holds whether one is north or south of the Equator, so cyclonic circu-lations have positive vorticity in the northern hemi-sphere but negative vorticity in the southern hemisphere.

A second type of vorticity is that related to the rota-tion of the Earth itself, called **planetary vorticity**. Be-cause the Earth rotates counterclockwise (when looking down on the North Pole), its mo-tion has a positive vorticity. This vorticity increases as one moves toward the pole, as seen in FIG-URE 8.15. Note that near the Equator, an individual rotates through a very small distance, while at the pole the same individual rotates through a much larger distance. Objects on the Earth, including air and water, acquire the same rotation rate as the Earth at the point they are located.

planetary vorticity Rotation imparted to an object by the rotation of the Earth.

More important, as an air parcel moves, it must maintain its *absolute vorticity*—the sum of its relative vor-ticity and planetary vorticity. For example, if a cyclone moves toward the Equator, its planetary vorticity de-creases. Since its absolute vorticity cannot change, its relative vorticity has to increase. In other words, the cy-clone "speeds up" as it moves equatorward.

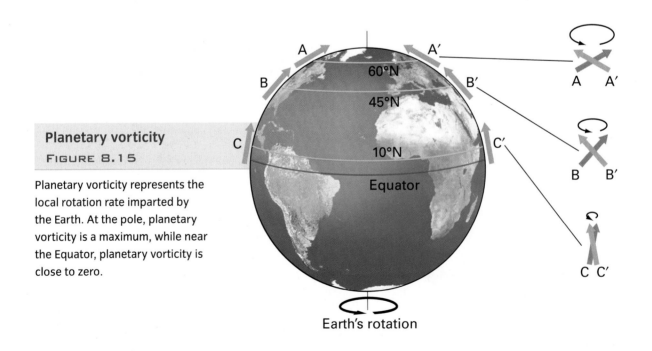

Planetary vorticity
FIGURE 8.15

Planetary vorticity represents the local rotation rate imparted by the Earth. At the pole, planetary vorticity is a maximum, while near the Equator, planetary vorticity is close to zero.

This dancer from San Miguel de Allende, Mexico, increases her rate of rotation by bringing her arms into her body and becoming narrower. In the same way, an air column's rotation, or vorticity, can increase by the column's stretching and becoming narrower.

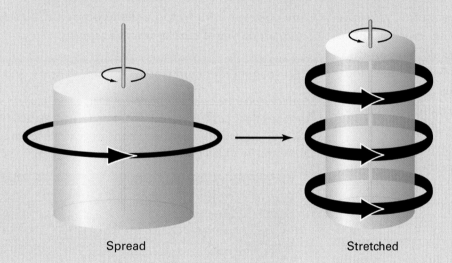

Spread   Stretched

There are other ways to generate relative vorticity. One involves the stretching and shrinking of circulating air columns. You may have seen figure skaters increase their rotation rate during spins by pulling their arms to their sides or holding them above their heads. Dancers do the same thing, as in FIGURE 8.16. By narrowing the area over which their body mass is spread, they can rotate faster. Similarly, a rotating air column that is elongated narrows and hence acquires positive vorticity.

Within the atmosphere, a similar process occurs. Diverging air in the exit region of jet streaks produces a vertical ascent of the air above the surface and an increase in the surface cyclonic circulation. The vertical motion narrows and elongates the air column, which adds to the relative vorticity of the midlatitude cyclone.

High-level divergence is only one way to generate midlatitude cyclones. Other processes can produce an elongation of the air column and hence can alter the vorticity of the atmosphere as well, leading to the formation of midlatitude cyclones. For example, as air flows over a mountain ridge, it tends to form low pressures on the lee side of the mountain as it descends, as

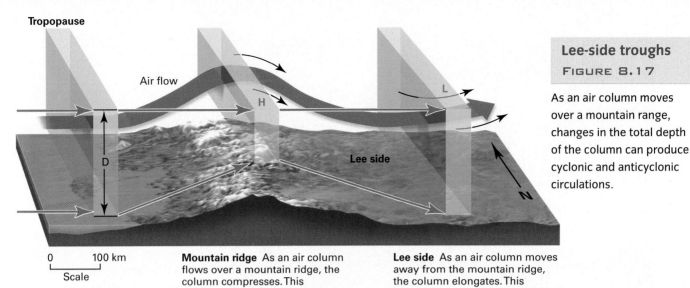

Tropopause

Air flow

D

Lee side

N

0   100 km
Scale

Mountain ridge As an air column flows over a mountain ridge, the column compresses. This generates anticyclonic vorticity with clockwise circulations and high pressures.

Lee side As an air column moves away from the mountain ridge, the column elongates. This generates cyclonic vorticity with counterclockwise circulations and low pressures.

shown in FIGURE 8.17. The descending air represents an elongation of the air column, much as the ascending air did in the previous example. This produces a **lee-side trough** on the downwind side of the mountain range, which can develop into a midlatitude cyclone or help reintensify a weakening midlatitude cyclone that passes over the mountain range.

Another way to generate vorticity is by a change in temperature from one location to another. For example, in winter, land surfaces in Asia and North America tend to be much colder than neighboring ocean regions. This is particularly true along the eastern seaboard of continents, where warm water currents offshore flow poleward from the subtropics. In these regions, warm moist air situated over the currents can become unstable, producing convection. This rising convective air can generate cyclonic vorticity in much the same way as the ascending air in the exit region of the jet streak. Hence, these regions are also breeding grounds for midlatitude cyclones and are called *storm-generation regions*. One such storm is shown in FIGURE 8.18. These storms can continue to move out over the Atlantic, where they intensify and eventually make landfall along western Europe. They can also continue to move up the eastern seaboard, where their strong cyclonic circulations bring cold ma-

rine polar air from Newfoundland down into New England, producing storms known as *nor'easters*.

Storm generation off North Carolina FIGURE 8.18

Surface isobars show the formation of a large midlatitude cyclone off the coast of North Carolina on January 22, 2007. Cyclonic circulation around the closed low generates both cold and warm fronts (blue and red lines).

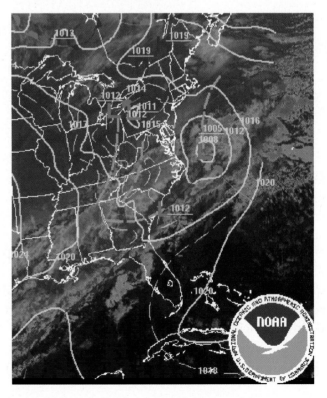

lee-side trough

Low-pressure region found on the downwind (lee) side of a mountain chain, which can subsequently generate midlatitude cyclones.

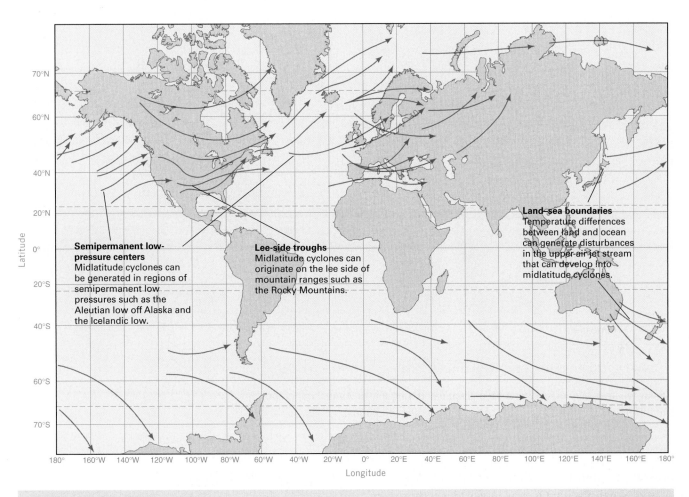

Paces annotations on map:

Semipermanent low-pressure centers
Midlatitude cyclones can be generated in regions of semipermanent low pressures such as the Aleutian low off Alaska and the Icelandic low.

Lee-side troughs
Midlatitude cyclones can originate on the lee side of mountain ranges such as the Rocky Mountains.

Land–sea boundaries
Temperature differences between land and ocean can generate disturbances in the upper-air jet stream that can develop into midlatitude cyclones.

Paths of midlatitude cyclones FIGURE 8.19

This world map shows typical paths of midlatitude cyclones (blue).

CYCLONE TRACKS AND CYCLONE FAMILIES

Many processes can generate the cyclonic circulations necessary for initiating traveling midlatitude cyclones, as seen in FIGURE 8.19. These surface pressure patterns are linked to the upper-air pressure patterns via vertical circulations, convergence/divergence, and vorticity. As the upper-air disturbances develop and grow, the midlatitude cyclones develop and grow as well. The link between the upper-air jet stream disturbances and the midlatitude cyclones and anticyclones at the surface also explains why these midlatitude cyclones move over time. As the upper-air disturbances move along the jet stream, their vertical circulations move with them. In turn, the associated surface pressure patterns also move with them. In this sense, the surface midlatitude cyclones are "dragged" along by the upper-air wind patterns. These upper-air winds are called **steering winds** because they tend to steer the midlatitude cyclones in a particular direction.

Because midlatitude cyclones tend to form in certain areas, they tend to travel common paths, called **storm tracks**, as they develop, mature, and dissolve. For example, in the northern hemisphere, midlatitude cyclones are heavily concentrated in the neighborhood of the Aleutian and Icelandic lows. These cyclones commonly form in

steering winds Upper-air wind patterns associated with the jet stream that tend to steer the direction in which midlatitude cyclones travel.

storm tracks Common paths that cyclonic storms tend to follow, usually associated with the location of the jet stream.

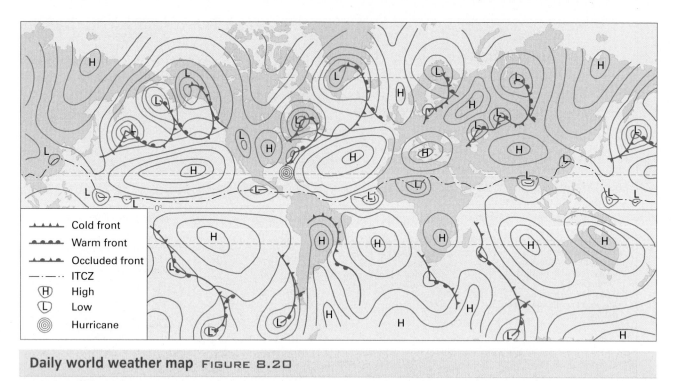

Daily world weather map FIGURE 8.20

A daily weather map of the world for a given day during July or August might look like this map, which is a composite of typical weather conditions.

Legend:
- ▲▲▲ Cold front
- ●●● Warm front
- ▲●▲● Occluded front
- —·—·— ITCZ
- Ⓗ High
- Ⓛ Low
- ◎ Hurricane

succession, traveling as a chain across the North Atlantic and North Pacific Oceans. A world weather map, such as in FIGURE 8.20, shows several such **cyclone families**. Each midlatitude cyclone moves northeastward along the storm track, deepening in low pressure and eventually occluding. For this reason, intense cyclones arriving at the western coasts of North America and Europe are usually occluded, while those arriving on the eastern seaboard after originating on the lee-side of the Rocky Mountains are still intensifying.

In the southern hemisphere, storm tracks are more nearly along a single lane, following the parallels of latitude. Three such cyclones are shown in Figure 8.21. This track is more uniform because of the uniform pattern of ocean surface circling the globe at these latitudes. Only the southern tip of South America projects southward to break the monotonous expanse of ocean.

> **cyclone families**
> Succession of cyclonic storms that follow one after the other along the same track.

COLD AIR OUTBREAKS

Another distinctive weather feature of midlatitude weather systems is the occasional penetration of powerful tongues of cold polar air from the midlatitudes into very low latitudes. These tongues are known as **cold air outbreaks**. The leading edge of a cold air outbreak is a cold front with squalls, which is followed by unusually cool, clear weather with strong, steady winds. The cold air outbreak is best developed in the Americas. Outbreaks that move southward from the United States into the Caribbean Sea and Central America are called *northers* or *nortes*, whereas those that move north from Patagonia into tropical South America are called *pamperos*. FIGURE 8.21 shows one such outbreak over North America. A severe polar outbreak may bring subfreezing temperatures to the low latitudes of both regions and damage tropical crops such as citrus and coffee.

> **cold air outbreaks**
> Equatorward movement of continental polar (cP) air masses into the low-latitude regions of the subtropics.

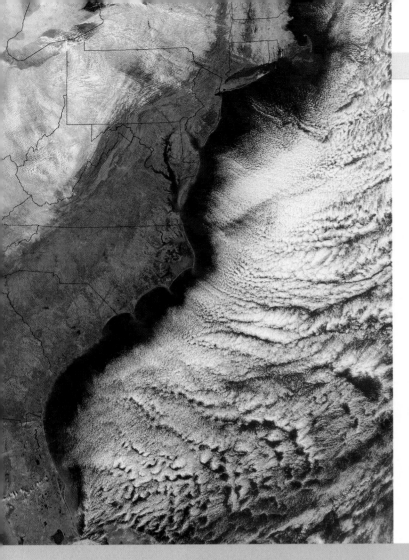

Cold air outbreak FIGURE 8.21

This image shows the eastern seaboard during a cold air outbreak on February 28, 2002. White regions over the central and northeastern portion of the United States are associated with snow cover. To the south, clear, cold conditions are found. This cold air outbreak brought subfreezing cP air from Canada down into Florida.

CONCEPT CHECK **STOP**

What role do upper-air convergence and divergence play in generating midlatitude cyclones?

HOW are the generation, development, and movement of a midlatitude cyclone linked to upper-air disturbances?

What other processes can generate midlatitude cyclones?

What is happening in this picture ?

The photograph shows a line of cumulus clouds advancing from left to right. The clouds formed when warm, moist air was pushed aloft.

Can you explain why the warm air was forced upward?

The clouds mark an advancing front. What type of front might this be?

SUMMARY

1 Air Masses

1. Air masses are distinguished by the latitudinal location and surface type of their source regions. Latitudinal location determines the temperature of the air mass, while the surface type (continental or marine) determines the amount of moisture in the air mass.

2. The United States is affected by the movement of four types of air masses: continental polar (cP), maritime polar (mP), continental tropical (cT), and maritime tropical (mT).

3. Fronts are the boundaries between air masses. They include **cold fronts**, where cold air masses are advancing into regions of warmer air, and **warm fronts**, where warm air masses are advancing into regions of cooler air. In the **occluded front**, a cold front overtakes a warm front, pushing a pool of warm, moist air above the surface.

2 Midlatitude Anticyclones and Cyclones

1. The traveling anticyclone is typically a midlatitude system in which divergence of air at the surface produces descending air from above. Warmed adiabatically, the descending air has a low relative humidity that inhibits condensation and precipitation. The anticyclone is thus a fair-weather system.

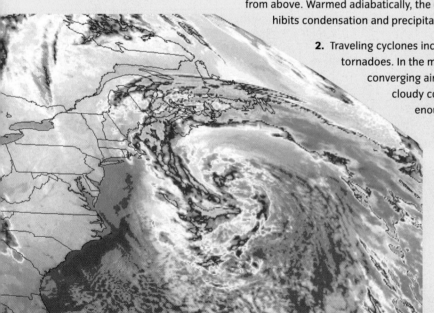

2. Traveling cyclones include **midlatitude cyclones**, tropical cyclones, and tornadoes. In the midlatitudes, traveling cyclones are associated with converging air near the surface, which produces ascent and cloudy conditions. On occasion, the convergence is strong enough to produce very heavy rainfall.

3. Midlatitude cyclones form at the boundary between cool, dry polar air masses and warm, moist tropical air masses. Circulations around midlatitude cyclones produce the movement of air masses, resulting in fronts. Distinct weather features are associated with the passage of midlatitude cyclones and the related fronts.

3 Midlatitude Cyclones and Upper-Air Disturbances

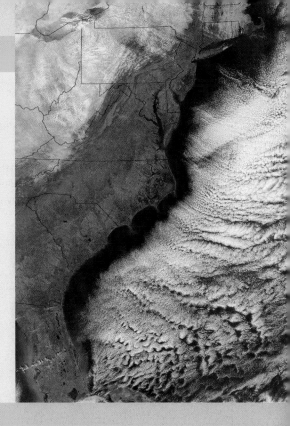

1. The development and movement of midlatitude cyclones are largely controlled by the development and movement of upper-air disturbances in the jet stream. Vertical motions associated with these disturbances generate high- and low-pressure centers at the surface that develop into anticyclones and cyclones.

2. In addition to upper-air disturbances, flow over mountain ranges and air–sea temperature differences can also generate the vertical motion necessary to produce cyclonic **vorticity**. Cyclonic vorticity can result in the generation of midlatitude cyclones as well.

3. Because vertical motions link upper-air disturbances with midlatitude cyclones at the surface, movement of upper-air disturbances with the jet stream produces movement of the midlatitude cyclones. Hence, midlatitude cyclones tend to follow **storm tracks** associated with the steering winds of the jet stream. A set of storms following the same track is called a **cyclone family**.

KEY TERMS

- **weather systems** p. 218
- **air mass** p. 219
- **front** p. 222
- **cold front** p. 222
- **warm front** p. 223
- **occluded front** p. 223

- **cyclonic storm** p. 226
- **midlatitude cyclone** p. 228
- **jet streak** p. 231
- **relative vorticity** p. 234
- **planetary vorticity** p. 234
- **lee-side trough** p. 236

- **steering winds** p. 237
- **storm tracks** p. 237
- **cyclone families** p. 238
- **cold air outbreaks** p. 238

CRITICAL AND CREATIVE THINKING QUESTIONS

1. What is an air mass? What two features are used to classify air masses? Compare the characteristics and source regions for mP and cT air mass types.

2. Identify three types of fronts, and draw a cross section through each. Show the air masses involved, the contacts between them, and the direction of air mass motion.

3. Sketch two weather maps, showing a midlatitude cyclone in open and occluded stages. Include isobars on your sketch. Identify the center of the cyclone as a low. Lightly shade areas where precipitation is likely to occur.

4. Explain how the development of midlatitude cyclones is linked to upper-air disturbances. Be sure to describe the vertical and horizontal winds in the upper atmosphere that produce these links as well as the winds and pressure patterns at the surface that result from these links.

5. What role does elongation of the air column play in producing cyclonic circulations? How is this related to the vorticity of the air column?

6. Identify key generation regions for storms that affect the United States. Are there different generation regions for the western portion of the United States than for the eastern United States? How do these storms move from the generation region to the areas they affect?

SELF-TEST

1. A(n) _____ is a large body of air with fairly uniform temperature and moisture characteristics.
- a. air source
- b. air resource
- c. climate zone
- d. air mass

2. The diagram shows the source regions that give global air masses their characteristics. Label the following sources: (a) cA, (b) cAA, (c) cP, (d) cT, (e) mE, (f) mP, and (g) mT.

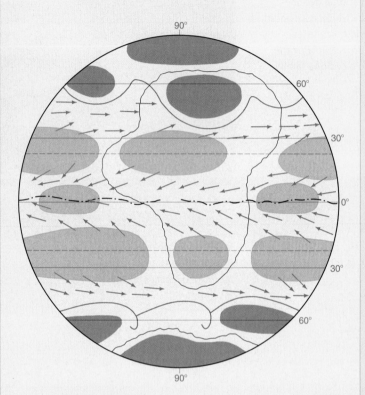

3. _____ air masses generally possess the lowest moisture content.
- a. Maritime tropical
- b. Continental polar
- c. Maritime tropical
- d. Continental tropical

4. An air mass that develops over equatorial waters would be expected to be warm and moist.
- a. True
- b. False

5. The _____ air mass generally does not influence North America.
- a. mE
- b. mP
- c. mT
- d. cP

6. A(n) _____ is a center of high pressure and is generally responsible for fair weather.
- a. anticyclone
- b. cyclone
- c. trade wind
- d. midlatitude storm front

7. A _____ forms when a cold air mass penetrates a warm air mass.
- a. warm front
- b. occluded front
- c. cold front
- d. stationary front

8. In _____, convergence and uplift typically cause condensation and precipitation, whereas descent in _____ causes the air to warm, producing clear conditions.
- a. cyclones; warm fronts
- b. anticyclones; cyclones
- c. anticyclones; warm fronts
- d. cyclones; anticyclones

9. A _____ forms between two high-pressure cells.
- a. ridge
- b. siphon of low pressure
- c. low-pressure trough
- d. high-pressure ridge

10. In middle and high latitudes, the dominant form of weather system is the _____, a large inspiral of air that repeatedly forms, intensifies, and dissolves along the polar front.
- a. warm front
- b. midlatitude anticylone
- c. midlatitude cyclone
- d. cold front

11. The diagram shows the progression of a midlatitude cyclone across the United States. At each stage, shade regions where you expect to find precipitation. Use arrows to indicate the wind direction in front of and behind the cold front and the warm front.

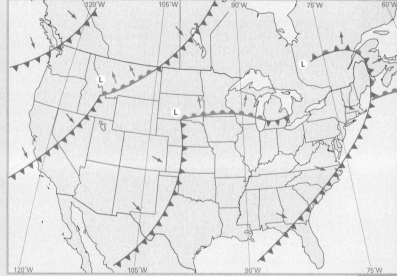

12. An occluded front occurs within a midlatitude cyclone
_____.

    a. on every occasion of warm front formation

    b. on every occasion of cold front formation

    c. whenever a cP air mass meets a mE air mass

    d. when a cold front has overtaken a warm front

13. Prevailing easterlies aloft propel midlatitude cyclones.

    a. True   b. False

14. The leading edge of a polar outbreak is _____.

    a. a warm front that develops into an occluded front

    b. clear weather

    c. clear weather followed by warm and occluded fronts

    d. a cold front with squalls

15. On the diagram, circle three regions in which storm generation occurs. Label the process that generates midlatitude cyclones in each of these regions.

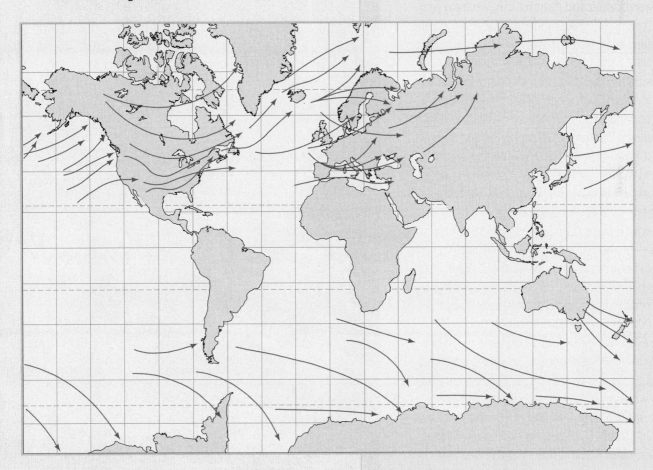

Tropical Weather Systems

As farmers were hurrying to finish harvesting rice across southeast Asia, the 2008 North Indian cyclone season—which runs from about May through November—was about to start with one of the deadliest storms on record. In the Bay of Bengal, east of India, on April 27, 2008, a low pressure system was intensifying. As the winds circling the central low pressure rose above 39 km/hr (63 mph), the storm was officially named Tropical Storm Nargis. One day later, the winds pushed past 119 km/hr (74 mph), and the storm was upgraded to cyclone (or hurricane) status. Initially it tracked northwestward towards India and Bangladesh, where a year earlier, Cyclone Sidr had made landfall, killing over 10,000 people. However, history would not repeat itself.

On April 29, Cyclone Nargis shifted course, moved northeast, and then almost directly eastward. It continued to intensify, with winds eventually reaching 215 km/hr (135 mph). Following its eastward track, the hurricane made landfall on May 2 and moved parallel to the low-lying coast of Mayanmar, wreaking havoc as it went. As Nargis slowly made its way along the coast, the on-shore winds—which were still in excess of 130 km/hr (80 mph)—pushed a surge of water 6 m (20 ft) high far inland, flooding an area the size of Connecticut. Heavy rain poured down. Entire villages were submerged or torn asunder. In all, it is estimated that over 80,000 people died and over 50,000 went missing. Even these numbers may be low—in one township alone 80,000 people may have perished.

In recent years, we have seen the extreme devastation wrought by tropical cyclones such as Cyclone Nargis in Mayanmar and Hurricane Katrina in the United States. Such events remind us that we are still very much at the mercy of the weather.

NATIONAL GEOGRAPHIC

A boy sits in a boat in Labutta, Myanmar, five days after Cyclone Nargis. Behind him are the remains of his village.

Tropical Cyclone Structure

Explain how easterly waves influence tropical weather.

Describe the characteristic structure of tropical cyclones.

Explain how measurements of tropical cyclones are made.

So far, we have discussed weather systems of the midlatitudes and poleward. Weather systems of the tropical and equatorial zones show some basic differences from those of the midlatitudes. Upper-air winds are often weak, so air-mass movement is slow and gradual. Air masses are warm and moist, and different air masses tend to have similar characteristics, so fronts are not as clearly defined. Without the strong temperature gradients across unlike air masses, there are no large, intense upper-air disturbances. On the other hand, the high moisture content leads to intense convective activity in low-latitude maritime air masses. Because these air masses are very moist, only slight convergence and uplift are needed to trigger precipitation.

EASTERLY WAVES

One of the simplest forms of tropical weather systems is an **easterly wave**—a slowly moving trough of low pressure within the belt of tropical easterlies (the trade winds). These waves occur in latitudes 5°–30° N and S over oceans, but not over the Equator itself. FIGURE 9.1 shows circulations and weather features associated with an easterly wave.

> **easterly wave**
>
> A traveling surface low-pressure system in the tropics that moves from east to west.

An easterly wave passing over the West Indies FIGURE 9.1

A zone of weak low pressure is at the surface, under the axis of the wave. The wave travels westward at a rate of 300–500 km (about 200–300 mi) per day. Rainy weather associated with the passage of the wave may last a day or two.

Airflow diverges on the western side of the wave axis. This divergence causes subsidence and fair weather.

Airflow converges on the eastern, or rear, side of the wave axis. This convergence causes the moist air to be lifted, producing scattered showers and thunderstorms, indicated by yellow shading.

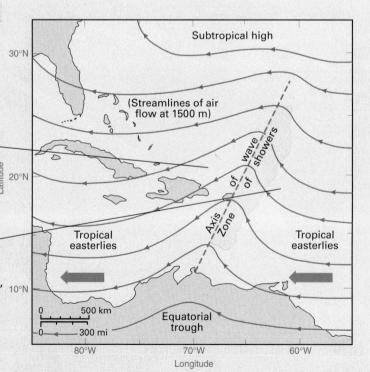

TAIWAN

Typhoons Chaba and Aere FIGURE 9.2

This satellite image shows Typhoon Aere, situated over Taiwan, and Typhoon Chaba, situated over the western Pacific, on August 24, 2004. Typhoon Chaba became one of the 15 most intense tropical cyclones on record and surpasses any hurricane in the Atlantic.

TROPICAL CYCLONES

Although easterly waves occur fairly regularly and generally involve only moderate changes in winds and rainfall, at times they grow into *cyclonic storms* involving intense circulations around low-pressure centers. Over the globe, cyclonic storms generally take three forms: traveling midlatitude cyclones, tornadoes, and tropical cyclones.

The **tropical cyclone** is the most powerful and destructive type of cyclonic storm (FIGURE 9.2). It is known as the *hurricane* in the western hemisphere, the *typhoon* in the western Pacific off the coast of Asia, and the *cyclone* in the Indian Ocean. This type of storm typically develops over oceans between 10°–20° N and S latitudes and no closer than 5° from the Equator.

tropical cyclone

Intense traveling cyclone of tropical and subtropical latitudes, accompanied by high winds and heavy rainfall.

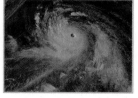

A tropical cyclone can originate as an easterly wave or weak low, which then intensifies and grows into a deep, circular low. It can also form as upper-air disturbances in the subtropical jet stream move south into the tropical regions. Once formed, the storm moves westward through the trade-wind belt, often intensifying as it travels. It can then curve poleward and eastward, steered by winds aloft. Tropical cyclones can penetrate well into the midlatitudes, as many residents of the southern and eastern coasts of the United States have experienced.

Tropical cyclones grow from *tropical depressions*, which are cyclones with winds below 17 m/s (39 mph). They intensify into *tropical storms* with winds of 18–33 m/s (40–74 mph). When winds exceed 33 m/s (74 mph), a tropical storm becomes a tropical cyclone.

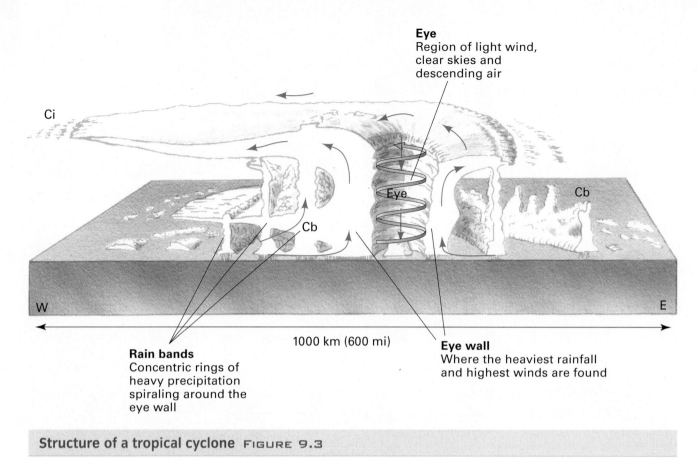

Eye
Region of light wind, clear skies and descending air

Ci

Eye

Cb

Cb

W

E

Rain bands
Concentric rings of heavy precipitation spiraling around the eye wall

1000 km (600 mi)

Eye wall
Where the heaviest rainfall and highest winds are found

Structure of a tropical cyclone FIGURE 9.3

In this schematic diagram, cumulonimbus (Cb) clouds in concentric rings rise through dense stratiform clouds. Cirrus clouds (Ci) fringe out ahead of the storm.

An intense tropical cyclone is an almost-circular storm center of extremely low pressure (**FIGURE 9.3**). Because of the very strong pressure gradient, winds spiral inward at high speed. Convergence and uplift are intense, producing very heavy rainfall. The storm gains its energy through the release of latent heat as the intense precipitation forms. The storm's diameter may be 150–500 km (about 100–300 mi). Wind speeds can range from 30 to 50 m/s (about 65 to 135 mph) and sometimes much higher. Barometric pressure in the storm center commonly falls to 950 mb (28.1 in. Hg) or lower. **FIGURE 9.4** on pages 250–251 shows how satellites, planes, and other instruments collect data that can help us understand what happens inside a tropical cyclone.

Another characteristic feature of a well-developed tropical cyclone is its central eye, in which clear skies and calm winds prevail. The eye is a cloud-free vortex produced by the intense spiraling of the storm. Around the eye, the wind is spinning so fast that it cannot converge into the center. Here in the eye, air descends from high altitudes and is adiabatically warmed, causing re-evaporation of cloud droplets. As the eye passes over a site, calm prevails, and the sky is clear. It may take about half an hour for the eye to pass, after which the storm strikes with renewed ferocity but with winds in the opposite direction.

Wind speeds and precipitation are highest along the cloud wall surrounding the eye. However, precipitation can also be intense in the rain bands that extend in concentric circles away from the hurricane, as visualized by special satellites described in *What a Scientist Sees: Rainfall in a Tropical Cyclone.*

Rainfall within a Tropical Cyclone

This set of satellite images shows two views of Hurricane Emily in the Gulf of Mexico on July 20, 2005.

In part A, we see a satellite image of clouds, with the storm's eye at the center. Although this image gives a sense of the scale of the hurricane, it does not identify where the winds are most intense or where rainfall is heaviest.

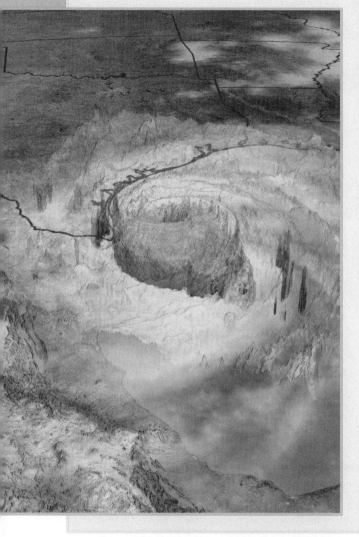

Part B is an image produced using a special radar aboard a satellite that maps water droplets, providing a three-dimensional image of the tropical cyclone. Note the very intense rainfall near the eye (represented by the red colors), as well as the local rainfall in the bands to the northwest and southeast of the eye.

CONCEPT CHECK **STOP**

What are the weather characteristics of an easterly wave?

What are the characteristic features of tropical cyclones?

HOW do scientists measure pressure, winds, and precipitation within tropical cyclones?

Visualizing

To understand how hurricanes work and to improve forecasts, researchers need detailed information from the heart of the storms. During the 2005 hurricane season, the most active on record, scientists investigated hurricanes from top to bottom (this one shown in cross section) with satellites, airplanes, and new kinds of instrumented probes.

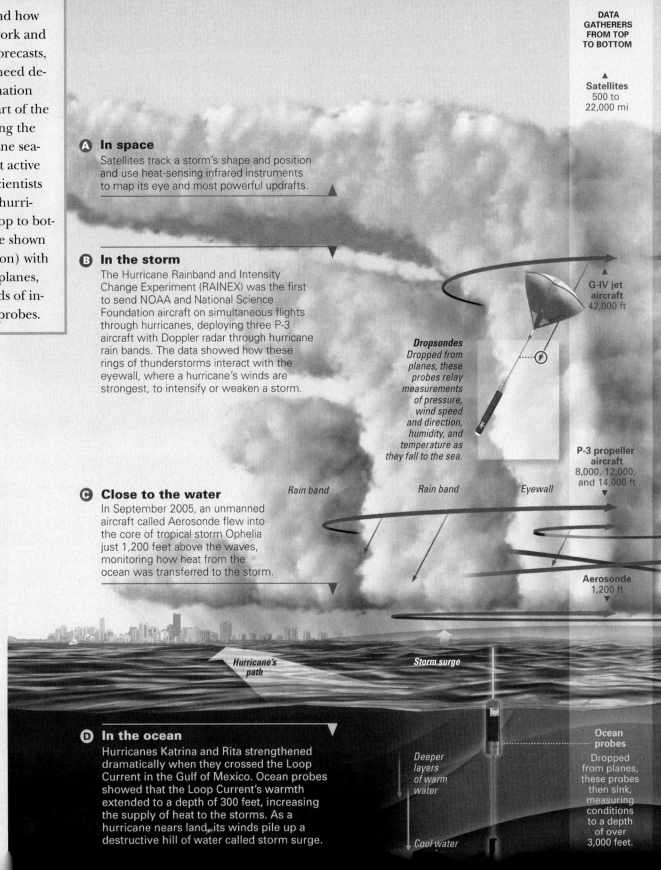

DATA GATHERERS FROM TOP TO BOTTOM

▲ **Satellites**
500 to 22,000 mi

A **In space**
Satellites track a storm's shape and position and use heat-sensing infrared instruments to map its eye and most powerful updrafts.

B **In the storm**
The Hurricane Rainband and Intensity Change Experiment (RAINEX) was the first to send NOAA and National Science Foundation aircraft on simultaneous flights through hurricanes, deploying three P-3 aircraft with Doppler radar through hurricane rain bands. The data showed how these rings of thunderstorms interact with the eyewall, where a hurricane's winds are strongest, to intensify or weaken a storm.

▲ **G-IV jet aircraft**
42,000 ft

Dropsondes
Dropped from planes, these probes relay measurements of pressure, wind speed and direction, humidity, and temperature as they fall to the sea.

P-3 propeller aircraft
8,000, 12,000, and 14,000 ft
▼

Rain band   *Rain band*   *Eyewall*

C **Close to the water**
In September 2005, an unmanned aircraft called Aerosonde flew into the core of tropical storm Ophelia just 1,200 feet above the waves, monitoring how heat from the ocean was transferred to the storm.

Aerosonde
1,200 ft
▼

Hurricane's path   *Storm surge*

D **In the ocean**
Hurricanes Katrina and Rita strengthened dramatically when they crossed the Loop Current in the Gulf of Mexico. Ocean probes showed that the Loop Current's warmth extended to a depth of 300 feet, increasing the supply of heat to the storms. As a hurricane nears land, its winds pile up a destructive hill of water called storm surge.

Deeper layers of warm water

Cool water

Ocean probes
Dropped from planes, these probes then sink, measuring conditions to a depth of over 3,000 feet.

Clues to Intensity

An image of Hurricane Rita based on infared data from NASA's Tropical Rainfall Measuring Mission (TRMM) satellite reveals a pair of chimney clouds, called hot towers, reaching more than 11 miles high. First observed in 1998, hot towers may indicate that the storm is about to intensify.

A

G-IV jet
Soon to be equipped with Doppler radar, NOAA's jet flies over and around developing hurricanes.

B

In Hurricane Rita the RAINEX experiment documented a phenomenon called eyewall replace-ment, in which a second eyewall **(1)** forms around the eye. The inner eyewall collapses **(2)** and temporarily weakens the storm. The outer eyewall then contracts and takes its place **(3)**, strengthening the storm again.

Eyewall

1

2

3

Doppler radar

P-3 hurricane hunters
Two of NOAA's planes were aided by the National Science Foundation's P-3, which carried a Doppler radar with four times more resolution than the standard radar.

C

10 feet

Aerosonde
Small enough to be launched from the back of a pickup truck, the 28-pound plane flew in winds that topped 78 mph, relaying data every half second.

D

SOURCES: PETER BLACK AND JOSEPH CIONE, NOAA ATLANTIC OCEANOGRAPHIC AND METEOROLOGICAL LABORATORY; SHUYI CHEN AND NICK SHAY, ROSENSTIEL SCHOOL OF MARINE AND ATMOSPHERIC SCIENCE

IMAGES: NASA GODDARD SPACE FLIGHT CENTER SCIENTIFIC VISUALIZATION STUDIO (TOP); ROSENSTIEL SCHOOL OF MARINE AND ATMOSPHERIC SCIENCE (MIDDLE AND BOTTOM)

REPORTING BY BRENNA MALONEY; DESIGNED BY JUAN VELASCO; ILLUSTRATIONS BY ROBERT KINKAID AND RAYMOND WONG

Category 5

Loop Current

An image of the Loop Current three days before Katrina's landfall shows how the storm intensified as it traveled over warmer waters (red).

Development and Movement of Tropical Cyclones

Why do tropical cyclones become so intense? The intensification is due to positive feedback within the environment itself, described in FIGURE 9.5. In the case of tropical cyclones, the positive feedback between the ocean and atmosphere allows the atmosphere to draw massive amounts of energy stored in the ocean.

As an easterly wave moves over the warm ocean waters of the low latitudes, the convergence at the surface

Process Diagram

Development and intensification of tropical cyclones FIGURE 9.5

Tropical cyclones are intense wind and rain events. The intensification of these tropical cyclones involves positive feedback loops between the ocean and the atmosphere.

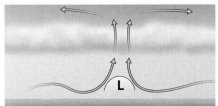

A Starting the engine
Tropical cyclones begin when low-level air flow is disturbed—by an easterly wave or the equatorward intrusion of an upper-air disturbance. Either can initiate the convection needed to start a hurricane. Once convection begins, a low-pressure center forms near the surface.

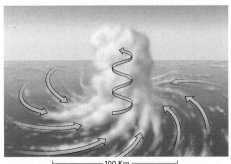

— 100 Km —

B Feeding it some fuel
The low-pressure center produces inspiraling air from the tropical ocean. This warm, moist air converges.

C Feeding it more
As warm, moist air rises, it expands and cools adiabatically. Once the air cools to the dew point temperature, condensation begins, releasing tremendous latent heat into the surrounding air. This heating accelerates the upward flow of air.

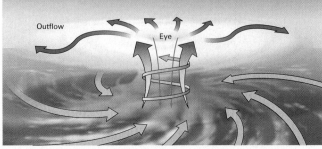

Outflow | Eye

— 500 Km —

D Running it wide open
Convection grows "explosively," accelerating air flow vertically and lowering surface pressures even more. The lowering pressure induces stronger inspiraling of warm, moist air. As this air rises, its water vapor condenses, releasing more latent heat. This enhances convection further, leading to even lower pressures. Around the center of the hurricane, convection and winds are most intense. However, because the air is spinning so fast, it never reaches the center. Here, calm prevails with descending air producing a clearing of clouds characteristic of the hurricane eye.

Cyclone paths FIGURE 9.6

Tropical cyclones form over warm, tropical oceans and are typically carried westward by easterly trade winds. Many eventually turn poleward and eastward. This image shows all known tropical cyclone tracks from 1985–2004—strong cyclones are in browns and yellows.

results in convective lifting. This lifting subsequently results in condensation, which releases latent heat that warms the surrounding air and produces even greater convection. As the convection intensifies, it further lowers the pressure at the surface. The result is enhanced convergence of warm, moist air, which supplies even more latent heat to the system as the air is lifted. Hence, a positive feedback loop forms in which intensified convergence leads to intensified convection, which leads to even more convergence. As the low pressure at the surface continues to drop, the inspiraling winds intensify, producing the ferocious winds described in the chapter opening.

Another reason for the very intense winds is the latitude at which tropical cyclones are located. Consider the winds around two identical low-pressure centers, one at 55° N and one at 15° N. Ignoring the effects of friction, the pressure gradient force on moving air parcels in both of these systems has to be balanced by the same Coriolis force. However, at low latitudes the Coriolis force is weaker than at high latitudes, so

the only way for the low-latitude winds to generate the same Coriolis force is to reach a higher speed. Thus, wind speeds in tropical cyclones can become much higher than in traveling cyclones of the midlatitudes.

TROPICAL CYCLONE TRACKS

tropical cyclone track Path of movement of a tropical cyclone, typically characterized by westward movement at low latitudes, followed by poleward and eastward movement in higher latitudes.

Tropical cyclones occur only during certain seasons. For hurricanes of the North Atlantic, the season typically runs from June through November, with maximum frequency in late summer or early autumn. In the southern hemisphere, the season is roughly the opposite. These periods follow the annual migrations of the ITCZ to the north and south with the seasons, and correspond to periods when ocean temperatures are warmest.

Most of the storms originate at 10°–20° N and S latitude and tend to follow known **tropical cyclone tracks** (FIGURE 9.6). In the northern hemisphere, they most often travel westward and northwestward

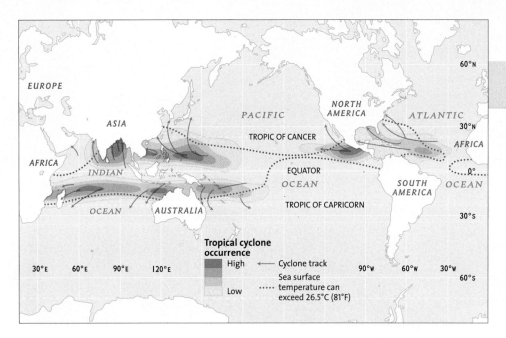

Tropical cyclones are most likely to occur in areas of greatest heating. Dashed lines show where the sea-surface temperature can be greater than 26.5°C (80°F). Cyclones last until they move over cooler waters or hit land.

through the trade winds, and then turn northeast at about 30°–35° N latitude into the zone of the westerlies. Here their intensity lessens, especially if they move over land. In the trade-wind belt, the cyclones travel 10–20 km (6–12 mi) per hour. In the zone of the westerlies, their speed is more variable.

As you can see in FIGURE 9.7, tropical cyclones always form over ocean regions in the low latitudes (but not within 5° of the Equator). Here, the conditions necessary for tropical cyclones to form are most likely to be met. These conditions include warm ocean temperatures greater than 26.5° C (80° F), which supply latent and sensible heat needed to drive the tropical cyclone. In addition, the overlying atmosphere has to have an unstable environmental lapse rate so that convection can easily be initiated by passing easterly waves or upper-air disturbances.

There also need to be weak winds with little change in wind speed or direction with height so that the large uniform vertical structure of the tropical cyclone can develop. If this uniform structure is disrupted, the tropical cyclone will quickly die out. Finally, the atmosphere must have a high water content through the bottom 5 km of the atmosphere—if warm, moist surface air mixes with dry air above, the relative humidity of the air will decrease and the latent heat release needed to sustain the tropical cyclone will be shut down.

What factors prevent the formation of tropical cyclones? Because weak winds are needed, moderate to strong winds above the surface will tend to disrupt the uniform structure of the tropical cyclone. Also, descent of air from above will inhibit convection and the vertical circulations necessary to release latent heat. Finally, if the low is too close to the Equator, the Coriolis force will be too weak to deflect air, and it will rapidly converge into the low-pressure center, re-equilibrating the pressures at the surface.

In the western hemisphere, hurricanes originate in the Atlantic off the west coast of Africa, in the Caribbean Sea, or off the west coast of Mexico. In the Indian Ocean, cyclones originate both north and south of the Equator, moving north and west to strike India, Pakistan, and Bangladesh, as well as south and west to strike the eastern coasts of Africa and Madagascar. Typhoons of the western Pacific also form both north and south of the Equator, moving into northern Australia, Southeast Asia, China, and Japan. Curiously, tropical cyclones almost never form in the South Atlantic or

Atlantic tropical cyclone gallery for 2005 FIGURE 9.8

2005 was the most active year on record for Atlantic hurricanes. Shown here are the 27 named storms that occurred during the official season of June 1 to November 30. Katrina was the costliest Atlantic storm on record. Wilma was the most intense, at one point observed with a central pressure of 882 mb (26.05 in. Hg) and winds of 83 m/s (185 mph).

Arlene–June 11, 2005 Bret–June 28 Cindy–July 5 Dennis–July 7

Emily–July 14 Franklin–July 28 Gert–July 24 Harvey–Aug. 4

Irene–Aug. 16 Jose–Aug. 22 Katrina–Aug. 28 Lee–Aug. 31

Maria–Sept. 5 Nate–Sept. 7 Ophelia–Sept. 14 Philippe–Sept. 19

Rita–Sept. 21 Stan–Oct. 4 Tammy–Oct. 5 Vince–Oct. 9

Wilma–Oct. 19 Alpha–Oct. 22 Beta–Oct. 29 Gamma–Nov. 18

Delta–Nov. 25 Epsilon–Dec. 5 Zeta–Jan. 3, 2006

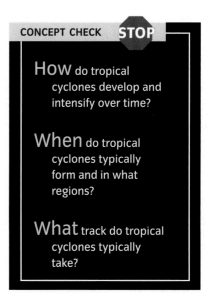

U.S. LANDFALL TRACKS

southeast Pacific regions. As a result, South America is not threatened by these severe storms.

Once formed, tropical cyclones are given names for convenience as they are tracked by weather forecasters. Male and female names are alternated in an alphabetical sequence that is renewed each season. Different sets of names are used within distinct regions, such as the western Atlantic, western Pacific, or Australian regions. Names are reused, but the names of storms that cause significant damage or destruction are retired from further use.

To track tropical cyclones, we now use satellite images. Within these images, tropical cyclones are often easy to identify by their distinctive pattern of inspiraling bands of clouds and a clear central eye. FIGURE 9.8 shows a gallery of satellite images of the Atlantic tropical cyclones from 2005.

CONCEPT CHECK STOP

How do tropical cyclones develop and intensify over time?

When do tropical cyclones typically form and in what regions?

What track do tropical cyclones typically take?

Impacts of Tropical Cyclones

LEARNING OBJECTIVES

Explain the various impacts of tropical cyclones.

Define storm surge and explain why it occurs.

Describe how tropical cyclone activity has been changing over the last 50 years.

T ropical cyclones can be tremendously destructive storms, with intense rainfall and very strong winds. The effects of wind, sea-level rise, and rain can cause devastation across very large areas. This devastation can occur along the coasts and farther inland along waterways and over mountain ranges, affecting thousands as the tropical cyclone moves along its track.

CHANGES IN WEATHER AND SEA LEVEL

The intensity of a tropical cyclone is based on the central pressure of the storm, mean wind speed, and height of the accompanying sea-level rise—also called the *storm surge*. Storms are then ranked from category 1 (weak) to category 5 (devastating) on the Saffir-Simpson scale, shown in FIGURE 9.9.

Saffir-Simpson scale of tropical cyclone intensity FIGURE 9.9

1. Minimal Damage
Winds 33–42 m/s (74–95 mph)
Storm surge 1.2–1.5 m (4–5 ft)

2. Moderate Damage
Winds 43–49 m/s (96–110 mph)
Storm surge 1.8–2.4 m (6–8 ft)

Small trees down, roof damage

3. Extensive Damage
Winds 50–58 m/s (111–130 mph)
Storm surge 2.7–3.6 m (9–12 ft)

Moderate to heavy damage to homes, many trees down

4. Extreme Damage
Winds 59–69 m/s (131–155 mph)
Storm surge 3.9–5.4 m (12–18 ft)

Major damage to all structures

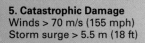

5. Catastrophic Damage
Winds > 70 m/s (155 mph)
Storm surge > 5.5 m (18 ft)

Severe damage to all structures

Wind damage FIGURE 9.10

High wind speeds, combined with tornado-force gusts, can generate damage to infra-structure as well as to living creatures caught in the path of a tropical cyclone. This photo captures the effects of the winds of Hurricane Allen at 45 m/s (100 mph) as it hit the Texas coast near Corpus Christi in 1980.

To be classified as a hurricane, a tropical cyclone must have sustained winds of over 33 m/s (74 mph) (FIGURE 9.10). However, sustained winds in the strongest storms can exceed 70 m/s (about 150 mph) with wind gusts of 90 m/s (about 200 mph) at times.

In addition, rainfall rates can be extremely high. During the passage of some tropical cyclones, 600 mm (2 ft) of rain or more can fall at a location. In some coastal regions, these storms provide much of the summer rainfall. Although this rainfall is a valuable water resource, it can also produce freshwater flooding, raising rivers and streams out of their banks. On steep slopes, soil saturation and high winds can topple trees and produce disastrous earthflows.

As an example, Hurricane Mitch, one of the deadliest Atlantic tropical cyclones in history, struck the heart of Central America, inundating it with rainfalls as high as 750 mm (30 in.). The monster category 5 storm attained a central low pressure of 905 mb with winds of 77.2 m/s (172 mph) in the western Caribbean on October 26, 1998. Claiming a toll of 9086 deaths, Mitch killed most of its victims with its torrential rains and subsequent downhill movements of mud, water, and debris. The damage was truly devastating. Honduras and Nicaragua suffered losses of about half their annual gross national products. Guatemala and El Salvador were also hard hit.

However, the most serious effect of tropical cyclones is usually coastal destruction by storm waves and very high tides, seen in FIGURE 9.11. Since atmospheric pressure at the center of the cyclone is so low, the sea level rises within the center of the storm. In addition, high winds create damaging surf and push water toward the coast on the side of the storm with onshore winds, raising the sea level even higher. Waves attack the shore at points far inland of the normal tidal range. Low pressure, winds, and the underwater shape of a bay floor can combine to produce a sudden rise of water level known as a **storm surge**, which carries ocean water and surf far inland. If high tide accompanies the storm, waters will be even higher. For ex-

> **storm surge** Rapid rise of the coastal water level accompanying the onshore arrival of a tropical cyclone.

ample, low-lying coral atolls of the western Pacific may be entirely swept over by wind-driven seawater, washing away palm trees and houses and drowning the inhabitants.

An example of the devastation that can be wrought by a storm surge occurred when an unnamed hurricane struck Galveston, Texas, on September 8, 1900 (FIGURE 9.12). The sudden storm surge generated by the severe hurricane flooded the low coastal city and drowned 6000–8000 people. This death toll was the largest yet experienced in a natural disaster within the United States. There was little warning of the 1900 hurricane. However, as was evident in New Orleans after Hurricane Katrina struck, even such warning may not have saved the city.

Storm surge and its effect on coastal areas FIGURE 9.11

As a tropical cyclone moves onshore, it can bring with it a devastating storm surge. By raising the sea level through a combination of high winds and low surface pressures, the storm surge can inundate low-lying areas and subject them to heavy surf.

Normal conditions
Structures above normal high tide are not subject to damage from continuously breaking waves.

Storm surge
A storm surge combined with a high tide can lift sea level so that structures are subjected to continuous pounding of heavy surf.

Galveston hurricane of 1900 FIGURE 9.12

On September 8, 1900, an unnamed hurricane struck Galveston, Texas, killing 6000–8000 people in the deadliest natural disaster in American history.

Normal conditions Under normal conditions, Galveston, a barrier island in the Gulf of Mexico, sits about 3 m (10 ft) above sea level.

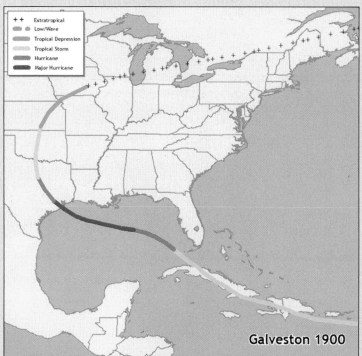

Galveston 1900

Legend:
- ++ Extratropical
- Low/Wave
- Tropical Depression
- Tropical Storm
- Hurricane
- Major Hurricane

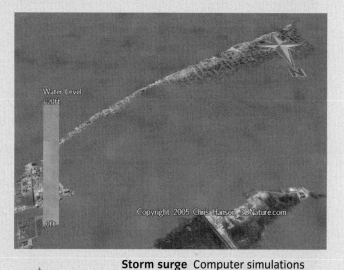

Storm surge Computer simulations indicate that the storm surge of 5 m (16 ft) that accompanied the 1900 hurricane completely inundated the city.

Estimated storm track This 1900 hurricane occurred before the practice of naming storms was established. Its track took it over open water, away from any weather stations, although reports from both Cuba and boats in the Gulf suggested the approach of a large storm. Today, satellites and "hurricane hunter" planes would give residents of Galveston more warning.

Devastation As this photo shows, the effects of the storm surge completely destroyed most structures in the city. Wind speeds at landfall were about 215 km/h (135 mph), which would be a category 4 storm using the Saffir-Simpson scale.

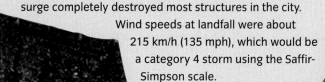

Major Atlantic hurricanes in 2004 FIGURE 9.13

Hurricane Ivan
September 2–17

Hurricane Jeanne
September 13–27

Hurricane Charley
August 5–15

Hurricane Frances
August 23–September 6

Four major hurricanes (category 3 or higher) made landfall in Florida in 2004, three of which passed within 40 km (25 mi) of the same location. Shown here are superimposed satellite images acquired at different times as the four hurricanes followed various tracks through Florida. The circle indicates the region that was hit three separate times.

IMPACTS ON COASTAL COMMUNITIES

Certainly, coastal residents of South Florida are particularly aware of the damage hurricanes can cause. In 1992, Hurricane Andrew struck the east coast of Florida near Miami. The second most damaging storm to occur in the United States, it claimed 26 lives and caused more than $35 billion in property damage, measured in today's dollars. In 2004, four major hurricanes crossed Florida—Charley, Frances, Ivan, and Jeanne (FIGURE 9.13). Taken together, the storms destroyed over 25,000 homes in Florida, with another 40,000 homes sustaining major damage.

Of course, South Florida has not been the only coastal region to suffer the devastating effects of hurricanes. In 2005, Hurricane Katrina laid waste to the city of New Orleans and much of the Louisiana and Mississippi Gulf coasts (FIGURE 9.14). Originating southeast of the Bahamas, the hurricane first crossed the South Florida peninsula as a category 1 storm, then moved into the Gulf of Mexico, where it intensified to a category 5 storm. Weakening somewhat as it approached the Gulf Coast, its eye came ashore at Grand Isle, Louisiana, with sustained winds of 56 m/s (125 mph) early on August 29.

The city of New Orleans is particularly vulnerable to hurricane flooding. Built largely on the floodplain of the Mississippi River, most of its land area has slowly sunk below sea level as underlying river sediments have compacted through time. In addition, it is a city surrounded by water. Levees protect the city from Mississippi River floods on the south, as well as from ocean waters along the saline Lakes Borgne on the east and Pontchartrain on the north, which are both connected to the Gulf of Mexico. Rain water falling into the sunken basin is pumped up and out of the city by discharge canals that run through the city itself and out to Lake Pontchartrain. There are also several canals and shipping channels connecting the river with the lakes and the Gulf.

Flooding in New Orleans FIGURE 9.14

This image shows the extent of flooding during and following the passage of Hurricane Katrina. The depth of the flooding can be gauged by the scale on the right.

In the aftermath of Hurricane Katrina, entire neighborhoods in New Orleans lay in ruins for months.

A This photo, taken a day after the passage of Hurricane Katrina, shows waters rushing through a broken levee, flooding homes across the neighborhood.

B Two weeks after the hurricane the devastation is nearly complete. The red object in the foreground is a barge.

Katrina's first assault was mounted from the east along one of these shipping channels, as a storm surge swept westward from Lake Borgne, overtopping and eroding levees and flooding east New Orleans and St. Barnard Parish. Penetrating deep into the city along the Intracoastal Waterway, the surge also overtopped and breached floodwalls and levees along the main canal connecting the Mississippi River with Lake Pontchartrain. To the north, water levels rose in Lake Pontchartrain, overtopping dikes and levees and filling the discharge canals to dangerously high levels. Eventually, large sections of the canal walls failed, allowing water to pour into the central portion of the city (FIGURE 9.15). Over the next two days, the water level rose until it equalized with the level of Lake Pontchartrain. Eighty percent of the city was covered with water at depths of up to 6 m (20 ft).

The result was devastation. Total losses were estimated at more than $100 billion. The official death toll exceeded 1300. The Gulf coasts of Mississippi and Louisiana were also hard hit, with a coastal storm surge as high as 8 m (25 ft) penetrating from 10 to 20 km (6–12 mi) inland. Adding insult to injury, much of New Orleans reflooded three weeks later because of Hurricane Rita, a category 3 storm that made landfall on September 24 at the Louisiana–Texas border.

CHANGES IN TROPICAL CYCLONE ACTIVITY

Tropical cyclone activity varies from year to year and decade to decade. In the Atlantic Basin, 13 strong hurricanes of category 3 or higher struck the eastern United States or the Florida Peninsula from 1947 to 1969, whereas only one hurricane struck the same region in the period 1970–1987. Although a number of strong storms, including category 4 or 5 hurricanes Gilbert (1988), Hugo (1989), and Andrew (1992) occurred in 1988–1992, Atlantic hurricane activity remained depressed until 1994 (FIGURE 9.16).

However, in 1995 a new phase of Atlantic hurricane activity began. Since 1995, sea-surface temperatures in the northern Atlantic during August–October have averaged about 0.5°C (about 1°F) warmer than 1970–1994, providing more latent heat to fuel the cyclones. High-level easterly winds, which can shear the tops off growing storms, weakened on average by about 2 m/s (about 4 mph) during the same period.

The result has been the present period of hurricane activity, unequalled in historic records. By 2005, the average number of named storms had increased to 13 per year, compared to 8.6 during 1970–1994. The number of hurricanes had increased from 5 to 7.7 per year with 3.6 major hurricanes, compared to 1.5 in prior years. In 2005 alone, there were 27 named storms, with a record 4 storms reaching category 5.

The most recent climatological studies suggest that this enhanced hurricane activity is linked to slowly changing cyclical patterns of the thermohaline circulation, sometimes called the *Atlantic multidecadal oscillation*. During periods of enhanced thermohaline circulation, as at present, hurricanes are more frequent. Because these periods of enhanced circulation appear to last for several decades, the outlook is for conditions that favor the development of Atlantic tropical cyclones to persist for another 15–20 years.

Hurricane tracks along the southeastern coast of the United States FIGURE 9.16

These figures show hurricane tracks for 1985–1994 A and 1995–2004 B. The thicker the line of the track, the more intense the hurricane. The bar chart in A shows sea-surface temperatures for both periods as well as sea-surface temperatures going back to 1944. Comparing A and B, you can see that the number and intensity of storms increases when sea-surface temperatures increase.

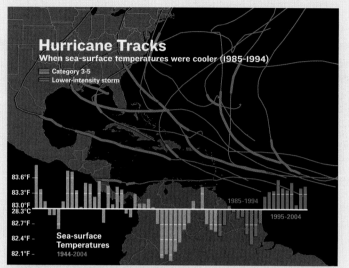

A Hurricane tracks for 1985–1994

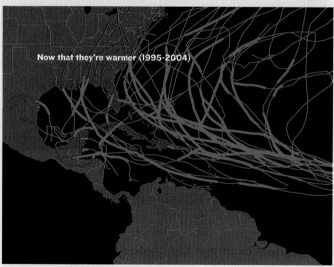

B Hurricane tracks for 1995–2004

CONCEPT CHECK **STOP**

What is a storm surge?

How does it form?

What other weather changes accompany the passage of a tropical cyclone?

What regions are usually most vulnerable to the passage of tropical cyclones?

Hurricane Eye

Although it is possible to track tropical cyclones from space, to take measurements of the pressure and winds within the tropical cyclone itself the National Weather Service employs "hurricane hunters"—planes designed to fly through high winds and heavy turbulence. The photo shown here was taken from one of these planes within the eye of Hurricane Katrina in 2005.

■ What characteristic features of a tropical cyclone do you see here?

■ Why is the sky so clear in the region of the eye?

SUMMARY

1 Tropical Cyclone Structure

1. **Easterly waves** are tropical weather systems that occur when a weak low-pressure trough develops in the easterly wind circulation of the tropical zones, producing convergence, uplift, and shower activity.

2. **Tropical cyclones** can be the most powerful of all cyclonic storms. They develop over very warm tropical oceans and can intensify to become vast inspiraling systems of very high winds with very low central pressures and heavy precipitation.

2 Development and Movement of Tropical Cyclones

1. Tropical cyclones form through a positive feedback between the ocean and atmosphere. During their formation, convergence of warm, moist air at the surface causes uplift, which leads to condensation, latent-heat release, and accelerated convection. The convection in turn lowers surface pressures, which causes even more convergence of warm, moist air.

2. Tropical cyclones form in the summer and fall when the waters are warmest over the tropical oceans. They can form in the central and western Atlantic, the eastern and western Pacific, as well as the Indian Ocean. Once they form, they tend to track westward in lower latitudes, then turn poleward before moving eastward in higher latitudes.

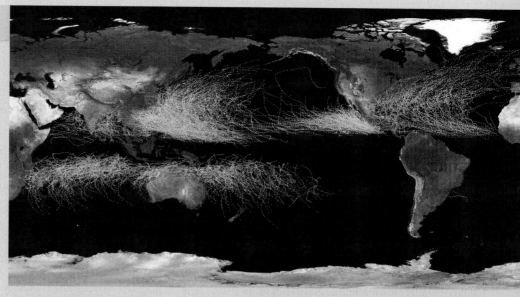

3 Impacts of Tropical Cyclones

1. As tropical cyclones move onto land, they bring heavy surf and storm surges of very high water. Tropical cyclones have caused many deaths and billions of dollars of destruction in coastal regions.

2. In addition to the **storm surge** and heavy surf, tropical cyclones can cause severe wind damage. The heavy rain that accompanies them can also produce intense flooding and earthflows, particularly over elevated terrain.

3. Over the last 10 years or so, the number and intensity of tropical cyclones has been increasing on a global scale. In 2005, the North Atlantic had a record number of tropical cyclones, as well as a record number of category 5 hurricanes. The increase in tropical cyclone activity in this region has been associated with rising sea-surface temperatures, which can feed more intense convection and stronger tropical cyclones.

KEY TERMS

CRITICAL AND CREATIVE THINKING QUESTIONS

1. If you were sitting on a tropical island in the western Atlantic, describe the weather you might experience as an easterly wave passes over your area. How long would the process of the easterly wave take? Other than precipitation, what other changes in weather might you experience?

2. On the diagram below, label the following components of a tropical cyclone: (a) eye; (b) eye wall; (c) rain band; (d) region of rising air; (e) region of descending air.

3. Tropical cyclones intensify because of a positive feedback between the ocean and atmosphere. Describe how this feedback works. Start with an initial disturbance in the surface pressure and describe why this disturbance grows over time. Also describe how the change in pressure affects the wind fields and the formation of precipitation that accompany tropical cyclones.

4. What background conditions are necessary for the development of a tropical cyclone? Give a typical track for the movement of a tropical cyclone in the northern hemisphere.

5. Why are tropical cyclones so dangerous? Explain the different types of damage they can cause. Overall, which type of damage is the most severe?

6. Wind speeds in tropical cyclones tend to be much higher than in midlatitude cyclones, even given the same pressure gradient force. Why? (*Hint:* Think of how latitude affects the Coriolis force.)

SELF-TEST

1. Tropical weather tends to be _____.
  a. based on westerly waves
  b. based on divergence of air masses
  c. convective in nature
  d. based on the horse-latitude divergence

2. Easterly waves predominantly occur within the _____ latitudinal zone(s).
  a. 0°–5° N and S
  b. 10°–40° N and S
  c. 5°–30° N and S
  d. 25° N to 25° S

3. Hurricanes and typhoons generally develop within the _____ latitudinal zones.
  a. 0°–10° N and S
  b. 10°–20° N and S
  c. 20°–30° N and S
  d. 30°–40° N and S

4. Hurricane season in the northern hemisphere runs from _____.
  a. November through May
  b. March through September
  c. June through November
  d. August through February

5. The _____ continent is rarely if ever threatened by hurricanes or typhoons.
   a. North American
   b. South American
   c. African
   d. Eurasian

6. The tools that weather forecasters use to track and study hurricanes are _____.
   a. satellites
   b. "hurricane hunter" planes
   c. ocean probes
   d. all of the above

7. Tropical cyclones may have diameters of up to about 500 km (about 300 mi) and wind speeds averaging over 33 m/s (74 mph).
   a. True
   b. False

8. On the map below, indicate where tropical cyclones tend to form and what tracks they take.

9. As tropical cyclones grow, the low-pressure center produces a convergence of _____ air; as this air rises, _____ is released, which results in increased convection and _____ pressures at the surface.
   a. warm, dry; longwave heat; lower
   b. cool, dry; latent heat; higher
   c. warm, moist; latent heat; lower
   d. cool, moist; longwave heat; lower

10. _____ is (are) an unfavorable condition for the initial stages of tropical cyclone formation.
   a. Weak upper-air winds
   b. Presence of subsiding or sinking air
   c. Warm ocean temperatures
   d. Presence of significant moisture to feed evaporation

11. A _____ is a sudden rise of water level caused by a tropical cyclone.
   a. storm surge
   b. high tide
   c. tsunami
   d. tidal flood

12. The intensity of tropical cyclones is categorized based on the _____.
   a. Fujita scale
   b. Richter scale
   c. decibel scale
   d. Saffir-Simpson scale

13. Hurricane activity in the Atlantic has _____ over the last decade because of _____.
   a. decreased; increased ocean salinity
   b. increased; stronger winds aloft
   c. increased; warmer ocean temperatures
   d. decreased; warmer land temperatures

14. The largest U.S. death toll in a natural disaster occurred from a hurricane that struck _____.
   a. Galveston, Texas, in 1900
   b. New Orleans, Louisiana, in 2005
   c. Miami, Florida, in 1992
   d. Orlando, Florida, in 2004

15. Destruction from tropical cyclones is *not* associated with _____.
   a. high winds
   b. heavy rainfall
   c. sea-level rise
   d. rapid temperature drops

Thunderstorms and Tornadoes

Lightning strikes an isolated house near Inverness, Scotland.

Lightning is one of the most awesome displays in nature, in terms of both its beauty and its power. Approximately 80 percent of all lightning flashes are between clouds, but the remaining 20 percent shoot from the clouds to the Earth's surface. In fact, the Earth is struck by lightning about 100 times every second. When it is, electricity flows between sky and ground, traveling at about a third of the speed of light and packing a punch of 60,000 to 100,000 amps—thousands of times as much as a household circuit.

A lightning bolt is actually an arc of atmospheric gases that is heated to as high as 30,000°C (50,000°F)—a temperature hotter than the surface of the Sun—as an electric current flows through it. The hot gas emits light as a flash and expands explosively to create the familiar crack of thunder.

It is no wonder that lightning is one of the most deadly natural phenomena. In an average year, more people are killed in the United States by lightning than by tornadoes or hurricanes. In the United States alone, lightning sets 10,000 forest fires and causes $100 million in property damage every year. Between 1940 and 1991, lightning killed 8,316 people.

Lightning also presents science with one of its greatest mysteries. We still do not know exactly what causes lightning. If we hope to find the answer one day, we must look in depth at some of our atmosphere's most elusive processes.

Thunderstorms

LEARNING OBJECTIVES

Describe the stages of development of thunderstorms.

Characterize the differences between air-mass thunderstorms and mesoscale convective systems.

Identify the weather conditions that are conducive to thunderstorms.

A thunderstorm is any storm that produces thunder and lightning (FIGURE 10.1). At the same time, **thunderstorms** can also produce high winds, hail, and tornadoes. They are typically associated with cumulus clouds that indicate the presence of rising, unstable air. It is this rising motion that produces the characteristic rainfall and lightning that accompany thunderstorms.

Thunderstorms can range from fairly isolated, short-lived storms, sometimes called *air-mass thunderstorms*, to massive, well-organized complexes of storms, called *mesoscale convective systems*. Next, we describe the different types of thunderstorms and the environmental conditions necessary for their formation.

thunderstorm
Any storm in which vertical motions are sufficient to cause lightning and thunder.

Thunderstorm over New Mexico FIGURE 10.1

This figure shows a severe thunderstorm over New Mexico during summer. Note the cumulus cloud structure, indicating significant vertical movement. Other characteristics of severe thunderstorms include the overshooting cloud top, the anvil cloud, and the shelf cloud at the bottom.

Overshooting top

Anvil cloud

Shelf cloud

Typical air-mass thunderstorm FIGURE 10.2

This figure shows an isolated air-mass thunderstorm over the Everglades in Florida during summer. Note the vertical extent of the clouds and the dark region beneath, indicating the presence of rain.

AIR-MASS THUNDERSTORMS

Air-mass thunderstorms are isolated thunderstorms generated by daytime heating of the land surface. They occur when surface heating makes the environmental lapse rate—the vertical change in temperature of the surrounding air—unstable with respect to the dry and moist adiabatic lapse rates. At that point, isolated air masses can begin to rise through the air column. As they do so, they cool adiabatically. If the air masses rise high enough to reach the lifting condensation level (also called the *convective condensation level*), they form cumulus clouds. Continued lifting can subsequently result in enough condensation that precipitation begins to fall.

Because these thunderstorms require significant surface heating, they are typi-cally found during spring and summer over land surfaces. They are also frequently associated with the intrusion of maritime tropical air into a given region. This warm, moist air tends to be unstable more often than dry continental air, giving rise to storms like the one seen in FIGURE 10.2.

One of the key differences between air-mass thunderstorms and other types of thunderstorms is that they are not associated with the interaction of two air masses along a front. Because of this lack of frontal activity, there is no process for organizing the thunderstorms, and they appear to rise haphazardly in different locations. Without the continued lifting supplied by frontal activity, these thunderstorms usually die out relatively quickly as they mix with the surrounding environmental air and become drier, cooler, and more stable.

air-mass thunderstorm Thunderstorm arising from daytime heating of the land surface, usually characterized by isolated cumulus and cumulonimbus clouds.

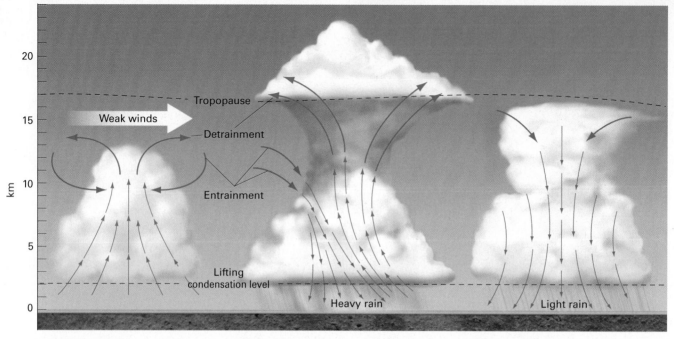

Cumulus stage
Vertical motions are limited by mixing with cool, dry environmental air. Continuous mixing moistens the environment, allowing subsequent convection to ascend higher.

Mature stage
In this most intense stage, there are well-organized updrafts and downdrafts. Updrafts can reach the tropopause and then spread out as an anvil cloud. Downdrafts that reach the ground can reinforce convective lifting by producing cold surface outflow that forces warm surface air aloft.

Dissipating stage
Dissipation occurs when cool, dry environmental air is entrained by the rising air parcels. Subsequent sensible and latent cooling, combined with frictional drag imparted by falling rain, result in significant downdrafts that inhibit vertical circulations necessary to sustain convection.

Stages in the development of an air-mass thunderstorm FIGURE 10.3

This figure shows the three development stages of an air-mass thunderstorm, which are the cumulus, mature, and dissipating stages. Each stage has characteristic vertical winds and precipitation.

Cumulus stage   The typical life cycle of an air-mass thunderstorm involves three stages of development (FIGURE 10.3). The first of these is the **cumulus stage**. In this stage, initial air parcels near the surface are heated and begin to rise. The first air parcels may reach the lifting condensation level and form isolated cumulus clouds. However, these mix with the surrounding dry environmental air, and the cloud water droplets evaporate. This process cools the temperature of the air parcel and prevents it from rising further.

At the same time, this mixing adds water vapor to the surrounding environmental air. Then, as the next air parcel rises through the moister environment, its cloud droplets evaporate more slowly. Eventually, after

> ▮ **cumulus stage**
> Stage of air-mass thunderstorm development in which surface heating produces isolated regions of significant updrafts, leading to the formation of cumulus clouds and precipitation.

enough air parcels have moistened the environmental air, cumulus clouds rising through this air have very little evaporation. Instead, condensation continues to occur, warming the air parcel and allowing the cloud to rise even higher.

As the air parcels continue to rise, condensation continues until water drops become large enough to fall. However, if the air is rising fast enough, the updrafts keep the water drops suspended in the atmosphere, where they continue to grow by collision and condensation. Eventually, they become large enough that the force of gravity overcomes the force applied to the drops by the updraft. At that point, the water drops start to fall to the ground.

Mature stage The cumulus stage is dominated by the presence of updrafts throughout the air column. It is the rising motion of the heated surface air that produces the updraft. At the same time, however, downdrafts also begin to form during this stage, marking the transition to the **mature stage**. One reason for the downdrafts is the drag exerted by the falling precipitation. Because the environmental lapse rate is unstable, air parcels that initially move down under the influence of this drag continue to descend (just as the surface air that initially rose by surface heating continues to rise).

Another process also initiates downdrafts during this stage of development. This process involves the mixing, or **entrainment**, of cold, dry environmental air into the cloud. Because this environmental air is cooler than the cloud, it begins to descend. In addition, the drier environmental air produces more evaporation of cloud water drops, thereby cooling the surrounding air. Again, this cooler air begins to descend toward the surface.

During this mature stage, the thunderstorm has formed into an organized convection cell and is at its most active. In one part of the cell, warm, moist air rises through the cooler, drier environmental air. As it does so, significant condensation occurs, releasing latent heat that allows the air parcel to continue to ascend. If enough latent heat is released, these air parcels can ascend 10–15 km to the tropopause. At that level, their ascent is limited by the strong temperature inversion of the tropopause. The cloud top then spreads laterally, forming an *anvil cloud* that extends downwind from the cumulus cloud.

In the other part of the cell, there are significant downdrafts. These downdrafts—initiated by the drag of falling precipitation as well as the evaporation of the precipitation as it falls through the cold, dry environmental air—can be as strong as the updrafts. The updraft region is tilted somewhat, allowing the downdrafts to reach the surface without being countered by the opposing updrafts. When these downdrafts hit the surface, they spread laterally, producing gusty, strong winds that blow in all directions.

The updrafts and downdrafts found during the mature stage are actually part of an organized convection cell that can allow the thunderstorm to grow. As the downdrafts sink and spread along the surface, they initiate the lifting of warmer surface air. This initial lifting of the warm unstable air results in enhanced convection, which releases more latent heat and produces more precipitation, thereby enhancing the downdrafts.

Dissipating stage The **dissipating stage** occurs when the stabilizing effects of entrainment overcome the destabilizing effects of convection. As continued mixing between the warm, moist rising air mass and the cool, dry surrounding environment occurs, widespread downdrafts form throughout the air column. These downdrafts inhibit the upward motion associated with convection. Without convection—and the associated condensation and latent heat release—the thunderstorm quickly dies out.

In all, air-mass thunderstorms can develop, mature, and dissipate over the course of an hour or so. Once the mature stage is reached—with its active regions of updrafts and downdrafts—the necessary environmental conditions that would allow them to overcome the stabilizing effect of entrainment are missing, and the storms eventually dissipate.

■ **mature stage**
Stage of air-mass thunderstorm development in which strong updrafts and downdrafts are present, resulting in heavy rainfall, gusty winds, and lightning.

■ **entrainment**
Process in which cool, dry environmental air mixes with warm, moist rising air within cumulus clouds.

■ **dissipating stage**
Stage of air-mass thunderstorm development in which strong downdrafts throughout the air column inhibit the convection and latent heat release needed to sustain the thunderstorm.

Occurrence Air-mass thunderstorms are initiated by surface heating during the day. Usually, these thunderstorms are associated with uneven heating, which allows certain air parcels to become warmer than others. One method for producing this uneven heating is through slope heating along mountainsides. As air at the lower elevations is heated, it begins to rise up the mountain. As it rises, it continues to be heated by the nearby mountain slope, which allows significant heating of the air parcel to occur. This warmer air tends to lift higher than air that is found over flat terrain.

Processes other than surface heating can also produce the lifting needed to initiate these types of thunderstorms (**FIGURE 10.4**). This lifting usually involves some type of forced convergence. For example, over Florida in the summertime, significant air-mass thunderstorms frequently form at the convergence of the sea breezes from the Atlantic and the Gulf of Mexico. This convergence forces air at the surface to rise, which initiates convection. In addition, surface convergence of the northeasterly and southeasterly trade winds along the intertropical convergence zone can also initiate upward motion that in turn generates air-mass thunderstorms. Finally, thunderstorms can develop where upper-air divergence occurs within a jetstream disturbance. This divergence aloft lifts air from the surface, again initiating convection.

Severe thunderstorms An air-mass thunderstorm may produce strong winds and rain, but only for a short duration. Usually, these thunderstorms start, mature, and dissipate within an hour or two.

Other thunderstorms, however, can persist for much longer and are called **severe thunderstorms**. By definition, severe thunderstorms must have winds greater than 26 m/s (58 mph). Alternatively, they must either have hail of a certain size (19 mm or 0.75 in. in diameter) or produce a tornado. Weather of this strength usually only accompanies a mature thunderstorm that has been active for many hours.

> ■ **severe thunderstorm**
> Thunderstorm in which the surface winds are greater than 26 m/s (58 mph), hail is more than 19 mm (0.75 in.) in diameter, or there is an accompanying tornado.

Severe thunderstorms persist longer than others because they develop an organized convection cell that allows for constant intake of warm, moist air, as seen in **FIGURE 10.5**. The only way to accomplish this development is to have air from *outside* the region begin to enter the thunderstorm. In that scenario, the downdrafts become so large that they spread out past the radius of the air column itself. As they spread, they force warm, moist air from surrounding regions to lift up and

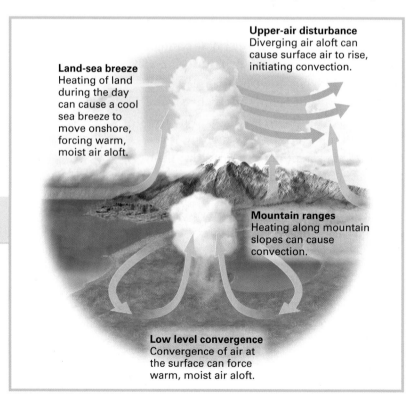

Processes that can generate air-mass thunderstorms FIGURE 10.4

This schematic shows various processes that can initiate the convection necessary to produce air-mass thunderstorms.

Upper-air disturbance Diverging air aloft can cause surface air to rise, initiating convection.

Land-sea breeze Heating of land during the day can cause a cool sea breeze to move onshore, forcing warm, moist air aloft.

Mountain ranges Heating along mountain slopes can cause convection.

Low level convergence Convergence of air at the surface can force warm, moist air aloft.

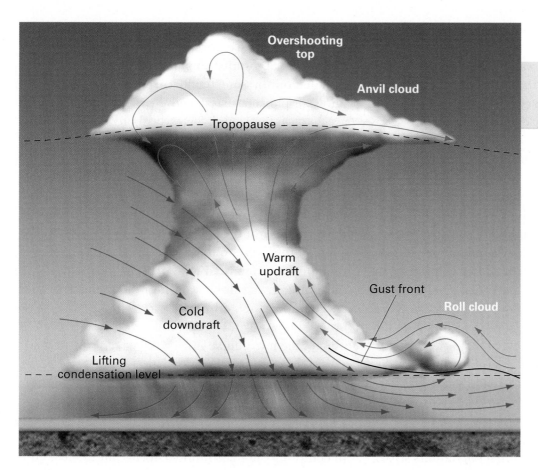

Overshooting top

Anvil cloud

Tropopause

Warm updraft

Gust front

Roll cloud

Cold downdraft

Lifting condensation level

Anatomy of a severe thunderstorm

FIGURE 10.5

A severe thunderstorm can maintain convection and precipitation for long periods of time if it can continuously incorporate warm, moist air. Gust fronts associated with downdrafts extending ahead of the storm system force warm, moist surface air aloft, which then flows back toward the thunderstorm to replace the descending, outflowing cold air.

over the colder, denser air. This warm, moist air then flows back toward the thunderstorm and becomes part of the updraft. As long as there is the presence of warm, moist air that can be incorporated into the thunderstorm, it can continue to grow.

At the same time, the strong downdrafts must not inhibit the vertical motions associated with this convection. The interference can be prevented if *wind shear*—the change in winds with height—is significant. In that case, cool, dry air is entrained only on the upwind side of the convection cell, while warm, moist convecting air is positioned toward the downwind side by the strong winds aloft. This orientation allows the downdrafts to reach the surface without subsequently shutting down convection.

As severe thunderstorms continue to grow, the vertical motions can become so strong that they overcome the inversion found at the tropopause, producing a characteristic *overshooting top* that can be seen by planes flying overhead. The air that does not continue to rise through the tropopause again spreads out horizontally,

being blown by the prevailing upper-air winds. The result is a horizontal cloud at the top of the thunderstorm called an *anvil cloud*. Where low-level, warm, moist air starts to rise over the cold downdraft air, condensation and cloud formation can also occur, resulting in a *roll cloud*.

The most severe of these thunderstorms are called *supercell thunderstorms*. These are massive thunderstorms with a single circulation cell comprising very strong updrafts and downdrafts. Because of their vertical extent, which can be up to 25 km (16 mi), they are affected differently by winds at the surface and winds aloft. If the background wind shear not only involves a change in wind speed with height but also a change in direction—typically in a counterclockwise direction—a rotation of the storm can occur, which is a precursor to the formation of tornadoes.

Most supercell storms arise in response to the suppression of smaller thunderstorms by a temperature inversion aloft. The air trapped near the surface gets warmer and moister as the day goes on. The sensible and latent heat associated with this air continues to build. Eventually, the air becomes so warm and moist that it is less dense than the warm, dry capping air above it. At this point, the air at the surface breaks through the inversion as a single, very large air parcel. This air then surges upward in a violent updraft. The process is similar to what occurs when a balloon is blown up too much and the surface of the balloon eventually breaks.

Microbursts

Another characteristic feature of many severe thunderstorms is the formation of a **microburst**—an intense downdraft or *downburst* that accompanies the gust front. Once the microburst hits the ground, it flows outward in all directions, producing intense, localized winds called *straight-line winds* or *plough winds*. A microburst is also often, but not always, accompanied by rain.

> **microburst**
> A strong localized downdraft from a thunderstorm that produces rapid changes in surface wind speeds.

The microburst itself can be so intense that it is capable of causing low-flying aircraft to crash (**FIGURE 10.6**). An aircraft flying through the microburst first encounters strong headwinds, which may cause a bumpy ride but do not interfere with the airplane's ability to fly. However, as the airplane passes through the far side of the microburst, it encounters a strong tailwind. The lift of the airplane's wings depends on the speed of the air flowing across them, and the tailwind greatly reduces the air speed, which causes a loss of lift. If the tailwind is strong enough, the airplane cannot hold its altitude and may crash.

Microbursts can be detected by special radar instruments that measure horizontal wind speeds. These instruments are quite expensive, however, and are installed only at some major airports. Training procedures for pilots have reduced the incidence of aviation accidents in the United States attributed to microbursts and associated wind shear. However, in 1994 a jet aircraft attempting to land at Charlotte, North Carolina, crashed during a violent thunderstorm, killing 37 people and injuring 20 others. A microburst was detected just as the airplane approached the airport, but the warning was not broadcast to the pilots on the radio channel they were following. Although they attempted to abort the landing, a pilot error occurred, and their aircraft was brought down by the severe tailwind of the microburst.

MESOSCALE CONVECTIVE SYSTEMS

> **mesoscale convective system** A relatively long-lived, large, and intense convective cell or cluster of cells characterized by exceptionally strong updrafts.

In contrast to air-mass thunderstorms, **mesoscale convective systems** (MCS) are thunderstorms in which multiple, organized thunderstorm cells form. Gener-

Anatomy of a microburst FIGURE 10.6

Schematic cross section showing a downdraft reaching the ground and producing a horizontal outflow.

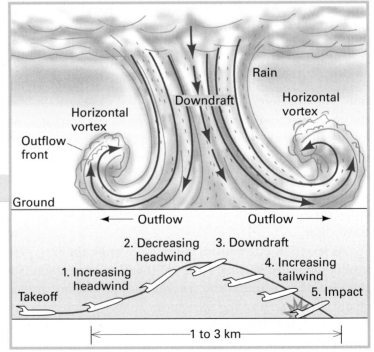

Mesoscale convective systems
FIGURE 10.7

Mesoscale convective systems are large storms with multiple, organized thunderstorm cells. They can span hundreds of kilometers. This photo shows a mesoscale convective system forming over the badlands of South Dakota.

ally, they are much larger than air-mass thunderstorms and can range from 25 km (16 mi) to hundreds of kilometers across, like the one in **FIGURE 10.7**. In addition, the vertical winds associated with these MCSs are also much stronger, and they can be accompanied by much more intense precipitation and winds.

The massive, interconnected cells of an MCS can form in a line, called a *squall line*, or in a circular shape, called a *mesoscale convective complex*. In either case, the process that allows these thunderstorms to organize into cells makes them different from the air-mass thunderstorms described earlier.

Mesoscale convective complexes

One type of MCS forms within **mesoscale convective complexes** (sometimes referred to as *MCCs*). Mesoscale convective complexes are circular formations of multiple thunderstorms that can span several hundred kilometers and are active during the night.

> **mesoscale convective complex** A large complex of thunderstorms, generally round or oval-shaped, characterized by long-lived convection and heavy rainfall that persists through the night.

An MCC usually starts as a group of individual air-mass thunderstorms formed during the day. After the daytime surface heating ceases, diverging upper-air flow in the region of a jet stream disturbance organizes the storm into an MCC. The upper-air divergence supplies the vertical lift needed to sustain the convective activity, while low-level flow of warm, moist air into the region feeds the convective updrafts. As with severe thunderstorms, this convection can be enhanced if the downdrafts within the storm complex generate additional lifting of the incoming low-level flow.

Because the convection in an MCC depends on divergence in the upper-level flow, the complex tends to move at a slow speed, following the motion of the upper-air divergence. New convective cells are added on the side fed by the low-level flow, while old ones die out on the opposite side.

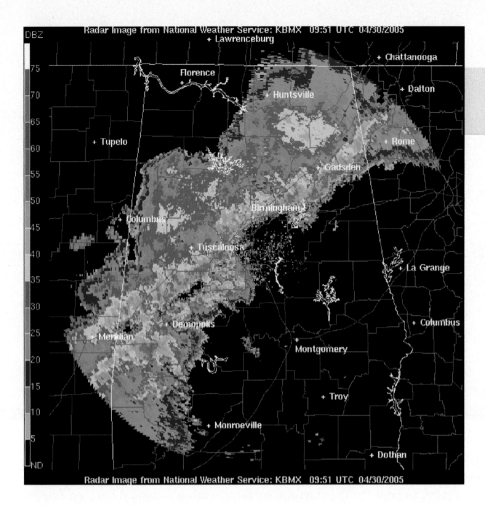

Radar image showing thunderstorms aligned along a squall line FIGURE 10.8

This radar image shows convective precipitation (in reds and yellows) associated with a line of thunderstorms extending across the state of Alabama on April 30, 2005. Gust fronts associated with these storms produced surface winds in excess of 27 m/s (60 mph), knocking down trees and power lines.

Squall lines and drylines

MCSs often form along a **squall line**. A squall line is a line of individual thunderstorms in different stages of development, as seen in FIGURE 10.8. Usually, squall lines form out ahead of an approaching cold front. However, the convection associated with the thunderstorms is not associated with uplift over the cold front itself. Instead, two different processes initiate convection and give rise to these thunderstorms.

> **squall line** Line of thunderstorms and strong winds that extends for several hundred miles.

The first process involves the interaction between winds at the surface and winds aloft (FIGURE 10.9). If we look at the surface winds and winds aloft associated with a well-developed traveling cyclone, we see that at the surface, the winds out ahead of the cold front are southerly. Aloft, however, the winds are southwesterly. These southwesterly winds aloft can produce an axis of divergence that aligns with the surface winds. In such a situation, lifting of surface air occurs parallel to the axis of divergence. The surface winds then feed warm, moist air all along this region of lifting. As the air is forced aloft, water vapor condenses out, releasing latent heat and further enhancing the vertical motions.

Another process that can produce squall lines is the formation of a **dry line** out ahead of an approaching cold front, seen in FIGURE 10.10. A dry line is a boundary between hot, dry, continental tropical (cT) air and warm, moist, marine tropical (mT) air. Both of these air masses are drawn into a region by the prevailing southerly winds out ahead of the cold front. As the hot, dry air encounters the warm, moist air ahead of it, the moister air starts to lift (because humid air is actually less dense than dry air). Severe convection can occur along this line because the cT air is warmer than the

> **dry line** Boundary separating hot, dry air from warm, moist air along which thunderstorms tend to form.

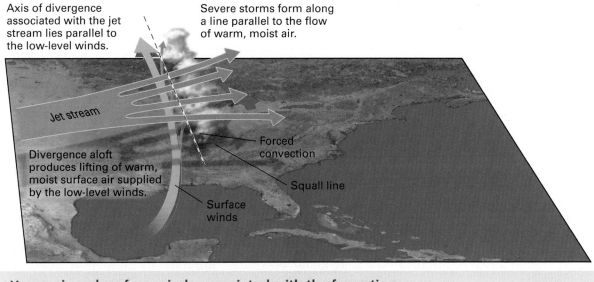

Axis of divergence associated with the jet stream lies parallel to the low-level winds.

Severe storms form along a line parallel to the flow of warm, moist air.

Jet stream

Divergence aloft produces lifting of warm, moist surface air supplied by the low-level winds.

Forced convection

Squall line

Surface winds

Upper-air and surface winds associated with the formation of a squall line FIGURE 10.9

The orientation of the upper-air winds and the surface winds associated with a traveling cyclone can produce the conditions necessary to form a squall line.

Dry line and squall line formation FIGURE 10.10

This schematic shows the dry line separating cT and mT air masses out ahead of a cold front. Squall lines can form along this dry line.

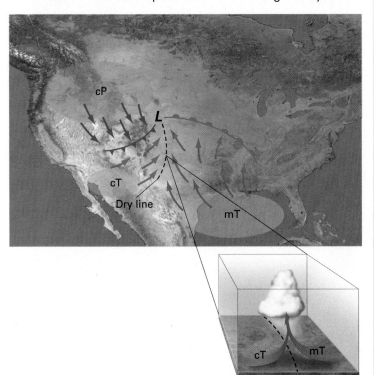

cP

L

cT

Dry line

mT

cT   mT

mT air. As it mixes with the warm, moist air ahead of it, the cT air warms the air parcels further, making them less stable. In contrast, an approaching cold front may initiate convection, but once the cold air begins to mix with the warmer air ahead of it, the overall temperatures tend to decrease and make the air more stable, which reduces convection.

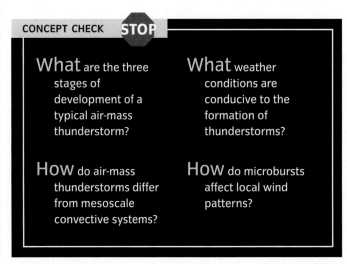

CONCEPT CHECK **STOP**

What are the three stages of development of a typical air-mass thunderstorm?

What weather conditions are conducive to the formation of thunderstorms?

How do air-mass thunderstorms differ from mesoscale convective systems?

How do microbursts affect local wind patterns?

Thunder and Lightning

Learning Objectives

Identify the different types of lightning.

Describe how charge builds up during thunderstorms.

Explain how lightning removes this charge buildup.

Describe the relation between thunder and lightning.

thunder Sound waves generated from the rapid expansion of superheated air around a lightning bolt.

lightning An electrical discharge produced by a thunderstorm, resulting in a flash of light extending from cloud to cloud or cloud to ground.

By definition, a thunderstorm requires the presence of thunder. **Thunder** is the result of a rapid flow of electricity through the atmosphere called an *electrical discharge*, much like a spark but on a much larger scale. Because of the heat associated with the electric discharge, air molecules become hot enough to emit visible light, which we see as **lightning**. At the same time, the heat causes a rapid expansion of the air molecules, which sends out a shock wave that we hear as thunder.

Charge separation during a thunderstorm FIGURE 10.11

The rapid vertical motions associated with a thunderstorm lead to charge separation within a cloud and between clouds. This charge separation produces positively charged and negatively charged regions. When the difference in charge between these regions grows large enough, lightning occurs.

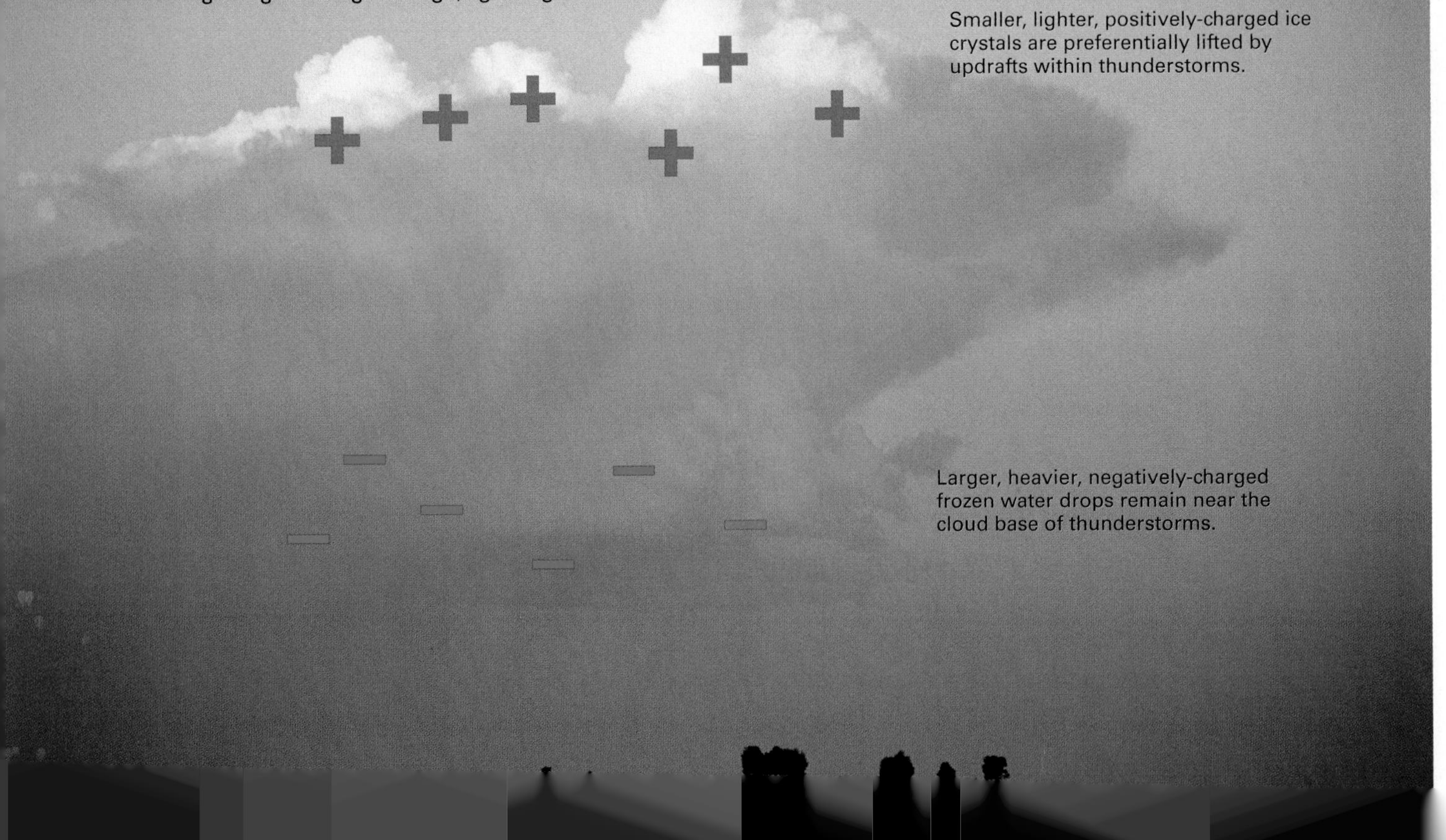

Smaller, lighter, positively-charged ice crystals are preferentially lifted by updrafts within thunderstorms.
Larger, heavier, negatively-charged frozen water drops remain near the cloud base of thunderstorms.

CHARGE SEPARATION

What causes this electrical discharge? It is the result of charge separation that occurs within clouds during the vertical ascent and descent of water and ice within the cloud (FIGURE 10.11). Charge separation is the process in which atoms and molecules in certain portions of the cloud acquire an unbalanced electrical field within them—termed an *electric charge*. In the case of a cloud, certain portions of the cloud build up an excessive negative charge, while other portions build up an excessive positive charge. If the difference between the negatively charged regions of the cloud (or ground) and the positively charged regions becomes large enough, an electrical discharge can occur between the two regions.

The cause of **charge separation** in a cloud is still a subject of active inquiry. In fact, there are at least a dozen hypotheses for how this process occurs. It appears, however that it is related to the fact that positively charged water molecules tend to accumulate on colder objects.

charge separation Process of cloud electrification in which convective transport of charged air parcels results in differences of charge between the top and bottom of a cloud.

One hypothesis is that, as a supercooled liquid water drop freezes, the frozen portion of the water—which is colder—acquires more positively charged water molecules, and hence a positive charge, while the liquid portion of the water acquires more negatively charged molecules, and hence a negative charge. Because water drops freeze from the outside in, the frozen outer shell of the water drops tends to be positively charged, while the interior liquid portion tends to be negatively charged.

As the water drop continues to freeze, it actually expands (which is why cans of soda left in a freezer too long will explode). As the water drop expands, the frozen shell eventually breaks apart into smaller ice fragments. Because these fragments are smaller than the drop itself, they are more easily transported aloft. Hence, the tops of clouds tend to be populated by small, positively charged ice crystals, while the lower portions tend to be populated by larger, negatively charged water drops.

Another hypothesis attributes charge separation to the collision of small, cold ice crystals with frozen water drops—called *graupel*—within the turbulent updrafts. Graupel, which grow as supercooled water drops freeze to their surface, tend to be slightly warmer than the surrounding air, because they are warmed by latent heat released as the water drops change phase. As the warmer graupel collide with the colder ice crystals, individual positively charged water molecules are transferred from the graupel to the ice crystals, once again increasing the positive charge of the lighter ice crystals, which are lofted upward by the updrafts. Removal of positively charged water molecules from the graupel also increases the negative charge of these heavier semi-frozen water drops, which remain closer to the cloud base.

As both of these processes continue, the charge between the top of the cloud and the bottom of the cloud builds up. Eventually, the difference in charge is great enough to trigger a lightning stroke between the two regions of the cloud. This stroke is called *sheet lightning* because it appears as a broad, dull flash throughout the cloud. However, it really represents a lightning stroke whose light is diffused by the cloud.

Whereas 80 percent of lightning takes place between different parts of the cloud, or between one cloud and another, about 20 percent of the time, lightning occurs between a cloud base and the ground. This lightning is called **cloud-to-ground lightning**.

In order for cloud-to-ground lightning to occur, there needs to be a charge separation within the ground as well as within the cloud. The ground charge separation usually occurs because of the buildup of a negative charge within the base of the cloud. As the cloud base passes over a given region, the negatively charged particles within the ground are repelled by the cloud base (much as a negatively charged magnet is repelled by another negatively charged magnet). This process induces a positive charge to the underlying ground.

If the charge difference between the bottom of the cloud and the underlying surface grows large enough, it can produce a **lightning stroke**.

cloud-to-ground lightning Lightning discharge between cloud and ground.

lightning stroke The process of electrical discharge between a cloud and the ground that results in a visible lightning flash.

The lightning stroke is just one part of a much more complicated process that effectively removes (by electrical discharges) the charge separation between the ground and the cloud, described in FIGURE 10.12. In the first step of this process, a **leader** of electrically charged molecules—called *ions*—extends from the cloud base to the ground. It takes about 100 milliseconds for the leader, which is a narrow path about 10 cm wide, to reach from the cloud base to the ground.

Electrons flow downward along the leader. As they near the surface, they rapidly accelerate toward the positively charged ground. In fact, they accelerate so quickly that they produce enough energy to heat the air to temperatures several times hotter than the Sun. This process generates the visible light we see as the **lightning flash**. Electrons at higher levels subsequently accelerate downward as well. The region of acceleration progresses upward along the *lightning channel* established by the leader. This process is like cars at a stoplight progressively accelerating once the light turns green so that the cars farthest from the stoplight accelerate last.

However, unlike cars at a stoplight, the region of acceleration for electrons moves up the leader very rapidly (about 10 milliseconds compared with the 100 milliseconds it took for the leader to reach the ground). The upward-moving region of accelerating electrons is called the **return stroke**. At each point that the electrons are accelerating, they produce a flash. Hence, in stop-action photos it appears the lightning is traveling from the ground to the cloud. However, the *electrons* are traveling downward from the cloud base to the ground, decreasing the negative charge of the cloud base.

leader Channel of ionized air extending from the cloud bottom to the ground that initiates the lightning stroke.

lightning flash The visible light produced by the rapid acceleration of electrons within the leader of the lightning stroke.

return stroke Intense light that propagates upward from the Earth to the cloud base in the last phase of each lightning stroke.

THUNDER

As the return stroke of lightning progresses upward, the energy released instantaneously heats the air, causing the air to expand rapidly. This expansion generates sound waves in the atmosphere that we refer to as *thunder*.

Thunder can be heard as a sharp crack as well as a low rumble. The sound of the thunder we hear depends on how close the lightning is to us as well as the type of lightning that produced the thunder.

In addition, since sound travels through the air at a rate that is about a million times slower than the speed of light, the light from a lightning stroke reaches our eyes before the sound of the lightning reaches our ears. The farther the lightning is from us, the larger is the difference between when we see the lightning and when we hear it.

In fact, the difference can tell us something about how far away the lightning strike is. As a rule of thumb, a lightning strike that is 1.6 km (1 mi) away will have thunder that lags the lightning by 5 seconds. Hence, if there is a 10-second delay between when you see the lightning and when you hear the thunder, you can estimate that the storm that generated the lightning is 3.2 km (2 mi) away.

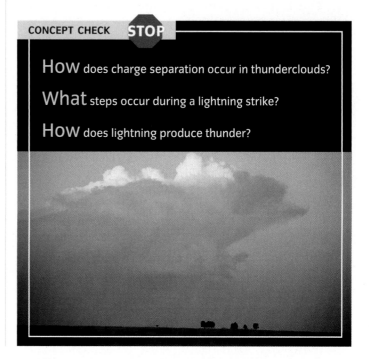

CONCEPT CHECK STOP

How does charge separation occur in thunderclouds?

What steps occur during a lightning strike?

How does lightning produce thunder?

Cloud-to-ground lightning FIGURE 10.12

VIEW THIS IN ACTION
in your WileyPLUS course

The lightning we see during thunderstorms is only one part of a complex process in which charge separation between a cloud base and the underlying ground is discharged.

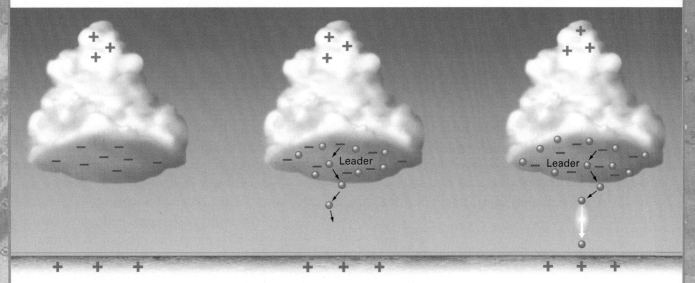

Charge separation between cloud base and ground
Once charge separation has occurred within the cloud, the cloud base acquires an overall negative charge. The negative charge of the cloud base repels negatively charged particles in the ground, inducing a positive charge to the ground.

Descent of leader from the cloud bottom to the surface
The strong negative charge at the cloud bottom is enough to force electrons from the cloud to the molecules below it. This process creates ions, which are negatively charged. These negatively charged ions in turn force electrons to move to the air molecules below them. In this way, a leader of ions extends from the cloud base to the ground.

Initiation of return stroke
Once the leader gets close enough to the ground, the lead electrons rapidly accelerate under the attraction of the positively charged ground. This rapid acceleration heats the surrounding air molecules to over 30,000°C, so they are now hot enough to radiate energy in the visible spectrum. We see this radiation, which we call *lightning*.

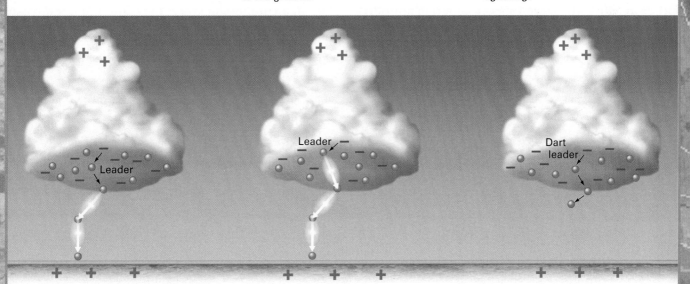

Ascent of the lightning stroke
As the electrons near the ground accelerate rapidly surfaceward, the electrons above them begin to accelerate as well. Hence the region of accelerating electrons moves up along the lightning path created by the leader. The lightning we see is associated with the region of rapidly accelerating electrons.

Discharge of cloud base
As the electrons progressively accelerate down the lightning channel, they effectively remove electrons from the cloud base to the ground. This discharges the negatively-charged cloud base. In addition, it also discharges the positively charged ground by increasing the electrons found there.

Establishment of dart leaders
The discharge is not complete after only one lightning stroke. Additional leaders, called *dart leaders*, begin to extend from the cloud base to the ground. These re-establish lightning channels along which return strokes will travel. Usually one lightning flash is made up of many separate return strokes that the human eye sees as a single flash.

Tornadoes

Describe the characteristic structure of tornadoes.

Explain the weather conditions necessary for tornado formation.

Identify which regions and times of year have high numbers of tornadoes.

 tornado is a small but intense cyclonic vortex in which air spirals at tremendous speed. It is associated with thunderstorms spawned by fronts in the midlatitudes of North America. Tornadoes can also occur inside tropical cyclones (hurricanes).

tornado A rapidly rotating column of air extending from the base of a thunderstorm that comes in contact with the ground.

TORNADO CHARACTERISTICS

A **tornado**, seen in FIGURE 10.13, appears as a dark funnel cloud hanging from the base of a dense cumulonimbus cloud. At its lower end, the funnel may be 100 to 450 m (about 300 to 1500 ft) in diameter. The base of the funnel appears dark because of the density of condensing moisture, dust, and debris swept up by the wind. Wind speeds in a tornado exceed speeds known in any other storm. Estimates of wind speed run as high as 100 m/s

(about 225 mph), although generally they are closer to 50 m/s (about 110 mph). As the tornado moves across the countryside, the funnel writhes and twists. Where it touches the ground, it can cause the complete destruction of almost anything in its path.

The center of a tornado is characterized by low pressures, which are typically 10–15 percent lower than the surrounding air pressures. Although the pressure difference inside and outside the tornado may not be as great as within the interior of a hurricane, it occurs over a very short distance. The result is a very large pressure gradient force that generates high wind speeds as the air rushes into the low-pressure center of the tornado. Because of their small size, tornadoes are not affected by the Coriolis force. Most tornadoes rotate in a counterclockwise direction, but a few rotate the opposite way.

Tornado FIGURE 10.13

A tornado is the rapidly spinning funnel seen descending from the cloud base of a supercell thunderstorm that extends many kilometers up.

A Wind shear

Change in speed of winds with height causes horizontal vortexes to form.

High speed winds aloft (1,000 to 3,000 m)

Low speed winds near surface (100–500 m)

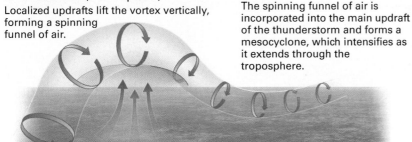

B Convection (warm updraft)

Localized updrafts lift the vortex vertically, forming a spinning funnel of air.

The spinning funnel of air is incorporated into the main updraft of the thunderstorm and forms a mesocyclone, which intensifies as it extends through the troposphere.

C Mesocyclone

Rotation at the bottom of the mesocyclone can induce circulations in the air below it that can become a tornado.

Tornado

Formation of mesocyclones and tornadoes FIGURE 1 0 . 1 4

The combination of vertical shear in the winds, along with regions of very strong convection, produce mesocyclones that extend through the troposphere. Rapidly rotating circulations extending from the bottom of the mesocyclones to the surface can then develop into tornadoes.

TORNADO DEVELOPMENT

The generation of tornadoes is usually associated with the presence of intense thunderstorm activity. This thunderstorm activity provides one of the key ingredients for the initial development of tornadoes—namely, very strong vertical circulations. The other key ingredient is the presence of significant changes in wind speed and direction with height, termed *wind shear*.

In regions where there is significant wind shear, spinning circulations aligned with the ground—*horizontal vortexes*—can form. Strong convection can then lift portions of the vortex, as shown in

> **mesocyclone**
> A vertical column of cyclonically rotating air that develops in the updraft of a severe thunderstorm cell.

FIGURE 1 0 . 1 4, which results in a vertical tower of slowly rotating air, called a **mesocyclone**.

Initially, the mesocyclone is fairly broad. However, as the convection extends the top of the mesocyclone to the tropopause, the mesocyclone stretches and narrows. As it does so, it begins to spin faster, much as ice skaters spin faster as they pull their arms close to their bodies. The winds associated with the mesocyclone begin to cause the air below it to spin as well. About a fifth of the time, a narrow, rapidly circulating vortex stretches from the base of the mesocyclone down to the ground. When it touches the ground, it officially becomes a tornado.

Inside a tornado FIGURE 10.15

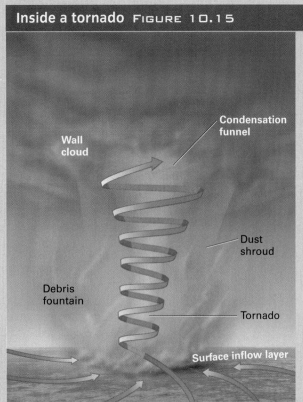

A Within a tornado, inspiraling air circulates around a low-pressure center. Condensation and debris within the tornado produce the characteristic image we see in photos.

B A large tornado can have multiple suction vortexes. Each of these vortexes is about 30 ft across and has very intense winds. The vortexes move slowly around the tornado center.

What happens inside the vortex of a tornado? At its center is low pressure, with strong, inspiraling air surrounding it (**FIGURE 10.15**). As air spirals into the low-pressure center, it expands and therefore cools adiabatically. If it cools enough, water vapor condenses, revealing the shape of the tornado. If the air does not cool enough to produce condensation, the only signature of the tornado is circulating debris and dust lifted by the high winds circulating around the tornado.

While some tornadoes consist of a single funnel (or vortex), some contain multiple vortexes. These tornadoes are called *multiple vortex tornadoes*. Within these tornadoes are separate suction vortexes, each only about 10 m (33 ft) in diameter. However, these vortexes have very intense circulations. When a multiple vortex tornado strikes an area, structures that are hit by the suction vortexes sustain maximum damage, while those a short distance away are left untouched. As a result, tornado damage seems to lie along very narrow tracks, even when the tornado itself is relatively large.

WEATHER CONDITIONS LEADING TO TORNADO FORMATION

One condition necessary for the formation of tornadoes is significant convective activity. Intense convection is typically associated with thunderstorms and supercell thunderstorms in particular. Because of this convection, these thunderstorms can generate tornadoes (although only about one percent of thunderstorms actually do so). However, other weather phenomena can also produce the convection needed to initiate tornadoes, including hurricanes and the passage of cold fronts.

The other condition necessary for tornado formation is the presence of significant wind shear. One way to generate wind shear is to have a very strong horizontal temperature gradient—for example, as found across a cold front. The temperature gradient creates a pressure gradient that gets stronger with elevation, creating the wind shear that initiates mesocyclones and tornadoes. Wind shear in the atmosphere can also be produced by thunderstorms themselves. For example, in the region of a gust front, the air near the surface is moving away from the thunderstorm; however, above the gust front, the air is rushing toward the thunder-

storm to sustain the convection. This change in wind direction with height results in wind shear.

The presence of strong convection and significant wind shear were both met on February 6, 2008, when a series of tornadoes struck the southeastern United States (Figure 10.16). During this outbreak, tornadoes were spread over five different states. There were over fifty reported fatalities, the most from tornadoes in over twenty years.

TORNADO ALLEY

Although the weather conditions necessary for tornado formation can occur in any month, they are most common in the spring and summer. During this time, the passage of midlatitude cyclones generates cold fronts that bring cold, dry air from the north down to the lower latitudes. In addition, the warming of the land surface, combined with the movement of warm, moist air from lower latitudes, can generate the strong temperature gradient needed to produce the wind shear and convection necessary for initiating tornadoes.

Geographically, these conditions are found most often over the central and southeastern United States,

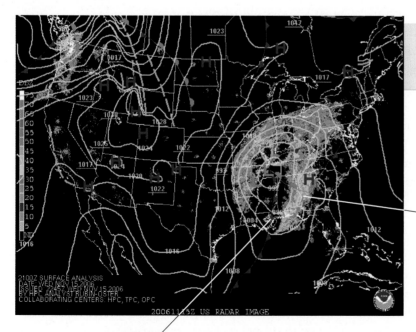

Weather map during tornado outbreak of February 6, 2008

FIGURE 10.16

This map shows the weather conditions during a February 6, 2008, tornado outbreak over the southeastern United States. During this outbreak, tornadoes formed across 5 different states, killing at least 55 people.

Squall line
The radar image shows regions of severe convection (in reds and yellows) forming along a squall line—the red dashed line—stretching across Florida, Alabama, and Georgia. Tornadoes formed throughout these states.

Cold front
Continental polar air moving down from the north met with tropical marine air from the Gulf of Mexico, setting up significant wind shear across the southeastern United States.

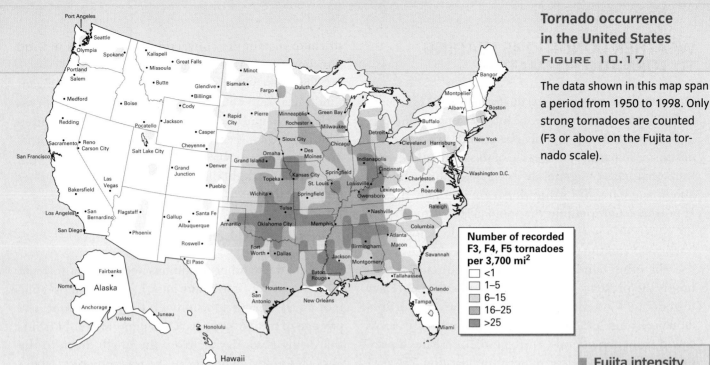

Tornado occurrence in the United States

FIGURE 10.17

The data shown in this map span a period from 1950 to 1998. Only strong tornadoes are counted (F3 or above on the Fujita tornado scale).

Number of recorded F3, F4, F5 tornadoes per 3,700 mi²
- <1
- 1–5
- 6–15
- 16–25
- >25

where cold, dry air from Canada comes in contact with warm, moist air brought up from the Gulf of Mexico by the prevailing flow around the Bermuda-Azores high. In addition, the landscape is generally flat across this region, which allows the two air masses to flow unimpeded, retaining their initial temperature and moisture characteristics. Because of this unique geography, tornadoes occur in greatest numbers in the central and southeastern United States—termed *tornado alley*—but are rare over mountainous and forested regions (**FIGURE 10.17**). They are almost unknown west of the Rocky Mountains and are relatively less frequent on the eastern seaboard. Tornadoes are a typically American phenomenon, being most frequent and violent in the United States. They also occur in Australia in substantial numbers and are occasionally reported from other midlatitude locations.

TORNADO DESTRUCTION

The large majority of tornadoes are relatively weak, lasting only a few minutes and covering only a few hundred meters. On the other hand, large tornadoes, which make up only about 5 percent of all tornadoes, can last for hours and spread destruction over hundreds of kilometers. Devastation from these tornadoes is often complete within the narrow limits of their paths. Only the strongest

buildings constructed of concrete and steel can withstand the extremely violent winds. One way scientists measure the strength of various tornadoes is to measure wind speeds. However, it is almost impossible to position an anemometer or wind vane in the path of a tornado, so direct wind-speed readings are very difficult to obtain. Readings of the pressures in the center of the tornado are also difficult to obtain, again because it is so difficult to place instruments in the proper position.

One way to measure tornado wind speeds is via Doppler radar. In addition to being used to estimate precipitation, Doppler radar can be used to estimate the horizontal velocity of raindrops (and debris) in a tornado, which gives us an estimate of the wind speeds. However, only the very largest tornadoes are visible on Doppler radar. Learn more about Doppler radar and tornadoes in *What a Scientist Sees: Doppler Radar Images of Tornadoes.*

Presently, the most commonly used measure of tornado intensity is the **Fujita intensity scale**, or the F-scale. This scale ranks tornadoes on a scale from 1 to 5, weakest to strongest. The scale is based on the severity of damage found in the tornado's wake. The more severe the damage, the more intense is the tornado. The method is not perfect, because damage is also affected

Fujita intensity scale Scale used to rate the intensity of a tornado by examining the damage caused to different types of human-made structures.

Doppler Radar Images of Tornadoes

In this photo, we see two different radar images of an F5 tornado that struck the Oklahoma City area on May 3, 1999. This tornado, and others that occurred at the same time, caused over $1.5 billion in damage and took 48 lives.

Like all radar, Doppler radar detects objects by sending out a pulse of radiation, which then reflects off the object and returns to the receiver. The strength of the reflection—the *reflectivity*—is stronger for solid objects like dust and debris than it is for liquids such as clouds and raindrops. Hence, localized regions of high reflectivity can indicate airborne debris picked up by a tornado. The radar image on the left shows a characteristic "hook echo" associated with the cloud drops and precipitation within the mesocyclone that spawned the F5 tornado. The white region is an area of very high reflectivity that could only have been generated by airborne debris lofted by the tornado itself. Typically, tornadoes are too far from radar sites to show the signature of the tornado this clearly, and the shape of the hook echo is usually the sole indicator of possible tornado activity.

In addition, by measuring the change in the wavelength of the radar signal that was emitted by the radar and reflected back to the radar site, it is possible to determine whether the reflecting objects are moving toward or away from the radar and how fast. In the right image, we see the velocities toward (green/white) and away from (red/orange) the radar. A shift from greens to the south and reds to the north in the end of the hook indicates a counterclockwise circulation associated with the mesocyclone. In our detailed image, there is also a distinct region with very strong winds toward (white) and away from (pinks) the radar in contact—a *tornado vortex signature*—indicating the high winds circulating around the tornado itself.

Reflectivity
Debris lofted into the atmosphere by the tornado reflects more strongly than surrounding water droplets.

Hook echo
This is a characteristic signature of mesocyclones that can generate tornadoes.

Velocity
Radial velocity toward (green/white) and away from (orange/red) the radar shows high winds moving in opposite directions—a *tornado vortex signature* characteristic of spinning winds in a tornado.

Reflectivity | Storm-relative Velocity

WSR-88D Imagery of Moore, Oklahoma Tornadic Supercell (3 May 1999)

NATIONAL GEOGRAPHIC

by the resilience of the structure. However, it is one of the few ways to directly compare one tornado to another.

In 2007, an *enhanced Fujita intensity scale* was adopted to remove some of the deficiencies in the earlier scale, particularly the tendency to overestimate wind speeds based on damage sustained by various structures. The new scale now incorporates 28 separate damage indicators, with up to 12 different "degree of damage" ratings for each. This scale allows scientists to compare tornadoes that passed through very different

regions (for example, industrial and rural areas) and better estimate the wind speed within each.

As mentioned, only about 5 percent of tornadoes reach 4 or 5 on the Fujita intensity scale, shown in FIGURE 10.18 on pages 290–291. However, these tornadoes are responsible for about 70 percent of all tornado deaths. On average, about 75 people in the United States die each year from tornadoes, more than from any other natural phenomena except flooding and lightning strikes.

F0: 18–32 m/s (40–72 mph); damage to chimneys or TV antennas; broken branches; shallow-trees uprooted; signs and sign boards damaged

Palm trees were uprooted as a weak, F0 tornado passed over Kuwait City's beach front on April 12, 2008.

F1: 33–50 m/s (73–112 mph); surface of roofs peeled away; windows broken; trees on soft ground uprooted; trailer houses moved or overturned

The roof of a restaurant was torn off but the walls were left standing after an F1 tornado hit Daleville, Alabama, on November 25, 2001.

F2: 51–70 m/s (113–157 mph); roofs torn off frame houses; weak structures or outbuildings demolished; large trees snapped or uprooted; cars blown off highways; walls badly damaged but standing

An F2 tornado hit Greymouth New Zealand on March 10, 2005, peeling the roof away from the steel beams of this video store and causing millions of dollars of damage to the city.

F3: 71–92 m/s (158–206 mph); roofs torn off well-constructed houses; some rural building completely demolished; steel-framed, warehouse-type structures badly damaged; cars lifted off the ground; most trees uprooted or leveled

This middle-school gymnasium—constructed of cinder block and concrete—was severely damaged when a tornado strunck Jackson, Mississippi on May 4, 2003.

The Fujita scale is used to rate the severity of tornadoes. It is based on the type of damage left in the wake of the tornado.

F4: 93–116 m/s (207–260 mph); well-constructed houses leveled, leaving piles of debris; structures with weak foundations lifted and blown away; cars blown large distances; large projectiles generated

The walls were torn from the foundation of this home when an F4 tornado struck Madisonville, Kentucky, on December 8, 2005.

F5: 117–142 m/s (261–318 mph); strong frame houses lifted clear off foundation and disintegrate; steel-reinforced concrete structures badly damaged; trees completely debarked; automobile-sized projectiles thrown 100 yards or more

This aerial photo shows the severity and extent of damage associated with an F5 tornado that hit Greensburg, Kansas, on May 4, 2007. This tornado destroyed more than 95 percent of the town.

GREENSBURG, KANSAS

Global Locator

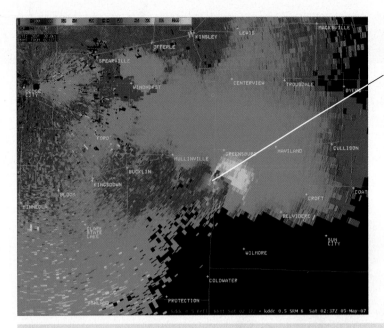

Velocity
Radial velocity toward (green/white) and away from (orange/red) the radar shows high winds moving in opposite directions—a tornado vortex signature characteristic of spinning winds in a tornado.

Doppler radar image leading to a tornado warning FIGURE 10.19

The National Weather Service can alert people to the potential danger of tornadoes by issuing severe storm alerts, tornado watches, tornado warnings, and tornado emergencies. Tornado warnings and emergencies are only announced if an actual tornado has been spotted or its signature has been seen on Doppler radar. In this image, we see the radar velocity for May 4, 2007. A tornado in the vicinity of the tornado signature hit Greensburg, Kansas. In response to this image and sightings from tornado spotters, a tornado emergency was declared 30 minutes before the tornado struck.

TORNADO FORECASTING

Although tornadoes are short-lived and generally confined to local regions, because of their ability to cause significant damage and loss of life, they are a priority when it comes to weather forecasting. However, their short life span, small size, and erratic path make them one of the most difficult weather phenomena to predict.

Generally, there are three stages in the prediction and forecasting of tornadoes. In the first stage, forecasters announce a severe storm outlook. These outlooks are made throughout the day and highlight regions across the country that may experience severe storms over the next 24 hours.

In the next stage, a *tornado watch* is issued. Watches are issued only for specific regions (covering 10,000 square miles) in which severe weather patterns have organized into possible tornado-producing circulations. These watches are usually for a fixed 3–6 hour period.

In the final stage, a *tornado warning* or *tornado emergency* is issued. A tornado warning is only issued if a tornado has been spotted or if the signature of a tornado is seen on a Doppler radar screen (FIGURE 10.19). Warnings are for very specific regions (usually covering only 1000 square miles) and for a fairly short period, usually 30–60 minutes. However, they indicate a very high probability of tornado activity in the region. Because of the use of Doppler radar for detecting tornadoes and issuing warnings, the average time between a tornado warning and the arrival of the tornado has increased from 5 minutes to 11 minutes. This may not seem like much, but for events as severe as these, any extra time to find shelter can save many lives.

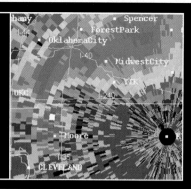

What are the characteristic sizes and wind speeds of tornadoes?

What weather conditions are necessary for tornado formation?

When and where do tornadoes most frequently strike?

How is the intensity of tornadoes measured?

What is happening in this picture **?**

This image shows dust being blown across grasslands by a gust front.

◾ Based on your understanding of gust fronts and their relation to thunderstorms, where do you think the thunderstorm itself is located?

◾ If you were sitting on the road seen here, would you expect clear or rainy conditions over the next 15–30 minutes?

1 Thunderstorms

1. **Thunderstorms** are intense local storms associated with a tall, dense cumulonimbus cloud in which there are very strong updrafts of air. They can range from fairly isolated, short-lived storms, sometimes called **air-mass thunderstorms**, to massive, well-organized complexes of storms, called **mesoscale convective systems**.

2. Air-mass thunderstorms develop along three stages. In the **cumulus stage**, surface heating produces isolated regions of significant updrafts, leading to the formation of cumulus clouds and precipitation. In the **mature stage**, organized convection forms within the thunderstorm, producing strong updrafts and downdrafts, heavy rainfall, gusty winds, and lightning. Eventually, the thunderstorm reaches the **dissipating stage**, in which strong downdrafts throughout the air column inhibit convection and latent heat release needed to sustain the thunderstorm.

3. Mesoscale convective systems are thunderstorms in which there is the formation of multiple, organized thunderstorm cells. The organization of these multiple cells can occur along a **squall line** or as part of a **mesoscale convective complex**. These thunderstorms can be sustained for long periods of time by divergence of the winds aloft. In addition, squall lines can be formed by the presence of a **dry line**, in which warm, moist air encounters hot, dry air.

2 Thunder and Lightning

1. By definition, thunderstorms are associated with the presence of thunder. **Thunder** in turn is generated by lightning. Hence, lightning and thunder are integral components of any thunderstorm.

2. **Lightning** requires charge separation within a cloud, as lighter, colder ice crystals, which tend to be positively charged, move upward and heavier, warmer ice particles, which tend to be negatively charged, move downward in the cloud. This effect concentrates positive charge in the upper part of the cloud and negative charge in the lower part. The electrical potential builds until an electrical discharge occurs, creating cloud-to-cloud lightning, termed *sheet lightning*.

3. About 20 percent of the time, the charge in the cloud is discharged through **cloud-to-ground lightning**. This lightning occurs as a complex process in which the negative cloud-base charge induces a positive charge in the underlying ground. A **leader** of ions extends from the cloud base to the ground, establishing a lightning channel. The **lightning stroke**, which produces the light we actually see, is associated with rapidly accelerating electrons moving toward the surface within this lightning channel.

4. The rapidly accelerating electrons within the lightning channel heat the surrounding air molecules to 30,000°C. The heating causes the air to expand rapidly, generating a shock wave that we hear as thunder.

3 Tornadoes

1. Occasionally, thunderstorms can also produce tornadoes. **Tornadoes** are small but intense cyclonic vortexes in which air spirals at tremendous speed. Only about one percent of thunderstorms generate tornadoes; however, when they do, they can be extremely destructive and cause high numbers of fatalities.

2. Tornadoes are generally 100–450 m (about 300–1500 ft) across, although sometimes they can span several kilometers. The wind speeds within a tornado range from 50 to 100 m/s (110–225 mph).

3. The generation of tornadoes requires the presence of strong convection and wind shear—significant changes in wind speed and direction with height. When these two are present, a portion of the rolling vortex associated with the wind shear can be lifted by convective activ-

ity, producing a vertical vortex termed a **mesocyclone**. Under certain circumstances, a narrow, rapidly circulating vortex stretches from the base of the mesocyclone down to the ground, which we call a tornado.

4. The majority of tornadoes last only a few minutes and produce relatively little damage. However, other tornadoes can last from 30 minutes to a few hours and generate massive damage across their track. The intensity of tornadoes is usually gauged by the damage that they cause. This intensity is based on the **Fujita intensity scale**, which rates tornadoes on a scale of 1 to 5.

KEY TERMS

CRITICAL AND CREATIVE THINKING QUESTIONS

1. How do air-mass thunderstorms develop? What are the three stages of their development?

2. How do mesoscale convective systems differ from air-mass thunderstorms? How are these thunderstorms produced?

3. What process produces charge separation in clouds? What region of the cloud is negatively charged? What region is positively charged?

4. In cloud-to-ground lightning, what are the stages of discharge that ultimately produce a lightning flash?

5. Describe a tornado. What are the typical wind speeds of tornadoes? What is their typical size?

6. What meteorological conditions are necessary for tornado formation? Where and when do tornadoes typically occur?

1. The two conditions that promote thunderstorm development are _____.
   a. warm, moist air and a stable environmental lapse rate
   b. cool, dry air in collision with a colder air mass
   c. warm, moist air and an unstable environmental lapse rate
   d. cold, wet air and a stable environmental lapse rate

2. A very intense downdraft of air produced by a thunderstorm is called a _____.
   a. tornado
   b. hurricane
   c. microburst
   d. sleet storm

3. The three stages of air-mass thunderstorm development include _____.
   a. cumulus, mature, and dissipating
   b. inititiation, mature, and closing
   c. early, middle, and late
   d. onset, growth, and decay

4. A characteristic feature we would *not* expect to find in a severe thunderstorm is _____.
   a. anvil cloud formation
   b. a gust front
   c. an eye wall
   d. strong updrafts and downdrafts

5. On the diagram below, label the following features of a severe thunderstorm: (a) anvil cloud, (b) overshooting top, (c) updrafts, (d) downdrafts, (e) gust front, (f) roll cloud.

6. Squall lines can form along a dry line separating these two air masses: _____ and _____.
   a. cP; mT
   b. mT; cT
   c. cP; cT
   d. mT; mP

7. Cloud-to-ground lightning is initiated by _____ between the cloud base and the ground.
   a. static bonding
   b. water transfer
   c. wind shear
   d. charge separation

8. On the photo below, indicate which regions you would expect to be positively charged and which regions you would expect to be negatively charged. Do not forget to consider the ground as well as the cloud.

9. A lightning flash is the visible light emitted as electrons accelerate down the _____.
   a. leader
   b. vortex
   c. gust front
   d. downdraft

10. A _____ is a small but intense cyclonic vortex with very high wind speeds.
    a. hurricane
    b. tornado
    c. typhoon
    d. cyclone

11. Tornadoes occur as parts of cumulonimbus clouds traveling in advance of a warm front.
    a. True
    b. False

12. Tornadoes are typically associated with regions of
    _____.
    a. strong wind shear
    b. intense convection
    c. advancing cold fronts
    d. All of the above

13. Tornadoes typically form in the _____ season over the _____ portion of the United States.
    a. winter and fall; northeastern and northwestern
    b. winter; southwestern
    c. spring and summer; southeastern and central
    d. summer; southwestern and northwestern

14. Tornadoes are typically _____ wide.
    a. 1–5 m
    b. 100–450 m
    c. 5–10 km
    d. 50–100 km
    e. greater than 500 km

15. On the figure below, label the following features of a tornado: (a) wall cloud, (b) dust shroud, (c) condensation funnel, (d) debris fountain.

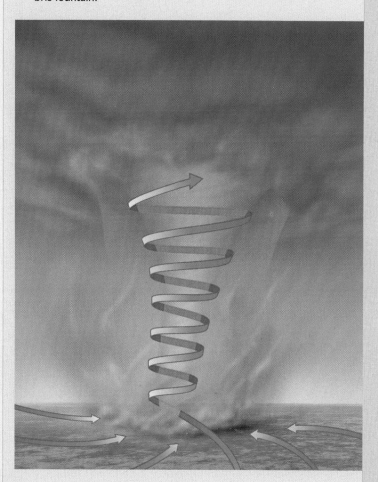

The Global Scope of Climate

Starting in the 1300s, much of Europe fell under the spell of an unusually cool period—depicted by Pieter Brueghel the Elder in this painting of a village in Flanders in 1564—that lasted up until the mid-1800s. Glaciers descended from mountains, overrunning farms and villages in England, Switzerland, and Denmark. In winter, rivers large and small froze over, including the Thames in England, which at one point was covered in more than 2 meters (6 feet) of ice. Farther north, whole colonies in Iceland and Greenland died out. The colder temperatures even played a part in American history. The famous picture of George Washington crossing an ice-packed Delaware River on December 25, 1776, as part of a surprise attack during the Revolutionary War would be difficult to replicate today—the Delaware River rarely freezes in today's climate.

How much colder was the climate then than it is today? Most estimates put the temperatures of the northern hemisphere at less than 1°C (or 1.8°F) colder than at present—yet this small change in the climate had an enormous influence across large portions of the globe. Here we see that the climate of a given region not only affects the weather of that region, but also the social, cultural, and even political character of the region as well.

298

Hunters in the Snow (1565) by Pieter Brueghel the Elder.

Factors Controlling Climate

LEARNING OBJECTIVES

Define climate.

Identify the factors that affect climate.

Climate is the average weather of a region. The primary driving force for weather is the flow of solar energy received by the Earth and atmosphere. Because that energy flow varies on daily cycles with the planet's rotation and on annual cycles with its revolution in orbit, it imposes these cycles on temperature and precipitation.

Climate controls include latitude, elevation, and proximity to oceans, as well as others. These controls influence the annual and seasonal temperatures and precipitation for a given region.

Latitude
Insolation varies with latitude. The annual cycle of temperature at any place depends on its latitude. Near the Equator, temperatures are warmer and the annual range is low. Toward the poles, temperatures are colder and the annual range is greater. Ellesmere Island, Nunavut, Canada

Coastal-continental location
Ocean surface temperatures vary less with the seasons than land surface temperatures, so coastal regions show a smaller annual variation in temperature. Aerial view of Cornwall, England

Elevation
High-elevation stations show cooler temperatures than sea-level stations, because the atmosphere cools with elevation at the average environmental temperature lapse rate. Kluane National Park, Canada

Annual and monthly air temperatures
Warm air can contain more moisture than cold air, so colder regions generally have lower precipitation than warmer regions. Also, precipitation will tend to be greater during the warmer months of the temperature cycle. Mount Des Voeux, Tavenui Island, Fiji Islands

However, climate includes other time cycles as well, ranging from the cycles of several years' duration, such as the cycles imparted by El Niño, to cycles of hundreds of thousands of years, like the cycles of continental glaciation.

In addition to recognizing the time cycles imposed by the flow of solar energy to a given region, a few other principles are very helpful in understanding the global scope of climate. As shown in FIGURE 11.1, three major factors influence the annual cycle of air temperature experienced at a location: latitude, coastal versus continental location, and elevation.

For precipitation, these same three factors are important, but precipitation is also affected by annual and daily air temperatures, prevailing air masses, relation to mountain barriers, position of persistent high- and low-pressure centers, and prevailing wind and ocean currents.

Keeping these key ideas in mind as you read this chapter will help make the climates we discuss easier to understand and explain.

Visualizing

Climate controls FIGURE 11.1

Air masses
Air masses that come from continental regions are drier, while air masses from marine locations are moister and hence support more precipitation. In regions where two prevailing air masses collide, fronts will form, which can lead to precipitation. Baffin Island, Canada

Topography
Location on a mountain can greatly affect the amount of precipitation. On the windward side, the forced uplift of air over the mountains produces condensation and precipitation. On the leeward side, adiabatic warming of the air produces hot, dry conditions. Winery in Paso Robles, California

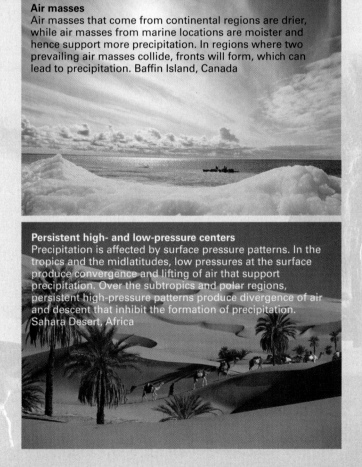

Persistent high- and low-pressure centers
Precipitation is affected by surface pressure patterns. In the tropics and the midlatitudes, low pressures at the surface produce convergence and lifting of air that support precipitation. Over the subtropics and polar regions, persistent high-pressure patterns produce divergence of air and descent that inhibit the formation of precipitation. Sahara Desert, Africa

Prevailing wind and ocean currents
On the western portion of midlatitude continents, the prevailing westerly winds bring warm, moist air off the ocean and onto the continent, resulting in higher precipitation. On the eastern portion, these same winds bring dry air from the continental locations without enough moisture to produce significant precipitation. Redwood National Park, California

CONCEPT CHECK STOP

What characteristics do we use to describe climate in different regions?

What factors affect climate in different regions?

Temperature and Precipitation Regimes

LEARNING OBJECTIVES

Describe the characteristics of temperature regimes at different latitudes and coastal/continental locations.

Explain the role latitude plays in determining the location of precipitation regimes.

Explain how air masses affect precipitation regimes.

P atterns of mean monthly temperatures are referred to as *temperature regimes*— distinctive types of annual temperature cycles related to latitude and location.

TEMPERATURE REGIMES

FIGURE 11.2 shows annual cycles of air temperature for different temperature regimes. The equatorial regime (Douala, Cameroon, 4° N) is uniformly very warm with temperatures close to 27°C (81°F) year round because insolation is nearly uniform throughout the year. In contrast, temperatures in the tropical continental regime (In Salah, Algeria, 27° N) change from very hot when the Sun is high, near one solstice, to mild at the opposite solstice. The tropical west-coast regime at Walvis Bay, Namibia (23° S)—nearly the same

Temperature regimes
FIGURE 11.2

This figure shows typical patterns of mean monthly temperatures observed at stations around the globe. Each of these *temperature regimes* is labeled according to its latitude zone: equatorial, tropical, midlatitude, and sub-arctic. Some labels also describe the location of the station in terms of its position on a land-mass—"continental" for a continental interior location and "west coast" or "marine" for a lo-cation close to the ocean.

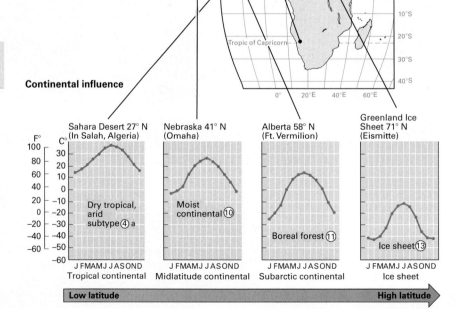

latitude as In Salah—has only a weak annual cycle and no extreme heat because of the moderating effects of its maritime location. This moderating effect persists poleward, into the midlatitude west-coast regime—Monterey, California (36° N) and Sitka, Alaska (57° N). In continental interiors, however, the midlatitude continental regime of Omaha, Nebraska (41° N), and the subarctic continental regime of Fort Vermilion, Alberta (58° N), show large annual variations in mean monthly temperature of about 30°C (54°F) and 40°C (72°F), respectively. The ice sheet regime of Greenland (Eismitte, 71° N) experiences severe cold all year.

Overall, we find that (1) annual variation in insolation, determined by latitude, provides the basic control on temperature, and (2) the effect of location—maritime or continental—moderates that variation.

PRECIPITATION REGIMES

Global precipitation patterns are largely determined by air masses and their movements, which in turn are produced by global air circulation patterns. Before taking a detailed look at global precipitation patterns, let's look at FIGURE 11.3, which shows the general patterns expected for a hypothetical supercontinent that has most of the features of the Earth's continents but is simplified. The map recognizes and defines five classes of annual precipitation: wet, humid, subhumid, semiarid, and arid.

Beginning with the equatorial zone, the figure shows a wet band stretching across the continent. This band is produced by convective precipitation over the equatorial lows near the intertropical convergence zone. Note that the wet band widens and is extended poleward into the tropical zone along the continent's eastern coasts. This region is kept moist by the influence of the trade winds, which move warm, moist mT air masses and tropical cyclones westward onto the continental coast. Farther poleward, humid conditions continue along the east coasts into the midlatitude zones. In these regions, subtropical high-pressure cells tend to move mT air masses from the east onto the continent

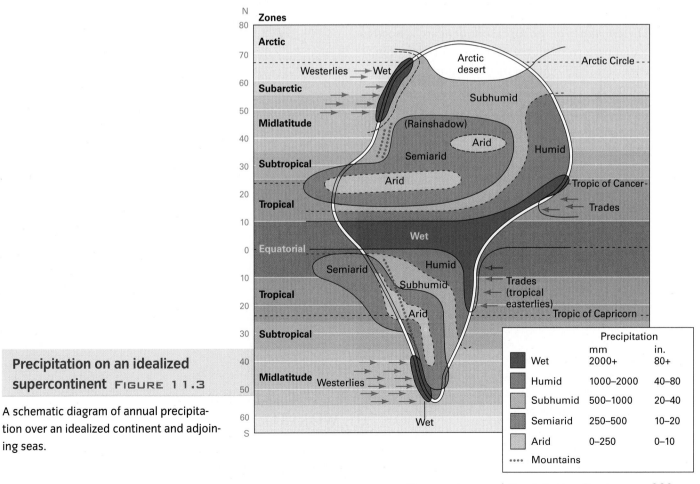

Precipitation on an idealized supercontinent FIGURE 11.3

A schematic diagram of annual precipitation over an idealized continent and adjoining seas.

	Precipitation	
	mm	in.
Wet	2000+	80+
Humid	1000–2000	40–80
Subhumid	500–1000	20–40
Semiarid	250–500	10–20
Arid	0–250	0–10
···· Mountains		

in the summer, whereas in winter, midlatitude cyclones bring cyclonic precipitation from the west.

In the arctic zone, shown on the continent as arctic desert, precipitation remains low because air temperatures are low, and only a small amount of moisture is contained in cold air.

Another important feature of the hypothetical continent is the pattern of arid and semiarid regimes that stretches from tropical west coasts to subtropical and midlatitude continental interiors. In the tropical and subtropical latitudes, the arid pattern is produced by dry, subsiding air in persistent subtropical high-pressure cells. The aridity continues eastward and poleward into semiarid continental interiors, which remain relatively dry because they are far from source regions for moist air masses. Rain-shadow effects provided by coastal mountain barriers are also important in maintaining inland aridity.

Yet another obvious feature of the supercontinent is the pair of wet bands along the west coasts of the midlatitude and subarctic zones produced by the eastward movement of moist mP air masses onto the continent driven by the prevailing westerlies.

These features of the supercontinent are echoed in the actual pattern of global precipitation, shown in FIGURE 11.4. This map of mean annual precipitation shows *isohyets*—lines drawn through all points having the same annual precipitation. Using the same logic that we used to explain the precipitation patterns of the hypothetical continent, we can recognize seven global precipitation regions as follows:

1. *Wet equatorial belt.* This zone of heavy rainfall, over 2000 mm (80 in.) annually, straddles the Equator and includes the Amazon River Basin in South America, the Congo River Basin of equatorial Africa, much of the African coast from Nigeria west to Guinea, and the East Indies. In this zone, the warm temperatures and high-moisture content of the mE air masses favor abundant convective rainfall.

2. *Trade-wind coasts.* Narrow coastal belts of high rainfall, 1500 to 2000 mm (about 60 to 80 in.), extend from near the Equator to latitudes of about 25° to 30° N and S on the eastern sides of every continent or large island. Examples include the eastern coast of Brazil, Central America, Madagascar, and northeastern Australia. The rainfall of these coasts is supplied by moist mT air masses from warm oceans that are brought

World precipitation
FIGURE 11.4

This global map of mean annual precipitation uses *isohyets*—lines drawn through all points having the same annual precipitation— labeled in mm (in.).

MEAN ANNUAL PRECIPITATION OF THE WORLD

Millimeters 0   100   300   500   1000   2000   5000 above 5000

Inches 0   4   12   20   40   80   200 above 200

Isohyets labeled in millimeters, inches (approximate) in parentheses.

over the land by the trade winds and encounter coastal hills and mountains, producing heavy orographic rainfall.

3. *Tropical deserts.* In contrast to the wet equatorial belt are the zones of tropical deserts lying approximately on the Tropics of Cancer and Capricorn. These are hot, barren deserts with less than 250 mm (10 in.) of rainfall annually and, in many places, with less than 50 mm (2 in.). They are located under the large, stationary subtropical cells of high pressure, in which the subsiding cT air mass is adiabatically warmed and dried.

4. *Midlatitude deserts and steppes.* Farther northward, in the interiors of Asia and North America between latitude 30° and latitude 50°, are great deserts, as well as vast expanses of semiarid grasslands known

as *steppes.* Annual precipitation ranges from less than 100 mm (4 in.) in the driest areas to 500 mm (20 in.) in the moister steppes. Dryness here results from remoteness from ocean sources of moisture.

Located in regions of prevailing westerly winds, these arid lands typically lie in the rain shadows on the lee side of coastal mountains and highlands. In the southern hemisphere the dry steppes of Patagonia, lying on the lee side of the Andean chain, are roughly the counterpart of the North American deserts and steppes.

5. *Moist subtropical regions.* On the southeastern sides of the continents of North America and Asia, in latitude 25° to 45° N, are the moist subtropical regions, with 1000 to 1500 mm (about 40 to 60 in.) of rainfall annually. Smaller areas of the same type are found in Uruguay, Argentina, and southeastern Australia. These regions are positioned on the moist western sides of the oceanic subtropical high-pressure circulations, which bring moist mT air masses from the tropical ocean onto the continent.

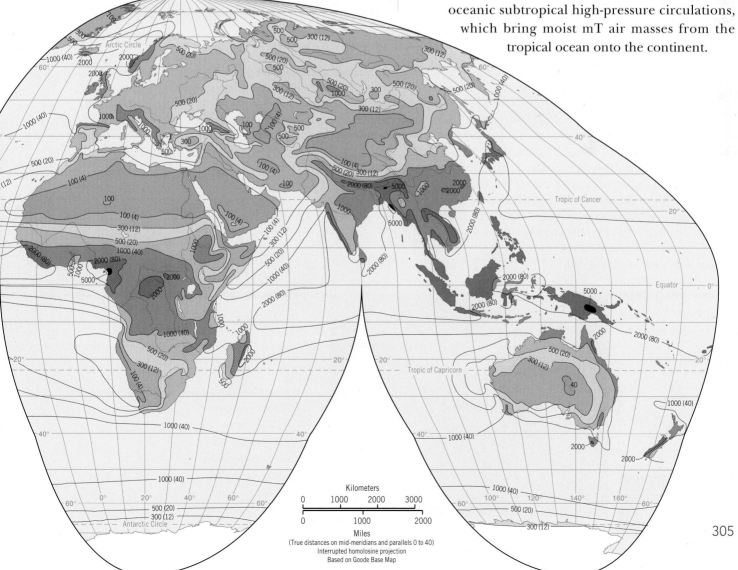

Kilometers
0   1000   2000   3000

Miles
0   1000   2000

(True distances on mid-meridians and parallels 0 to 40)
Interrupted homolosine projection
Based on Goode Base Map

305

6. *Midlatitude west coasts.* Another wet location is on midlatitude west coasts of all continents and large islands lying between latitudes about 35° and 65° in the region of prevailing westerly winds. In these zones, abundant orographic precipitation occurs as a result of forced uplift of mP air masses. Where the coasts are mountainous, as in British Columbia, southern Chile, Scotland, and South Island of New Zealand, the annual precipitation is over 2000 mm (79 in.).

7. *Arctic and polar deserts.* A seventh precipitation region is formed by the arctic and polar deserts. Northward of the 60th parallel, annual precipitation is largely under 300 mm (12 in.), except for the west-coast belts. Cold cP and cA air masses cannot contain much moisture, and, consequently, they do not yield large amounts of precipitation.

SEASONALITY OF PRECIPITATION

Total annual precipitation is a useful quantity in establishing the character of a climate type, but it does not account for the seasonality of precipitation. The variation in monthly precipitation through the annual cycle is a very important factor in climate descriptions. If there is a pattern of alternating dry and wet seasons instead of a uniform distribution of precipitation throughout the year, we can expect that the natural vegetation, soils, crops, and human use of the land will all be different. It also makes a great deal of difference whether the wet season coincides with a season of higher temperatures or with a season of lower temperatures. If the warm season is also wet, growth of both native plants and crops is enhanced. If the warm season is dry, the stress on growing plants is great, and irrigation is required for crops (see *What a Scientist Sees: Drought and Vegetation*).

What a Scientist Sees

Drought and Vegetation

Rainfall is important for supporting agriculture and livestock. When the rains don't come, regions can go into a prolonged drought. With ready access to irrigation, the effects of drought can be minimized—but even in a developed country such as the United States, sustained droughts can be devastating. The maps show summertime drought conditions during 2002. The top map shows how rainfall up through August 2002 ranked within each state's 112-year record. Regions in the southwestern United States had some of the lowest rainfall years recorded, while those in the upper portion of the Mississippi Valley had rainy conditions.

The bottom image is an estimate of vegetation growth during August 2002 based on satellite observations of reflected solar radiation. Brown colors in this image represent decreased vegetation growth compared with normal years, while green colors represent enhanced vegetation growth. The geographic correspondence is very good between regions with the most severe drought and those with the sharpest drop in vegetation, particularly over Colorado, Utah, and Arizona.

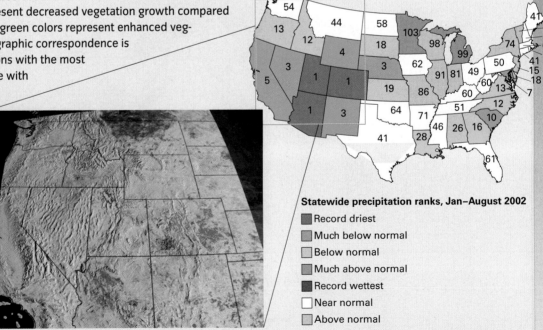

Statewide precipitation ranks, Jan–August 2002

- Record driest
- Much below normal
- Below normal
- Much above normal
- Record wettest
- Near normal
- Above normal

Precipitation regimes FIGURE 11.5

Seasonal patterns of precipitation can be relatively uniform, ranging from very dry ④ to very wet ①; can have a maximum in summer during the high-Sun period ②; or can have a maximum in the winter during the low-Sun period ⑦.

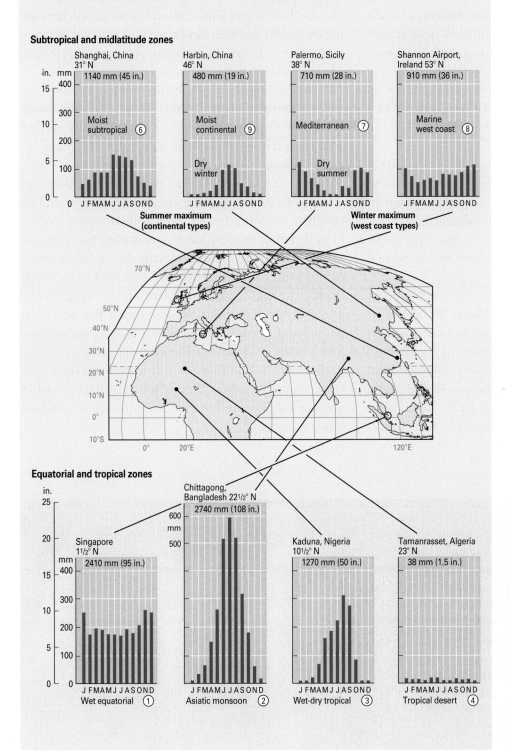

Monthly precipitation patterns can be grouped into three types: (1) uniformly distributed precipitation; (2) a precipitation maximum during the summer (or season of high Sun), in which insolation is at its peak; and (3) a precipitation maximum during the winter or cooler season (season of low Sun), when insolation is least. Note that the uniform pattern can include a wide range of possibilities from little or no precipitation in any month to abundant precipitation in all months.

FIGURE 11.5 shows a set of monthly precipitation diagrams selected to illustrate the major types that occur over the globe. Two stations show the uniformly distributed pattern described above—Singapore, a wet equatorial station near the Equator (1 1/2° N), and Tamanrasset, Algeria, a tropical desert station very near the Tropic of Cancer at 23° N. In Singapore, rainfall is abundant in all months, but some months have somewhat more rain than others. Tamanrasset has so little rain in any month that it scarcely shows on the graph.

Chittagong, Bangladesh (22 1/2° N), and Kaduna, Nigeria (10 1/2° N), both show patterns of the second type—that is, a wet season at the time of high Sun (summer solstice) and a dry season at the time of low Sun (winter solstice). Chittagong is an Asian monsoon station, with a very large amount of precipitation falling during the high-Sun season. Kaduna, an African station with about half the total annual precipitation, shows a similar pattern and is also of the wet-dry tropical type. Both stations experience their wet season when the intertropical convergence zone (ITCZ) is nearby and their dry season when the ITCZ has retreated to the other hemisphere.

The summer precipitation maximum also occurs at higher latitudes on the eastern sides of continents. Shanghai, China (31° N), shows this pattern nicely in the subtropical zone. The same summer maximum persists into the midlatitudes. For example, Harbin, in eastern China (46° N), has a long, dry winter with a marked period of summer rain.

In contrast to these patterns are cycles with a winter precipitation maximum. Palermo, Sicily (38° N), is an example of the Mediterranean type of climate, named for its prevalence in the lands surrounding the Mediterranean Sea. This type experiences a very dry summer but has a moist winter. Southern and central California are also regions of this climate type. In Mediterranean climates, summer drought is produced by subtropical high-pressure cells, which intensify and move poleward during the high-Sun season. These cells extend into the regions of Mediterranean climate, providing the hot, dry weather associated with cT air masses while blocking the passage of other, moister, air masses from the oceans. In the low-Sun season, the subtropical high-pressure cells move equatorward and weaken, allowing frontal and cyclonic precipitation to penetrate into Mediterranean climate regions.

The dry-summer, moist-winter cycle is carried into higher midlatitudes along narrow strips of west coasts. Shannon Airport, Ireland (53° N), has this marine west-coast type of climate, although the difference between summer and winter rainfall is not as marked. Summers at Shannon Airport have less rainfall for two reasons. First, the blocking effects of subtropical high-pressure cells tend to extend poleward into the region, keeping moister air masses and cyclonic storms away. Second, the cyclonic storms that produce much of the winter precipitation are reduced in intensity during the high-Sun season because the temperature and moisture contrasts between polar and arctic air masses and tropical air masses are weaker in summer. The weaker contrasts are caused by increased high-latitude insolation. Without the strong temperature contrast, disturbances in the jet stream do not tend to grow as large, so neither do the underlying midlatitude cyclones.

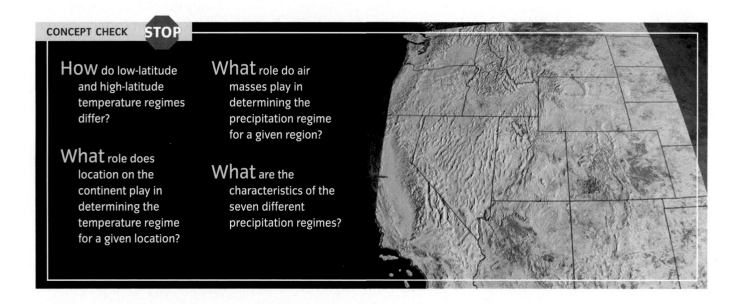

CONCEPT CHECK STOP

How do low-latitude and high-latitude temperature regimes differ?

What role does location on the continent play in determining the temperature regime for a given location?

What role do air masses play in determining the precipitation regime for a given region?

What are the characteristics of the seven different precipitation regimes?

Climate Classification

Define climograph, and describe how the climograph is used to characterize regional climates.

Explain how latitude determines the three broad climate groups.

Describe the difference between dry and moist climates.

Mean monthly values of air temperature and precipitation can describe the climate of a weather station and its nearby region quite accurately. To study climates from a global viewpoint, climatologists classify these values into distinct climate types. This classification requires developing a set of rules to use in examining monthly temperature and precipitation values. By applying the rules, a climatologist can use each station's data to determine the climate to which it belongs.

This textbook recognizes 13 distinctive climate types that are designed to be understood and explained by air-mass movements and frontal zones—that is, by the weather various regions experience throughout the year.

An air mass is classified according to the general latitude of its source region, which determines the temperature of the air mass and its surface type—land or ocean—within that region, which controls the moisture content. Since the air-mass characteristics control the two most important climate variables—temperature and precipitation—we can explain climates using air masses as a guide. In addition, where differing air masses are in contact, frontal zones will form. The position of these

frontal zones changes with the seasons. The seasonal movements of air masses and frontal zones therefore influence annual cycles of temperature and precipitation in certain locations as well, particularly in the mid- and high-latitudes.

The rules that define the climate types in this text are based on an analysis of how the amount of moisture held in the soil varies throughout the year, which is determined by air temperature and rainfall. Our discussion here will not focus on the specific climate rules but will instead focus on showing how the classification follows quite naturally from an understanding of the processes that produce variations of temperature and precipitation around the globe.

FIGURE 11.6 shows a schematic diagram of air-mass source regions used in conjunction with the

Climate groups and air-mass source regions
FIGURE 11.6

Using the map of air-mass source regions, we can identify five global bands associated with three major climate groups. Within each group is a set of distinctive climates with unique characteristics that are explained by the movements of air masses and frontal zones.

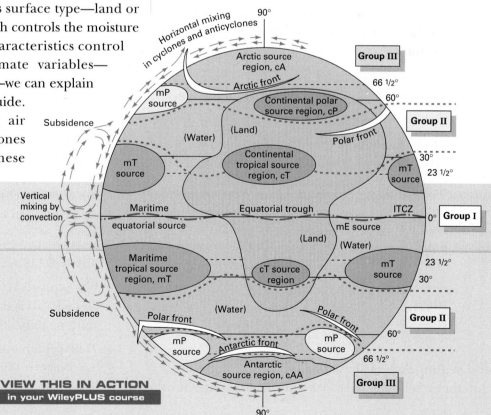

VIEW THIS IN ACTION in your WileyPLUS course

13 climate types described in the text. We have subdivided this diagram into global bands that contain three broad groups of climates: low-latitude (Group I), midlatitude (Group II), and high-latitude (Group III), described briefly as follows.

- *Group I: Low-latitude climates.* The region of low-latitude climates (Group I) is dominated by the source regions of continental tropical (cT), maritime tropical (mT), and maritime equatorial (mE) air masses. These source regions are related to the three most obvious atmospheric features that occur within their latitude band—the two subtropical high-pressure belts and the equatorial trough at the intertropical convergence zone (ITCZ). Air of polar origin occasionally invades regions of low-latitude climates. Easterly waves and tropical cyclones are important weather systems in this climate group as well.

- *Group II: Midlatitude climates.* The region of midlatitude climates (Group II) lies in the polar-front zone—a zone of intense interaction between unlike air masses. In this zone, tropical air masses moving poleward and polar air masses moving equatorward are in contact. Midlatitude cyclones are normal features of the polar front, and this zone may contain as many as a dozen midlatitude cyclones around the globe.

- *Group III: High-latitude climates.* The region of high-latitude climates (Group III) is dominated by polar and arctic (including antarctic) air masses. In the arctic belt of the 60th to 70th parallels, continental polar air masses meet arctic air masses along an arctic-front zone, creating a series of eastward-moving midlatitude cyclones. In the southern hemisphere, there are no source regions in the subantarctic belt for continental polar air—just a great single oceanic source region for maritime polar (mP) air masses. The pole-centered continent of Antarctica provides a single great source of the extremely cold, dry antarctic air mass (cAA). These two air masses interact along the antarctic-front zone.

Within each of these three climate groups are a number of climate types (or simply, climates)—four

low-latitude climates (Group I), six midlatitude climates (Group II), and three high-latitude climates (Group III)—for a total of 13 climate types. In this textbook, the climates are numbered for ease of identification on maps and diagrams. We will still refer to each climate by name because the name describes the general nature of the climate and also suggests its global location. For convenience, however, we also include the climate number next to the name.

OVERVIEW OF CLIMATES

In presenting the climates, we make use of a pictorial device called a **climograph**. It shows the annual cycles of monthly mean air temperature and monthly mean precipitation for a location, along with some other useful information. **FIGURE 11.7** shows an example of a climograph for New Delhi and Simla, both in India. At the top of the climograph, the mean monthly temperature is plotted as a line graph. At the bottom, the mean monthly precipitation is shown as a bar graph. The annual range in temperature

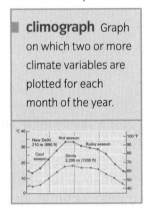

climograph Graph on which two or more climate variables are plotted for each month of the year.

and the total annual precipitation are stated on every climograph as well. Most climographs also display dominant weather features, which are shown using picture symbols. For both these regions, the two dominant features are the seasonal movement of the subtropical high and equatorial trough (ITCZ). Many of our climographs also include a small graph of the Sun's declination in order to help show when solstices and equinoxes occur.

Although the next chapter provides a thorough examination of the 13 climates in our system, here we give a brief description of each and provide their locations (**FIGURE 11.8**).

Low-latitude climates (Group I)

- *Wet equatorial* ①. Warm to hot with abundant rainfall, this is the steamy climate of the Amazon and Congo basins.

- *Trade-wind coastal* ②. This warm to hot climate has a very wet rainy season and occurs in coastal regions that are influenced by trade winds or a monsoon circulation. The climates of Vietnam, Bangladesh and coastal India are good examples.

- *Wet-dry tropical* ③. This climate is warm to hot, with very distinct wet and dry seasons. The monsoon region of central India falls into this type, as does much of the Sahel region of Africa.

- *Dry tropical* ④. The climate of the world's hottest deserts—extremely hot in the high-Sun season, a little cooler in the low-Sun season, has little or no rainfall. The Sahara desert, Saudi Arabia, and the central Australian desert have this type of climate.

Midlatitude climates (Group II)

- *Dry subtropical* ⑤. This desert climate is not quite as hot as the dry tropical climate, because it is found farther poleward. This type includes the hottest part of the American southwest desert.

- *Moist subtropical* ⑥. This is the climate of the southeastern regions of the United States and China—hot and humid summers, with mild winters and ample rainfall year-round.

- *Mediterranean* ⑦. Hot, dry summers and rainy winters mark this climate. Southern and central California, as well as the lands of the Mediterranean region—Spain, southern Italy, Greece, and the coastal regions of Lebanon and Israel—are prime examples.

- *Marine west-coast* ⑧. This is the climate of the Pacific Northwest—coastal Oregon, Washington, and British Columbia. Warm summers and cool winters, with more rainfall in winter, are characteristics of this climate.

- *Dry midlatitude* ⑨. This dry climate is found in midlatitude continental interiors. The steppes of central Asia and the Great Plains of North America are familiar locales with this climate—warm to hot in summer, cold in winter, with low annual precipitation.

- *Moist continental* ⑩. This is the climate of the eastern United States and lower Canada—cold in winter, warm in summer, with ample precipitation throughout the year.

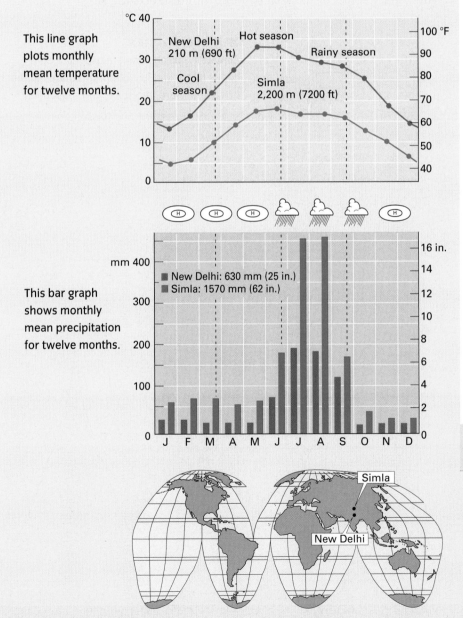

This line graph plots monthly mean temperature for twelve months.

This bar graph shows monthly mean precipitation for twelve months.

Climograph for New Delhi and Simla
FIGURE 11.7

A climograph combines graphs of monthly temperature and precipitation for an observing station. This example features two stations plotted together to compare their climates. Climographs also show total annual precipitation and annual temperature range (omitted here). Many include weather features using picture symbols.

▲ **Tropical desert**

The climate here is hot and dry all the time.

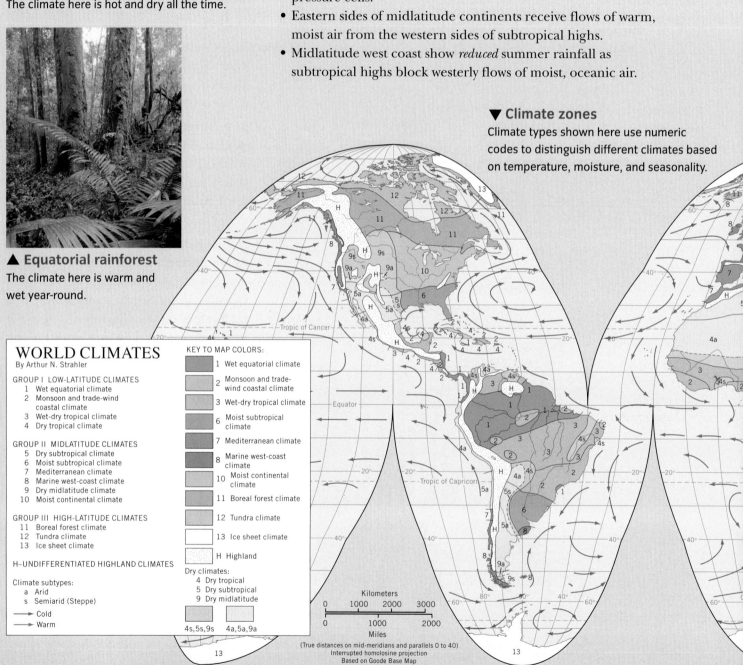

▲ **Equatorial rainforest**

The climate here is warm and wet year-round.

PRECIPITATION

Seasonal precipitation patterns depend on movements of air masses.

- Equatorial regions are wet year-round.
- Trade winds bring rain to equatorial and tropical east coasts.
- Tropical deserts underlie the descending air of subtropical high pressure cells.
- Eastern sides of midlatitude continents receive flows of warm, moist air from the western sides of subtropical highs.
- Midlatitude west coast show *reduced* summer rainfall as subtropical highs block westerly flows of moist, oceanic air.

▼ **Climate zones**

Climate types shown here use numeric codes to distinguish different climates based on temperature, moisture, and seasonality.

WORLD CLIMATES
By Arthur N. Strahler

GROUP I LOW-LATITUDE CLIMATES
1 Wet equatorial climate
2 Monsoon and trade-wind coastal climate
3 Wet-dry tropical climate
4 Dry tropical climate

GROUP II MIDLATITUDE CLIMATES
5 Dry subtropical climate
6 Moist subtropical climate
7 Mediterranean climate
8 Marine west-coast climate
9 Dry midlatitude climate
10 Moist continental climate

GROUP III HIGH-LATITUDE CLIMATES
11 Boreal forest climate
12 Tundra climate
13 Ice sheet climate

H–UNDIFFERENTIATED HIGHLAND CLIMATES

Climate subtypes:
a Arid
s Semiarid (Steppe)

→ Cold
→ Warm

KEY TO MAP COLORS:
1 Wet equatorial climate
2 Monsoon and trade-wind coastal climate
3 Wet-dry tropical climate
6 Moist subtropical climate
7 Mediterranean climate
8 Marine west-coast climate
10 Moist continental climate
11 Boreal forest climate
12 Tundra climate
13 Ice sheet climate
H Highland

Dry climates:
4 Dry tropical
5 Dry subtropical
9 Dry midlatitude

4s,5s,9s   4a,5a,9a

Kilometers
0   1000   2000   3000

0   1000   2000
Miles

(True distances on mid-meridians and parallels 0 to 40)
Interrupted homolosine projection
Based on Goode Base Map

The world map of climates shows the actual distribution of climate types on the continents. This map, based on data collected at a large number of observing stations, is simplified because the climate boundaries are uncertain in many areas where observing stations are thinly distributed.

TEMPERATURE

Seasonal temperature patterns depend on latitude, location, and elevation.

- Latitude: Temperatures drop from the equator toward the poles.
- Location: Continental interiors experience a greater range in temperature with the seasons than coastal locations.
- Elevation: Temperatures drop with elevation.

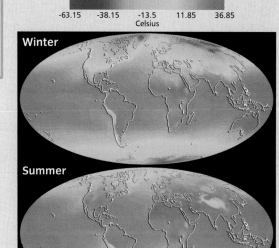

Temperature

-63.15   -38.15   -13.5   11.85   36.85
Celsius

Winter

Summer

▲ Temperature

Temperatures vary seasonally and with latitude as the Earth offers first one, then the other, hemisphere to more direct sunlight. Temperatures are modified by ocean currents and vegetation and are depressed by altitude.

▼ Precipitation

Rainfall hugs the Equator but migrates alternately north–south to the summer hemispheres, when land is warmer than surrounding seas and rising hot air draws in moisture.

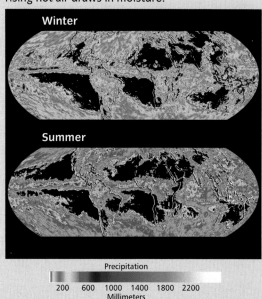

Winter

Summer

Precipitation

200   600   1000   1400   1800   2200
Millimeters

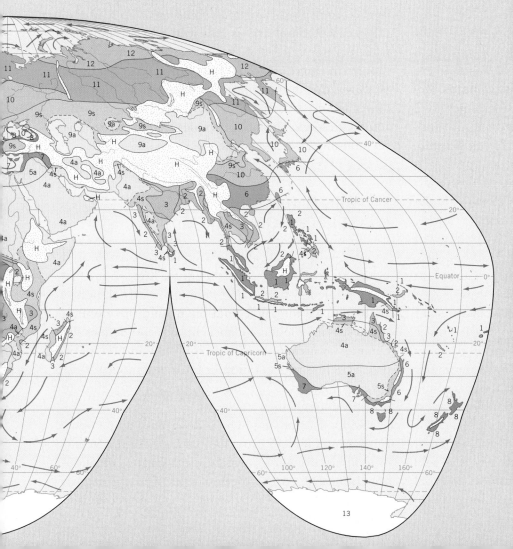

High-latitude climates (Group III)

- *Boreal forest* ⑪. Short, cool summers and long, bitterly cold winters characterize this snowy climate. Northern Canada, Siberia, and central Alaska are regions of boreal forest climate.

- *Tundra* ⑫. Although this climate has a long, severe winter, temperatures on the tundra are somewhat moderated by proximity to the Arctic Ocean. This is the climate of the coastal arctic regions of Canada, Alaska, Siberia, and Scandinavia.

- *Ice sheet* ⑬. The bitterly cold temperatures of this climate, restricted to Greenland and Antarctica, can drop below –50°C (–58°F) during the sunless winter months. Even during the 24-hour days of summer, temperatures remain well below freezing.

DRY AND MOIST CLIMATES

All but two of the 13 climate types we've discussed are classified as either **dry climates** or **moist climates**. Dry climates are those in which total annual evaporation of moisture from the soil and from plant foliage exceeds annual precipitation by a wide margin. Generally speaking, the dry climates do not support permanently flowing streams. The soil is dry much of the year, and the land surface contains only sparse plant cover—scattered grasses or shrubs—or simply lacks plant cover. Moist climates are those with sufficient rainfall to maintain the soil in a moist condition through much of the year and to sustain the year-round flow of the larger streams. Moist climates support forests or prairies of dense, tall grasses.

> ■ **dry climate**
> Climate in which evaporation is limited by low soil moisture.
>
> ■ **moist climate**
> Climate in which precipitation provides enough soil moisture to support evapotranspiration throughout the year.

Within the dry climates there is a wide range of degree of aridity, from very dry deserts nearly devoid of plant life to moister regions that support a partial cover of grasses or shrubs. We will refer to two dry climate subtypes: (1) semiarid (or steppe) and (2) arid. The *semiarid* (steppe) subtype, designated by the letter *s*, is found next to moist climates. It has enough precipitation to support sparse grasses and shrubs. The *arid* subtype, indicated by the letter *a*, ranges from extremely dry climates to climates that are almost semiarid.

In addition, two of our 13 climates cannot be accurately described as either dry or moist climates. These are the wet-dry tropical (3) and Mediterranean (7) climate types. Instead, they show a seasonal alteration between a very wet season and a very dry season. This striking contrast in seasons gives a special character to the two climates, and so we have singled them out for special recognition as wet-dry climates.

One climate type that isn't usually included in broad climate classification schemes is the highland climate, which occupies mountains and high plateaus. They tend to be cool to cold because air temperatures in the atmosphere normally decrease with altitude. Highland climates are usually wetter at higher locations, because rainfall increases as air is forced over the mountain ranges through orographic precipitation. However, the annual temperature cycle and the times of wet and dry seasons in highland areas are usually driven by the same climate processes that influence the surrounding lowland climate types.

KÖPPEN CLIMATE CLASSIFICATION SYSTEM

An alternative classification of climate types was devised by the Austrian climatologist Vladimir Köppen in 1918 and modified by Geiger and Pohl in 1953. Designed to capture variability of vegetation around the globe, it uses a system of letters to label climates and defines the climates by mean annual precipitation and temperature as well as precipitation in the driest month. **FIGURE 11.9** gives a brief overview of this climate classification scheme.

Köppen's climate system is descriptive and empirical, relying on rules using annual and monthly values of temperature and precipitation at stations throughout the world. At the highest level of classification, there are five major climate types, which are recognized by capital letter codes, A–E. Subtypes are denoted by additional codes using large and small letters.

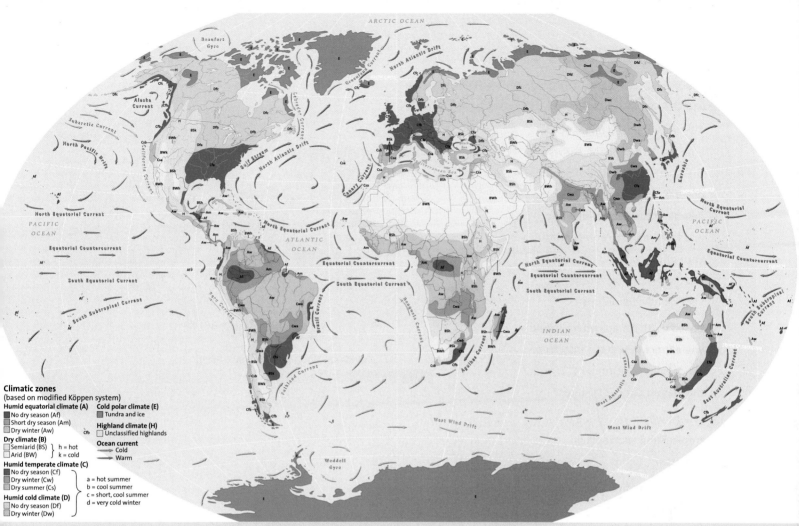

Climatic zones
(based on modified Köppen system)

Humid equatorial climate (A)
- No dry season (Af)
- Short dry season (Am)
- Dry winter (Aw)

Dry climate (B)
- Semiarid (BS) } h = hot
- Arid (BW)   } k = cold

Humid temperate climate (C)
- No dry season (Cf)
- Dry winter (Cw)
- Dry summer (Cs)

Humid cold climate (D)
- No dry season (Df)
- Dry winter (Dw)

Cold polar climate (E)
- Tundra and ice

Highland climate (H)
- Unclassified highlands

Ocean current
- → Cold
- → Warm

a = hot summer
b = cool summer
c = short, cool summer
d = very cold winter

Köppen climate classification system FIGURE 11.9

Although it is now more than fifty years old, the Köppen-Geiger-Pohl climate classification system still finds many uses. Originally devised by the Austrian climatologist Vladimir Köppen in 1918, it was designed to bring out the relationship between vegetation cover type and climate. Over the years, Köppen's classification was modified several times to enhance the correspondence between climate and vegetation, with the last principal revision by Geiger and Pohl in 1953.

The **Köppen climate system** is easy and convenient to use, given monthly weather data, but it is not related to the underlying processes that differentiate the Earth's climates. In comparison, the classification used in this chapter is designed to be understood and explained by air-mass movements and frontal zones—that is, by the weather various regions experience throughout the year. It is a more natural system that is easier to learn and understand.

Köppen climate system Classification of climates based upon annual temperatures and precipitation and their seasonality.

CONCEPT CHECK STOP

What is a climograph?

Where are the three broad climate groups located?

How many climate types are there, and what are they called?

This image, based on many years of satellite data, shows the fractional amount of cloud cover a region receives over the course of the year, with white and blue regions representing very low cloud amounts and reds and pinks representing very cloudy conditions.

- Based on this map, where would you expect to find the intertropical convergence zone (ITCZ) and the wet equatorial climates?

- Where would you find the dry tropical and subtropical climates?

- What are conditions usually like over the poles?

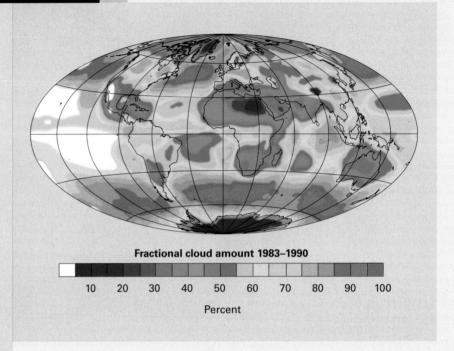

Fractional cloud amount 1983–1990

10	20	30	40	50	60	70	80	90	100

Percent

SUMMARY

1 Factors Controlling Climate

1. **Climate** is the average weather of a region. Because temperature and precipitation are measured at many stations worldwide, we can use the combined annual patterns of monthly averages of temperature and precipitation to assign climate types.

2. Climate is usually characterized by the average temperature and precipitation as well as their seasonality. Various climate factors control these characteristics, including a location's latitude, elevation, relation to the coast, and dominant air masses.

2 Temperature and Precipitation Regimes

1. Temperature regimes are characterized by the annual temperature and seasonal variations in temperature. These characteristics depend on latitude, which determines the annual pattern of insolation, and on location—continental or maritime—which enhances or moderates the annual insolation cycle.

2. Global precipitation patterns are largely determined by air masses and their movements, which in turn are produced by global air circulation patterns. Precipitation regimes are also influenced by the location of high- and low-pressure patterns associated with these global circulations.

3. Seasonality of precipitation falls into three patterns: uniform (ranging from abundant to scarce); high-Sun (summer) maximum; and low-Sun (winter) maximum. The combination of seasonality in precipitation and overall annual rainfall amount results in seven general precipitation patterns over the globe.

3 Climate Classification

1. There are three groups of climate types, arranged by latitude. Low-latitude climates (Group I) are dominated by mE, mT, and cT air masses and are largely related to the global circulation patterns that produce the ITCZ, trade winds, and subtropical high-pressure cells. Midlatitude climates (Group II) lie in the polar-front zone and are strongly influenced by eastward-moving midlatitude cyclones in which mT, mP, and cP air masses interact. High-latitude climates (Group III) are dominated by polar and arctic (antarctic) air masses. Midlatitude cyclones mixing mP and cP air masses along the arctic-front zone provide precipitation in this region.

2. **Dry climates** are those in which precipitation is largely evaporated from soil surfaces and transpired by vegetation, so that permanent streams cannot be supported.

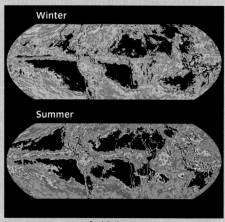

Within dry climates, there are two subtypes: arid (driest) and semi-arid or steppe (a little wetter). In **moist climates**, precipitation exceeds evaporation and transpiration, providing for sustained year-round stream flow. In wet-dry climates, strong wet and dry seasons alternate.

KEY TERMS

- climograph p. 310
- dry climate p. 314
- moist climate p. 314
- Köppen climate system p. 315

CRITICAL AND CREATIVE THINKING QUESTIONS

1. Discuss the use of monthly records of average temperature and precipitation to characterize the climate of a region. Why are these measures useful?

2. Why are latitude and location (maritime or continental) important factors in determining the annual temperature cycle of a station?

3. List and describe the important climate control factors and how they can potentially influence temperature and precipitation in a given location.

4. Describe three temperature regimes, and explain how they are related to latitude and location.

5. Identify seven important features of the global map of precipitation, and describe the factors that produce each.

6. Sketch a hypothetical supercontinent with a shape and features of your own choosing. It should stretch from about 70° N to 40° S latitude. Add some north-south mountain ranges, positioned where you like. Then select four locations on the supercontinent, describing and explaining the annual cycles of temperature and precipitation you would expect at each.

1. Climate is _____ .

   a. based on vegetation types alone

   b. a model of reality that is too simplistic to be useful

   c. based on temperature and precipitation averages over decades

   d. a reflection of solar insolation values and placement in relation to the prime meridian

2. _____ are the two major factors that influence the annual cycle of air temperature experienced at a station.

   a. Latitude and longitude

   b. Longitude and coastal versus continental location

   c. Latitude and coastal versus continental location

   d. Annual insolation and longitude

3. Warm air can contain more moisture than cold air. This affects all of the following climate characteristics except the characteristic that _____ .

   a. colder regions have lower precipitation than warmer regions

   b. precipitation is usually greater during warmer months

   c. the maximum precipitation is most likely to be in the colder months

   d. air temperature has an important effect on precipitation

4. A maritime influence will affect the temperature regime experienced at a location by _____ .

   a. always creating a uniformly warm temperature pattern year-round

   b. creating a strong annual temperature cycle

   c. enabling extreme variability in temperatures

   d. moderating the variation of temperature throughout the year

5. On the accompanying figure, draw lines connecting the monthly temperature plots with their respective locations on the globe.

Marine influence

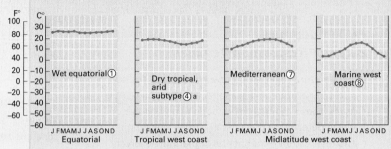

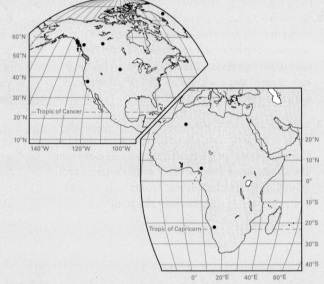

Continental influence

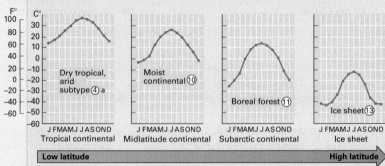

6. _____ is (are) a factor that does not always contribute to aridity.
a. The rain-shadow effect
b. Being far from source regions for moist air masses
c. Warm temperatures
d. Subsidence in subtropical high-pressure cells

7. Isohyets are lines that show areas of equal _____.
a. barometric pressure   c. precipitation
b. air temperature   d. elevation

8. On the accompanying figure, label each air mass with its appropriate name and acronym.

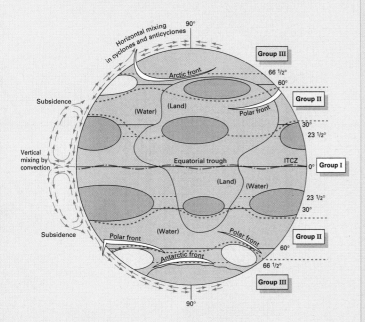

9. The tropical desert climates are caused by _____.
a. the presence of large mountain ranges
b. cold temperatures that reduce the ability of the air to hold moisture
c. a rate of precipitation that outweighs the rate of evaporation
d. stationary subtropical high-pressure cells

10. _____ represents a uniform precipitation pattern.
a. Maximum precipitation in high-Sun season
b. No precipitation in any months
c. Minimum precipitation in high-Sun season
d. An equal amount of precipitation in high- and low-Sun seasons with drought in between

11. On the accompanying figure, draw lines connecting the monthly precipitation plots with their respective locations on the globe.

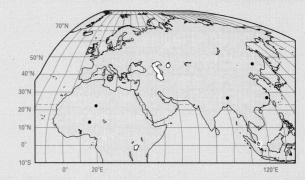

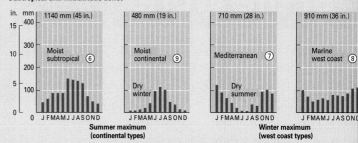

Equatorial and tropical zones

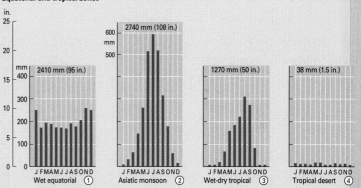

12. _____ is an air mass that is not usually part of the low-latitude climates.
a. cT   b. mE   c. cP   d. mT

13. The region of midlatitude climates lies in the _____.
a. polar-front zone   c. tropical zone
b. ITCZ   d. arctic zone

14. A(n) _____ shows the annual cycles of monthly mean air temperature and monthly mean precipitation for a location.
a. thermograph   c. isotherm
b. isohyet   d. climograph

15. _____ is (are) a geographic feature that is common in dry climates.
a. No permanently flowing streams
b. Extremely hot nights
c. A complete lack of precipitation
d. Strong surface winds

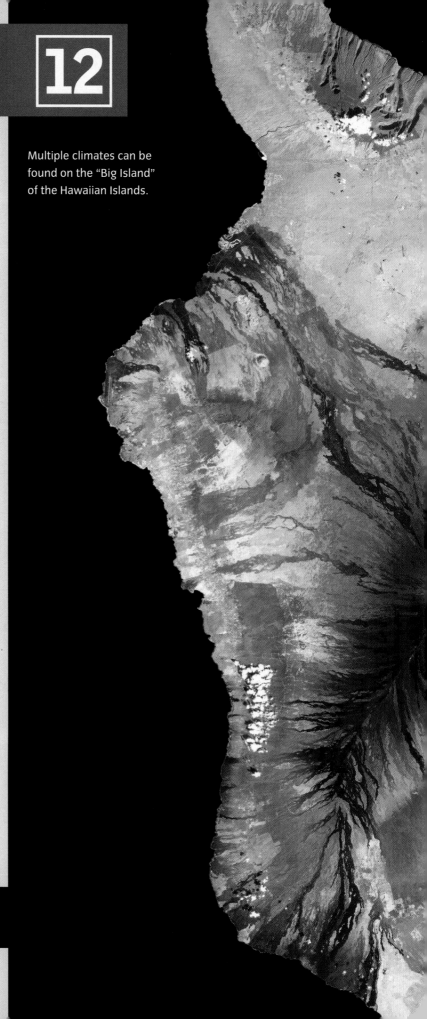

Climates of the World

The Hawaiian Islands have the distinction of being one of the few places in the world where you can experience the climates of warm tropical beaches and chilly snowfields within a few hours' drive. This image, taken from NASA's Aqua satellite, shows the broad spectrum of climates found on the "Big Island" of Hawaii. On the eastern edge, the prevailing trade winds bring ample rain to the region, creating a *coastal trade-wind climate*, indicated by green vegetation in this region. On the northwestern portion, in the lee side of the mountain range, the lighter brown regions are semi-arid and constitute a *dry subtropical climate*. Over the southern portion, where there is green on both the east and west sides of the island, rainfall is intermittent with the seasons, constituting a *wet-dry climate*. Near the center of the island lies Mauna Loa and Mauna Kea (whose name means "White Peak"). These peaks are 4169 m (13,677 ft) and 4205 m (13,796 ft) above sea level. At these altitudes, snow sheets can form during part of the year, resulting in a climate that is like the *dry steppe* regions in winter.

In this chapter, we investigate the different types of climates that are found across the globe. In addition, we describe the physical processes that give rise to very similar climates, despite the fact that they may be separated by thousands of miles.

Multiple climates can be found on the "Big Island" of the Hawaiian Islands.

NATIONAL GEOGRAPHIC

Low-Latitude Climates

LEARNING OBJECTIVES

Identify the features of low-latitude climates.

Describe wet equatorial, coastal trade-wind, wet-dry, monsoon, and dry tropical climates.

The low-latitude climates lie for the most part between the Tropics of Cancer and Capricorn, occupying all of the equatorial zone (10° N to 10° S), most of the tropical zone (10–15° N and S), and part of the subtropical zone. The low-latitude climate regions include the equatorial trough of the intertropical convergence zone (ITCZ), the belt of tropical easterlies (northeast and southeast trades), and large portions of the oceanic subtropical high-pressure belt. There are four low-latitude climates: wet equatorial ①, trade-wind coastal ②, wet-dry tropical ③, and dry tropical ④. We now look at each one in detail.

WET EQUATORIAL CLIMATE

The **wet equatorial climate** ① region lies between 10° N and 10° S and includes the Amazon lowland of South America, the Congo Basin of equatorial Africa, and the East Indies, from Sumatra to New Guinea (**FIGURE 12.1**).

> **wet equatorial climate** ① Moist climate of the equatorial zone with a large annual water surplus and uniformly warm temperatures throughout the year.

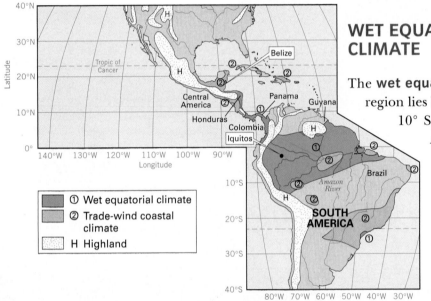

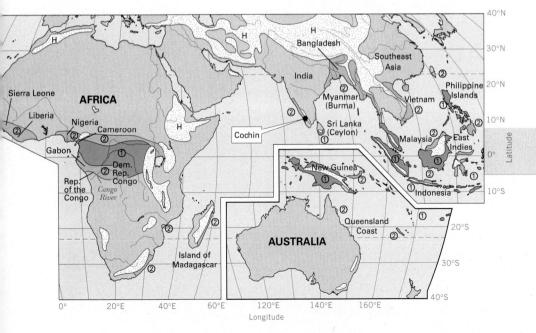

Low-latitude climates

FIGURE 12.1

World map of wet equatorial ① and trade-wind coastal climates ②.

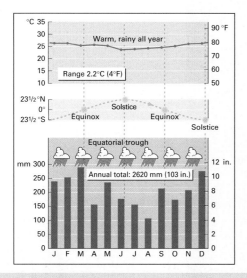

Wet equatorial climate FIGURE 12.2

Iquitos, in Peru (lat. 3° S), is located in the upper Amazon lowland, close to the Equator. Temperatures differ very little from month to month, and there is abundant rainfall throughout the year, as this climograph shows.

The wet equatorial climate ① is controlled by the ITCZ and is dominated by warm, moist maritime equatorial (mE) and maritime tropical (mT) air masses that produce heavy convective rainfall (FIGURE 12.2). A large amount of precipitation falls every month, and the annual total often exceeds 2500 mm (about 100 in.). There is a seasonal rainfall pattern, however, with heavier rain when the ITCZ migrates into the region. Temperatures are uniform throughout the year, with mean monthly and mean annual temperatures close to 27°C (81°F). Read *What a Scientist Sees: Rainforests and Agriculture* for a closer look at this climate. Typically, the monthly air temperature varies between 26° and 29°C (79° and 84°F) for stations at low elevation in the equatorial zone.

Rainforests and Agriculture

Rainforests are home to an incredible range of plants and animals. A 16-km² (6-mi²) area of rainforest in Panama, for example, contains about 20,000 insect species, whereas all of France has only a few hundred.

In the past, native peoples farmed the rainforests by cutting down the vegetation in a small area, then burning it. This may seem irrevocably destructive, but scientists recognize that this slash-and-burn method doesn't threaten the rainforest ecosystem. Burning the vegetation releases trapped nutrients, returning some of them to the soil and making them available to growing crops. Once the site is abandoned, the rainforest can reestablish itself.

Presently, however, modern methods that use heavy machinery are a threat. Large areas of land are cleared, and when these are abandoned, seed sources are so far away that the original forest species cannot take hold. A rainforest ecosystem that has been cleared in this way will never return to its earlier state.

The satellite image on the left shows a scene from 1975 before large-scale deforestation began. The image on the right shows the same region in 2001. As one region becomes depleted, heavy machinery brings deforestation into ever more remote locations. Depleted plots that are abandoned, however, are no longer able to recover the character and diversity of the original forest.

What a Scientist Sees

A **Belize city** Trade winds bring frequent showers to the coastal city of Belize City, Belize.

Trade-wind coastal climate ② FIGURE 12.3

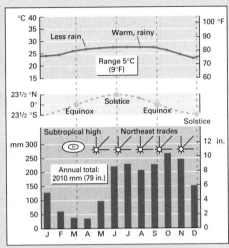

B **Belize climograph** This climograph for Belize, a Central American east-coast city (lat. 17° N), is exposed to the tropical easterly trade winds. Rainfall is abundant from June through November, when the ITCZ is nearby. Easterly waves are common in this season, and on occasion a tropical cyclone brings torrential rainfall. Following the December solstice, rainfall is greatly reduced, with minimum values in March and April, when the ITCZ is farthest away.

THE TRADE-WIND COASTAL CLIMATE

The trade-wind coastal climate ② occurs between 5° and 25° N and S and is found along the east sides of Central and South America, the Caribbean Islands, Madagascar (Malagasy), southeast Asia, the Philippines, and northeast Australia.

Like the wet equatorial climate ①, the **trade-wind coastal climate** ② has abundant rainfall. The climate is influenced by moisture-laden maritime tropical (mT) and maritime equatorial (mE) air masses that are moved onshore across narrow coastal zones by trade winds. As the warm, moist air passes over coastal hills and mountains, the orographic effect touches off convective shower activity (**FIGURE 12.3**).

> ■ **trade-wind coastal climate**
> ② Moist climate of low latitudes showing a rainfall peak in the high-Sun season and a short period of reduced rainfall in the middle of the low-Sun season.

Unlike the wet equatorial climate, however, the trade-wind coastal climate rainfall always shows a stronger seasonal pattern. In the high-Sun season ("summer," depending on the hemisphere), the ITCZ is nearby, so monthly rainfall is greater. Shower activity is also intensified by easterly waves, which are more frequent when the ITCZ is nearby.

In the low-Sun season, when the ITCZ has migrated to the other hemisphere, the region is dominated by subtropical high pressures, so there is less monthly rainfall. However, during these periods there is still some rainfall, which is supported by the persistent trade winds blowing in from the ocean.

Temperatures in the trade-wind coastal climate ② are warm throughout the year. The warmest temperatures occur in the high-Sun season, just before the

ITCZ brings clouds and rain, with minimum temperatures at the time of low Sun.

THE WET-DRY TROPICAL CLIMATE

As we move farther poleward, the seasonal rainfall and temperature cycle becomes stronger, and the trade-wind coastal climate ② grades into the **wet-dry tropical climate** ③ (FIGURE 12.4).

The wet-dry tropical climate ③ region lies between 5° to 20° N and S in Africa and the Americas

and 10° to 30° N in Asia. In Africa and South America, the climate occupies broad bands poleward of the wet equatorial and trade-wind coastal climates. Because these regions are farther away from the ITCZ, less rainfall is triggered by the ITCZ during the rainy season, and in the low-Sun season subtropical high pressure dominates more strongly.

The wet-dry tropical climate ③ has a very dry season at low Sun and a very wet season at high Sun. During the low-Sun season, when the equatorial trough is far away, dry continental tropical (cT) air masses

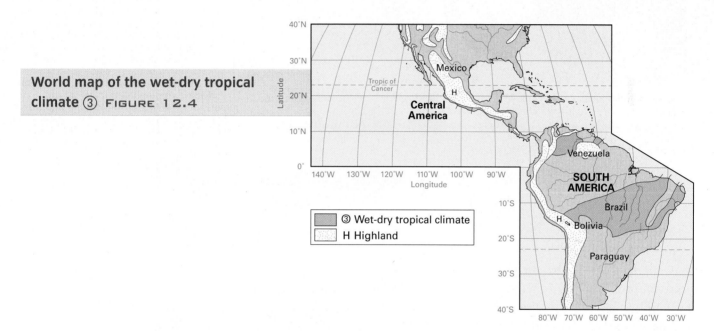

World map of the wet-dry tropical climate ③ FIGURE 12.4

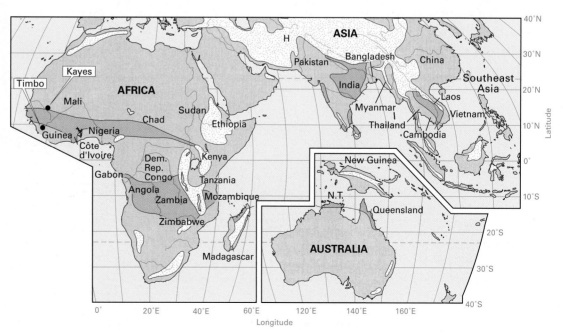

A **Kindia, Guinea** This busy market town, located about 150 km (about 100 mi) from Timbo, is typical of the wet-dry tropical climate region of West Africa.

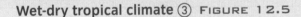
Wet-dry tropical climate ③ FIGURE 12.5

B **Wet-dry tropical climate** Timbo, in Guinea (lat. 10° N), in West Africa, has a rainy season that begins just after the March equinox and peaks when the ITCZ has migrated to its most northerly position. Monthly rainfall decreases as the low-Sun season arrives and the ITCZ moves south. December through February are practically rainless, when subtropical high pressure dominates the climate, and stable, subsiding continental tropical (cT) air pervades the region. In February and March, insolation increases, so air temperature rises sharply. When the rains set in, the cloud cover and evaporation of rain make temperatures drop.

prevail (FIGURE 12.5). In the high-Sun season, when the ITCZ is nearby, the climate is dominated by moist maritime tropical (mT) and maritime equatorial (mE) air masses. Cooler temperatures in the dry season give way to a very hot period before the rains begin.

The strong seasonality in precipitation found in the wet-dry climate results in significant periods of low rainfall, so vegetation must be drought resistant to survive. In some regions, this precipitation pattern produces a savanna of grasses with widely scattered trees.

Monsoon coastal climate FIGURE 12.6

Cochin, on the lower peninsula of India (lat. 10° N), shows a peak of rainfall during the rainy monsoon and a short dry season at the time of low Sun. Air temperatures have a weak annual cycle, cooling a bit during the rains, so the range of temperature is small.

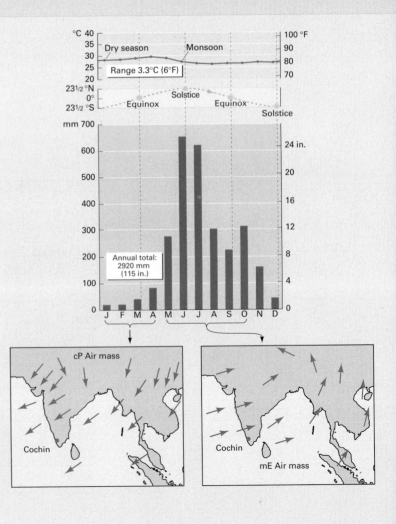

MONSOON CLIMATES

In our climate system, the Asian and Indian monsoons are seen in both the trade-wind coastal ② and wet-dry tropical climates ③. During the wet season, southwesterly winds bring mE and mT air from the Indian Ocean across the west coasts of India and Indochina, producing a rainy season similar to that of trade-wind coasts when the ITCZ is near. FIGURE 12.6 presents a climograph for Cochin, on the west coast of India, that shows the effect of this monsoon flow quite clearly. During the low-Sun season, high pressures form over the interior land masses as the continents cool more than the surrounding oceans. This high-pressure system produces an outflow of cold, dry air cT from central Asia across India and Southeast Asia from the northeast. Skies clear, and precipitation drops abruptly.

In central India and Indochina, mountains provide a barrier from the warm, moist mE and mT air flows, creating a rain-shadow effect. Less rainfall occurs during the rainy season, and the dry season is drier still, placing these areas firmly in the wet-dry tropical climate.

THE DRY TROPICAL CLIMATE

The **dry tropical climate** ④ is found in the center and east sides of subtropical high-pressure cells (FIGURE 12.7). Air descends and warms adiabatically, inhibiting condensation, so rainfall is very rare and occurs only when unusual weather conditions move moist air into the region. Skies are clear most of the time,

> ■ **dry tropical climate** ④ Climate of the tropical zone with high temperatures and low rainfall.

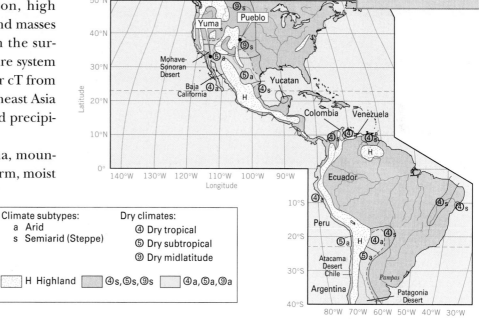

Climate subtypes:
a Arid
s Semiarid (Steppe)

Dry climates:
④ Dry tropical
⑤ Dry subtropical
⑨ Dry midlatitude

☐ H Highland ▨ ④s,⑤s,⑨s ☐ ④a,⑤a,⑨a

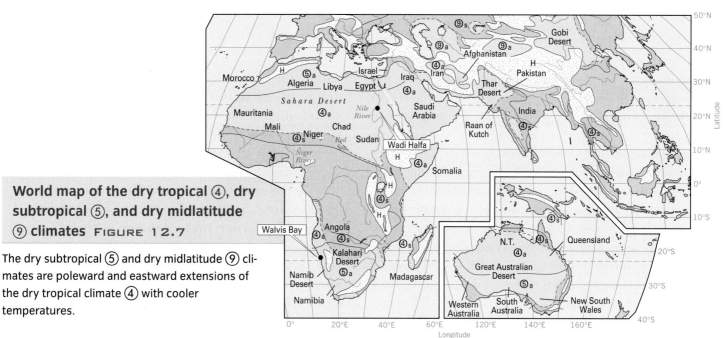

World map of the dry tropical ④, dry subtropical ⑤, and dry midlatitude ⑨ climates FIGURE 12.7

The dry subtropical ⑤ and dry midlatitude ⑨ climates are poleward and eastward extensions of the dry tropical climate ④ with cooler temperatures.

A **Wadi Halfa** This aerial view of Wadi Halfa, Sudan, situated on the Nile River, shows the town's flat-roofed residences with walled courtyards, laid out in a rectangular pattern.

Dry tropical climate ④ FIGURE 12.8

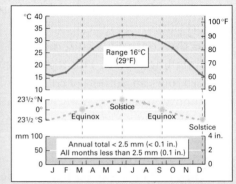

B **Dry tropical climate ④ Dry desert.**
Wadi Halfa is on the Nile River in Sudan at lat. 22° N, almost on the Tropic of Cancer. There is a strong annual temperature cycle with a very hot period at the time of high Sun. Daytime maximum air temperatures are frequently between 43° and 48°C (about 110° to 120°F) in the warmer months. There is a comparatively cool season at the time of low Sun. There is too little rainfall to show on the climograph. Over a 39-year period, the maximum rainfall recorded in a 24-hour period at Wadi Halfa was only 7.5 mm (0.3 in.).

so the Sun heats the surface intensely, keeping air temperatures high. During the high-Sun period, heat is extreme. During the low-Sun period, temperatures are cooler. Given the dry air and lack of cloud cover, the daily temperature range is very large.

The driest areas of the dry tropical climate ④ are near the Tropics of Cancer and Capricorn (FIGURE 12.8). Rainfall increases as we move from the tropics toward the Equator. Continuing in this direction, we encounter regions that have short rainy seasons when the ITCZ extends close enough to produce rainfall, until finally the climate grades into the wet-dry tropical ③ type.

Nearly all of the dry tropical climate ④ areas lie between latitudes 15° to 25° N and S. The largest re-

gion is the Sahara–Saudi Arabia–Iran–Thar desert belt of North Africa and southern Asia, which includes some of the driest regions on Earth. Another large region is the desert of central Australia. The west coast of South America and southern Africa, including portions of Ecuador, Peru, and Chile, also exhibits the dry tropical climate ④ (FIGURE 12.9). In those areas, temperatures are strongly influenced by cold ocean currents and the upwelling of deep, cold water. This cooling enhances the descent of air, making these some of the driest regions in the world. The cool water also moderates coastal zone temperatures, reducing the seasonality of the temperature cycle.

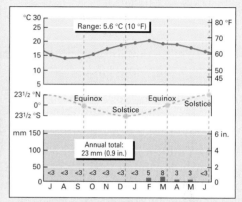

A **Walvis Bay, Namibia** The city of Walvis Bay is situated along the desert coast of western Africa near the Tropic of Capricorn.

Dry tropical climate, western coastal desert subtype

FIGURE 12.9

B **Dry tropical climate ④ Western coastal desert subtype.** Walvis Bay, Namibia (lat. 23° S), is a desert station on the west coast of Africa. The monthly temperatures are remarkably cool for a location that is nearly on the Tropic of Capricorn. Because of its coastal location, the annual range of temperatures is also small—only 5°C (9°F). Coastal fog is a persistent feature of this climate.

CONCEPT CHECK **STOP**

What are the four climate types of the low-latitude climate group? Give an example of each type.

What are each climate type's location and characteristic features?

What main climate characteristic distinguishes the low-latitude climate types?

Midlatitude Climates

LEARNING OBJECTIVES

Identify the features of midlatitude climates.

Describe dry subtropical, moist subtropical, Mediterranean, marine west-coast, dry midlatitude, and moist continental climates.

The midlatitude climates almost fully occupy the land areas of the midlatitude zone and a large proportion of the subtropical latitude zone. They also extend into the subarctic latitude zone, along the western fringe of Europe, reaching to the 60th parallel. Unlike the low-latitude climates, which are about equally distributed between northern and southern hemispheres, nearly all of the midlatitude climate area is in the northern hemisphere. In the southern hemisphere, the land area poleward of the 40th parallel is so small that the climates are dominated by a great southern ocean.

In the northern hemisphere, the midlatitude climates lie between two groups of very unlike air masses, which interact intensely. Tongues of maritime tropical (mT) air masses enter the midlatitude zone from the subtropical zone, where they meet and conflict with tongues of maritime polar (mP) and continental polar (cP) air masses along the polar-front zone.

The midlatitude climates include the poleward halves of the great subtropical high-pressure systems and much of the belt of prevailing westerly winds. As a result, weather systems, such as midlatitude cyclones and their fronts, characteristically move from west to east. This global airflow influences the distribution of climates from west to east across the North American and Eurasian continents.

There are six midlatitude climate types. They span the range from those with strong wet and dry seasons to those with uniform precipitation. Temperature cycles for these climate types are also quite varied. We now examine each climate type in more detail.

■ **dry subtropical climate** ⑤ Dry climate of the subtropical zone, transitional between the dry tropical climate and the dry midlatitude climate.

THE DRY SUBTROPICAL CLIMATE

The **dry subtropical climate** ⑤ is simply a poleward extension of the dry tropical climate ④. It is caused by somewhat similar air-mass patterns, but the annual temperature range is greater for the dry subtropical climate ⑤. Although the great summer heat is comparable to the dry tropical climate ④, the low Sun brings a winter season unseen in the tropical deserts. The lower latitude portions of the dry subtropical climate have a distinct cool season, and the higher latitude portions have a cold season (**FIGURE 12.10**). The cold season occurs at a time of low Sun and is caused in part by the invasion of cold continental polar (cP) air masses from higher latitudes. Midlatitude cyclones occasionally move into the subtropical zone in the low-Sun season, producing precipitation. There are both arid (*a*) and semiarid (*s*) subtypes in this climate.

This climate type is found in a broad band of North Africa, connecting with the Near East. Southern Africa and southern Australia also contain this climate. A band of dry subtropical climate ⑤ occupies Patagonia, in South America, and in North America, the Mojave and Sonora Deserts of the American southwest and northwest Mexico are of this type.

A **Yuma, Arizona** Spanish mission-style architecture is a highlight of this Arizona desert city.

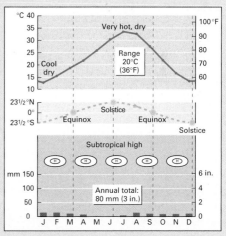

B **Dry subtropical climate** ⑤ Yuma, Arizona (lat. 33° N), has a strong seasonal temperature cycle, with a dry hot summer, and freezing temperatures in December and January. The annual temperature range is 20°C (36°F). Precipitation totals about 80 mm (3 in.) and is small in all months but has peaks in late winter and late summer. The August maximum is caused by the invasion of maritime tropical (mT) air masses, which bring thunderstorms to the region. Higher rainfalls from December through March are produced by midlatitude wave cyclones following a southerly path. Two months, May and June, are nearly rainless.

MOIST SUBTROPICAL CLIMATE

Circulation around subtropical high-pressure cells provides a flow of warm, moist air onto the eastern side of continents. This flow of maritime tropical (mT) air also dominates the **moist subtropical climate** (FIGURE 12.11) ⑥. There is abundant rainfall in the summer, much of which is convective, fed by the onshore movement of mT air masses circling around the subtropical highs. Occasional tropical cyclones add to this summer precipitation. Summer temperatures are warm, with persistent high humidity.

There is also plenty of winter precipitation, produced by midlatitude cyclones.

■ **moist subtropical climate** ⑥ Moist climate of the subtropical zone, characterized by a moderate to large annual water surplus and a strong seasonal temperature cycle.

Moist subtropical climate ⑥ FIGURE 12.11

A Charleston, South Carolina These old homes on the Charleston waterfront display a local architecture well adapted to the climate of the region. The upper porches were used for outdoor living and sleeping in the hot and humid summer weather.

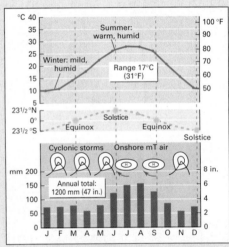

B Moist subtropical climate ⑥ Charleston, South Carolina (lat. 33° N), located on the eastern seaboard, has a mild winter and a warm summer. There is ample precipitation in all months but a definite summer maximum. Total annual rainfall is abundant—1200 mm (47 in.). There is a strongly developed annual temperature cycle, with a large annual range of 17°C (31°F). Winters are mild, with the January mean temperature well above the freezing mark.

Continental polar (cP) air masses frequently invade these climate regions in winter, bringing spells of subfreezing weather, but no winter month has a mean temperature below 0°C (32°F).

The moist subtropical climate ⑥ is found on the eastern sides of continents between latitude 20° to 35° N and S (FIGURE 12.12). In South America, it includes parts of Uruguay, Brazil, and Argentina. In Australia, it consists of a narrow band between the eastern coastline and the eastern interior ranges. Southern China, Taiwan, and southernmost Japan are included—where this climate is characterized by a strong monsoon effect, with much more rainfall in the summer than in the winter—as is most of the Southeast of the United States, from the Carolinas to east Texas.

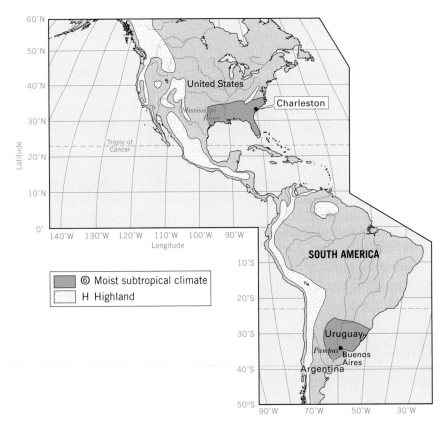

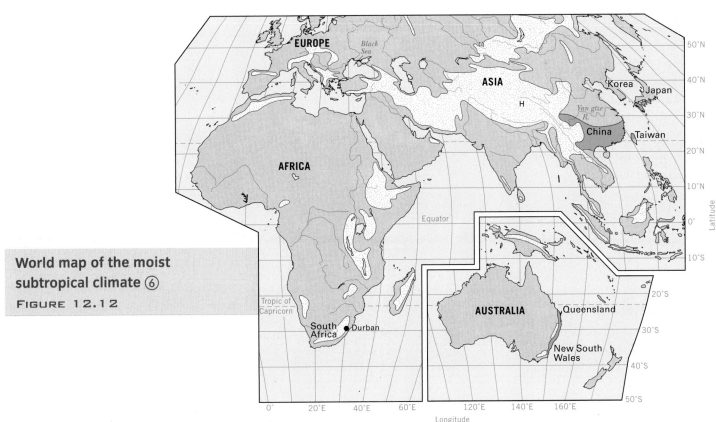

World map of the moist subtropical climate ⑥

FIGURE 12.12

THE MEDITERRANEAN CLIMATE

The **Mediterranean climate** ⑦ is unique in having a wet winter and a very dry summer (FIGURE 12.13), caused by its location along the west coasts of continents, just poleward of the dry, eastern side of the subtropical high-pressure cells. When these cells move poleward in summer, they enter the Mediterranean climate region. Dry continental tropical (cT) air then dominates, producing the dry summer season. In winter, the moist mP air mass invades with cyclonic storms and generates ample rainfall.

The Mediterranean climate ⑦ is found between latitude 30° and 45° N and S. In the southern hemisphere, it occurs along the coast of Chile, in the Cape Town region of South Africa, and along the southern and western coasts of Australia. In North America, it is found in central and southern California. In Europe, this climate type surrounds the Mediterranean Sea, giving the climate its distinctive name.

> **Mediterranean climate ⑦** Climate type of the subtropical zone characterized by a very dry summer and a mild, rainy winter.

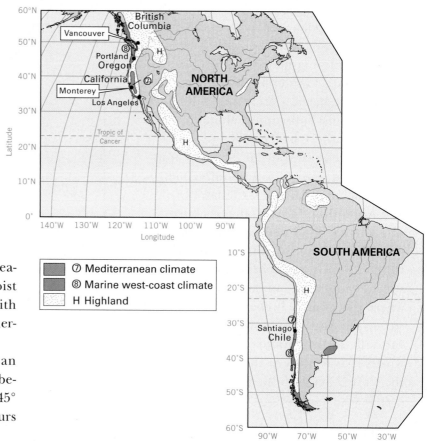

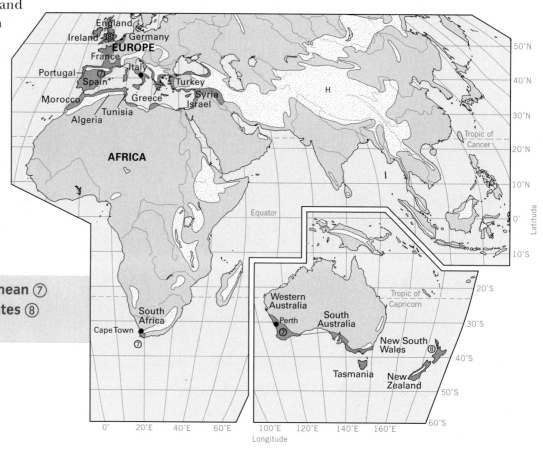

World map of the Mediterranean ⑦ and marine west-coast climates ⑧

FIGURE 12.13

- ⑦ Mediterranean climate
- ⑧ Marine west-coast climate
- H Highland

The Mediterranean climate ⑦ spans arid to humid climates, depending on location. Generally, the closer an area is to the tropics, the stronger the influence of the subtropical high pressure, and thus the drier the climate. The temperature range is moderate, with warm to hot summers and mild winters (FIGURE 12.14). Coastal zones between latitude 30° and 35° N and S, such as southern California, show a smaller annual range, with very mild winters.

Mediterranean climate ⑦ FIGURE 12.14

A **Monterey, California** As evident in this photo of Monterey harbor, the west coasts of the Mediterranean climate ⑦ often support extensive fisheries.

B **Mediterranean climate** ⑦ Monterey, California (lat. 36° N), has a very weak annual temperature cycle because of its closeness to the Pacific Ocean. The summer is very dry. Fogs are frequent. Rainfall drops to nearly zero for four consecutive summer months but rises to substantial amounts in the rainy winter season.

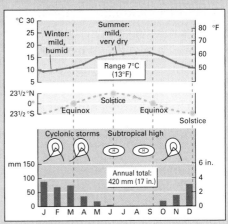

A **Vancouver, BC** Situated on the western coast of Canada, Vancouver is a major deep-water seaport. The nearby Coast Mountains are lined with conifer forests.

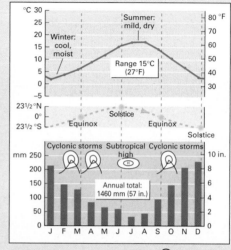

B **Marine west-coast climate** ⑧ Vancouver, British Columbia (lat. 49° N), has a large annual total precipitation, with most precipitation falling in winter. The annual temperature range is small, and winters are very mild for this latitude. Evergreen needleleaf forest is typical of this climate.

THE MARINE WEST-COAST CLIMATE

The **marine west-coast climate** ⑧ occupies midlatitude west coasts. These locations receive the prevailing westerlies from a large ocean and have frequent cyclonic storms involving cool, moist mP air masses (FIGURE 12.15). Where the coast is mountainous, the orographic effect causes large amounts of precipitation annually.

Precipitation is plentiful in all months, but there is often a distinct winter maximum. In summer, the rainfall is reduced because the influence of the subtropical high pressure extends poleward into the region. However, because of the prevailing westerlies, these locations still receive some pre-

■ **marine west-coast climate** ⑧ Cool moist climate of west coasts in the midlatitude zone, usually with abundant precipitation and a distinct winter precipitation maximum.

cipitation during the high-Sun season, unlike the Mediterranean climates further to the south (FIGURE 12.16). The annual temperature range is comparatively small for midlatitudes. The marine influence keeps winter temperatures milder than at inland locations at equivalent latitudes.

In North America, the marine west-coast climate ⑧ occupies the western coast from Oregon to northern British Columbia. Western Europe, the British Isles, Portugal, and much of France fall under this climate. In the southern hemisphere, it includes New Zealand and the southern tip of Australia, as well as the island of Tasmania and the Chilean coast south of 35° S. The general latitude range of this climate is 35° to 60° N and S.

Influence of latitude on Mediterranean and marine west-coast climates FIGURE 12.16

VIEW THIS IN ACTION in your WileyPLUS course

Although both the Mediterranean and marine west-coast climates tend to be located on the western portion of continents, they have very different climate and vegetation, principally because of differences in annual and seasonal rainfall. The differences have to do with the latitude of each with respect to the subtropical highs and polar jet stream.

July

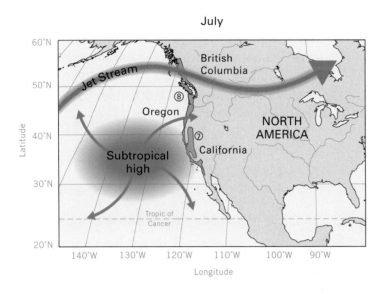

July During summer, both the subtropical high pressure center and the polar jet stream are shifted north. Midlatitude cyclones tend to move north as well, impacting regions farther north in Canada. In addition, the intensification of the Hawaiian high tends to produce divergence and subsidence over the Mediterranean climate region. This leads to substantial decreases in precipitation. In contrast, the anticyclonic flow around the Hawaiian high also leads to westerly flow from the oceans into the northwestern portion of the United States. The relatively moist air produces orographic precipitation as it rises over the Cascade mountains. As a result, the marine west-coast climates in this region experiences some rainfall in summer while the Mediterranean regime is dry.

January

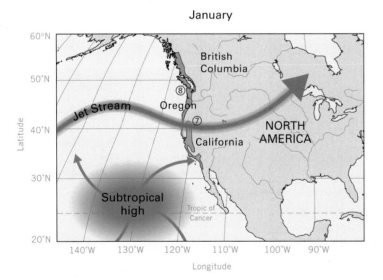

January During winter, the southward movement of the subtropical high and polar jet stream allows midlatitude cyclones to impact regions along the western coast of the U.S. This produces wintertime precipitation in the form of snow and rain, particularly along the windward side of the mountain ranges found over this portion of the continent. Hence, both Mediterranean and marine west-coast climates have substantial precipitation during this time.

⑦ Mediterranean climate
⑧ Marine west-coast climate

DRY MIDLATITUDE CLIMATE

The **dry midlatitude climate** ⑨ is almost exclusively limited to the interior regions of North America and Eurasia, where it lies within the rain shadow of mountain ranges on the west or south. The ranges effectively block the eastward flow of maritime air masses, and continental polar (cP) air masses dominate the climate in winter (**FIGURE 12.17**). In summer, a dry continental air mass of local origin dominates, but occasionally maritime air masses invade, causing convective rainfall.

> **dry midlatitude climate** ⑨ Dry climate of the midlatitude zone with a strong annual temperature cycle and cold winters.

The annual temperature cycle is strongly developed, with a large annual range. Summers are warm to hot, but winters are cold to very cold.

The largest expanse of the dry midlatitude climate ⑨ is in Eurasia, stretching from the southern republics of the former Soviet Union to the Gobi Desert and northern China. True arid (*a*) deserts and extensive areas of highlands can be found in the central portions of this region. In North America, the dry western interior regions, including the Great Basin, Columbia Plateau, and the Great Plains, are of the semiarid (*s*) subtype, which can sustain agriculture. A small area of dry midlatitude climate ⑨ is found in southern Patagonia, near the tip of South America. The latitude range of this climate is 35° to 55° N.

Dry midlatitude climate ⑨ FIGURE 12.17

A **Pueblo, Colorado** Located on the Arkansas River, Pueblo is the gateway to the southern Rocky Mountains. Pictured here is a pedestrian walkway along the river.

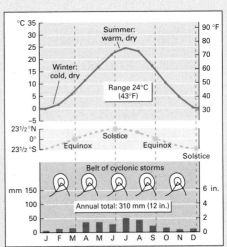

B **Dry midlatitude climate** ⑨ Pueblo, Colorado (lat. 38° N), just east of the Rocky Mountains, has a marked maximum of rainfall in the summer months. Total annual precipitation is 310 mm (12 in.), and most of this is convectional summer rainfall, which occurs when moist maritime tropical (mT) air masses invade from the south and produce thunderstorms. In winter, snowfall is light. The temperature cycle has a large annual range, with warm summers and cold winters. January, the coldest winter month, has a mean temperature just below freezing.

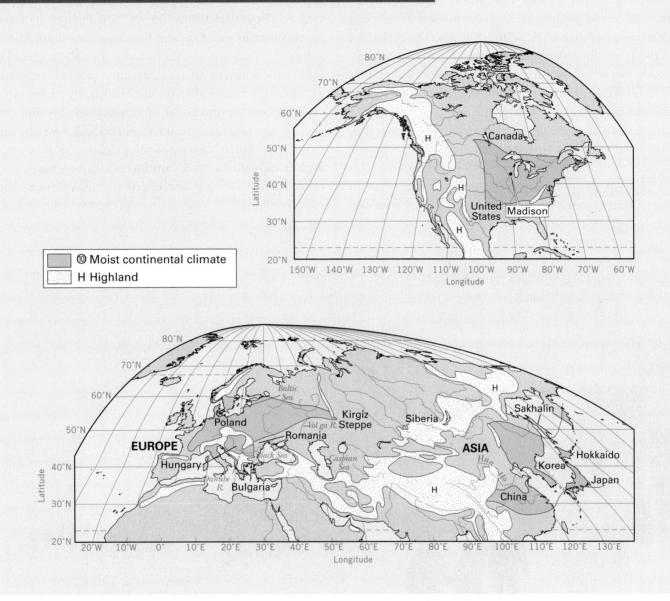

⑩ Moist continental climate
H Highland

MOIST CONTINENTAL CLIMATE

The **moist continental climate** ⑩ is located in central and eastern parts of North America and Eurasia in the midlatitudes (FIGURE 12.18). It lies in the polar-

moist continental climate ⑩ Moist climate of midlatitude zones with strongly defined winter and summer seasons and adequate precipitation throughout the year.

front zone—the battleground of polar and tropical air masses. Seasonal temperature contrasts in the moist continental climate are strong, and day-to-day weather is highly variable. There is ample precipitation throughout the year, which increases in summer when maritime tropical (mT) air masses invade (FIGURE 12.19). Cold winters are dominated by continental polar (cP) and continental arctic (cA) air masses from subarctic source regions. The influence of the prevailing air masses produces a much stronger seasonality in temperature than is found in the marine west-coast climate regions, which lie at about the same latitude (FIGURE 12.20).

The moist continental climate ⑩ is restricted to the northern hemisphere. In North America the moist continental climate ⑩ lies between latitudes 30° and 55° N, covering most of the eastern half of the United States from Tennessee to the north as well as the southernmost strip of eastern Canada. In Asia, it is found in northern China, Korea, and Japan between latitudes 30° and 55° N. These regions have more summer rainfall and drier winters than North America. This is an effect of the monsoon circulation, which moves moist maritime tropical (mT) air across the eastern side of the continent in summer and dry continental polar air southward through the region in winter. In most of central and eastern Europe, the moist continental climate ⑩ lies in a higher latitude belt (45° to 60° N) and receives precipitation from mP air masses coming from the North Atlantic.

Moist continental climate ⑩ FIGURE 12.19

A Madison, Wisconsin The snowy winters here offer plenty of opportunity for outdoor recreation.

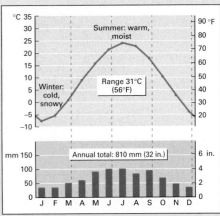

B Moist continental climate ⑩ Madison, Wisconsin (lat. 43° N), has cold winters—with three consecutive monthly means well below freezing—and warm summers, making the annual temperature range very large. There is ample precipitation in all months, and the annual total is large. There is a summer maximum of precipitation when the maritime tropical (mT) air mass invades, and thunderstorms form along moving cold fronts and squall lines. Much of the winter precipitation is snow, which remains on the ground for long periods.

Influence of location on marine west-coast and moist continental climates FIGURE 12.20

Although both the marine west-coast and moist continental climates tend to be located at the same latitudes, they have very different climates, with regard to both temperature and precipitation. Some of these differences have to do with the location of each with respect to the seasonal change in pressure centers found over the nearby oceans.

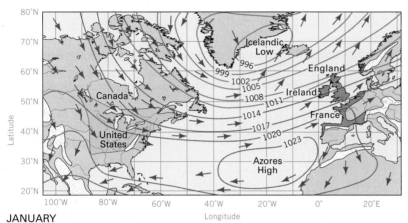

JANUARY

January During winter, the subtropical high over the Atlantic (the Azores High) weakens, shifts south, and is replaced in the North Atlantic by the Icelandic Low, which has shifted south as well. The cyclonic circulation around this low pressure center brings cP air from Canada down into the central and northeastern United States. This cP air produces cold, dry conditions over moist continental climates. In contrast, the same cyclonic circulation brings mT air from the subtropical Atlantic into southern Europe. This air, which tends to be moist and warm, produces wetter, milder wintertime conditions over the marine west-coast climate.

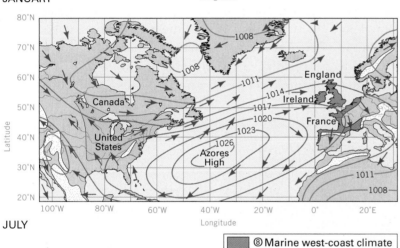

JULY

July During summer, the Icelandic Low weakens, shifts north, and is replaced over the north Atlantic by the Azores High. The anticyclonic circulation around this high pressure center brings mT air from the subtropical Atlantic into the central and northeastern United States. This mT air tends to make summertime conditions in the moist continental climate regions both hot and humid. In contrast, the same anticyclonic circulation brings mP air from the north Atlantic into Europe. This air, which tends to be humid but cool, results in some precipitation during summer but also produces much milder conditions than found over the moist continental climate regimes.

⑧ Marine west-coast climate
⑩ Moist continental climate
Highland

CONCEPT CHECK STOP

What are the six climate types of the midlatitude climate group? Give an example of each type.

What are each climate type's location and characteristic temperature and precipitation features?

How does precipitation in subtropical and midlatitude climates over the eastern portion of continents differ from that in central and western portions of continents?

High-Latitude Climates

LEARNING OBJECTIVES

Define the features of high-latitude climates.

Describe the boreal forest, tundra, and ice sheet climates.

igh-latitude climate are climates of the northern hemisphere, occupying the northern subarctic and arctic latitude zones. However, they also extend southward into the midlatitude zone as far south as about the 47th parallel in eastern North America and eastern Asia. One of these, the ice sheet climate ⑬, is present in both hemispheres in the polar zones.

The high-latitude climates coincide closely with the belt of prevailing westerly winds that circles the subarctic region. In the northern hemisphere, this circulation sweeps maritime polar (mP) air masses, formed over the northern oceans, into conflict with continental polar (cP) and continental arctic (cA) air masses on the continents. Upper-air disturbances form in the westerly flow, bringing lobes of warmer, moister air poleward into the region in exchange for colder, drier air that is pushed equatorward. As a result of these processes, traveling cyclones—*polar lows* and *polar cyclones*—are frequently produced along the arctic-front zone.

BOREAL FOREST CLIMATE

The **boreal forest climate** ⑪ is a continental climate with long, bitterly cold winters and short, cool summers (**FIGURE 12.21**). It occupies the source region for cP air masses, which are cold, dry, and stable in the winter. Very cold cA air masses

boreal forest climate ⑪ Cold climate of the subarctic zone in the northern hemisphere with long, extremely severe winters and several consecutive months of frozen ground.

Boreal forest climate ⑪ **FIGURE 12.21**

A Yakutsk, Siberia Well-dressed pedestrians make their way down a snowy street in winter.

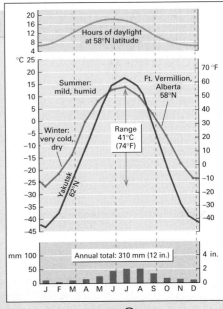

B Boreal forest climate ⑪ Extreme winter cold and a very large annual range in temperature characterize the climate of Fort Vermilion, Alberta (lat. 58° N). The very great annual temperature range is typical for North America. Monthly mean air temperatures are below freezing for seven consecutive months. The summers are short and cool. Precipitation has a marked annual cycle with a summer maximum, but the total annual precipitation is small. A snow cover remains over solidly frozen ground through the entire winter. We can also see temperature data for Yakutsk, a Siberian city at latitude 62° N. There is an enormous annual range, as well as extremely low means in winter months—January reaches about –42°C (–44°F).

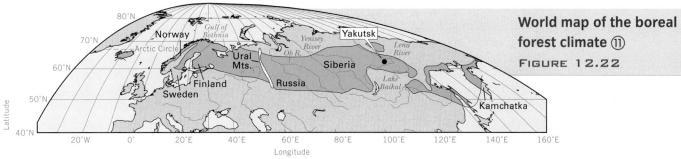

① Boreal forest climate

World map of the boreal forest climate ⑪

FIGURE 12.22

commonly invade the region. The annual range of temperature is greater than the range of any other climate and is greatest in Siberia, Russia.

Precipitation in the boreal forest climate increases substantially in summer, when maritime air masses penetrate the continent with traveling cyclones, but total annual precipitation is low. Although much of the boreal forest climate does receive enough precipitation to support forests of needle-leaf trees, large areas in western Canada and Siberia have low annual precipitation and are therefore cold and dry.

The boreal forest climate in North America stretches from central and western

Alaska, across the Yukon and Northwest Territories to Labrador on the Atlantic coast (FIGURE 12.22). In Europe and Asia, it reaches from the Scandinavian Peninsula eastward across all of Siberia to the Pacific. This climate type ranges from latitude 50° to 70° N.

TUNDRA CLIMATE

tundra climate ⑫
Cold climate of the arctic zone with eight or more months of frozen ground.

The **tundra climate** ⑫ occupies arctic coastal fringes and is dominated by polar (cP, mP) and arctic (cA) air masses. Winters are long and severe (FIGURE 12.23). The nearby ocean water moderates winter temperatures so they don't fall

Tundra climate ⑫ FIGURE 12.23

A Upernavik, Greenland A chunk of glacial ice floats past the village of Upernavik, situated on the tundra-covered slopes of a small island in Baffin Bay.

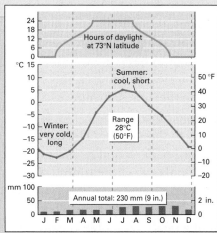

B Tundra climate ⑫ Upernavik, located on the west coast of Greenland (lat. 73° N), has short mild periods, with above-freezing temperatures—equivalent to a summer season in lower latitudes. The long winter is very cold, but the annual temperature range is not as large as that for the boreal forest climate to the south, such as at Fort Vermilion. Total annual precipitation is small. In July, the sea-ice cover melts and the ocean water warms, raising the moisture content of the local air mass, increasing precipitation.

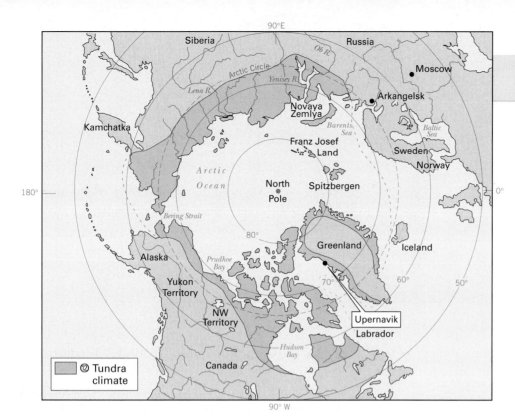

⑫ Tundra climate

to the extreme lows found in the continental interior. There is a very short mild season, which many climatologists do not recognize as a true summer. Although most portions of the tundra receive little rainfall because they are far from the polar front and generally under the influence of the polar high, coastal regions along the eastern portion of the continents do receive some precipitation as moist maritime polar (mP) air masses are brought in off the ocean by the prevailing polar easterlies.

The tundra climate ⑫ rings the Arctic Ocean and extends across the island region of northern Canada (**FIGURE 12.24**). It includes the Alaskan north slope, the Hudson Bay region, and the Greenland coast in North America. In Eurasia, this climate type occupies the northernmost fringe of the Scandinavian Peninsula and Siberian coast. The Antarctic Peninsula also belongs to this climate. The latitude range for this climate is 60° to 75° N and S, except for the northern coast of Greenland, where tundra occurs at latitudes greater than 80° N.

The term *tundra* describes both an environmental region and a major class of vegetation. Because of the cold temperatures experienced in the tundra and northern boreal forest climate zones, the ground is typically frozen to great depth. This perennially frozen ground, or *permafrost*, prevails over the tundra region. Normally, a top layer of the ground, 0.6–4 m (2–13 ft) thick, will thaw each year during the mild season.

Trees in the tundra are stunted because of the seasonal damage to roots by freezing and thawing of the soil layer and to branches exposed to the abrading action of wind-driven snow. In some places, a distinct tree line—roughly along the 10°C (50°F) isotherm of the warmest month—separates the forest and tundra.

THE ICE SHEET CLIMATE

The **ice sheet climate** ⑬ coincides with the source regions of arctic (A) and antarctic (AA) air masses, situated on the vast, high ice sheets of Greenland and Antarctica and over polar sea ice of the Arctic Ocean (**FIGURE 12.25**). The mean annual temperature is much lower than that of any other climate, with no monthly mean above freezing. Strong temperature in-

■ **ice sheet climate**
⑬ Severely cold climate found on the Greenland and Antarctic ice sheets.

versions, caused by radiation loss from the surface, develop over the ice sheets. In Antarctica and Greenland, the high surface altitude of the ice sheets intensifies the cold. Strong cyclones with blizzard winds are frequent. Precipitation, almost all occurring as snow, is very low, but the snow accumulates because of the continuous cold. The latitude range for this climate is 65° to 90° N and S.

Throughout this chapter, our focus has been on showing how the climate classifications presented here follow naturally from a consideration of the weather and climate processes that influence temperature and precipitation around the globe. However, both temperature and precipitation also control the amount of moisture held in the soil and how much it varies throughout the year. Hence, we expect different climates to support very different types of natural and planted vegetation. FIGURE 12.26 gives examples of the vegetation found in some of the different climate types discussed in this chapter.

Ice sheet climate ⑬ FIGURE 12.25

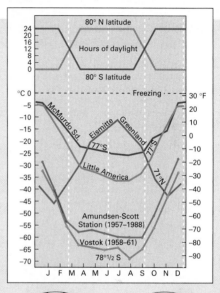

A **Antarctica** Snow and ice accumulate at higher elevations here in the Dry Valleys region of Victoria Land, Antarctica. Most of Antarctica is completely covered by ice sheets of the polar ice cap.

B **Ice sheet climate** ⑬ Temperature graphs for five ice sheet stations. Eismitte is on the Greenland ice cap; the other stations are in Antarctica. Temperatures in the interior of Antarctica are far lower than at any other place on Earth. A low of −88.3°C (−127°F) was observed in 1958 at Vostok, about 1300 km (about 800 mi) from the South Pole at an altitude of about 3500 m (11,500 ft). At the Pole (Amundsen-Scott Station), July, August, and September have averages of about −60°C (−76°F). Temperatures are considerably higher, month for month, at Little America in Antarctica because it is located close to the Ross Sea and is at a low altitude.

CONCEPT CHECK **STOP**

What are the three climate types of the high-latitude climate group? Give an example of each type.

What are each climate type's location and characteristic temperature and precipitation features?

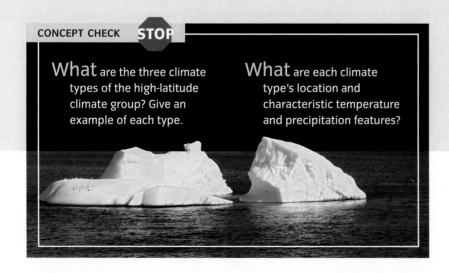

▲ Savanna woodland
In wet-dry tropical regions, coarse grasses occupy the open space between the rough-barked and thorny trees. There may also be large expanses of grassland. In the dry season, the grasses turn to straw, and many of the tree species shed their leaves to cope with the drought.

Dry Climates The soil in dry climates is dry much of the year, and the land surface contains only sparse plant cover—scattered grasses or shrubs—or simply lacks a plant cover. However, even in these regions, vegetation can be quite varied.

▲ Nomadic grazing
The semiarid steppes bordering many of the world's deserts often support nomadic grazing cultures. These Shahsavan tribespeople near Tabriz, Iran, are packing their possessions in preparation for a move to summer pastures.

▲ Desert vegetation
The odd-looking plants are Joshua trees, which are abundant here in the Mojave Desert—a dry subtropical climate. Most areas of this desert have fewer plants than shown here.

◀ Wheat harvest
Wheat is a major crop of the semiarid, dry midlatitude steppe lands, but wheat harvests are at the mercy of rainfall variations from year to year. Good spring rains mean a good crop, but if spring rains fail, so does the wheat crop. This aerial view shows combines bringing in a Kansas wheat harvest.

Climate types and vegetation FIGURE 12.26

Moist Climates Moist climates are those with sufficient rainfall to maintain the soil in a moist condition through much of the year and to sustain the year-round flow of the larger streams. Moist climates support forests or prairies of dense tall grasses.

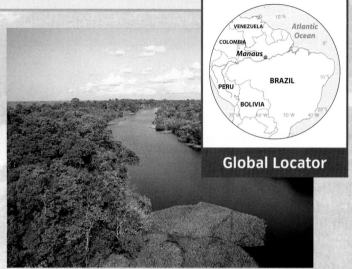

Global Locator

▲ Old-growth rainforests

Some of the world's tallest trees are found in the marine west-coast climates. Here, abundant water—in the form of precipitation as well as fog arriving from the nearby ocean—provides moisture for growth. In addition, the moderating influence of the nearby ocean keeps temperatures mild throughout the year. These conditions allow forests, such as this one in the Jedediah Smith Redwoods State Park in California, to thrive.

▲ Wet equatorial and trade-wind coastal climate vegetation

The wet equatorial and trade-wind coastal climates are quite uniform in temperature and have a high annual rainfall. Streams flow abundantly throughout most of the year, and the riverbanks are lined with dense forest vegetation. These factors create a special environment—the low-latitude rainforest. This picture shows the rainforest of the western Amazon lowland, near Manaus, Brazil. The river is a tributary of the Amazon.

▲ Moist subtropical forest

A mix of broadleaf deciduous trees and shrubs (including oaks, hickories, and poplars) and occasional pines are quite typical of forests in this climate zone. Broadleaf evergreen trees and shrubs, such as the mountain laurel shown here, can also be found.

▲ Lakes in a boreal forest

Much of the boreal forest consists of low but irregular topography, formed by continental ice sheets during the last glaciation. Low depressions scraped out by the moving ice are now occupied by lakes. Alaska's Mulchatna River is in the foreground of this aerial photo.

This image shows the migration of caribou across the arctic tundra of northern Alaska. These caribou travel over 1000 km between their summering and wintering ranges as they search for food and suitable calving grounds.

- Given that this picture was taken around November, in which direction do you think the caribou are headed?

- What does the distance they travel tell you about the seasonal shifts in climate patterns over the tundra region?

SUMMARY

1 Low-Latitude Climates

1. **Wet equatorial climates** ① are warm to hot with abundant rainfall. This is the steamy climate of the Amazon and Congo basins.

2. **Trade-wind coastal climates** ② are warm to hot with very wet rainy seasons. They occur in coastal regions that are influenced by trade winds. The climates of Vietnam and Bangladesh are good examples. The monsoon region of coastal India also represents a trade-wind coastal climate.

3. **Wet-dry tropical climates** ③ are warm to hot with very distinct wet and dry seasons. The monsoon region of interior India falls into this type, as does much of the Sahel region of Africa.

4. **Dry tropical climates** ④ describe the world's hottest deserts—extremely hot in the high-Sun season and a little cooler in the low-Sun season, with little or no rainfall. The Sahara Desert, Saudi Arabia, and the central Australian desert are of this climate.

2 Midlatitude Climates

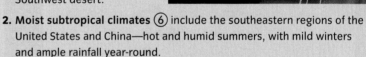

1. **Dry subtropical climates** ⑤ also include desert climates, but they are found farther poleward than the dry tropical climate ④, and so aren't as hot. This type includes the hottest part of the American Southwest desert.

2. **Moist subtropical climates** ⑥ include the southeastern regions of the United States and China—hot and humid summers, with mild winters and ample rainfall year-round.

3. **Mediterranean climates** ⑦ are marked by hot, dry summers and rainy winters. Southern and central California, Spain, southern Italy, Greece, and the coastal regions of Lebanon and Israel are prime examples.

4. **Marine west-coast climates** ⑧ have warm summers and cool winters, with more rainfall in winter. Climates of this type include the Pacific Northwest—coastal Oregon, Washington, and British Columbia.

5. **Dry midlatitude climates** ⑨ are dry climates found in midlatitude continental interiors. The steppes of central Asia and the Great Plains of North America are familiar locales with this climate—warm to hot in summer, cold in winter, and low annual precipitation.

6. **Moist continental climates** ⑩ are found in the eastern United States and lower Canada—cold in winter, warm in summer, and ample precipitation throughout the year.

3 High-Latitude Climates

1. **Boreal forest climates** ⑪ are snowy climates with short, cool summers and long, bitterly cold winters. Northern Canada, Siberia, and central Alaska are regions of boreal forest climate.

2. **Tundra climates** ⑫ have a long, severe winter. Temperatures on the tundra are somewhat moderated because they are near the Arctic Ocean. This is the climate of the coastal arctic regions of Canada, Alaska, Siberia, and Scandinavia.

3. **Ice sheet climates** ⑬ are bitterly cold. Temperatures of this climate, restricted to Greenland and Antarctica, can drop below −50°C (−58°F) during the sunless winter months. Even during the 24-hour days of summer, temperatures remain well below freezing.

KEY TERMS

CRITICAL AND CREATIVE THINKING QUESTIONS

1. Why is the annual temperature cycle of the wet equatorial climate ① so uniform? The wet-dry tropical climate ③ has two distinct seasons. What factors produce the dry season? The wet season?

2. Sketch the temperature and rainfall cycles for a typical station in the trade-wind coastal climate ②. What factors contribute to the seasonality of the two cycles?

3. Why is the dry tropical climate ④ dry? How does the dry subtropical climate ⑤ differ from the dry tropical climate ④?

4. Both the moist subtropical ⑥ and moist continental ⑩ climates are found on eastern sides of continents in the midlatitudes. What are the major factors that determine their temperature and precipitation cycles? How do these two climates differ?

5. Both the Mediterranean ⑦ and marine west-coast ⑧ climates are found on the west coasts of continents. Why do they experience more precipitation in winter than in summer? How do the two climates differ?

6. Both the boreal forest ⑪ and tundra ⑫ climate are climates of the northern regions, but the tundra is found fringing the Arctic Ocean and the boreal forest is located farther inland. Compare these two climates from the viewpoint of coastal-continental effects.

7. Suppose South America were turned over. That is, imagine that the continent was cut out and flipped over end-for-end so that the southern tip was at about 10° N latitude and the northern end (Venezuela) was positioned at about 55° S. The Andean chain would still be on the west side, but the shape of the land mass would now be quite different. Sketch this continent and draw possible climate boundaries, using your knowledge of global air circulation patterns, frontal zones, and air-mass movements.

SELF-TEST

1. In contrast to the marine west-coast climate, the dry midlatitude climate is characterized by _____.
   a. uniform temperatures all year round
   b. a high annual variation in solar insolation
   c. no annual variation in solar insolation
   d. a very strong temperature cycle

2. The tropical desert climates are caused by _____.
   a. an abundance of moisture
   b. their placement in relationship to the prime meridian
   c. their distance from the Equator and the Arctic Circle
   d. stationary subtropical cells of high pressure

3. The _____ air masses have no influence on low-latitude climates.
   a. cT   c. cP
   b. mE   d. mT

4. The _____ type climate is not a member of the low-latitude climate group.
   a. wet equatorial   c. wet-dry tropical
   b. moist subtropical   d. dry tropical

5. The _____ air mass is drawn onshore over Asia during the summer monsoon season.
   a. mP   c. cT
   b. cP   d. mT

6. The region of midlatitude climates lies in the _____, a zone of intense interaction between air masses with significantly different temperature and moisture characteristics.
   a. polar-front zone   c. tropics
   b. ITCZ   d. subtropics

7. The _____ air masses are responsible for enhanced summertime precipitation in the moist continental climate.
   a. mP   c. cP
   b. mT   d. cA

8. Boreal forest climates have short, cool summers and long, bitterly cold winters.
   a. True   b. False

9. High-latitude climate types are dominantly found in the _____ hemisphere.
   a. western   c. northern
   b. eastern   d. southern

10. Given the surface pressures and winds and upper-air jet stream location: (1) indicate the climate regimes associated with the two shaded regions; (2) determine the time of year; (3) indicate whether the shaded regions experience wet or dry conditions.

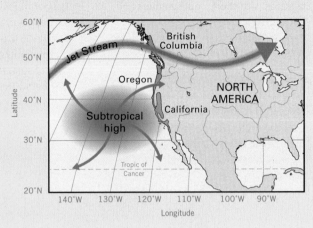

11. Wet equatorial climates _____.
    a. are significantly influenced by the ITCZ
    b. are dominated by warm, moist mE and mT air masses
    c. show only small seasonal rainfall variations
    d. all of the above

12. The areas shaded green on this map represent which climate regime?
    a. moist subtropical   c. Mediterranean
    b. dry subtropical   d. ice sheet

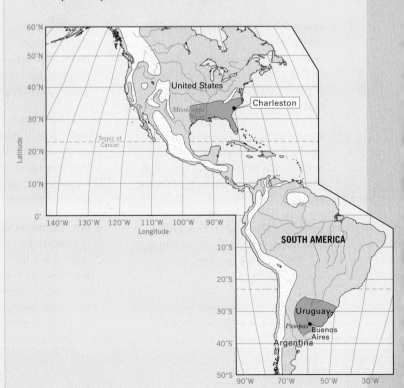

13. Dry tropical climates are generally located in the center and on the _____ of subtropical high-pressure cells.
   a. west sides
   c. south sides
   b. north sides
   d. east sides

14. The _____ climate is renowned for its very scarce precipitation during the summer season only.
   a. moist subtropical
   c. Mediterranean
   b. dry subtropical
   d. ice sheet

15. Identify which climographs represent: (a) the dry midlatitude climate at Pueblo, Colorado; (b) the tundra climate of Upernavik, Greenland; and (c) the wet-dry tropical climate of Timbo, Guinea.

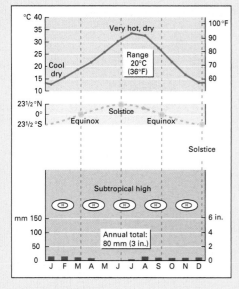

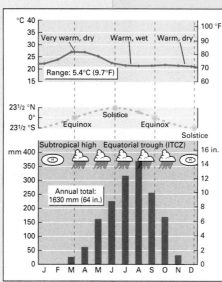

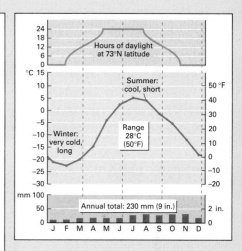

Climate Variability

A swarm of Mastigias jellyfish in land-locked Jellyfish Lake, Rock Islands, Palau.

On the tropical island of Eil Malk in Palau, in the Pacific Ocean, is a very special saltwater lake—home to more than 10 million jellyfish. In the distant past, Jellyfish Lake was connected to the ocean. A geological upheaval isolated a section of the marine lagoon, creating a land-locked lake and trapping the *Mastigias* jellyfish.

Over the years, the jellyfish adapted to make the most of their changed climate and environment. Free from any predators, their stings became unnecessary. They adapted to feeding on quickly reproducing algae, which they house in their own bodies. The algae rely on photosynthesis, so during the day, clouds of *Mastigias* follow the Sun's rays around the lake.

During the strong La Niña climate event of 1998–1999—to the great surprise of divers—the entire jellyfish population mysteriously vanished. The jellyfish had not died, though—the population reappeared once normal climate conditions resumed. The explanation is that the jellyfish life cycle had adapted to the local climate. The *Mastigias* begin life as polyps, transform into larvae, and then finally develop into their familiar adult form. Because the water was abnormally hot during that La Niña, the jellyfish remained in their polyp state. Once the normal climate returned, the polyps began maturing again, restoring the population.

The *Mastigias* jellyfish are only one of many animal and plant species around the globe that have adapted to variations in the climate conditions around them.

NATIONAL GEOGRAPHIC

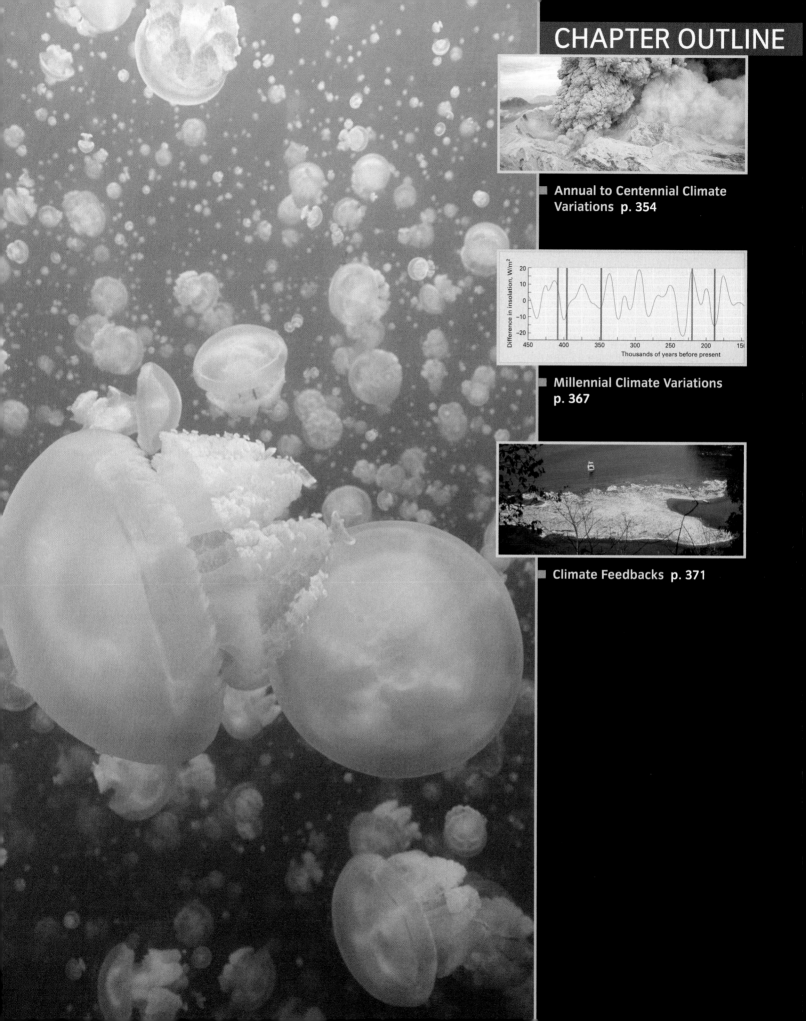

Annual to Centennial Climate Variations

LEARNING OBJECTIVES

Identify different processes that can produce global-scale climate variations.

Describe how El Niño conditions differ from La Niña conditions.

Characterize the climate changes associated with the North Atlantic oscillation.

Describe how changes in the thermohaline circulation can affect climate in the North Atlantic.

I s our global climate changing? Yes. The climates of the Earth are not fixed, but change over time on several scales. The processes that affect climate are always in motion and are often interconnected in complex ways. As these processes change and interact, they change climate as well. In this chapter, we look at the natural processes that affect climate as well as the feedback mechanisms that affect these processes. Of course, human activity has also changed the Earth's climates in various ways—but we will hold that part of our story for Chapter 16.

HISTORICAL RECORD

One of the most obvious descriptors of climate is temperature, and scientists have developed many techniques for determining past temperatures over hundreds to thousands of years. **FIGURE 13.1** shows how the mean annual surface temperature of the northern hemisphere has varied over the last three centuries. A lot of year-to-year and decade-to-decade variability occurs. On the scale of a century or so, the temperature cycled through a change of about 0.6°C from 1700 to about 1850. Then for about a hundred years, the temperature rose, only to fall briefly from about 1950 to 1970. Since then, the temperature has continued its upward rise.

In fact, the last decade is the warmest of the past thousand years, at least in the northern hemisphere, as seen in **FIGURE 13.2**. At this scale, the cycles we see in Figure 13.1 are part of a longer rhythm of ups and downs that has persisted over the millennium. Except for the steep upward climb in the 20th century, the graph shows a decreasing trend. The period of decreasing temperature starting about 1400 and carrying through to about 1850 is known as the *Little Ice Age.* In some of the colder periods of this era, rivers and harbors in Europe and North America froze, impeding

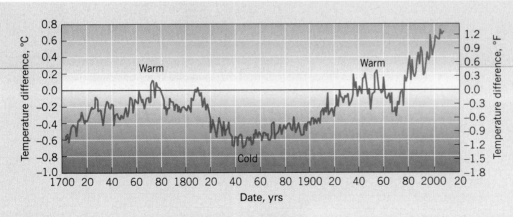

Three centuries of northern hemisphere temperatures
FIGURE 13.1

A reconstruction of the departures of northern hemisphere temperatures from the 1950–1965 mean, based on analyses of tree rings sampled along the northern tree limit of North America. Original data are from 1700 to 1975. For completeness, observed mean annual surface temperatures of the Earth from 1975 to 2000 are shown in a different color.

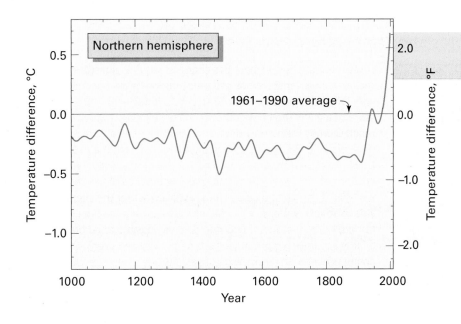

Compared with the reference temperature of
the 1961–1990 average, northern hemisphere
temperatures followed a trend slightly downward until the beginning of the 20th century.
The line shown is a 40-year moving average obtained from thermometers, historical data, tree
rings, corals, and ice cores.

trade and transportation. Alpine glaciers grew, and snow lingered on the ground for extra weeks, reducing the growing season.

On an even longer time scale, the Earth was much colder during the last glacial stage of the most recent ice age. This stage started about 120,000 years ago and ended about 12,000 years ago. Thus, climate change over time is the norm, not the exception.

How do scientists determine what the surface temperatures were hundreds or thousands of years ago? They use **proxy data**—information preserved over long time periods that can be used to reconstruct past temperature. For example, the growth of some species of trees in stressful environments is strongly affected by temperature. Each year, a tree produces a growth ring that is thicker in warm years and thinner in cold years (see **FIGURE 13.3**). By counting rings and measuring their thicknesses, scientists can reconstruct a history of temperature. Some corals also grow in annual increments, and the chemical composition of their growth layers can be related to water temperature, producing

proxy data Information used for indirect measurement of past climate characteristics, such as the thickness of annual tree rings or the chemical composition of annual coral growth rings.

a similar temperature record. Ice cores can also provide a temperature record that scientists can "read" by examining the chemical composition of air bubbles trapped within annual layers of snow, as discussed in *What a Scientist Sees: Ice Cores and Temperature Records*.

Tree rings FIGURE 13.3

The width of annual tree rings varies with growing conditions and in some locations can indicate air temperature.

Ice Cores and Temperature Records

Global Locator

Understanding how climate has changed requires a historical record of how temperature and precipitation have changed over very long time periods. How do scientists know what the temperature of a given region was thousands of years ago? Much of the record comes from proxy data, such as the chemical composition of layers of ice in this ice core being extracted from the Antarctic ice sheet at Vostok, Antarctica (A). Each year new snow falls, covering the old snow pack and trapping bubbles of air. Chemical analysis of the ice and air bubbles reveals whether the climate was relatively wet or dry, or cold or warm, during the time it was buried. From this kind of information, it is possible to reconstruct a fairly accurate picture of global temperature and precipitation changes over many thousands of years (B).

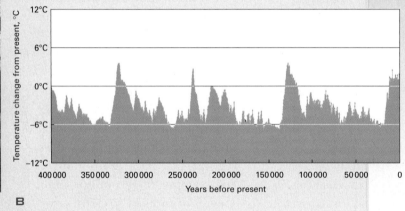

A

B

VOLCANOES

Volcanic eruptions are spectacular events, with lava flows and showers of rocks and ash that can melt mountain glaciers and trigger mud flows (FIGURE 13.4). In addition to these local effects, a strong eruption can change the climate around the globe. Volcanic gases are rich in sulfur dioxide, a gas that combines with water to form tiny drops—*aerosols*—of liquid sulfuric acid. These particular aerosols strongly reflect sunlight.

In a large and powerful eruption, many tons of sulfur dioxide are released and lifted high into the atmosphere, where they form aerosols and become trapped in the stratosphere. Because they are above the level of cloud formation, the sulfate aerosols are not washed out of the atmosphere by precipitation. Instead, they spread widely, forming a thin layer that reflects a small portion of the sunlight reaching the Earth. This layer raises the Earth's albedo and acts to lower global temperatures.

Eruption of Mount Pinatubo, Philippine Islands, April 1991 FIGURE 13.4

Volcanic eruptions like this can inject particles and gases into the stratosphere, influencing climate for several years afterward.

The last large eruptions to produce significant changes in the global climate were those of Mount Agung in Bali in 1963, El Chichón in Mexico in 1982, and Mount Pinatubo in the Philippines in 1991 (FIGURE 13.5). For these eruptions, the decrease in incoming solar radiation produced a subsequent drop in global temperatures of as much as 0.2–0.3°C (0.36–0.54°F). Other large eruptions include those at Mount Tambora and Krakatoa in Indonesia in 1815 and 1883. The Mount Tambora eruption produced the "year without a summer," during which there was snowfall in August in New England and parts of Europe. Because the sulfuric acid aerosols eventually leave the stratosphere after a year or two, the effect of volcanic eruptions on climate is generally short compared with other climate-change processes.

The impact of a particular volcanic eruption on climate depends on the amount of sulfur dioxide it injects into the stratosphere, which is determined by the magnitude of the eruption. Another factor is location. Eruptions occurring in equatorial regions often have a greater effect on climate because the aerosols they release can be carried by stratospheric circulation into both hemispheres. However, if the eruption occurs at higher latitudes, where the tropopause is lower, it does not need to be as large to inject aerosols into the stratosphere. For example, the eruption of the Laki volcano in Iceland in 1783, although weak, was followed by the lowest average winter temperatures on record in the eastern United States.

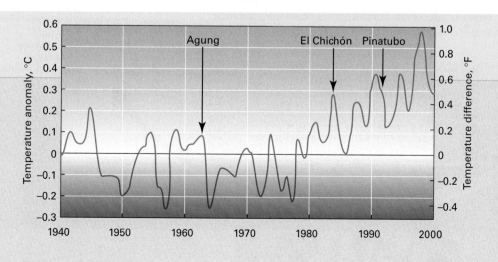

Impact of volcanoes on global temperatures

FIGURE 13.5

Global temperatures have generally shown an increase over the last 60 years of about 1°C. However, following major volcanic eruptions, temperatures decreased by 0.2–0.3°C (0.36–0.54°F) for 2–3 years. Note that other temperature decreases also occur that are not related to volcanic eruptions.

EL NIÑO AND LA NIÑA

Another phenomenon that can produce a significant change in global climate is **El Niño**—a period in which the waters of the eastern equatorial Pacific are warmer than normal. For hundreds of years, Peruvians occupying the arid Pacific coast of their country looked forward to an unusually wet winter that, come spring, would make their desert land bloom. These wet winters, occurring on a 3- to 8-year cycle, were thought to be gifts of the Christ child, El Niño.

However, the blooming desert was also associated with a dying fishery. The coast of Peru is normally an upwelling zone where cold bottom waters rich in nutri-

El Niño Phenomenon in which the waters of the eastern tropical Pacific warm significantly.

ents come to the surface. The nutrients stimulate the phytoplankton that form the base of the oceanic food chain, creating a rich and abundant population of fish. However, during these El Niño winters, a warm, poleward-flowing, sterile ocean current invaded the coastal waters, and fish populations plummeted. Thus, the name El Niño also became associated with the warm, poleward-flowing current that brought misfortune to the Peruvian fishing fleets.

Meanwhile, scientists were observing a shift in the atmospheric circulation of the equatorial Pacific Ocean that they called the *southern oscillation*. Normally, the eastern equatorial Pacific is a region of cooler air tem-

El Niño and the southern oscillation FIGURE 13.6

VIEW THIS IN ACTION
in your WileyPLUS course

Maps of temperatures, pressures, and ocean currents in the tropical Pacific and eastern Indian Ocean in November during normal and El Niño years. The interaction of the ocean and atmosphere in this region—termed the El Niño /Southern Oscillation or ENSO—results in a series of positive feedbacks that allows initial changes in the fields to grow over time.

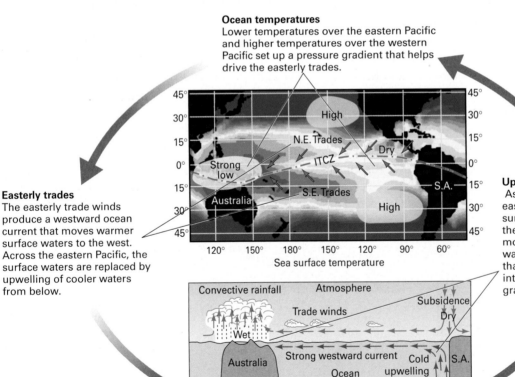

Ocean temperatures
Lower temperatures over the eastern Pacific and higher temperatures over the western Pacific set up a pressure gradient that helps drive the easterly trades.

Easterly trades
The easterly trade winds produce a westward ocean current that moves warmer surface waters to the west. Across the eastern Pacific, the surface waters are replaced by upwelling of cooler waters from below.

Upwelling
As cold water upwells in the eastern Pacific, it lowers the surface temperatures there. Over the western Pacific, the westward movement of the warm surface waters raises the temperatures in that region. These two processes intensify the initial temperature gradient.

peratures and drier weather with higher surface pressures, while the western region is warmer and wetter with lower surface pressures. The pressure gradient between the two regions helps drive the easterly trade winds that blow westward.

Every few years, however, this situation changed. The trade winds failed and so did the rains in the western Pacific. This condition was marked by changes in barometric pressures observed at Tahiti, in the central Pacific, and at Darwin, in the western Pacific on the northern tip of the Australian continent. During normal years, the pressure at Darwin was lower than at Tahiti. In the abnor-

mal dry years, though, the pressure at Darwin was higher than normal and the pressure at Tahiti was lower than normal, thereby reducing the pressure difference between the two. Scientists called the record of weakening and strengthening pressure differences between the two locations the southern oscillation.

In about the 1950s, however, atmospheric scientists and oceanographers recognized that El Niño and the southern oscillation were actually part of the same phenomenon that we now refer to as the **El Niño/Southern Oscillation**, or **ENSO**. FIGURE 13.6 shows the phenomenon in more detail.

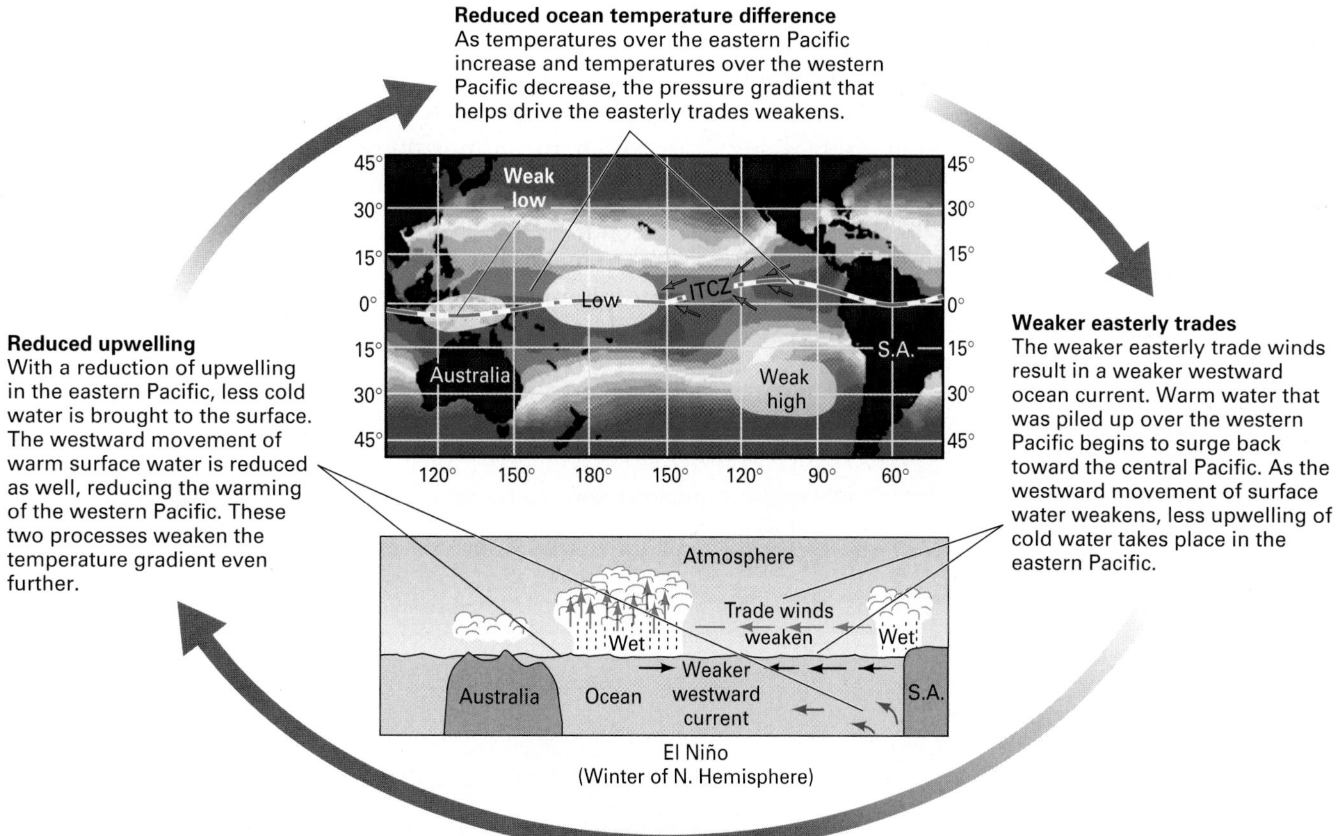

■ **El Niño/Southern Oscillation (ENSO)** A coupled system of ocean and atmospheric circulation that produces cyclic changes in winds, currents, temperatures, and rainfall patterns in the equatorial Pacific region.

Reduced ocean temperature difference
As temperatures over the eastern Pacific increase and temperatures over the western Pacific decrease, the pressure gradient that helps drive the easterly trades weakens.

Reduced upwelling
With a reduction of upwelling in the eastern Pacific, less cold water is brought to the surface. The westward movement of warm surface water is reduced as well, reducing the warming of the western Pacific. These two processes weaken the temperature gradient even further.

Weaker easterly trades
The weaker easterly trade winds result in a weaker westward ocean current. Warm water that was piled up over the western Pacific begins to surge back toward the central Pacific. As the westward movement of surface water weakens, less upwelling of cold water takes place in the eastern Pacific.

El Niño
(Winter of N. Hemisphere)

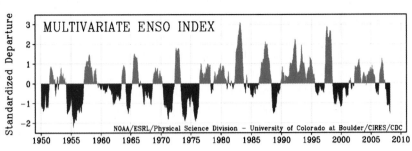

A These striking images show sea-surface temperature changes in the eastern tropical Pacific during an El Niño year and a La Niña year. Blue/green tones indicate cooler than normal temperatures, and red tones indicate warmer than normal temperatures.

B The time-series above shows monthly values for the multivariate ENSO index, which uses data such as pressure, air and sea-surface temperatures, winds, and cloudiness over the equatorial Pacific to monitor El Niño conditions (red) and La Niña conditions (blue) over the last 50 years.

El Niño and La Niña sea-surface temperature anomalies FIGURE 13.7

In the normal circulation pattern, easterly trade winds are driven by a gradient of high pressure and cool water temperatures in the eastern Pacific and low pressure and warmer water temperatures in the western Pacific. Surface water, blown by the trade winds, piles up in the west. In an El Niño event, upwelling along the South American coast does not reach the surface, and the surface waters remain warm. With the reduced temperature and pressure gradient, trade winds weaken or stop, and warm ocean water surges eastward.

At other times, the upwelling along the Peruvian coast becomes more intense, significantly cooling sea-surface temperatures over the eastern Pacific. The normal circulation pattern becomes more intense, with increased surface pressures in the eastern Pacific, reduced surface pressures in the western Pacific, and stronger trade winds. This condition is referred to as **La Niña** (the girl-child).

What causes the ENSO phenomenon? One view is that the cycle is a natural oscillation caused by the way in which the atmosphere and oceans are coupled through temperature and pressure changes. Each ocean–atmosphere state has positive feedbacks that tend to make that state stronger. Thus, the ocean and atmosphere tend to stay in one state or the other until something occurs to reverse the state.

La Niña Situation in which the waters of the eastern tropical Pacific cool significantly.

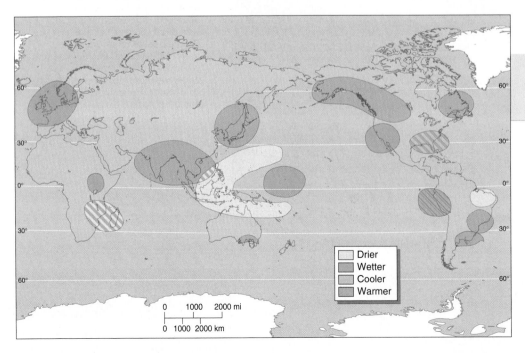

Drier
Wetter
Cooler
Warmer

0 1000 2000 mi

0 1000 2000 km

Climate changes during El Niños

FIGURE 13.8

El Niño events drastically alter local and regional climate, even in many areas far from the Pacific Ocean. As a result, some areas are drier, some wetter, some warmer, and some cooler than usual.

El Niño's impact on rainfall FIGURE 13.9

During an El Niño, shifts in the tropical convection produce changes in rainfall over wide regions, with some regions experiencing significantly more precipitation (reds) and others experiencing significantly less (blues).

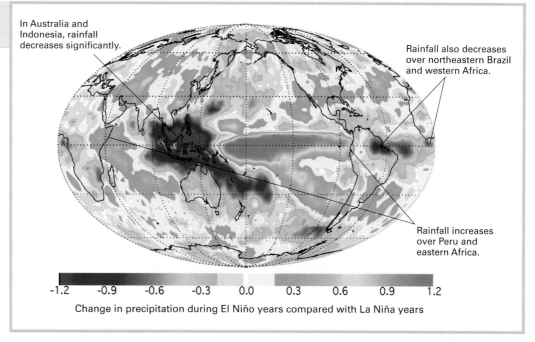

In Australia and Indonesia, rainfall decreases significantly.

Rainfall also decreases over northeastern Brazil and western Africa.

Rainfall increases over Peru and eastern Africa.

-1.2   -0.9   -0.6   -0.3   0.0   0.3   0.6   0.9   1.2

Change in precipitation during El Niño years compared with La Niña years

One process that can cause a reversal of the ENSO state is another, negative feedback mechanism in the system. The upwelling of cool water along the eastern tropical Pacific also generates a broad but weaker downwelling of water to either side of the Equator. During an El Niño, when upwelling weakens, so does the downwelling. The weaker downwelling allows the thermocline in this region to lift, like the crest of a wave. Like a wave, the lifted thermocline propagates westward. Once it reaches the margins of the continents in the western Pacific, the wave crest of the lifted thermocline is reflected back, moving slowly eastward along the Equator. Eventually it reaches the eastern equatorial Pacific, where it lifts the thermocline back toward the surface. This lifting of the thermocline cools the surface waters and sets the stage for the return of normal conditions or even a La Niña. Because this mechanism turns an El Niño into a La Niña, it is a type of negative feedback. It takes many months for the slow-moving wave to do its work, however, so most El Niños last for a year or more (FIGURE 13.7).

Global effects of El Niño

The El Niño phenomenon is considered one of the most important influences on natural climate variability, producing changes across the globe (FIGURE 13.8). For example, the winter of 1997–1998 experienced one of the strongest El Niños of the century. In equatorial and tropical regions, torrential rains drenched Peruvian and Ecuadorian coastal ranges, producing mudflows, debris avalanches, and extensive river flooding. Large portions of Australia and the East Indies went rainless for months, and forest fires burned out of control in Sumatra, Borneo, and Malaysia. In east Africa, Kenya experienced rainfall 1000 mm (40 in.) above normal.

In North America, heavy rainfalls and flooding occurred in Florida and on the Gulf Coast, spreading across the southern United States. In the west, California was hit hard, with some locations experiencing 2–3 times their normal annual precipitation. However, in Mexico and Central America, extreme drought occurred. FIGURE 13.9 shows how precipitation can change during El Niño and La Niña.

How do changes in pressures, winds, and currents in the equatorial Pacific affect the climate of these far distant locations? Outside the tropics, El Niños and La Niñas can produce significant climatic changes via **teleconnections**, which are far-reaching changes in the general circulation of the atmosphere.

teleconnections Large-scale changes in the general circulation of the atmosphere that allow changes in atmospheric circulations in one region to affect the climate of regions far from the original location.

For El Niños, changes in surface pressure and sea-surface temperatures alter the regions of convection and produce changes in vertical motions aloft. Under normal conditions, the convection in the ITCZ produces an upper-air pressure gradient between the Equator and the poles, resulting in the formation of westerly jet streams. As the convection in the equatorial Pacific changes, the strength and location of jet streams also change. The subtropical jet intensifies and shifts southward (FIGURE 13.10), bringing storms farther south, over California, and farther east, over Florida. In the northwestern United States, the weather is warmer and drier, because most storms now track to the south. In contrast, the polar jet stays well to the north, sheltering Canada and the northeastern United States from storms and polar air outbreaks.

The southerly position of the subtropical jet during an El Niño also influences weather patterns in the subtropical and tropical Atlantic. There, the intensified upper-air winds associated with the jet stream disrupt the formation of tropical cyclones, and the number of hurricanes in the western Atlantic decreases. In contrast, during La Niña years the number of hurricanes increases as the jet stream shifts north and tropical storms grow unimpeded by strong upper-air winds.

Just how large an impact can El Niño and La Niña events have? During the 1997–1998 El Niño, powerful winter storms devastated parts of California. Monstrous tornadoes ripped through Florida. A record ice storm left millions without power in Quebec and the northeastern United States. All told, the modified weather patterns associated with the El Niño of 1997–1998 caused property damage estimated at $33 billion worldwide and killed an estimated 2,100 people.

The El Niño of 1997–1998 was rapidly followed by the La Niña of 1998–1999. The result was heavier monsoon rains in India and more rain in Australia. In the United States, winter conditions were colder than normal in the northwest and upper midwest as storms that had previously gone through California now shifted north. The eastern mid-Atlantic region endured drought through spring and early summer. Finally, the Atlantic hurricane season of 1998, which included the monster storm Hurricane Mitch, was the deadliest of the past two centuries.

PACIFIC DECADAL OSCILLATION

In addition to ENSO, another source of climate variability is found over the north Pacific Ocean. There, changes in the atmospheric pressure pattern—called the *North Pacific pressure pattern*—can produce climate changes across parts of Eurasia, Alaska, and the western United States. These changes interact with ENSO-related changes, either strengthening them or weakening them. The changes in the North Pacific pressure pattern can last from weeks to decades. The decadal changes—seen in FIGURE 13.11—are called the *Pacific decadal oscillation (PDO)*.

Midlatitude weather FIGURE 13.10

During an El Niño, the subtropical jet stream tends to shift equatorward and intensify. Because the jet streams act to steer storms in particular directions, midlatitude cyclones tend to move south across California and into the southeastern United States during El Niño years.

Normal

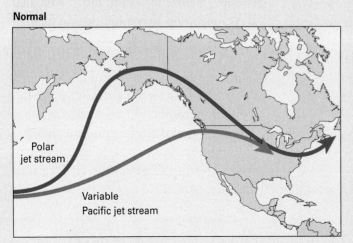

Polar jet stream

Variable Pacific jet stream

El Niño

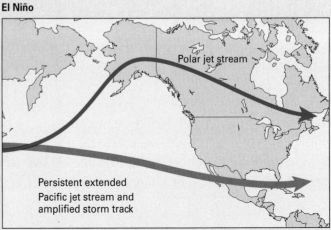

Polar jet stream

Persistent extended Pacific jet stream and amplified storm track

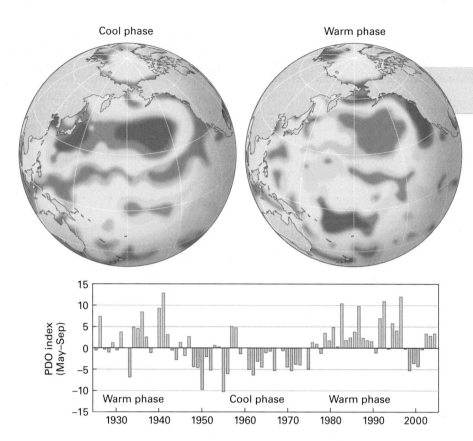

Cool phase   Warm phase

The changes in sea-surface temperature that are part of the Pacific decadal oscillation can produce variations in the overlying sea-level pressures, which produce changes in prevailing winds and storm tracks. In the warm phase, storm systems move south, leaving the northwestern portion of the United States warm and dry, while the southwest receives more rainfall than normal. During cool conditions, with opposite sea-surface temperatures and sea-level pressures, the northwestern United States becomes cooler and wetter, while drought comes to California and the American southwest. Along the western coast of North America, changes in the PDO can also produce variations in the abundance of salmon in fisheries as well as changes in plankton in the coastal waters of the eastern North Pacific, leading to changes in sea bird and sea lion populations.

NORTH ATLANTIC OSCILLATION

Another source of interannual-to-decadal climate variability is the **North Atlantic oscillation (NAO)**, which is associated with changes in atmospheric pressures and sea-surface temperatures over the mid- and high latitudes of the Atlantic (**FIGURE 13.12**).

The North Atlantic oscillation is predominantly an atmospheric phenomenon, although it may be linked with changes in both the ocean and possibly snow and ice cover over Europe and Greenland. The NAO is partly related to larger-scale variations in the surface pressure gradient between the polar cap and the mid-latitudes in both the Atlantic and Pacific Ocean basins, which is termed the Arctic oscillation (AO). Changes in these pressure gradients can significantly influence the climate in North America, Europe, and Asia.

Unlike the ENSO variations, shifts in the North American oscillation can occur over the course of weeks, seasons, and even decades. Since about 1975, the positive phase has dominated, bringing wetter but milder conditions to northern Europe and dry conditions to southern Europe.

North Atlantic oscillation (NAO)
A large-scale climate signal over the mid- and high latitudes of the North Atlantic characterized by changing pressure differences between polar (Iceland) and subtropical (Azores) surface pressures.

The North Atlantic oscillation involves a change in the wind patterns over the mid- and high latitudes of the northern hemisphere. During periods when the circulation intensifies, storm systems shift north, and cold arctic air is confined to the high latitudes. When this circulation is weak, storm systems shift south, and cold-air outbreaks are more prevalent.

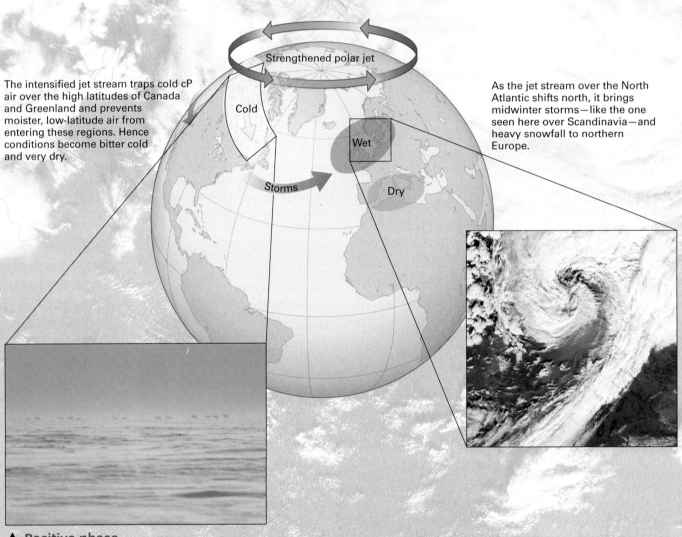

Strengthened polar jet

The intensified jet stream traps cold cP air over the high latitudes of Canada and Greenland and prevents moister, low-latitude air from entering these regions. Hence conditions become bitter cold and very dry.

Cold

Wet

Storms

Dry

As the jet stream over the North Atlantic shifts north, it brings midwinter storms—like the one seen here over Scandinavia—and heavy snowfall to northern Europe.

▲ **Positive phase**
Pressures over the Icelandic region and over much of the Arctic are lower than normal, with higher than normal pressures across the subtropical Atlantic and into southern Europe. The change in pressure enhances the typical north-south pressure gradient, increasing the strength of the jet stream as well as the surface westerlies. The European Atlantic region and central Europe experience cool summers and wet, mild winters, while southern Europe stays warm and dry.

Changes in the North Atlantic ▶ oscillation

Year-to-year changes in the wintertime index for the NAO are shown by the bars. The black lines represent the slower changes associated with the 5-year mean value of the NAO. The NAO contains both substantial year-to-year variability as well as slower changes over the course of decades. For example, a reversal of the NAO index started around 1975 as the NAO made the transition from negative (blue) toward positive (red) values. Since the late 1980s, the NAO has tended to remain in a strong positive phase.

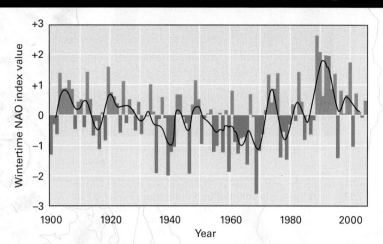

The weakened jet stream allows cold cP air to move to lower latitudes as cold-air outbreaks. These can produce hazardous conditions like those seen here in Tennessee.

As the midwinter storms shift south, they can bring wintery conditions to the Mediterranean. Here, snow covers the dome of the Rock in Jerusalem, Israel after a storm in January 2008.

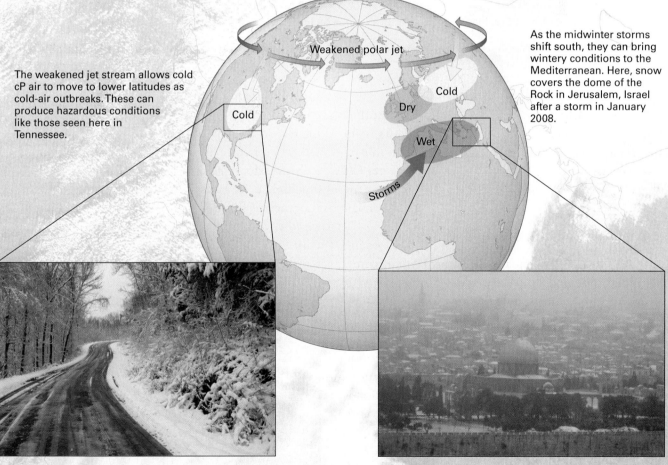

▲ Negative phase

The normal pressure gradient between low and high latitudes weakens, resulting in a decrease in the strength of the jet stream and prevailing westerlies. Atlantic and northern to central Europe see more storms and cold-air outbreaks, while southern Europe and North Africa are wetter. Over the eastern seaboard of North America, winters are colder and snowier.

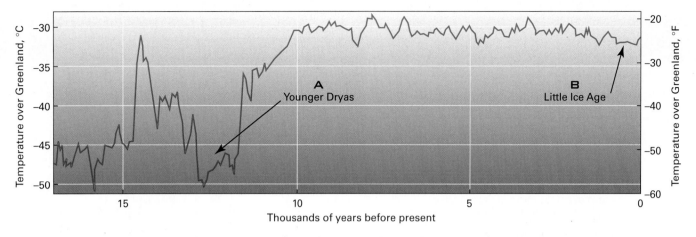

A Following the end of the last continental glaciation (around 12,000 years ago), it is believed that there was a massive rush of fresh water into the Atlantic after the collapse of a natural ice dam in Canada. This infusion of fresh water effectively shut down the thermohaline circulation and decreased temperatures over the North Atlantic for the next 1000 years.

B Historical periods of colder climates in Europe and North America from about 1300 A.D. to about 1800 A.D. are thought to have resulted from a slowdown of the thermohaline circulation.

Temperatures in Greenland FIGURE 13.13

This record of temperatures in Greenland is derived from ice cores on the continent.

ATLANTIC THERMOHALINE CIRCULATION

In the North Atlantic and northern Europe, another source of long-term climate variability results from changes in the underlying ocean circulation. In this region, a very slow circulation of waters in the deeper levels, called the *thermohaline circulation*, is driven by density differences between the equatorial and high-latitude seawater.

The thermohaline circulation transports warm, tropical waters from the eastern tropical Atlantic into the midlatitude and subpolar regions of the North Atlantic. An important part of this transport is the **Gulf Stream extension**. Although the Gulf Stream would exist even without the thermohaline circulation, the strength and northward extent of this current is modified by the strength of the thermohaline circulation. During periods in which the thermohaline circulation is weaker than normal, less heat is transported into the high latitudes of the Atlantic. This in turn produces lower temperatures over northern Europe, Greenland, and parts of Canada. These changes can last for decades or even as long as several centuries (FIGURE 13.13).

Why does the strength of the thermohaline circulation vary? The thermohaline circulation is driven by

the sinking of dense, cold, salty water at the northern margins of the Atlantic Ocean basin. During certain periods of high rainfall or significant melting of glaciers, fresh water can enter the North Atlantic, decreasing its salinity and hence its density. Because the water is less dense, it will not sink as deeply or as rapidly. Without this loss of sinking surface water, the Gulf Stream extension, which feeds warmer replacement water into the high latitudes, does not extend as far to the north.

Gulf Stream extension The extension of the Gulf Stream off of the South Carolina coast and into the northern Atlantic and the seas near Greenland and Iceland.

CONCEPT CHECK STOP

What four different processes can produce annual to multidecadal climate variability over the globe?

How do tropical and midlatitude climates change during an El Niño?

What climate changes are found during the positive phase of the North Atlantic oscillation? During the negative phase of the North Atlantic oscillation?

How do temperatures in the high latitudes of the North Atlantic change when the thermohaline circulation weakens?

Millennial Climate Variations

For the last 2–3 million years, the Earth has experienced an **ice age**, during which the Earth's climate has alternated between *glacial* and *interglacial* periods. During glacial periods, global temperature falls about 5°C (9°F), and glaciers expand to cover vast areas of land. During interglacial periods, climate warms, and the glaciers melt back.

> **ice age** Geologic time period during which glaciations alternate with inter-glaciations in rhythm with climate changes.

What caused the Earth to enter its present ice age? Scientists are not sure, but evidence suggests that the very slow motion of the Earth's continental plates is responsible. This motion brought Eurasia and North America northward to surround the Arctic Ocean. By restricting polar oceanic circulation and providing abundant land surfaces at high latitudes, the plate movement encouraged the growth of glaciers and sea ice.

The last interglacial period started about 120,000 years ago with an abrupt warming (**FIGURE 13.14**). After about 20,000 years, temperatures began to drop again, and the Earth entered its latest glacial period. About 20,000 years ago, temperatures began to rise, and for the past 12,000 years or so global temperatures have remained fairly high.

What causes the Earth's climate to shift between glacial and interglacial periods? To answer that question, we need to look carefully at how the energy of the Sun reaches the Earth, given the slow changes in patterns of motion of the Earth around the Sun.

CHANGES IN THE EARTH'S ROTATION AND REVOLUTION

Does the Sun's output change enough to cause climate changes? Variability in the Sun's intensity occurs over many time scales. Over the course of the Earth's 4-billion-year history, the Sun's intensity has increased approximately 30 percent. During recent millennia, however, variation has been much smaller. For example, the Sun has an 11-year cycle related to variations in solar eruptions (sunspots) on its surface, which changes the overall insolation reaching the Earth's surface by about 1.3 W/m², or about 0.1 percent.

Global temperatures of the ice-age years
FIGURE 13.14

This record of the Earth's global surface temperature, expressed as a difference from present temperature, shows glaciations and interglaciations to 800,000 years ago.

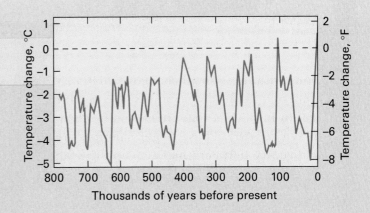

Changes in the Earth's rotation and orbit FIGURE 13.15

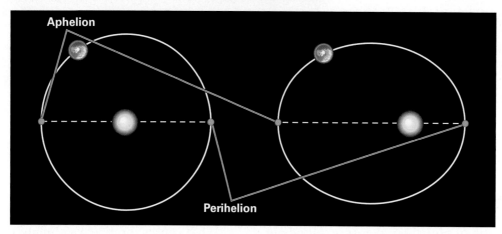

Change in the shape of the Earth's orbit Over time, the orbital ellipse of the Earth around the Sun becomes more and less circular (less and more eccentric). The change in shape changes the distance of the Earth from the Sun during perihelion and aphelion. The change in orbit occurs over a period of about 100,000 years.

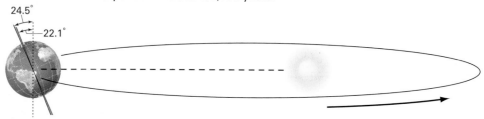

Change in the angle of the Earth's axis of rotation The axis of the Earth's rotation varies its angle from the perpendicular to the orbital plane. This angle varies from 22.1° to 24.5° within a period of about 41,000 years.

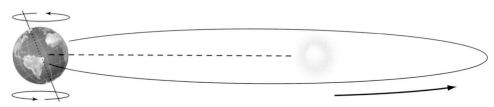

Change in the direction of the Earth's axis of rotation The Earth's axis of rotation slowly revolves, tracing a circle over a period of about 26,000 years.

Along with these small changes, which may have an effect on global temperatures, are much larger changes in insolation that are caused by changes in *Earth–Sun geometry*. These larger changes—related to the distance between the Earth and the Sun and the timing of maximum insolation during the year—appear to shift the Earth's climate between glacial and interglacial climates. The three components to these changes are discussed in **FIGURE 13.15**.

Eccentricity   In the 1930s, Serbian astrophysicist Milutin Milanković described the three main processes that change the Earth–Sun geometry, and hence insolation, over the time scales of 20,000 to 100,000 years. The first process is a cyclic change in **eccentricity**, which describes the shape of the Earth's orbit around the Sun.

Over the course of about 100,000 years, the shape of the Earth's orbit around the Sun goes from nearly circular (less eccentric) to slightly elliptical (more eccentric). In turn, this change in shape changes the distance between the Earth and Sun at aphelion and perihelion. At its greatest eccentricity, the Earth's orbit has a larger Earth–Sun distance at aphelion and a smaller Earth–Sun distance at perihelion. Because global insolation is related to Earth–Sun distance, a more eccentric (more elliptical) orbit makes insolation stronger at perihelion and weaker at aphelion.

Obliquity
The second component of change in Earth–Sun geometry involves a change in the **obliquity** of the Earth's axis of rotation. Although the Earth's axis of rotation is presently tilted at about 23.5° from a right angle to the Earth–Sun orbital plane, the tilt actually varies from about 22° (low obliquity) to 24.5° (high obliquity) on a cycle of about 41,000 years. Because the tilt of the Earth's axis of rotation causes the seasons, an increased tilt (high obliquity) will accentuate seasonal differences, especially at high latitudes, making summers warmer and winters colder.

Precession
The third component of change in Earth–Sun geometry involves **precession** of the axis of rotation. During precession, the axis of rotation changes direction with time. Like the wobbling of a spinning top, the axis itself rotates slowly in a circle. For the Earth, the axis of rotation circles a full 360° over the course of 26,000 years.

One of the important effects of precession is that it changes whether a given hemisphere experiences summer during perihelion, aphelion, or somewhere in between. At present, the northern hemisphere winter solstice, on December 21, occurs about 12 days before perihelion, on January 3. Some thousands of years from now, perihelion will coincide with the summer solstice.

eccentricity A measure of the elliptical shape of the Earth's orbit around the Sun.

obliquity A measure of the angle of tilt of the Earth's axis of rotation with respect to the plane of its orbit around the Sun.

precession Change in the direction of the Earth's axis of rotation over time.

The Milankovič curve
How do the factors of Earth–Sun geometry act to affect climate? When they interact to produce substantially higher summer insolation at high latitudes in the northern hemisphere, the increased warmth can trigger the melting of ice sheets and polar sea ice, thus starting an interglacial period. When the factors interact to produce lower summer insolation, the result can be a return of continental ice sheets. To understand how these factors interact, we will examine them individually.

First, let's consider the conditions that are optimal for generating interglacial periods. In any year, insolation is greatest at perihelion, when the Sun is nearest to the Earth. Thus, the high latitudes receive the most summer insolation when perihelion occurs near the time of the summer solstice, which is controlled by the 26,000-year cycle of the precession of the Earth's axis of rotation. In addition, orbital eccentricity acts to change the intensity of insolation at perihelion and aphelion on about a 100,000-year cycle. When the orbit is most eccentric, perihelion insolation is strongest. Finally, obliquity of the Earth's axis affects the distribution of heat at the poles. When obliquity is greatest, the axis is most tilted and polar heating will be greatest.

In other words, summer insolation in the northern hemisphere is greatest when (1) perihelion and summer solstice occur together, (2) axial obliquity is a maximum, and (3) orbital eccentricity is a maximum.

Now let's consider the conditions that are optimal for generating low summer insolation in the northern hemisphere, leading to glacial periods. First, the high latitudes receive the least summer insolation when the summer solstice occurs near the time of aphelion. In addition, when the orbit is most eccentric, insolation at aphelion is weakest because the Earth is further from the Sun. Finally, when obliquity is least, the axis is tilted to a lesser degree, and polar heating in summer will be weaker. In other words, summer insolation in the northern hemisphere is least when (1) aphelion and summer solstice occur together, (2) axial obliquity is a minimum, and (3) orbital eccentricity is a *maximum*.

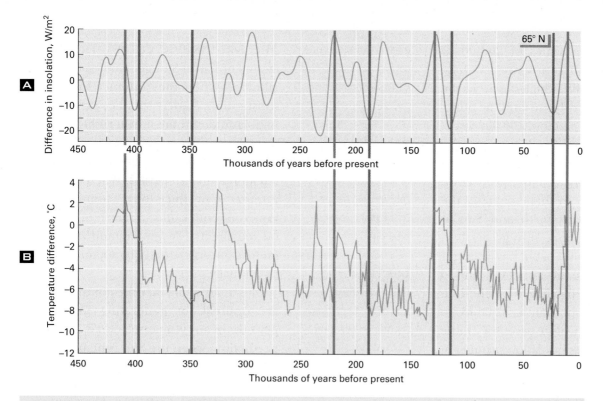

The Milanković curve FIGURE 13.16

A The vertical axis shows fluctuation in summer daily insolation at latitude 65° N for the last 500,000 years. The curve was first calculated from mathematical models of the variation in Earth–Sun geometry by Milutin Milanković. Zero represents the present value. B Global temperature changes over the last 500,000 years. Red lines (indicating warm periods) and blue lines (indicating cool periods) show how variations in the global temperatures vary with changes in the Milanković curve.

FIGURE 13.16 shows the *Milanković curve* —a plot of the difference in summer insolation at 65° N latitude given the cycles of motion of the Earth's orbit and its axis of rotation. This calculation translates the interaction of these cycles into heat flows. The magnitude of the cycles is quite strong, with a difference of nearly 45 W/m^2 between high and low values. This difference is about 20 percent of the annual average rate of insolation received at that latitude.

Comparing this curve with estimates of global temperatures based on ancient ice cores and deep lake sediment cores shows that the peaks at about 12,000, 130,000, 220,000, and 410,000 years correspond well with the rapid melting of ice sheets and a shift from a glacial to an interglacial climate. Conversely, decreases in insolation at about 25,000, 120,000, 170,000, 350,000, and 400,000 years ago correspond with glacial periods and minimum global temperatures.

One of the reasons for the abrupt shift from glacial climate to interglacial and back again is related to positive ice–albedo feedback. As we will see in the following section, when the area of land and ocean covered by ice and snow decreases, so does the surface albedo, which enhances warming. If the climate cools and snow and ice increase, the albedo also increases, which enhances cooling.

CONCEPT CHECK STOP

HOW have global temperatures changed over the last 150,000 years?

What three factors of Earth–Sun geometry are important for producing global temperature variations?

HOW does each factor influence the amount of summer insolation that the northern hemisphere receives?

HOW do all three factors combined influence the amount of insolation received in the high-latitude northern hemisphere?

Climate Feedbacks

This section is about the connections between climate and atmosphere in which a change in climate caused by an atmospheric factor is amplified or reduced by a feedback process. In discussing these processes, we will refer to them simply as "feedbacks," rather than "feedback loops" or "feedback mechanisms," following the usage of many scientists now studying climate change.

The two most important factors in considering variability in climate are forcing of climate, which produces initial changes in the climate system, and climate feedback, which determines the magnitude of the resultant change. The preceding discussions focused on natural forcing of climate variability, including the role of volcanoes, ocean–atmosphere interactions, and changes in solar insolation. Feedbacks act to modify the effects of changes in these natural forcing factors. Feedbacks can either amplify (*positive feedback*) or dampen (*negative feedback*) the effects of a change in the system.

longwave–temperature feedback Negative feedback in which an increase in temperature of a surface is moderated by increasing longwave emission of heat.

LONGWAVE–TEMPERATURE FEEDBACK

The most important type of negative feedback in the global climate system is the **longwave–temperature feedback**, shown in FIGURE 13.17. If the surface temperature increases—for example, by increased solar insolation—according to Stefan–Boltzmann's law the amount of

Longwave–temperature feedback FIGURE 13.17

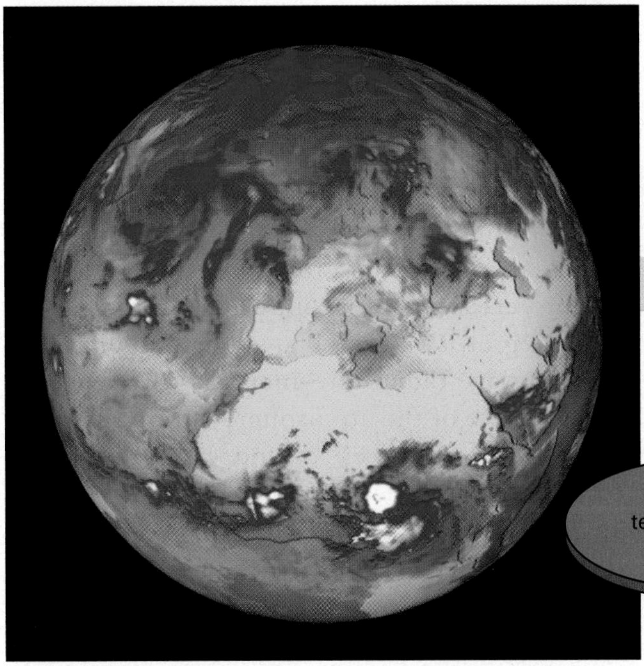

A As a region acquires excess heat, either through solar radiation, sensible heat, or latent heat, its temperature increases. As the temperature increases, the amount of longwave radiation the region emits also increases. Here, the deserts of Africa and the Middle East emit large amounts of longwave radiation (yellows) in response to their very high surface temperatures.

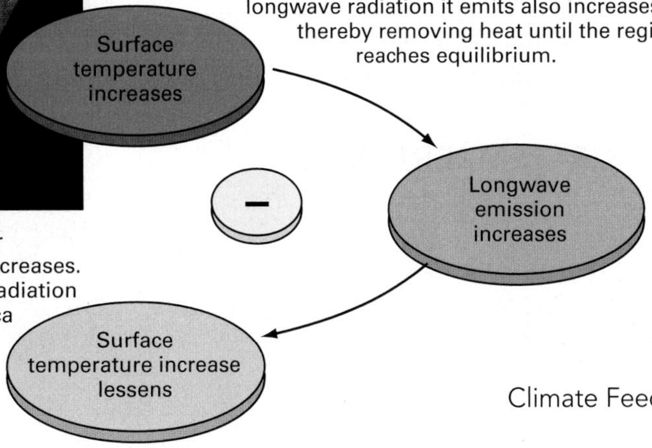

B This diagram shows the principle behind the longwave–temperature feedback. As a region acquires excess heat, its temperature increases. However its temperature increase is moderated, because the amount of longwave radiation it emits also increases, thereby removing heat until the region reaches equilibrium.

Surface temperature increases

Longwave emission increases

Surface temperature increase lessens

Climate Feedbacks 371

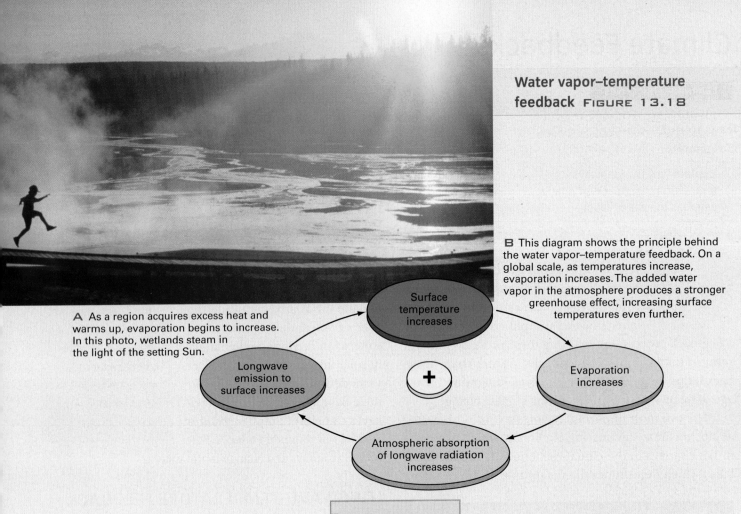

A As a region acquires excess heat and warms up, evaporation begins to increase. In this photo, wetlands steam in the light of the setting Sun.

B This diagram shows the principle behind the water vapor–temperature feedback. On a global scale, as temperatures increase, evaporation increases. The added water vapor in the atmosphere produces a stronger greenhouse effect, increasing surface temperatures even further.

Surface temperature increases

Longwave emission to surface increases

Evaporation increases

Atmospheric absorption of longwave radiation increases

+

radiation it emits will increase. This increase in emitted radiation removes energy from the surface, thus moderating the temperature increase. Because the increase in temperature is reduced, the feedback is negative.

In the same way, an initial decrease in temperature is moderated by this mechanism. With the decreased surface temperature, the longwave energy loss is reduced and the surface does not cool as much. Thus, the negative feedback acts to resist any change in temperature, whether an increase or decrease.

WATER VAPOR–TEMPERATURE FEEDBACK

The most important positive feedback in the global climate system is the **water vapor–temperature**

water vapor–temperature feedback Positive feedback in which an increase in temperature results in increased evaporation, thereby enhancing the natural greenhouse effect and subsequently increasing temperatures further.

feedback, shown in FIGURE 13.18. If the surface temperature increases, some of the excess energy is transferred to the atmosphere as latent heat through increased evaporation. The subsequent increase in atmospheric water vapor—a greenhouse gas that strongly absorbs longwave radiation—increases the temperature of the atmosphere, strengthens the greenhouse effect, and warms the surface further. Because the feedback acts to reinforce the initial temperature increase, it is a positive feedback.

When the surface temperature decreases, the water vapor–temperature feedback reduces the greenhouse effect, making the surface temperature decrease further. This positive feedback acts to accentuate any change in temperature, whether up or down.

ICE–ALBEDO FEEDBACK

Another important type of positive feedback is the **ice–albedo feedback** (FIGURE 13.19), in which an initial warming leads to the melting of snow and ice, uncovering low-albedo surfaces of soil or water. Because low-albedo surfaces absorb more solar radiation, surface warming is enhanced. This feedback is important at high latitudes, and can double the amount of temperature change that is experience elsewhere.

Ice–albedo feedback also enhances a decrease in surface temperature. With lower temperatures, more land and ocean are covered with snow and ice, increasing albedo and reducing absorption of solar energy. This feedback is extremely important for the rapid decreases in temperature during glaciations, allowing glaciers to expand over vast areas of the high latitude northern hemisphere.

> **■ ice–albedo feedback** Positive feedback in which an increase in temperature results in a melting of ice and snow, thereby decreasing the albedo of the surface and subsequently increasing temperatures further.

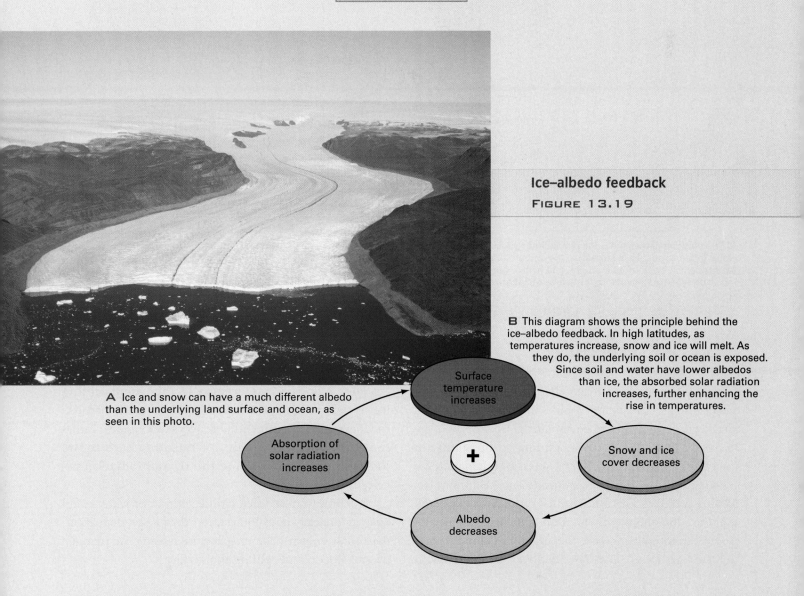

Ice–albedo feedback
FIGURE 13.19

A Ice and snow can have a much different albedo than the underlying land surface and ocean, as seen in this photo.

B This diagram shows the principle behind the ice–albedo feedback. In high latitudes, as temperatures increase, snow and ice will melt. As they do, the underlying soil or ocean is exposed. Since soil and water have lower albedos than ice, the absorbed solar radiation increases, further enhancing the rise in temperatures.

A Low clouds tend to reflect much of the incoming solar radiation, lowering temperatures at the surface.

B High clouds tend to be transparent to solar radiation, letting much of it reach the surface. However, they are efficient absorbers of longwave radiation and tend to enhance the natural greenhouse effect.

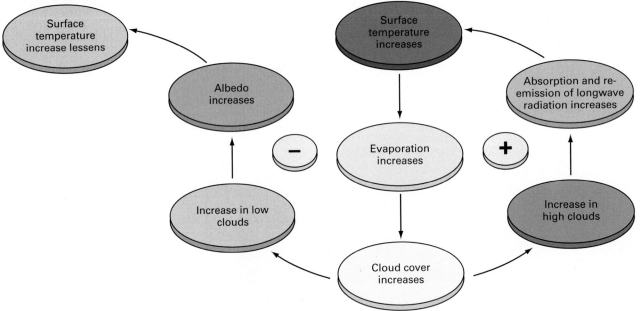

C The diagram shows the principle behind opposing cloud-cover feedback. As temperatures increase and the amount of water in the atmosphere increases, the amount of clouds is expected to increase. Low clouds, which have a high albedo, tend to reflect away more sunlight, and hence represent a negative feedback. High clouds, which are transparent to solar radiation but strongly absorb longwave radiation, intensify the greenhouse effect and represent a positive feedback.

Cloud-cover feedback FIGURE 13.20

CLOUD-COVER FEEDBACK

Cloud cover provides another feedback mechanism that affects the global climate. Unlike the types of feedback already described, cloud cover can act as either a negative or positive feedback, depending on the nature of the clouds (FIGURE 13.20). Given an initial increase in surface temperature, evaporation will increase, adding water vapor to the atmosphere. It is expected that the subsequent condensation of the excess water vapor will eventually increase the amount of clouds covering the Earth's surface. The feedback associated with the increased cloud cover can differ dramatically depending on the height at which the clouds form.

High clouds, which are usually thin clouds of fine ice crystals, tend to provide a positive feedback, because they serve as very efficient absorbers of longwave radiation. In turn, they reemit this radiation back to the Earth's surface, increasing the overall radiation received there.

In contrast, low, thick clouds tend to provide a negative feedback. Their liquid water droplets reflect abundant solar energy back to space, increasing the planet's overall albedo and cooling the surface.

Which feedback is stronger—negative or positive? Climatologists are not sure at the present time. It is very difficult to quantify the size and even the sign (positive or negative) of the overall feedback associated with clouds. The role of clouds represents one of our greatest uncertainties in determining how the climate ultimately responds to different climate influences.

BIOLOGICAL FEEDBACK

The Earth's biosphere also provides several types of important feedback that affect the global climate. One type involves the ocean **biological pump**, and it provides a positive feedback that enhances global warming, shown in FIGURE 13.21. Photosynthesis by phytoplankton—tiny floating plants in the ocean—removes CO_2 from the atmosphere and converts it into organic matter. When these organisms die, the organic matter falls to the ocean floor, where it is buried in the sediments at the bottom, effectively removing the carbon from the climate system.

biological pump
A process in which CO_2 from the atmosphere is taken up by marine phytoplankton and is then removed to the deep ocean as the phytoplankton die and sink.

When marine plants have all the nutrients they need, the biological pump is working at 100 percent efficiency, and the CO_2 in the atmosphere is around 165 ppm (parts per million). When no nutrients are available, the biological pump is working at 0 percent efficiency, and CO_2 is around 720 ppm. Presently, the concentration of CO_2 is more than 370 ppm, which suggests that the biological pump is not fully efficient.

What controls the efficiency of the biological pump? Typically, the photosynthetic efficiency of phytoplankton is limited by scarce nutrients, particularly iron. One of the major sources of iron is dust from the continents, brought in rainfall from storms that originate on land. Generally, when global temperatures are higher, the number of storms decreases, less iron is carried to the ocean, and the biological pump slows.

The number of storms decreases with increasing temperatures because the ice–albedo feedback warms the high latitudes more than the low latitudes, decreasing the temperature difference between them. It is this temperature difference that allows jet-stream disturbances to grow into large, traveling storm systems.

Biological pump FIGURE 13.21

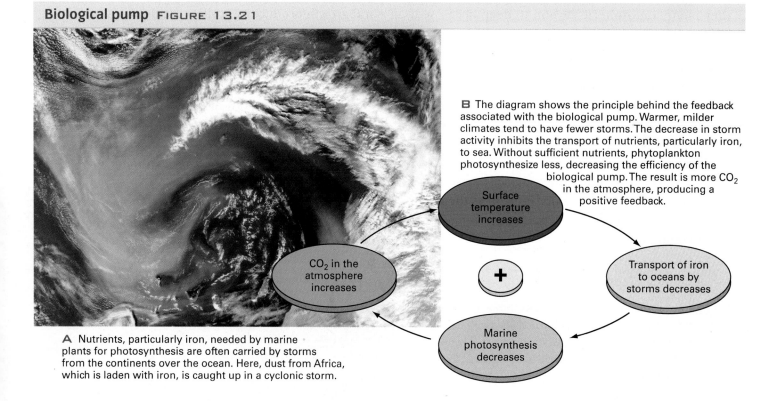

A Nutrients, particularly iron, needed by marine plants for photosynthesis are often carried by storms from the continents over the ocean. Here, dust from Africa, which is laden with iron, is caught up in a cyclonic storm.

B The diagram shows the principle behind the feedback associated with the biological pump. Warmer, milder climates tend to have fewer storms. The decrease in storm activity inhibits the transport of nutrients, particularly iron, to sea. Without sufficient nutrients, phytoplankton photosynthesize less, decreasing the efficiency of the biological pump. The result is more CO_2 in the atmosphere, producing a positive feedback.

Surface temperature increases

Transport of iron to oceans by storms decreases

CO_2 in the atmosphere increases

+

Marine photosynthesis decreases

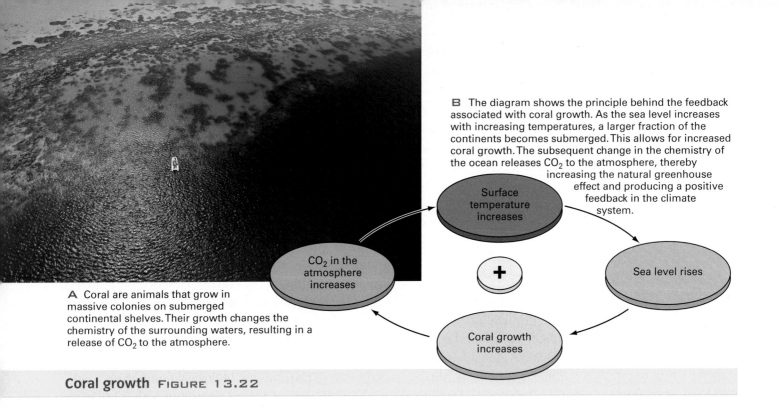

B The diagram shows the principle behind the feedback associated with coral growth. As the sea level increases with increasing temperatures, a larger fraction of the continents becomes submerged. This allows for increased coral growth. The subsequent change in the chemistry of the ocean releases CO_2 to the atmosphere, thereby increasing the natural greenhouse effect and producing a positive feedback in the climate system.

A Coral are animals that grow in massive colonies on submerged continental shelves. Their growth changes the chemistry of the surrounding waters, resulting in a release of CO_2 to the atmosphere.

Coral growth FIGURE 13.22

As the temperature difference decreases, the number of storms that form in the midlatitudes also decreases.

To review: Increasing global temperatures reduce the number of storms, which decreases the iron available to phytoplankton, which slows the rate at which they pump atmospheric carbon to deep ocean sediments, which causes CO_2 to accumulate in the atmosphere more rapidly, which further increases the warming. Thus, the biological pump a provides positive feedback to surface temperature.

Coral Growth
Another type of biological feedback to consider is associated with the growth of coral reefs (FIGURE 13.22). As coral grows over large portions of the flat, shallow, submerged surfaces in the warm tropical and equatorial oceans, it changes the chemistry of the surrounding ocean waters, releasing CO_2 to the atmosphere.

Coral bleaching FIGURE 13.23

Coral grows in warm waters of the tropics. However, if the water becomes too warm or too acidic, coral heads sicken and expel their symbiotic algae, turning white in a process known as coral bleaching. If conditions continue to deteriorate, the coral can die off completely.

For periods when global temperatures are higher and sea level rises, the area over which coral can grow will increase. The subsequent release of CO_2 to the atmosphere will lead to an increase in greenhouse gases and a further increase in surface temperatures. Thus, coral growth produces a positive feedback on surface temperature.

At the same time, however, coral growth is very sensitive to water temperature and water chemistry. If the water gets too warm or too acidic (which may occur as more and more CO_2 is absorbed by the ocean), coral begins to die off, which is called *coral bleaching*, seen in FIGURE 13.23. As the coral dies off, the chemistry of the ocean reverts back to its original state, and less

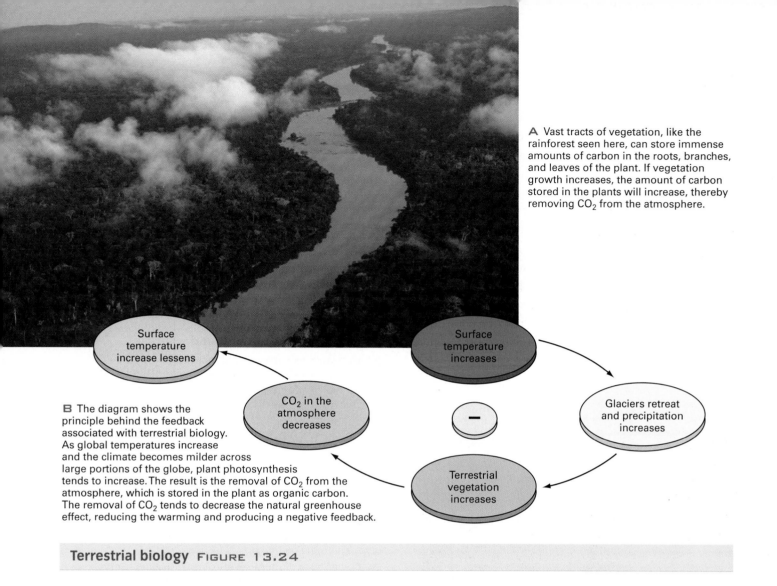

A Vast tracts of vegetation, like the rainforest seen here, can store immense amounts of carbon in the roots, branches, and leaves of the plant. If vegetation growth increases, the amount of carbon stored in the plants will increase, thereby removing CO_2 from the atmosphere.

Surface temperature increase lessens

CO_2 in the atmosphere decreases

Surface temperature increases

Glaciers retreat and precipitation increases

Terrestrial vegetation increases

B The diagram shows the principle behind the feedback associated with terrestrial biology. As global temperatures increase and the climate becomes milder across large portions of the globe, plant photosynthesis tends to increase. The result is the removal of CO_2 from the atmosphere, which is stored in the plant as organic carbon. The removal of CO_2 tends to decrease the natural greenhouse effect, reducing the warming and producing a negative feedback.

Terrestrial biology FIGURE 13.24

CO_2 is released to the atmosphere. A massive die-off of coral could also result in changes in the ecological food web of those regions where coral serves as the "infrastructure" for complex biological ecosystems.

Terrestrial biology

A final type of biological feedback involves land biomass—plant growth on land, shown in FIGURE 13.24. Generally, low temperatures inhibit plant growth, as do dry, arid conditions. If global temperatures increase, ice sheets will shrink, enhanced by the ice–albedo feedback, which will provide more land area for plant growth. In addition, the atmosphere will contain more water, as described in the discussion of water vapor–temperature feedback. If this results in more rainfall over more land area, it will en-

hance global vegetation growth. The enhancement of terrestrial vegetation growth from both of these processes removes CO_2 from the atmosphere, providing a negative feedback in the climate system.

As with coral growth feedback, there is a natural limit to which the land–biomass feedback will operate. If temperatures become too high, plants restrict their photosynthesis in order to conserve water. Long-term growth can suffer. In addition, higher temperatures can increase evaporation from the soil, limiting the soil moisture available to plants for photosynthesis. As with coral bleaching, if terrestrial plant growth slows, less atmospheric CO_2 is removed, leaving more in the atmosphere and providing a positive feedback.

OCEAN FEEDBACK

The Earth's vast ocean also provides several important types of feedback in the global climate system. First, the large heat capacity of the ocean serves as a negative feedback when considering temperature changes. If insolation increases, for example, much of the added heat will go into warming oceans rather than raising air temperatures. Because of the ocean's larger heat capacity, it will experience a smaller temperature change compared with land surfaces and hence will moderate the rise and fall of global temperatures.

Second, the solubility of CO_2 in ocean water decreases with temperature, providing a positive feedback. *Solubility* is the property that describes how much gas a volume of liquid can contain, based on temperature and pressure. As discussed in FIGURE 13.25,

solubility decreases with temperature, so a warmer global ocean will take up less CO_2, resulting in a higher overall concentration in the atmosphere. The higher concentration of CO_2 will increase the greenhouse effect, raising temperatures further.

GEOLOGICAL FEEDBACK

On very long time scales, feedbacks associated with the interaction of the atmosphere and the Earth's surface itself can occur through the process of **chemical weathering**. During this process, CO_2 dissolves in rainwater and groundwater to form a weak acid

> **chemical weathering**
> Process in which exposed land surfaces undergo chemical changes as they interact with carbon dioxide, water, and oxygen.

Ocean CO₂ uptake FIGURE 13.25

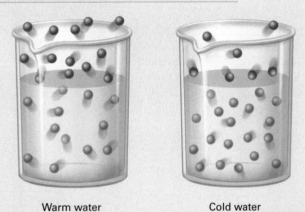

Warm water   Cold water

A The ability of the ocean to absorb CO_2 changes with its temperature. As temperatures increase, as on the left, the amount of dissolved CO_2 the ocean can contain decreases. As a result, more CO_2 remains in the atmosphere above.

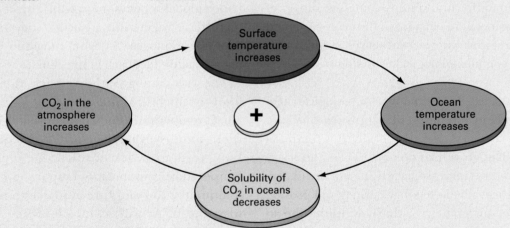

B The diagram shows the principle behind the feedback associated with ocean CO_2 uptake. As temperatures increase, the amount of dissolved CO_2 that can be contained within the oceans decreases. The CO_2 that cannot be dissolved into the ocean remains in the atmosphere, increasing the natural greenhouse effect. The process represents a positive feedback.

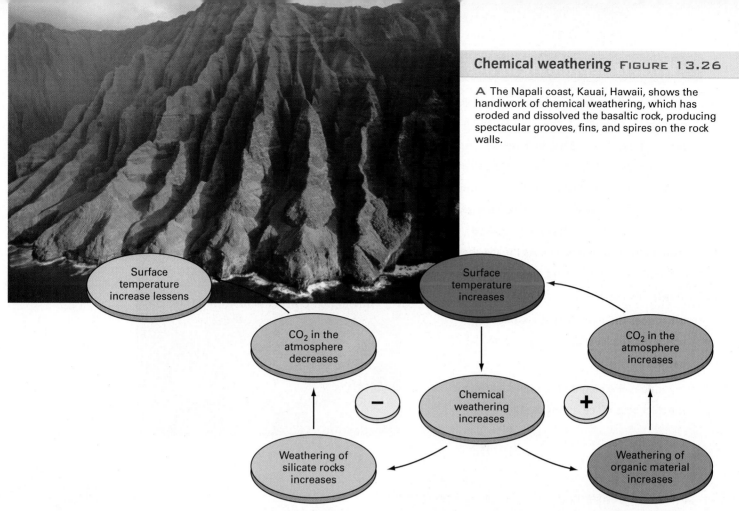

Chemical weathering FIGURE 13.26

A The Napali coast, Kauai, Hawaii, shows the handiwork of chemical weathering, which has eroded and dissolved the basaltic rock, producing spectacular grooves, fins, and spires on the rock walls.

B The diagram shows the principle behind the feedback associated with chemical weathering. As surface temperatures increase, air temperature and precipitation increase, leading to an increase in chemical weathering. If this weathering involves silicate minerals, atmospheric CO_2 will be incorporated into the dissolved rocks and washed into the ocean, thereby decreasing the natural greenhouse effect. If this weathering involves organic material, its carbon will be converted to CO_2 and released to the atmosphere, thereby increasing the natural greenhouse effect.

that reacts with the natural minerals of rocks and soils. The result is the formation of carbonate compounds that remove CO_2 from the atmosphere. Increasing temperatures speed chemical weathering, producing a negative feedback by removing more CO_2 from the atmosphere and reducing the greenhouse effect (**FIGURE 13.26**). Because most chemical weathering is very slow, this feedback works on a time scale of thousands to millions of years, matching climate changes on the scale of ice ages or longer.

In contrast to chemical weathering, the weathering of organic matter provides a positive feedback. Organic matter is broken down by bacteria, which respire CO_2. Moreover, as the temperature rises, so does bacterial activity and the release of CO_2, stimulating the greenhouse effect. There is presently widespread concern that any increases in global temperatures will speed the decomposition of vast volumes of organic matter in polar and arctic soils, significantly enhancing global temperatures even further through this positive feedback.

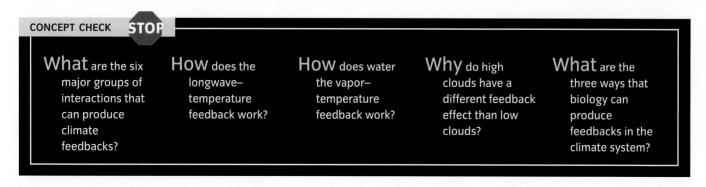

CONCEPT CHECK STOP

What are the six major groups of interactions that can produce climate feedbacks?

How does the longwave–temperature feedback work?

How does water the vapor–temperature feedback work?

Why do high clouds have a different feedback effect than low clouds?

What are the three ways that biology can produce feedbacks in the climate system?

This image shows a fjord—a deep, lake-filled canyon—in Greenland. The walls reach approximately 1000 m (3300 ft) in height, and the deep, water-filled channel drops 350 m (1155 ft) below sea level. Fjords were formed during the last glacial period, when massive "rivers" of ice flowed down the exposed mountainside to the ocean. As the ice sheets retreated, they left behind deep gorges that slowly filled as the sea level rose.

■ What does the depth of the channel say about the sea level during the last glacial period?

■ What does the height of the fjord say about the thickness of ice during that time?

SUMMARY

1 Annual to Centennial Climate Variations

1. Climate variability occurs on many time and space scales. From 1000 A.D. to 1900, temperatures moved slowly downward. Since 1900, northern hemisphere temperatures have risen, fallen, and risen again. The last decade has been the warmest of the past thousand years. **Proxy data**, derived from tree rings, coral growth increments, and ice-core layers, reveal past climate variations.

2. Strong volcanic eruptions can inject sulfuric acid aerosols into the stratosphere, lowering the Earth's albedo and cooling surface temperatures several tenths of a degree.

3. An oscillation of the ocean-atmosphere system, called **ENSO**, brings El Niño and La Niña conditions to the equatorial Pacific region. During an **El Niño**, upwelling along the Peruvian coast ceases and eastern Pacific trade winds weaken. Equatorial rainfall shifts eastward, and Malaysia and northern Australia are dry. During **La Niña**, upwelling increases, easterly trades intensify, and south Asia is wet. During strong El Niños, jet streams and storm tracks change positions, bringing storms as well as drought to different areas of the world.

4. The **North Atlantic oscillation (NAO)** describes a cyclic variation in atmospheric pressures and sea-surface temperatures between mid- and high latitudes of the North Atlantic region, changing the climate of Europe, North America, and western portions of Asia. The **Pacific decadal oscillation (PDO)** of air and sea-surface temperatures in the North Pacific also affects climate in Eurasia, Alaska, and the western United States.

5. The *thermohaline circulation* of the ocean is driven by the sinking of dense, cold, salty water along the northern

margins of the Atlantic. The sinking pulls warmer water northward in the **Gulf Stream extension**, warming northern Europe, Greenland, and parts of eastern Canada. High rainfall or increased inflows of fresh river water to the northern Atlantic could slow or stop the thermohaline circulation, causing regions surrounding the high-latitude North Atlantic to cool.

2 Millennial Climate Variations

1. About 2–3 million years ago, the Earth entered an **ice age** of alternating glacial (cold) and interglacial (warm) periods. The most recent glacial period lasted from about 120,000 to 12,000 years ago, and we are presently in an interglacial period.

2. The change between glacial and interglacial periods is triggered by changes in the solar heating of the northern hemisphere. The changes are caused by slow changes in the shape of the Earth's orbit and the direction of the Earth's axis of rotation. The **precession** of the Earth's axis of rotation determines whether northern hemisphere summer occurs near perihelion or aphelion. The **eccentricity** of the Earth's orbit determines whether perihelion is closer to or farther from the Sun. Finally, the **obliquity** of the rotational axis determines the extent to which the high latitudes of the northern hemisphere are exposed to the warming rays of the Sun.

3. The *Milanković curve* shows changes in northern hemisphere summer heating as a result of changes in the Earth–Sun geometry over the past 500,000 years. Peaks of heating trigger interglacials, whereas decreased heating brings on glacial periods.

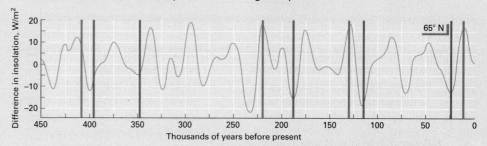

3 Climate Feedbacks

1. Climate responds to forcing factors, such as volcanic eruptions or the rhythms of ocean–atmosphere cycles, but the strength of the response depends on climate feedbacks. Positive feedbacks strengthen a change, whereas negative feedbacks weaken a change.

2. The most important type of negative feedback in the global climate system is the **longwave–temperature feedback**. When an increase in surface temperature occurs, longwave surface emissions increase, resulting in a loss of energy that offsets the initial heating. This process moderates the temperature increase associated with the initial heating.

3. The most important type of positive feedback is the **water vapor–temperature feedback**. As temperatures of the Earth's surface and lower atmosphere increase, evaporation increases, which increases the water vapor in the atmosphere. Because water vapor is a highly effective greenhouse gas, increased water vapor results in an increased greenhouse effect, thereby further increasing the temperatures of both the surface and the atmosphere.

4. **Ice–albedo feedback** is a type of positive feedback that occurs when warming causes snow and ice to melt, revealing dark soil or water that absorbs more solar energy, causing even warmer temperatures. **Cloud-cover feedback** is positive for high, thin clouds that absorb longwave radiation, increasing the greenhouse effect. However, this feedback is negative for low, thick clouds that reflect more solar energy back to space.

5. Biological feedback includes the ocean's **biological pump**, which produces a positive feedback. Coral growth, which increases with moderate ocean warming and increasing areas of shallow water habitat, creates another type of positive biological feedback. Plant growth on land, however, provides a negative feedback as higher temperatures stimulate plant growth and removal of CO_2 from the atmosphere.

6. The ocean's vast bulk and high heat capacity resists heating or cooling, producing a negative feedback. However, with increasing temperature, CO_2 becomes less soluble in water, reducing the amount of CO_2 taken up each year, which provides a positive feedback. Chemical weathering causes CO_2 to combine with minerals at the Earth's surface, reducing CO_2 levels on a geologic time scale and yielding a negative feedback. However, higher temperatures stimulate bacteria to digest organic matter in soils, releasing CO_2 to the atmosphere and producing a positive feedback.

KEY TERMS

CRITICAL AND CREATIVE THINKING QUESTIONS

1. Describe how global air temperatures have changed in the recent past. Identify some factors or processes that influence global air temperatures on this time scale.

2. Compare El Niño and La Niña conditions by drawing up contrasting lists of changes and climate effects that characterize each situation.

3. A friend of yours is thinking of moving to Spain. Looking at the climate of the region for the period 1970–2000, when the NAO was predominantly positive, she feels the winter climate should be sunny and mild with few storms. How would this climate change if the NAO became predominantly negative, as it was during the 1950–1970s?

4. Explain how the Earth's orbit around the Sun changes with time. Describe the motions of the Earth's axis of rotation. How do these motions interact to affect climate? Use the terms eccentricity, obliquity, and precession in your answer.

5. On the accompanying figure, identify which orbit would be most conducive to producing interglacial periods. Draw in the position and orientation of the Earth during a northern hemisphere summer that would correspond with interglacial periods. Also draw in the position and orientation of the Earth during a northern hemisphere summer that would correspond with glacial periods.

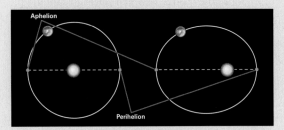

6. Name three important types of positive feedback associated with global climate variability, and describe how they work. Do the same for three important types of negative feedback.

7. How does water, as vapor and as clouds, influence global climate? How might water in these forms act to enhance or dampen climatic warming?

SELF TEST

1. Volcanic eruptions tend to _____ global temperatures by increasing concentrations of _____ .
   a. lower; greenhouse gases   c. raise; ash
   b. lower; aerosols   d. raise; ozone

2. El Niño is the name given to the warm phase of ocean temperatures in the eastern equatorial Pacific that produce a weakening of _____ .
   a. ocean currents in the northern Atlantic Ocean
   b. pressure patterns over the Arctic Circle
   c. wind-flow patterns along the ITCZ
   d. ocean currents in the southern Pacific Ocean

3. El Niños change the climate in midlatitudes by changing the _____ .
   a. extent of snow cover   c. location of monsoons
   b. direction of gyre circulations   d. location of jet streams

4. On the accompanying figure, draw in the surface winds, surface pressures, and surface currents that correspond to normal conditions. Also show on the figure how each changes during an El Niño event.

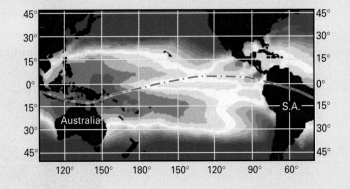

5. The negative phase of the North Atlantic oscillation is associated with _____.
a. wetter conditions over southern Europe
b. increased cold-air outbreaks
c. drier conditions over northern Europe
d. all of the above

6. Changes in the North Atlantic oscillation pressure patterns can occur on time scales of _____.
a. weeks   c. years
b. months   d. all of the above

7. Shifts in the thermohaline circulation can occur on time scales of _____.
a. weeks   c. years
b. months   d. decades or longer

8. The change in Earth–Sun geometry that does *not* influence the change in climate from glacial to interglacial periods is _____.
a. obliquity   c. eccentricity
b. heliocity   d. precession

9. Presently the Earth is in a(n) _____ period that started about _____ years ago.
a. interglacial; 12,000   c. interglacial; 250,000
b. glacial; 120,000   d. glacial; 1 million

10. Global temperature changes during glacial periods can be up to _____ lower than at present.
a. 1°C   c. 5°C
b. 2°C   d. 10°C

11. According to the longwave–temperature feedback, _____ in global temperatures result in _____ emissions of longwave radiation, thereby _____ the temperature change.
a. decreases; increased; moderating
b. increases increased; enhancing
c. decreases; decreased; moderating
d. increases; decreased; enhancing

12. According to the water vapor–temperature feedback, _____ in global temperatures result in _____ in water vapor in the atmosphere, which will _____ the temperature change.
a. increases; increases; enhance
b. increases; decreases; enhance
c. decreases; increases; moderate
d. decreases; decreases; moderate

13. Imagine that the amount of cloud cover increases globally. For the clouds in photos A and B, indicate which cloud type will tend to produce decreased surface temperatures. Also indicate which cloud type will tend to produce increased surface temperatures.

A

B

14. _____ is a biological process that does *not* provide important feedbacks for global climate variability.
a. The biological pump
b. Jellyfish reproduction
c. Coral growth
d. Terrestrial vegetation growth

15. If ocean waters become too _____ and too _____, coral tends to die off, which is called *coral bleaching*.
a. cold; basic   c. warm; acidic
b. warm; basic   d. cold; acidic

Human Interaction with Weather and Climate | 14

Venice is sinking. January 2001 saw the worst sustained flooding in the city's history—a combination of higher-than-normal tides, a drop in atmospheric pressures over the Adriatic Sea, and strong winds pushing seawater onshore. An eighth of the picturesque Italian city was underwater.

Venice was originally built on low-lying islands in a coastal lagoon. The city is now dropping about one millimeter a year due to the natural movement of the lithospheric plates underneath. Venice's plight worsened after World War II, however, when industries began pumping ground water from below the city. That caused Venice to sink a foot in two decades, leaving St. Mark's Square, the center of Venetian social life, just 2 inches above the normal high-tide level.

Although ground-water removal has now stopped in Venice, flooding is still a huge problem. Venice is not the only city in peril. Ground-water withdrawal affects the Thai city of Bangkok—now the world's most rapidly sinking city—in addition to regions in California, Texas, and Mexico City. New Orleans lies completely below sea level and continues to sink at almost a meter a century—one of the contributors to the devastation that accompanied the passage of Hurricane Katrina in 2005.

Venice provides a dramatic reminder of how human interaction with the world around us can affect the impacts that weather and climate have on our lives and livelihood.

High water in Saint Mark's Square, Venice.

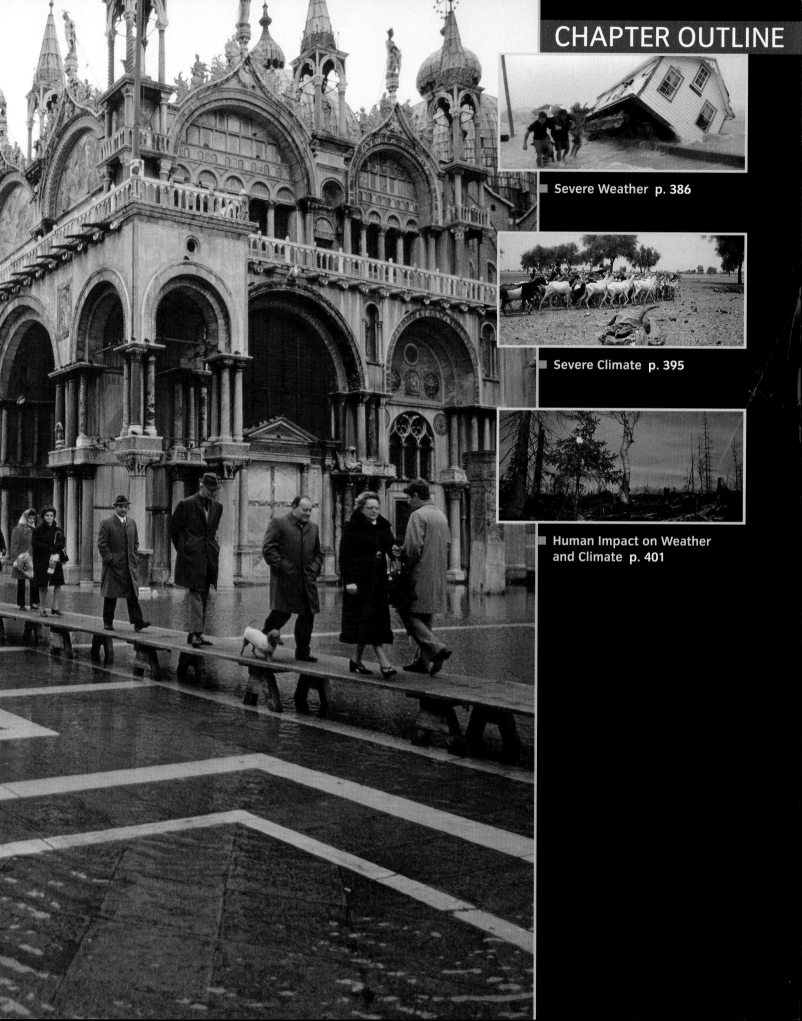

Severe Weather

LEARNING OBJECTIVES

Describe the effects of heat waves.

Identify the weather conditions that lead to heat waves and cold waves.

Describe the conditions that lead to river flooding and coastal flooding.

Almost any weather system influences our activities to some extent, which is what makes the study of weather so compelling. **Severe weather**, however, is hazardous to humans and can destroy property and infrastructure. Hurricanes, tornadoes, severe thunderstorms, and ice storms are examples we have discussed in previous chapters. Heat waves, cold waves, and flooding are three additional significant weather-related events.

severe weather Weather event that has the potential to cause substantial destruction or loss of life.

HEAT WAVES

A **heat wave** is a persistent period of high temperatures. In the United States, a heat wave is a period of 3 consecutive days with temperatures above 32.2°C (90°F). In northern Europe, where temperatures are typically cooler during the summer and air conditioning is less prevalent, a heat wave is defined as 5 consecutive days of temperatures above 25°C (77°F), which would be considered mild in the southeastern United States. Extremely high temperatures can occur anywhere though, as happened in Europe in 2003 (**FIGURE 14.1**).

heat wave Persistent period of significantly elevated temperatures, usually lasting 3–5 days.

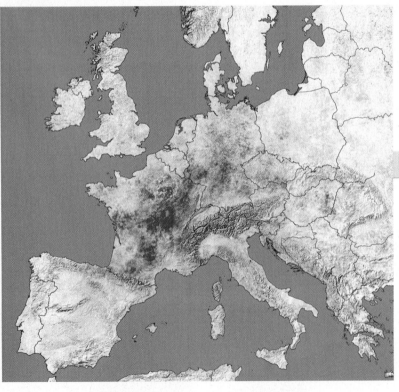

2003 heat wave FIGURE 14.1

In the summer of 2003, a heat wave hit Europe. High temperatures in southern France were above 36°C (97°F) for an entire week. In a 2-week period in August, more than 14,000 excess deaths were recorded in France alone. This image shows the departures in the 2003 temperatures during July 20–August 20, compared with the average.

-10   -5   0   5   10
Land surface temperature difference, °C

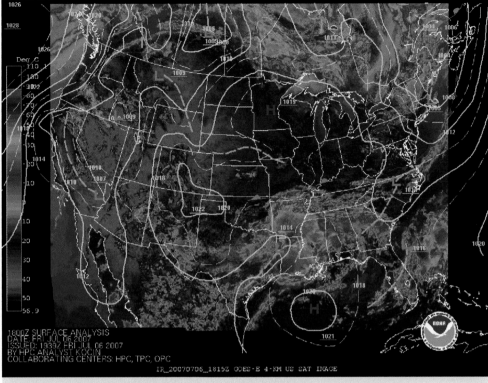

Heat wave in the southwestern United States FIGURE 14.2

This surface map shows weather conditions on July 6, 2007. On that day, the temperature hit 46.67°C (116°F) in Las Vegas, Nevada, which was 0.55°C (1°F) short of the all-time high set in 2005. The yellow isobars indicate a high-pressure center situated near the Oklahoma panhandle, producing a southerly flow of hot, dry air from Mexico. Satellite measurements show extremely high temperatures (purple tones) extending from Baja California into Montana.

Another component of many heat waves is high relative humidity. One of the ways our bodies remove excess heat is through the evaporation of perspiration from the skin, which removes latent heat. When the relative humidity is high, less evaporation occurs because the surrounding atmosphere is already relatively moist. Excess heat builds up in our bodies and we feel hotter.

The *heat index* indicates how hot temperatures feel based on both the actual temperature and the relative humidity. The heat index is given in degrees Celsius (or Fahrenheit) but can be very different from the actual temperature—up to 15–20°C (30–40°F) different!

Heat waves often occur when a high-pressure system stalls over a given region. In the southwestern United States, persistent periods of high pressure centered over New Mexico and Texas can produce searing temperatures (**FIGURE 14.2**). In this situation, a related upper-air high-pressure center, situated directly over Nevada, produces strong subsidence, inhibiting the formation of clouds and allowing large amounts of insolation to reach the surface. Subsidence also produces adiabatic warming of the air as it descends toward the surface.

In addition, the anticyclonic circulation of winds around the surface high-pressure center brings hot, dry cT air from Mexico northward into the region, increasing temperatures even further. The intrusion of this dry cT air from the south, combined with the adiabatic warming, produces a hot, but dry, heat wave.

Over the eastern portion of the United States, heat waves are accompanied by high relative humidity, which increases the heat index far above the actual temperature. This type of heat wave struck Chicago, Illinois, in mid-July 1995. Over the course of 5 days (July 12–16), approximately 600 people died from exposure to high temperature and humidity levels.

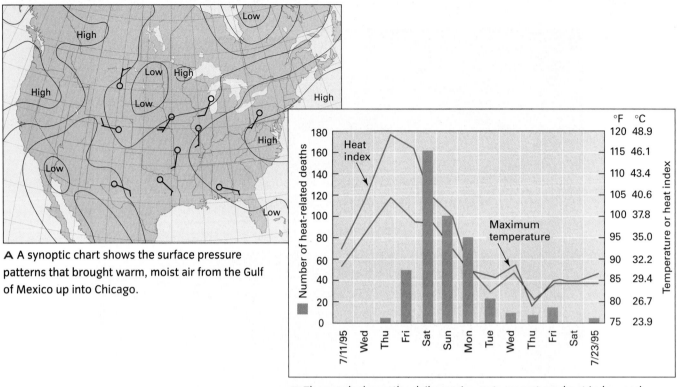

A A synoptic chart shows the surface pressure patterns that brought warm, moist air from the Gulf of Mexico up into Chicago.

B The graph shows the daily maximum temperature, heat index, and related deaths for the heat wave.

1995 heat wave in Chicago FIGURE 14.3

In mid-July 1995, temperatures in the central United States began to rise, eventually peaking at 39.5°C (103°F) on July 13 in Chicago. Over 600 heat-related deaths were recorded during and following this heat wave.

Like the Las Vegas heat wave, the Chicago heat wave was associated with high-pressure centers both at the surface and aloft. At the surface (FIGURE 14.3), the anticyclonic circulation of winds around the Bermuda high brought hot, humid air from the southeastern United States into the Great Lakes region. This southerly flow also prevented a lake-breeze circulation from forming and bringing in cooler air from the Great Lakes. Aloft, a high-pressure center at 500 mb was situated directly over Illinois, producing descending air that inhibited the formation of cloud cover.

The combination of warm, moist air from the south and strong solar heating during the day sent temperatures soaring. Although strong surface heating usually creates convection that mixes surface air, the descending air from the upper-level high produced an inversion at 850 mb. By suppressing convection, the inversion trapped the hot, humid air at the surface.

As suggested by the number of excess deaths in France and Chicago, heat waves are dangerous

weather events. Exposed to high heat, a victim first suffers *heat exhaustion*, in which excessive perspiration leads to dehydration. Eventually, perspiration decreases, and the body redirects blood flow to the skin and away from vital organs. Headaches, nausea, and dizziness set in—the precursors to a shutdown of internal temperature regulation. At this point, the victim suffers from *heat stroke* (also called hyperthermia). Without the means to remove excess heat, the body experiences a rise in core temperature. Fluids that cannot be removed by perspiration accumulate in the appendages, leading to swelling. If the internal temperature continues to rise, brain damage and eventual death can occur.

The same synoptic conditions that produce heat waves—including the presence of inversions and high relative humidity—also lead to high levels of pollution in the atmosphere. These pollutants, such as ozone and nitrogen oxides, can produce irritation of the lungs and eventually corrode lung tissue.

COLD WAVES

A **cold wave** is a period of relatively cold temperatures that begins with a rapid fall during a 24-hour period. The drop in temperature can be as rapid as 16–22°C (30–40°F) over the course of 6 hours. Temperatures must remain at least 8.33°C (15°F) below average over an extended area to constitute a true cold wave.

Cold waves usually involve the intrusion of a cold cP or cA air mass into a region following the passage of a cold front. Cold waves are particularly severe during winter, when the source regions for cP and cA air masses—Canada in North America, Greenland in Europe, and Siberia in Asia—are significantly colder than regions to the south.

After the initial temperature drop, cold temperatures typically persist as an anticyclone settles over the region. Skies are largely clear because of the descending air found above the high-pressure center. If snow falls during the passage of the cold front, its high albedo reflects most of the incoming solar radiation back to space. At night, longwave radiation escapes to space through the clear skies, and temperatures drop even further. Eventually, the cold wave breaks as another midlatitude cyclone moves through the region, bringing warmer air in behind the accompanying warm front.

In 2004, a cold wave hit the eastern United States (**FIGURE 14.4**), causing temperatures in Boston to reach −21.67°C (−7°F). The *wind chill*—which accounts for additional cooling associated with wind speed—indicated that temperatures felt even colder, reaching as low as −40°C (−40°F).

Cold waves can be just as hazardous as heat waves. Both hypothermia and frostbite can result. *Hypothermia* results when the body is unable to maintain its core temperature of 37°C (97.6°F). Initially, shivering sets in. This can become so severe that muscles stop functioning. Breathing becomes extremely shallow, and major organs do not receive necessary oxygen, eventually failing and leading to death. In addition, *frostbite* can occur as blood flow needed to maintain internal body temperatures is redirected from the exterior surfaces to the body's interior.

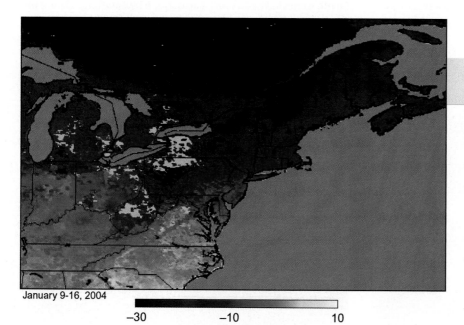

January 9-16, 2004

-30   -10   10
Minimum temperatures, °C

Northeast cold wave
FIGURE 14.4

In January 2004, record low temperatures covered much of the northeastern United States. This map shows the temperatures during the week of January 9–16. During that time, temperatures in Boston dropped to −21.67°C (−7°F). The coldest recorded temperature was on Mt. Washington, New Hampshire at −42°C (−43.6°F) with a wind chill of −75°C (−103°F).

Cold waves also affect infrastructure, agriculture, and transportation. In buildings without sufficient insulation, water pipes freeze and break. Gas and electric heaters, used improperly, start fires. The increased demand for electricity to heat homes can cause power outages. Crop damage occurs as water within plant cells freezes and expands, rupturing the cell walls and killing the plants. If the cold wave is accompanied by precipitation, road surfaces freeze. Ice accumulates on airplane wings, reducing the lift needed to keep an aircraft aloft. For regions without the proper equipment, highways and airports can be shut down for hours or even days.

FLOODING

Flooding is responsible for some of the most devastating effects of weather. There are three major categories of flooding—river floods, mud and debris flows, and the coastal flooding caused by ocean storms.

River floods After a heavy rainfall or snowmelt, stream flow increases. When the discharge of a river cannot be accommodated within its normal channel, the water spreads over the adjoining ground, causing flooding. Often, the flooded area is cropland or forest, but sometimes it is occupied by houses, factories, or transportation corridors (**FIGURE 14.5**).

Where does river flooding occur? Most rivers of humid climates have a **floodplain**—an area bordering the channel on one or both sides. Although the floodplains can be inundated annually without harmful consequences, once in 30–50 years we see examples in which higher discharges cause rare and disastrous floods that inundate ground well above the floodplain.

> **floodplain** A broad belt of low, flat ground bordering a river channel that floods regularly.

Another type of river flood is a *flash flood*, which occurs in streams draining small watersheds with steep slopes. After intense rainfall, the stream quickly rises to

---

River flooding FIGURE 14.5

A Harper's Ferry Historic town centers are often close to rivers and subjected to river flooding. Harper's Ferry, West Virginia, is at the junction of the Potomac and Shenandoah Rivers.

B Sainte Genevieve Large rivers, like the Mississippi, can flood extensive areas far from their banks. Pictured here is a home in Sainte Genevieve, Missouri, surrounded by sandbags during the flood of 1993.

a high level. The flood arrives as a swiftly moving wall of turbulent water, sometimes sweeping up great quantities of coarse rock debris, tree limbs and trunks, and soil. Flash floods often occur too quickly to warn people, so they can cause significant loss of life.

Cities and suburbs also affect the flow of streams and hence flooding (FIGURE 14.6). In these populated areas, it is more difficult for water to infiltrate the ground,

Urban water flow FIGURE 14.6

Impervious surfaces like this street in Bangkok increase surface runoff and hasten the flow of water, leading to increased flooding during heavy rainfall events.

which is largely covered by buildings, driveways, walks, pavements, and parking lots. In a closely built-up residential area, 80 percent of the surface may be impervious to water. Overland flow increases, making flooding more common during heavy storms for small regions lying largely within the urbanized area.

Cities and suburbs also have storm sewers, a system of large underground pipes that quickly carries storm runoff from paved areas directly to stream channels for discharge. The system shortens the time it takes runoff to travel to channels, while the proportion of runoff is increased by the expansion in impervious surfaces. Together, these factors produce higher flows in urban and suburban streams. Many rapidly expanding suburban communities are finding that low-lying, formerly flood-free, residential areas now experience periodic flooding as a result of urbanization.

> **mudflow** Mixture of water and soil that flows rapidly downhill, following a stream channel.

Mudflow and debris flood One of the most spectacular forms of flooding and also a potentially serious environmental hazard is the **mudflow**—a mud stream that pours swiftly down canyons in mountainous regions. In deserts, thunderstorms produce rain much faster than it can be absorbed by the soil. Without vegetation to protect soil slopes, the excess water runs off, picking up fine particles to form a thin mud that flows down to the canyon floors, gathering up additional sediment and becoming thicker and thicker as it goes. Great boulders are carried along, buoyed up in the mud. Roads, bridges, and houses in the canyon floor are engulfed and destroyed. A mudflow can severely damage property and even cause death as it emerges from the canyon and spreads out.

Another type of mudflow, found on the slopes of erupting volcanoes, is called a *lahar*. In a lahar, heavy rains or melting snow turn freshly fallen volcanic ash and dust into mud that flows downhill. Herculaneum, a city at the base of Mount Vesuvius, was destroyed by a mudflow during the eruption of 79 A.D. At the same time, the neighboring city of Pompeii was buried under volcanic ash.

Mudflows vary in consistency. They range from the thickness of concrete emerging from a mixing truck to a watery consistency similar to the floodwaters of a turbid river. The watery type of mudflow is called a *debris flood* or *debris flow*. Debris flows are common in southern California, with disastrous effects. They can carry anything from fine particles and boulders to tree trunks and limbs. On steep slopes in mountainous regions,

Debris flow FIGURE 14.7

Heavy rains saturate the soil of steep slopes, which then flow rapidly down narrow creek valleys and canyons, carrying mud, rocks, and trees. This debris flow, also known as an alpine debris avalanche, struck the village of Schlans, Switzerland, in November 2002.

these flows are called *alpine debris avalanches* (FIGURE 14.7). The intense rainfall of hurricanes striking eastern North America often causes debris avalanches and was responsible for most of the 11,000 deaths that occurred during the passage of Hurricane Mitch through Central America in 1998.

Coastal flooding and storm surges

Coastal flooding is associated with floodwaters moving landward from the ocean. Although heavy rainfall can produce coastal flooding at river mouths and estuaries, winds and tides are usually involved.

Strong winds can move massive volumes of water onshore, raising local sea levels by up to 4–6 m (15–20 ft) and overtopping natural and constructed coastal barriers. Once the water level lies above these barriers, waves can erode the foundation of buildings, beaches, and cliffs, causing severe damage to nearby structures and ecosystems (FIGURE 14.8).

Two processes can cause this rise in sea level, termed a *storm surge*. In some cases, the winds are strong enough to blow water directly onto the shore. This type of storm surge is typically associated with hurricanes approaching a coast. Strong onshore winds on one side of the storm push water toward the shore, where it can rise quickly and overwhelm protective barriers along the coast. On the other side of the storm, offshore winds move water away from the coast, lowering the local sea level. Because the distance between the onshore and offshore winds circling a hurricane's eye is typically only 25–50 km (15–30 mi), the exact location of the landfall can determine which coastal community is flooded and which is not. This type of surge is usually short-lived and produces very localized flooding.

More persistent storm surges are the consequence of winds blowing *along* the coast. This type of storm surge is produced by hurricanes moving parallel to the coast or by midlatitude storms that stall off the coast or move slowly along it. In this case, persistent winds drive an ocean current that slowly comes under the influence of the Coriolis force, veering to the right of the wind

Storm surge inundates Key West, Florida
FIGURE 14.8

This photo, taken September 25, 1998, shows the effects of the storm surge generated by Hurricane Georges as it swept a beach-front hotel off its foundation.

(in the northern hemisphere) and producing an *Ekman transport* of water. If the coast is to the right of the storm, the Ekman transport associated with the ocean current piles up water along the coast. The longer the storm lasts, the more water piles up along the coast. For this type of storm surge, it is not only the strength of the wind that matters but also its duration. Storms that move slowly parallel to the coast tend to produce larger storm surges than storms that move directly onshore (FIGURE 14.9).

Storm surges FIGURE 14.9

VIEW THIS IN ACTION
in your WileyPLUS course

During storm surges in the midlatitudes, winds are not typically strong enough to blow water directly onshore, as they can during hurricanes in the tropics. Instead, the persistent force applied by the winds, combined with the Coriolis force, can produce onshore movement of water and large increases in sea level along the coast.

A **Alongshore winds begin to set water in motion**
As winds blow along the shore, the frictional force applied to the underlying surface of the ocean causes the water to move in the direction of the winds.

B **Coriolis force causes currents to drift onshore**
In the northern hemisphere, the Coriolis force causes the currents to veer to the right. Eventually, the transport of water moves perpendicular to the direction of the winds. If the coast is to the right of the blowing winds, the underlying transport of water—called *Ekman transport*—is directly onshore.

3–5 m (10–15 ft)   40–50 km (25–31 mi)

C **Onshore currents lift local sea-levels**
As the transport of water moves toward the shore, it encounters the coast. Water begins to pile up, lifting local sea levels. The longer the winds blow parallel to the coast, the more water is transported to the coast, and the higher the sea level rises. Eventually, sea levels can rise high enough that inland structures are subjected to the destructive surface waves accompanying the storm system.

Process Diagram

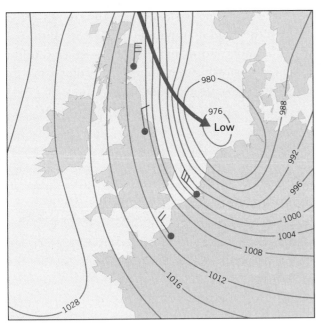

A The winds circling around the midlatitude cyclone produced massive flooding along the British coast as well as along the coasts of the Netherlands.

B The town of Abbenbrock, on an island off the coast of Holland, was submerged by the storm surge after a protective dyke broke.

1953 North Sea storm surge FIGURE 14.10

In 1953, a slow-moving midwinter storm traveled south over the North Sea.

FIGURE 14.10 shows the impact of a storm surge that occurred along the coast of the Netherlands and England in 1953. This storm moved slowly across the North Sea from the north. Because of the cyclonic circulation around the low-pressure center, winds were consistently northerly along the British coast. The underlying ocean water, initially forced southward by these winds, then drifted toward the coast. Over the course of 6 hours, local sea levels in some regions rose 5.6 m (18.5 ft) above the mean level. In the Netherlands alone, 1,365 km² (525 mi²) were flooded. In Britain, 1,600 km (990 mi) of coastline suffered flood damage. More than 2100 people died in the flooding.

CONCEPT CHECK STOP

What weather conditions contribute to heat waves?

What are the three types of flooding that can occur?

Why are cold waves associated with the presence of an anticyclone?

Severe Climate

LEARNING OBJECTIVES

Describe the impact of drought on human and natural systems.

Explain how changes in climate can produce changes in severe weather.

Identify regions in which the climate might shift from one regime to another.

Unlike the effects of gradual climate changes, drought, famine, and increases in the frequency of intense and destructive storms are examples of severe climate change that may occur too quickly for human systems to adapt.

DROUGHT

Climatic changes can take many forms. One of the most significant, in terms of its toll on both human health and livelihood, is **drought**. A drought is a persistent period of low precipitation. Unlike a flood or a heat wave, it does not simply arise from a single weather event. Instead, it comes from a systematic change in the number and intensity of weather events that affect a region—in other words, a change in the climate of the region.

Drought can be classified in different ways. A *meteorological* (or climatological) *drought* is simply the presence of below-normal rainfall over a given region or time period. An *agricultural drought,* on the other hand, occurs when soil moisture falls significantly below the level normally needed to sustain agriculture in a given region. The moisture level depends not only on how

drought A prolonged period of abnormally dry weather.

much precipitation is received, but also on how much evaporation occurs, which in turn is related to the temperature—higher temperatures lead to more evaporation and hence lower soil moisture. Finally, *hydrologic drought* represents lack of moisture in the surface or subsurface water supply (**FIGURE 14.11**). Because surface and subsurface reservoirs of water take time to adjust to changes in precipitation and evaporation, the impact of drought on the water supply tends to lag behind the impact on agriculture.

Human activities are affected differently by different types of drought. For instance, a drought during winter that seriously lowers reservoirs and produces shortages in drinking water may not affect the growth of crops during the summer at all.

Because drought can span broad areas and affect many different sectors of the economy, it can cause significant damage to a whole country. In the United States, a large-scale drought in 1988–89 resulted in a $15 billion loss in crops alone, with total damages in the range of $39–40 billion. Overall, drought costs the United States economy about $6–8 billion each year, more than floods, hurricanes, or any other weather-related events.

Drought in southern Colorado FIGURE 14.11
This dried and cracked reservoir bed shows the effects of a prolonged dry spell.

FAMINE

The greatest impact of drought on human health takes the form of a *famine*—a period in which a significant portion of a given population is underfed. Although drought is not the sole cause of famine, the lack of crop production during a drought can significantly decrease the amount of food available, as well as drive up its price. See *What a Scientist Sees: Drought and Famine in the Sahel.*

When famine strikes a region, disease rates climb. The weaker segments of a population—the young as well as the elderly—experience higher mortality. Fertility also declines. If the famine is severe, even relatively healthy people can die of starvation. Overall, famines killed 70 million people during the 20th century alone. In the mid–1800s, a series of famines—initiated by climate changes arising from a series of El Niño events in the eastern tropical Pacific—swept China, India, and Brazil, killing up to 50 million people.

Given the enormous cost in lives and livelihood associated with famine, new efforts are being made to both predict and mitigate the role of drought in producing famine. In the United States, the Famine Early Warning System has been established to monitor weather conditions, crop growth, and agricultural output around the world as a means to avert the most severe impacts of drought.

Drought and Famine in the Sahel

The climate of West Africa, including the adjacent semiarid southern belt of the dry tropical climate to the north, provides a lesson on the impact of climate on human and ecological systems. From 1968 through 1974 and again in the 1980s, countries of this perilous belt, called the Sahel, or Sahelian zone, were struck by severe drought. Both nomadic cattle herders and grain farmers share this zone. During the drought, grain crops failed, and foraging cattle could find no food to eat. In the worst stages of the Sahel drought, nomads were forced to sell their remaining cattle. Because the cattle were their sole means of subsistence, the nomads soon starved. Some 5 million cattle perished, and it has been estimated that 100,000 people died of starvation and disease in 1973 alone.

Periodic drought throughout past decades is well documented in the Sahel. The graph above the photo shows the percentages of departures from the long-term mean of each year's rainfall in the western Sahel from 1901 through 1995. Note the wide year-to-year variation. Since about 1950, the durations of periods of continuous departures both above and below the mean seem to have increased substantially. The period of sustained high-rainfall years in the 1950s contrasts sharply with a series of severe drought episodes starting in 1971.

Other historical periods of rainfall deficiency and excess—1820–1840, below normal; 1870–1895, above normal; 1895–1920, below normal—indicate that drought and wet periods are a normal phenomenon in the Sahel. Of course, that does not make them any less devastating.

Global Locator

NATIONAL GEOGRAPHIC

Bitterroot National Forest wildfire FIGURE 14.12

In 2000, a wildfire raged through the Bitterroot National Forest in Idaho and Montana. This fire was just one of the record number that occurred that year.

FIRE

Extended periods of drought can also increase the likelihood of large-scale fires, termed *wildfires* or *forest fires*. As vegetation begins to wither and die, it becomes more combustible. In addition, litter, in the form of fallen branches and leaves, dries out and can serve as fuel for fires.

Wildfires are typical of wet–dry climates in which there is enough precipitation during part of the year to sustain vegetation, but also an extended dry season that dehydrates the vegetation. Climates in which the dry season occurs during the high-sun season—such as Mediterranean climates—are particularly susceptible to wildfires.

During periods of severe drought, however, almost any vegetated region can sustain wildfires. In 2000, drought conditions spanned much of the western United States. Wildfires struck across this region, including Alaska, southern California, New Mexico, Idaho, Montana, Utah, Nevada, and Oregon. Overall, 7.2 million acres burned, making it the worst fire season in United States history (**FIGURE 14.12**).

One of the most common natural causes of wildfires is *dry lightning*, in which lightning occurs without any accompanying precipitation. As in a typical thunderstorm, dry lightning occurs when convection produces a charge separation between the top of the thunderstorm and the bottom, which is discharged through a lightning stroke.

Why doesn't precipitation accompany this convection? Actually, it does, but if the surrounding air is dry enough, the precipitation evaporates before it reaches the ground. In that situation, cloud-to-ground lightning initiates a fire, but no precipitation falls to help put out the fire.

Wildfires can be sustained or extinguished by meteorological events. The hot, dry conditions associated with heat waves make the vegetation more flammable, encouraging burning. High winds sustain and spread wildfires by providing oxygen and carrying and dropping firebrands. In California, high-pressure centers situated over the elevated inland plateau can produce strong, downslope winds—the *Santa Anas*—that bring warm, dry air to the coastal regions and greatly increase fire danger.

EXAMPLES OF CHANGES IN SEVERE WEATHER

In 2005, a record 28 named tropical storms tore through the North Atlantic. Of these, a record number developed into hurricanes (15) with a record number of category-5 storms (4). In addition, the most intense Atlantic tropical storm on record—Hurricane Wilma—formed that year. Since an average year brings 6 hurricanes and 2 major hurricanes of categories 3–5, the number and severity of storms that occurred during 2005 were far above normal.

Two significant contributors to the change in climate allowed these storms to form. First, ocean temperatures in the tropics and subtropics of the Atlantic and Gulf of Mexico were significantly warmer (FIGURE 14.13). This allowed developing storms to draw more latent and sensible heat from the ocean, setting the stage for them to grow ever larger. In the process, this latent and sensible heat was converted into kinetic energy that drove the high winds, strong convection, and heavy rainfall accompanying these storms.

Second, there was a change in the subtropical jet stream over the Atlantic, brought about by a weak La Niña event in the eastern tropical Pacific. During 2005, the subtropical jet stream weakened and shifted north, away from the tropics and subtropics of the North Atlantic. Without disruption by the jet's high-velocity, upper-level winds, tropical storms could expand vertically and grow into intense hurricanes.

In contrast, in 2006, the onset of an El Niño event shifted the jet stream farther south (FIGURE 14.14). As hurricanes started to develop, upper-level winds disrupted their development before they ever achieved hurricane status. Because of this change in the jet stream, only five hurricanes formed, and none made landfall in the United States.

Although we understand the climate conditions that spawned the storms of 2005, it is still unclear what led to the change in these conditions. The La Niña event is a natural phenomenon of the eastern equatorial Pacific that occurs every 3–7 years and is known to influence the tropical jet.

What about the abnormally warm waters of the Atlantic and Gulf of Mexico? Part of this temperature increase could be due to changes in the thermohaline circulation, which slows and accelerates every 40–60 years, leading to a warming and cooling of the North Atlantic. At the same time, temperatures in all ocean basins have been increasing slowly along with global temperatures over the last century. This warming has accelerated in the last 30 years, with ocean temperatures in the Gulf and tropical Atlantic increasing by 0.5°C (0.9°F).

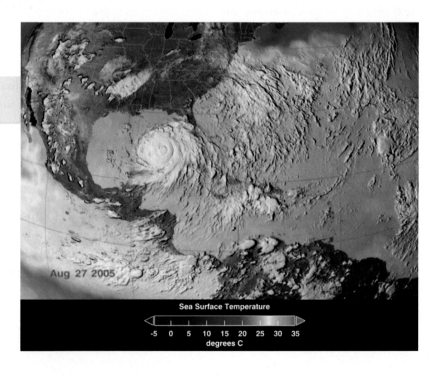

Atlantic Ocean temperatures in 2005

FIGURE 14.13

Hurricane Katrina sits poised over the warm waters of the Gulf of Mexico. Hurricanes can develop and intensify when ocean temperatures are higher than 26.5°C (80°F), shown here in red. During 2005, temperatures across much of the western tropical Atlantic were 0.5°–1°C (1°–1.8°F) warmer than normal, providing ample breeding grounds for hurricanes.

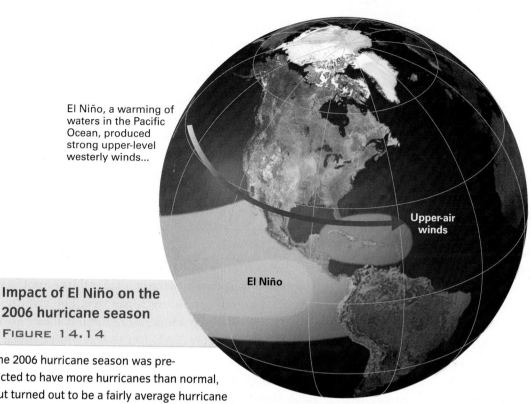

El Niño, a warming of waters in the Pacific Ocean, produced strong upper-level westerly winds...

Upper-air winds

El Niño

...which increased wind shear over the Atlantic hurricane basin, preventing tropical storms from fully developing.

Impact of El Niño on the 2006 hurricane season
FIGURE 14.14

The 2006 hurricane season was predicted to have more hurricanes than normal, but turned out to be a fairly average hurricane season. One of the reasons was the onset of an El Niño event in the tropical Pacific. Changes in ocean temperatures there affected the upper-air winds over the Atlantic, making them unfavorable for hurricane formation.

The global increase in temperature also coincides with an increase in emissions of heat-trapping gases to the atmosphere—carbon dioxide, methane, nitrous oxide, and others—arising from human activities. The addition of greenhouse gases has undoubtedly warmed the ocean and contributed in part to the record number and severity of hurricanes in 2005.

Changes in climate can also change the number of severe weather events in other parts of the globe. In 2007, the United Kingdom suffered severe floods (FIGURE 14.15). In certain areas, rainstorms brought a month's worth of rain in a single day. However, it was not simply one rainfall event that brought flooding—it occurred over the course of 2 months (June and July). The jet stream shifted south at the beginning of the summer and remained in that position for the next few months, steering a series of storms over the region. Like the Atlantic hurricanes of 2005, changes in the general circulation of the atmosphere brought major changes in severe weather events to yet another region.

Flooding in the British Isles FIGURE 14.15

Record rainfall in the summer of 2007 brought heavy flooding to regions of the United Kingdom. This flooding resulted from an increase in the number of severe rainfall events occurring throughout the summer.

SHIFTS IN CLIMATE REGIMES

Although climate-related changes in the number of severe weather events in a given region can occur during any given year, more systematic climate changes can also occur. For example, a climate shift is occurring presently in the high latitudes of the northern hemisphere, where temperatures have increased over 1.5°C (2.7°F) in the last 30 years. Tundra climates in those regions are shifting toward boreal forest climates. Wintertime land and sea ice is forming later in the season and melting earlier, which is affecting the migration patterns of caribou, the hunting season of polar bears, and oil production. In some areas, permafrost beneath the surface is thawing, creating shallow lakes and bogs and causing improperly constructed buildings and infrastructure to shift or collapse.

Other climate shifts can bring even more severe consequences. In the subtropics, regions that used to comprise the wet–dry tropical region—such as the Sahel in Africa, southern Brazil, and Venezuela—are slowly becoming more like a dry tropical region. The heavy but intermittent rains that came for a month or two each year are now not arriving at all. Although barely suitable for agriculture, in the past the climate of these regions could support populations through a combination of herding and hunting and gathering. Now, however, even these means of subsistence are in danger, threatening the local communities with famine.

In other regions, the climate is becoming warmer and wetter. Along the east coast of the United States, regions that previously had a moist continental climate are now more like a moist subtropical climate. With increases in temperature, humidity, and precipitation, these regions become susceptible to diseases that were typically confined to the hot, moist regions of Central America. For instance, malaria is spread by mosquitoes that need humid conditions to survive. The growth of the malaria parasite is temperature-limited and stops altogether for temperatures below 15.5°C (60°F). Malaria is predicted to spread widely where climates become both warmer and more humid (FIGURE 14.16). In 2003, eight cases of malaria were contracted internally within the United States, where the disease was once thought to be eradicated.

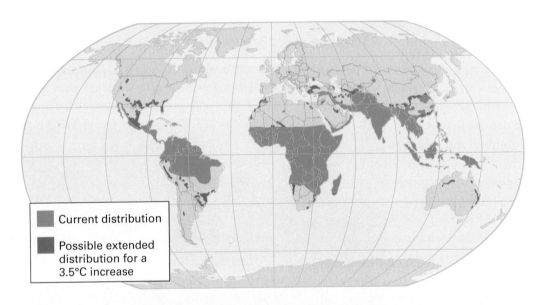

Current distribution

Possible extended distribution for a 3.5°C increase

Potential spread of malaria FIGURE 14.16

This map shows the increase in the geographic distribution of the parasite responsible for the spread of malaria that could accompany a 3.5°C increase in global temperatures.

CONCEPT CHECK **STOP**

HOW does drought affect human activities? How does it affect the natural system?

What weather events are affected by changes in climate and the general circulation of the atmosphere?

Why are arctic regions experiencing a shift in their climate?

Why are moist continental regions experiencing a shift in their climate?

Human Impact on Weather and Climate

LEARNING OBJECTIVES

Explain how inversions contribute to dangerous levels of pollution within the atmosphere.

Describe acid rain and how it forms.

Identify three types of land-use practices that affect local weather and climate.

Human activity can influence both the weather and climate of a particular region. For example, the release of chlorofluorocarbons (CFCs) through industrial processes has resulted in a significant decrease in the amount of stratospheric ozone over the high latitudes of the southern hemisphere. Because stratospheric ozone absorbs harmful ultraviolet radiation, the amount of ultraviolet radiation reaching the surface of these regions has increased.

Industrial, agricultural, and leisure activities have increased the emissions and concentrations of heat-trapping gases in the atmosphere—the *greenhouse gases,* which include carbon dioxide, methane, nitrous oxide, and others. These gases increase the amount of long-wave radiation that is absorbed by the atmosphere and is reemitted back to the Earth's surface, increasing the temperature of both the atmosphere and the surface.

In this chapter, we examine two more human influences on climate—one in which human activity combined with weather can produce life-threatening changes in the pollution of the atmosphere, and another in which human modification of the land surface on a large scale can affect climate across broad regions of the globe.

POLLUTION

Air pollution is largely the result of human activity. An **air pollutant** is an unwanted substance injected into the atmosphere from the Earth's surface by either natural or human activities. Air pollutants come as gases and aerosols—small bits of matter in the air, so small that they float freely with normal air movements—or as par-

> **air pollutant** An unwanted substance injected into the atmosphere from the Earth's surface by either natural or human activities; includes gases, aerosols, and particulates.

ticulates—larger, heavier particles that sooner or later fall back to the Earth.

Most pollutants are generated by the everyday activities of large numbers of people— for example, driving cars—or through industrial activities, such as fossil fuel combustion or the smelting of mineral ores to produce metals. The most common air pollutants are carbon monoxide, sulfur oxides, nitrogen oxides, and volatile organic compounds, which include evaporated gasoline, dry cleaning fluids, and incompletely combusted fossil fuels. Other pollutants are in the form of lead and particulates. Tropospheric ozone is also a pollutant. Near the surface, it forms through a complex interaction between naturally occurring and human-made organic chemicals, sunlight, and nitrogen oxides.

The combustion of fossil fuels is the most important source of all of these pollutants in the United States and accounts for 78 percent of emissions. Gasoline and diesel engine exhausts contribute most of the carbon monoxide, half the volatile organic compounds, and about a third of the nitrogen oxides.

Pollutants generated by a combustion process are contained within hot exhaust air emerging from a factory smokestack or an auto tailpipe. Because hot air rises, the pollutants are at first carried aloft by convection. However, the larger particulates soon settle under gravity and return to the surface as fallout. Particles too small to settle out are later swept down to Earth by precipitation in a process called *washout.* Through a combination of fallout and washout, the atmosphere tends to be cleaned of pollutants. Although a balance between input and output of pollutants is achieved in the long run, the quantities stored in the air at a given time fluctuate widely.

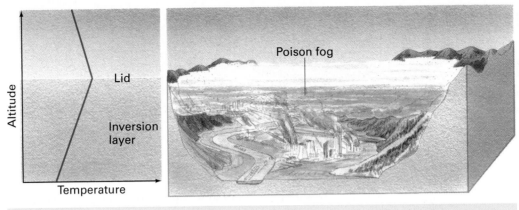

Low-level temperature inversion FIGURE 14.17

An inversion held air pollutants close to the ground, creating a poison fog accumulation at Donora, Pennsylvania, in October 1948.

Pollutants are also eliminated from the air over their source areas by wind. Strong, through-flowing winds will disperse pollutants into large volumes of cleaner air in the downwind direction. Strong winds can quickly sweep away most pollutants from an urban area, but during periods when winds are light or absent, the concentrations can rise to high values.

The concentration of pollutants over a source area rises to its highest levels when vertical mixing (convection) of the air is inhibited—for example, when the atmosphere has a stable environmental lapse rate. This stability is particularly strong if there is an *inversion*—a condition in which the temperature of the air increases with altitude.

Why does an inversion inhibit mixing? A heated air parcel, perhaps emerging from a smokestack or chimney, will rise as long as it is warmer than the surrounding air. As it rises, it is cooled according to the adiabatic principle. In an inversion, the surrounding air gets warmer, not colder, with altitude, so the parcel cools as it rises, while the surrounding air becomes warmer. Under these conditions, heated air moves only a short distance upward before its temperature matches that of the surrounding air, and uplift will stop. Pollutants in the air parcel will disperse at low levels, keeping concentrations high near the ground.

Two types of inversions can cause high air pollutant concentrations—low-level and high-level inversions. When a *low-level temperature inversion* develops over an urban area with many air pollution sources, pollutants are trapped under the "inversion lid." Heavy smog or highly toxic fog can develop.

An example is the tragedy that occurred in Donora, Pennsylvania, in late October 1948, when a persistent low-level inversion developed. The city occupies a valley floor hemmed in by steeply rising valley walls. The walls prevented the free mixing of the lower air layer with air of the surrounding region (FIGURE 14.17) and helped keep the inversion intact. Industrial smoke and gases from factories poured into the inversion layer for 5 days, increasing the pollution level. A poisonous fog formed and 20 people died, with several thousand persons stricken.

Another type of inversion is responsible for the smog problem experienced in the Los Angeles Basin and other California coastal regions, ranging north to San Francisco and south to San Diego (FIGURE 14.18). Off the California coast lies a persistent subtropical anticyclone that produces a layer of warm, dry air at upper elevations. Beneath it, a cold current of upwelling ocean bottom water runs along the coast, just offshore. Moist ocean air moves across this cool current and is chilled, creating a cool marine air layer.

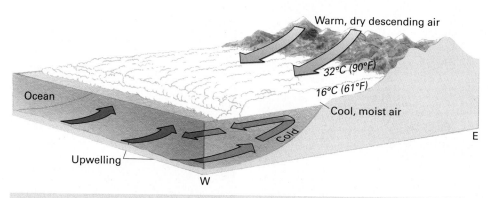

High-level temperature inversion FIGURE 14.18

A layer of warm, dry, descending air from a persistent fair-weather system rides over a cool, moist marine air layer at the surface to create a persistent temperature inversion, like this one on the California coast.

The Los Angeles Basin is a low, sloping plain lying between the Pacific Ocean and a massive mountain barrier on the north and east sides. Weak winds from the south and southwest move the cool marine air inland over the basin, where it is blocked by the mountain barrier. Because there is a warm layer above this cool marine air layer, the result is an inversion. Pollutants can accumulate in the cool air layer and produce smog. We describe this type of inversion, which persists to a higher level in the atmosphere, as a *high-level temperature inversion*.

Light or calm winds and stable air can also cause high concentrations of pollutants above a city. Stable air has a temperature profile that decreases with altitude but at a smaller rate than either the dry or moist adiabatic lapse rate. Some convection occurs in stable air, but the dispersion of pollutants is still inhibited. Under these conditions, a broad pollution dome can form over a city or region, and air quality will suffer (FIGURE 14.19). When there is a regional wind, the pollution from a large city will be carried downwind to form a pollution plume.

Pollution dome and pollution plume FIGURE 14.19

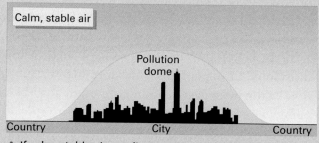

A If calm, stable air overlies a major city, a pollution dome can form.

B When wind is present, pollutants are carried away as a pollution plume.

A Forests Acid fallout from a nearby nickel smelter killed this lush forest in Monchegorsk, Russia, which then burned.

B Buildings Acid rain has eroded and eaten away the face of this stone angel in London, England.

Effects of acid rain FIGURE 14.20

Acid deposition

Perhaps you have heard about acid rain killing fish and poisoning trees (FIGURE 14.20). Acid rain is part of the phenomenon of *acid deposition*. It is made up of raindrops that have been acidified by air pollutants. Fossil fuel burning releases sulfur dioxide (SO_2) and nitric oxide (NO) into the air. The SO_2 and NO readily combine with oxygen and water in the presence of sunlight and dust particles to form sulfuric and nitric acid aerosols, which then act as condensation nuclei. The tiny water droplets created around these nuclei are acidic, and when the droplets coalesce in precipitation, the resulting raindrops or ice crystals are also acidic. Sulfuric and nitric acids can also form on dust particles, creating dry acid particles. These can be as damaging to plants, soils, and aquatic life as acid rain.

Acid deposition in Europe and North America has had a severe impact on some ecosystems. In Norway, acidification of stream water has virtually eliminated many salmon runs by inhibiting salmon egg development. Many lakes and streams in the northeastern United States have suffered as well. Forests, too, have been damaged by acid deposition. In western Germany, the impact has been especially severe in the Harz Mountains and the Black Forest.

LAND-COVER CHANGE

Human activity has greatly changed the face of the Earth over the last 5000 years. As human populations expanded, forests were clear-cut and converted to agriculture. With the development of irrigation, deserts bloomed with field crops. Sod-busting plows turned prairies into wheat fields. Grasslands became pastures. Cities and suburbs sprang up to add buildings, roads, and cultivated vegetation to the landscape.

What is the effect of land-cover change on weather and climate? Let's look at three human activities that can change climate: deforestation, irrigation, and urban development. These types of land-cover change can be seen in FIGURE 14.21.

Deforestation

Deforestation occurs when forests are converted to agricultural fields or grazing lands for long periods of time. This conversion reduces precipitation and increases surface temperature. In other words, the climate of the region becomes hotter and drier. Why does this occur?

A forest is a thick layer of vegetation, with leaves and branches extending many meters above the soil. As sunlight penetrates the forest canopy, solar heating is distributed throughout the thick forest layer. Transpiration by the many leaves of the forest also absorbs solar energy by converting liquid water to water vapor. Temperatures inside the forest are cool.

In contrast, cropland or grassland is a thin layer of vegetation over a bare soil surface. Solar energy is concentrated on the soil, which heats quickly. In addition, the top layer of soil dries out, cutting off evaporative cooling.

▲ Debris flow in Tegucigalpa, Honduras

After deforestation, soil is no longer held together by forest root systems, making it more susceptible to debris flow. The soil surrounding this neighborhood became saturated by Hurricane Mitch's deluge. The entire neighborhood flowed downhill, blocking the Choluteca River.

▲ Deforestation

Since humans took up agriculture, the Earth's forests have been diminishing. At present, about 13 million hectares (32 million acres) of forest is lost each year, largely to agriculture. More than half of this area is in South America, Africa, and equatorial Asia. Shown here is a land clearing effort in the state of Rondonia, in the Brazilian Amazon region. Most deforestation in the Amazon converts rainforest to cattle posture.

▲ Urban heat island

Phoenix, Arizona, has a significant heat island effect produced by urbanization. Nighttime temperatures in the city are typically 2–4°C (3.6–7.2°F) warmer than in the surrounding desert. Effects are strongest in the central city (indicated by reds and yellows at the top of the image) and weaker in residential and agricultural areas that have some vegetation cover (indicated by blues and greens bottom of image). To the left, dry, barren mountaintops also show higher temperatures than valleys, where cool nighttime air has descended.

▲ Irrigation

Irrigation can convert dry desert to productive cropland. Shown here is the Colorado River valley, near Parker, Arizona. The Colorado River is on the right. Canals along the edges of the fields provide irrigation water for crops.

The forest layer acts like a sponge, taking up rainwater in its leaves, branches, and roots and slowly releasing that rainwater to the air by transpiration. Without the forest to hold the rainwater, rainfall on crop or grazing land runs off into rivers and streams or sinks into the ground beyond the reach of plant roots. Because evapotranspiration is reduced, the air is drier and less water vapor is available to condense and form rain.

Another reason for reduced precipitation is reduced convection. More solar energy is reflected back to space by the dry soil surface of cropland, and the hotter soil of the cropland emits more longwave energy. The total energy available to warm and lift the air (net radiation) is lower, resulting in less convection in spite of higher surface temperatures. Coupled with the fact that drier air has to rise farther from the surface before condensation occurs, the loss of lift means fewer clouds and less precipitation.

How important is the effect of deforestation? Presently about 3–4 million hectares (about 12,000–15,000 mi^2) of Amazon rainforest is deforested each year. The removal of vegetation and accompanying overgrazing—termed *land degradation*—has made these regions hotter and drier. Using mathematical models, climatologists have estimated that clearing the Amazon rainforest would result in a reduction of precipitation in the region of about 1 mm per day. This is about 14 percent of the annual rainfall at Iquitos, Peru, for example. Mean temperatures would increase around 2°C (3.6°F). However, the effects of present deforestation are not strong enough to counteract the climate variations that the overall region naturally experiences.

Deforestation can also affect the impact of weather on a region. By removing the underlying support of natural masses of soil held together by forest root systems, deforestation can destabilize the soil and rock cover of steep hillsides and mountainsides. Saturated by heavy rains, the soil and rock can easily give way, producing mudflows and debris floods that travel far down the canyon floors and spread out, burying streets and houses in mud and boulders.

Irrigation Throughout the arid regions of the world, irrigation has greatly expanded the reach of agriculture. Based on both observations and models, arid

irrigated regions experience cooler surface air temperatures and more humid air than unirrigated arid regions. This is a natural result of the enhanced evapotranspiration by irrigated crops and moist soils.

As an example, the central valley of California is a semiarid area largely devoted to agriculture that is supported by water from streams and rivers draining the Sierra Nevada Mountains. In the irrigated regions of the San Joaquin Valley, daily maximum temperatures fell by 1.8–3.2°C (3.2–5.8°F) from 1887, when irrigation first began, to about 1980, when the expansion of the irrigated area ceased. The same decreasing temperature trend has been observed in other irrigated areas, including India, China, the Black Sea region, and the Great Plains. Daily maximum temperatures in the dry season are most affected, but in some cases there are decreases in the daily average temperature and minimum temperature as well.

Although irrigation acts to produce local cooling, irrigated agriculture actually increases the temperature of the globe, because it decreases the surface albedo. Under irrigation, dry soil and rock, which reflect more solar energy back to space, are replaced by vegetation and moist soil, which reflect less. Thus, the Earth absorbs more solar energy as a result, which leads to global warming. In addition, the increased evapotranspiration increases atmospheric water vapor concentrations and boosts the natural greenhouse effect.

Urbanization Cities tend to be warmer than the surrounding countryside—this is the *urban heat island* effect. The heat island is produced by several factors. First, evapotranspiration is lower because city surfaces of concrete, asphalt, and stone do not retain water that can be evaporated. Instead, the solar heat is directly absorbed, making city surfaces hotter.

Second, the vertical surfaces of buildings reflect part of the solar radiation back to the ground or to other buildings, where the radiation has another chance to be absorbed. This effect tends to trap solar radiation more effectively, significantly decreasing the city's albedo.

Third, the city releases significant amounts of heat of its own, produced by inefficiencies in space heating, refrigeration, air conditioning, and transportation. This release of *waste heat* is generally greater in winter than in summer for midlatitude cities, because of the need for winter space heating.

The urban heat island effect is felt largely at night, when air temperatures remain higher in the city than in the surrounding countryside. This is because urban building materials are good at conducting and storing daytime heat that is readily released at night. Compared to the countryside, then, the daily temperature range of the city is significantly smaller.

The presence of urban areas can also influence the weather in and around the city. Because of the elevated temperatures over urban areas, the environmental lapse rates tend to be more unstable than over the surrounding rural areas. As storms systems pass overhead, the unstable air over the city enhances convection and uplift, leading to greater condensation, cloud formation, and precipitation. The enhanced precipitation continues to fall over regions downwind of the city. In certain cases, urban and downwind areas can receive twice as much rainfall as upwind regions, raising the potential of flooding during heavy-rainfall periods.

CONCEPT CHECK STOP

How do inversions in the atmosphere contribute to high concentrations of pollutants? Describe two ways these inversions can form.

How do deforestation and urbanization affect regional climates? How do they affect weather and its impact on land and people?

What effects do acid rain have on natural and human structures?

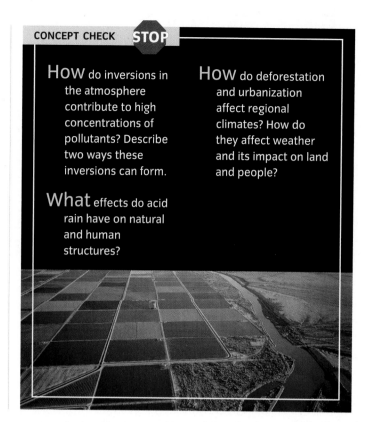

What is happening in this picture ?

This image shows the border between Haiti (to the left) and the Dominican Republic (to the right). These two countries occupy the same island in the Caribbean and have very similar climate conditions. However, they have very different regulations regarding land use of the forested regions.

- Based on this image, which country do you think has allowed unrestricted logging to occur?

- Which country would have greater likelihood of landslides and mudflows during the passage of tropical storms and hurricanes?

Human Impact on Weather and Climate 407

1 Severe Weather

1. **Severe weather** events are hazardous to humans and can destroy property and infrastructure. These events include tornadoes, hurricanes, midlatitude storm systems, as well as others.

2. **Heat waves** and **cold waves** are severe weather events associated with extreme temperatures. Heat waves are persistent periods of high temperatures usually accompanied by high relative humidity. Cold waves, which occur in the winter, are associated with the intrusion of cP or cA air from the high latitudes into the midlatitudes.

3. Severe weather events associated with excess precipitation can result in flooding. This flooding can take the form of river floods, in which water flow crests above the banks of rivers and streams. The excess precipitation can also produce **mudflows** and landslides, in which a mixture of water and soil flows rapidly downhill. Another type of flooding—coastal flooding—occurs when high winds push massive amounts of seawater onshore, inundating natural and human-made protective barriers.

2 Severe Climate

1. One of the most devastating forms of climate change involves **drought**—the persistent lack of precipitation and soil moisture. It is related to both a lack of rainfall as well as high temperatures that increase loss of soil moisture through evaporation. Drought can lead to agricultural failure, famine, and wildfires.

2. Climate change can also affect the number and intensity of severe storms in a given region. Changes in severe weather over a given region are particularly sensitive to changes in the location and strength of upper-air jet streams. Other severe weather events, like hurricanes, are sensitive to climate changes induced by changing sea-surface temperatures.

3. More persistent climate changes can produce a shift in the climate of a region as temperatures and precipitation change. Such shifts can be beneficial as well as detrimental. Some period of time is usually required before natural and human systems become accustomed to the new climate conditions.

3 Human Impact on Weather and Climate

1. Human activity can modify local and global weather and climate in many ways. Changes in the chemical composition of the atmosphere, such as the emission of CFCs or heat-trapping gases, can modify the amount of solar radiation that reaches the surface as well as the amount of longwave radiation that escapes to space.

2. Humans can also modify the local chemical composition of the atmosphere, leading to significant pollution. Most of the pollutants that affect us come from fossil-fuel combustion. The concentration of these pollutants can be exacerbated by local meteorological conditions, particularly inversions, which trap the pollutants near the surface.

3. Land-use change can also influence regional climate and weather features. Deforestation increases local air temperatures, reduces local precipitation, and makes regions more susceptible to mudflows and debris flows. Irrigation of arid lands provides abundant water for evapotranspiration, which lowers local air temperatures, particularly maximum temperatures observed during the day, and makes the air more humid. Urbanization increases local temperatures, largely because of reduced evapotranspiration from impervious urban surfaces, and can enhance precipitation by making air over the city regions more unstable.

- severe weather p. 386
- heat wave p. 386
- cold wave p. 389
- floodplain p. 390
- mudflow p. 391
- drought p. 395
- air pollutant p. 401

CRITICAL AND CREATIVE THINKING QUESTIONS

1. Describe how heat waves in the western portion of the United States differ from those in the eastern portion. Which heat waves typically have higher temperatures? Which typically have higher relative humidities? Why?

2. Imagine that you owned a farm in Oklahoma and that you grew wheat on the farm. Besides precipitation, what other climate variable might you be interested in if you were worried about drought affecting your crops? For what time of year would you be most interested in predicting the climate of your region?

3. Based on the figure shown here, what year had the highest overall precipitation in the Sahel? What year had the lowest? During 1994, rainfall was above average. Do you think this excess rainfall would be enough to alleviate the drought that occurred during the 1980s and 1990s?

4. Based on how the stability of the atmosphere changes with time of day, when do you think would be the most likely time to have severe pollution? When do you think would be the time when pollution is most easily distributed aloft?

5. Describe the most common forms of air pollution and the damage that air pollution can cause.

6. How and why do changes in land cover affect climate, both locally and globally? In your answer, describe the roles of changing surface characteristics and evapotranspiration as they affect temperature and precipitation.

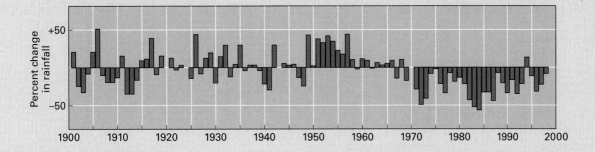

1. Heat waves in the eastern portion of the United States have high temperatures, as well as _____ relative humidities, which _____ the heat index.
   a. high; decreases
   b. high; increases
   c. low; decreases
   d. low; increases

2. Cold waves are associated with _____ that accompany _____ air masses.
   a. traveling cyclones; mT
   b. traveling cyclones; mP
   c. traveling anticyclones; cT
   d. traveling anticyclones; cP

3. A(n) _____ is a rapidly descending mixture of water and soil and is usually triggered by a(n) _____.
   a. landslide; earthquake
   b. landslide; thunderstorm
   c. mudflow; thunderstorm
   d. earthflow; earthquake

4. Flash floods are characteristic of streams draining _____ watersheds with _____ slopes.
   a. large; gentle
   b. small; steep
   c. small; gentle
   d. large; steep

5. Hurricane frequency and strength can be affected by changes in _____.
   a. ocean temperatures in the Atlantic
   b. upper-air winds
   c. ocean temperatures in the Pacific
   d. all of the above

6. Coastal flooding in the midlatitudes can be produced by _____ winds if the Coriolis force subsequently deflects the moving water masses _____.
   a. offshore; onshore
   b. offshore; offshore
   c. alongshore; onshore
   d. alongshore; offshore

7. On the accompanying pressure map, draw in the wind directions near the coast of England. Also draw in the direction of water transport associated with these winds.

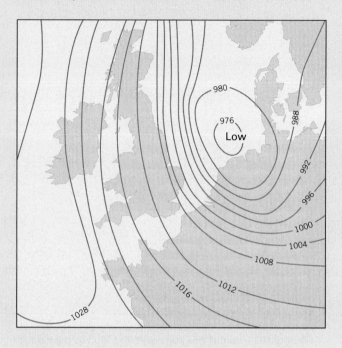

8. Droughts are only determined by the amount of precipitation that falls in a given region.
   a. True
   b. False

9. _____ is a type of drought classification.
   a. Meteorological
   b. Agricultural
   c. Hydrologic
   d. All of the above

10. One of the greatest impacts of drought on human health is _____.
    a. fire
    b. spread of disease
    c. famine
    d. loss of permafrost

11. _____ is *not* an example of the impact of climate change.
   a. Increase in number of extreme events
   b. Spread of disease
   c. Change of a region's climate type
   d. A mudflow arising from the passage of a thunderstorm

12. On the accompanying figure, indicate what levels of the atmosphere would tend to have trapped pollution. How warm would the surface have to become before the inversion disappeared? How warm would the surface have to be before it became unstable?

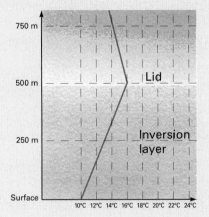

13. Acid deposition is produced by the release of sulfur dioxide and _____ into the air.
   a. carbon dioxide
   b. ozone
   c. nitric oxide
   d. sulfur perchlorate

14. Deforestation _____ local air temperatures and _____ global warming.
   a. increases; enhances
   b. increases; reduces
   c. decreases; enhances
   d. decreases; reduces

15. Irrigation and urbanization affect local temperatures primarily by changing surface _____.
   a. albedo
   b. interception
   c. evapotranspiration
   d. energy integration

Weather Forecasting and Numerical Modeling $\boxed{15}$

In the last 25 years, weather forecasting has vastly improved through the use of new technologies. These technologies now allow weather forecasters to make reliable predictions farther into the future, provide more accurate predictions of more types of weather phenomena than before, and present weather information in new and innovative ways.

One of the first technologies to be adopted by weather forecasters was radar. By placing ground-based radar equipment at sites across the country, forecasters can get up-to-the-minute measurements of local rainfall and wind speed. They can also use these images to identify conditions ideal for tornado formation, lightning strikes, and other severe weather phenomena.

Weather forecasters were also early adopters of computers. The original computers had no more computing power than today's calculators, although they could quickly do the multiple computations needed for simple numerical weather forecasts. As computers became faster, forecasts became more accurate, even as they focused on smaller and smaller regions. Computers for weather forecasting can now do 800 billion separate calculations a second.

Another technology that weather forecasters were quick to adopt was the use of satellite images taken from space. These images give the weather forecasters a new global view of the weather. Using the reflected and emitted radiation coming from the clouds themselves, forecasters can track storms over open ocean and across land. They can observe wind speeds in the upper atmosphere as well as at the surface. They can also watch storms evolve over time as they intensify, mature, and decay.

A portable Doppler radar is situated to capture the winds and rain associated with an approaching supercell thunderstorm.

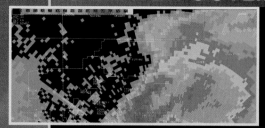

Local and Synoptic Weather Forecasting

Forecasting the weather is an ancient and honorable human activity. Weather predictions are found in ancient Egyptian, Chinese, and Greek texts. Without adequate ways to communicate weather activity from one region to another, early forecasters had only the weather in their own region as a tool for estimating how it might change over the next few hours or days. This is typically called *local weather forecasting* and can still be done today. With the advent of the telegraph, telephone, and wireless, it became possible to know the state of the weather in many locations simultaneously. This led to the development of **synoptic charts** like the one in FIGURE 15.1. The name refers to the fact that all the data on the chart are obtained at the same time.

> **synoptic chart**
> Map or chart showing meteorological data obtained simultaneously over a wide area.

By examining these charts, forecasters could determine the weather approaching from "over the horizon," which they could not see. They could also see relationships between the charts and take note of how weather systems developed over time. They could determine whether weather systems continued to move in the same direction, whether they intensified over time, and where there was heavy rainfall or snow. All of these allowed the forecaster to determine not only whether a region would be affected by an approaching storm, but also whether that storm would have strengthened in the intervening time and whether to expect snow, rain, or even hail.

The next important steps in weather forecasting occurred nearly simultaneously. The first was the development of *remote sensing* of weather either from space or from land. During World War II, *radar*, which stands for **ra**dio **de**tection **a**nd **r**anging, was developed to monitor incoming aircraft. The same technology was able to monitor cloud drops and rain drops in the atmosphere and could track storms from one region to the next. Similarly, as satellites were launched into space, beginning with TIROS-1 in 1960, they started taking pictures of the Earth. The weather features that the satellites were seeing, first in visible light and now with other

Early synoptic weather chart FIGURE 15.1

This early synoptic chart, produced on October 15, 1877, shows surface pressure patterns and temperatures across much of the northern hemisphere. Charts like this provided forecasters with their first look at weather features across broad regions of the world.

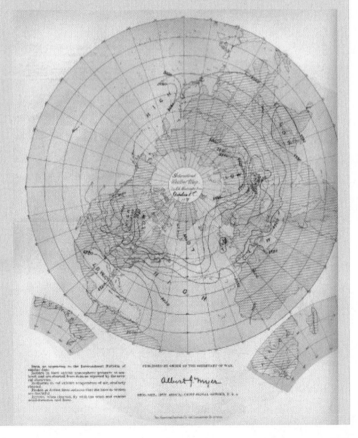

types of remote sensing techniques, could also be used to track the development of weather around the world. Presently, there are over a dozen satellites in orbit dedicated to monitoring weather and climate.

The other development that proceeded in parallel with remote sensing of weather was the advent of the computer—a device capable of carrying out many mathematical computations in a short time. Originally developed in the 1940s, modern computers have found their way into almost every aspect of our lives, including weather forecasting. Computer programs now depict and simulate the behavior of the whole atmosphere, including its interaction with ocean and land surfaces. These computer programs, called *atmospheric models,* attempt to predict the weather in the future based on equations that approximate the motion and energy transfer that govern the real atmosphere.

Wind direction

One way to predict changes in local weather is to use the local wind conditions and a weather map—or *surface chart*—to track where weather systems may go (**FIGURE 15.2**). Generally, weather systems and fronts will continue to move in the same direction and with the same speed as they did during the previous 6 hours—providing a *persistence* forecast. This method, however, does not foretell a shift in a storm track during the next 6 hours or an intensification of the storm.

It also is possible to predict the track of a storm by looking at the direction of the surface isobars. Low-pressure centers tend to move parallel to the surface isobars ahead of the cold front. Alternatively, the direction of the winds aloft can also provide information. These winds, typically found at about 500 mb—the *steering level*—tend to steer storms in a direction parallel

LOCAL WEATHER FORECASTING

Under certain scenarios, it is sometimes possible to predict changes in local weather based on some simple observations. This method can work well for short time periods (less than 12 hours) but tends to get worse as the forecasting period grows longer. For long-range forecasts, more computer-intensive and experience-based techniques are needed. Still, there are ways to predict near-future weather based on the surrounding weather conditions as well as those in the immediate vicinity.

Surface map FIGURE 15.2

The location and movement of frontal systems, along with current weather conditions for different observing sites, can be used to make short-term forecasts of local weather. **A** A map showing the location of frontal systems and current weather conditions for different observing sites across the United States for November 15, 2006. **B** A similar map for conditions 6 hours later.

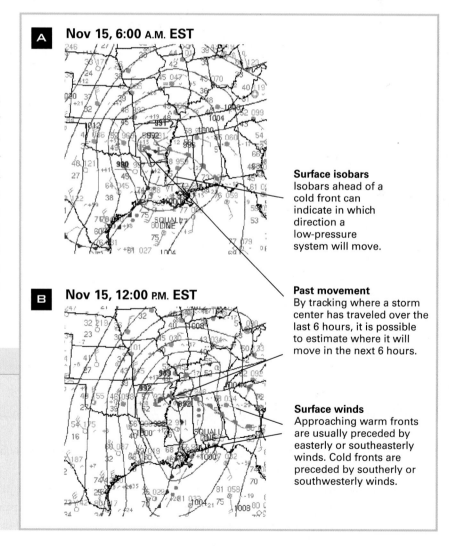

A Nov 15, 6:00 A.M. EST

B Nov 15, 12:00 P.M. EST

Surface isobars
Isobars ahead of a cold front can indicate in which direction a low-pressure system will move.

Past movement
By tracking where a storm center has traveled over the last 6 hours, it is possible to estimate where it will move in the next 6 hours.

Surface winds
Approaching warm fronts are usually preceded by easterly or southeasterly winds. Cold fronts are preceded by southerly or southwesterly winds.

Surface weather charts combine many different types of information, ranging from pressures and temperatures to wind speed and direction, cloud cover, and location of fronts. This table shows how information appears on a typical surface chart. As with many weather measurements taken in the United States, these are based on the English units of temperature (Fahrenheit) and wind speed (knots).

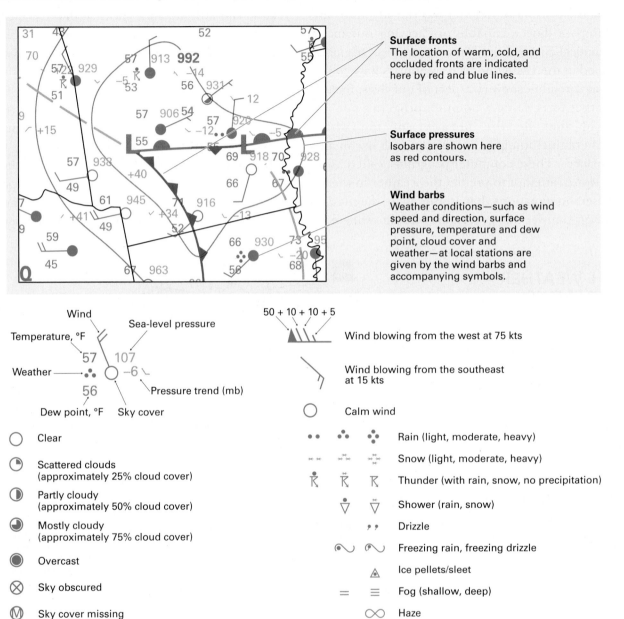

Surface fronts
The location of warm, cold, and occluded fronts are indicated here by red and blue lines.

Surface pressures
Isobars are shown here as red contours.

Wind barbs
Weather conditions—such as wind speed and direction, surface pressure, temperature and dew point, cloud cover and weather—at local stations are given by the wind barbs and accompanying symbols.

to the winds, although the storms typically move only half as fast as the winds themselves.

Other weather information can be gained from a surface chart. For example, it is possible to use the direction of the surface winds to determine whether a cold or warm front is approaching a given region. FIGURE 15.3 describes how surface winds and other weather conditions are depicted on surface charts. Approaching warm fronts are usually preceded by easterly or southeasterly winds. After the warm

A A line of cumulus clouds marks the advance of a cold front moving from left to right. The cold air pushes warmer, moister air aloft, triggering cloud formation.

B Cirrus and cirrostratus clouds mark the approach of a warm front. The higher cirrus clouds to the right and the lower cirrostratus clouds to the left indicate that the front is moving from left to right.

Clouds associated with approaching fronts FIGURE 15.4

front's passage, winds shift to southerly or southwesterly. In turn, these southerly or southwesterly winds typically precede the passage of a cold front, which follows behind the warm front. Once the cold front has passed, winds shift to a more northwesterly direction, followed by westerly winds as the storm system moves out of the region.

Given the movement and speed of pressure centers and fronts, it is possible to make a prediction of the weather in certain locations. For example, as a cold front passes, increased cloud cover and precipitation will be followed by a change in wind direction and a drop in temperature. Conversely, as a warm front passes, there may be increased precipitation ahead of the front but warmer temperatures behind it. It is also possible to forecast precipitation changes based on the rainfall patterns associated with the pressure centers themselves. For example, increased rainfall is associated with low-pressure centers, and clear skies are associated with high-pressure centers.

Cloud formation

Cloud formations can also indicate weather to come. Warm fronts tend to be preceded by the passage of cirrus and altostratus clouds, accompanied by easterly or southeasterly winds (FIGURE 15.4). If over time these clouds thicken and form lower in the atmosphere, precipitation is likely to occur as warmer air ahead of the warm front lifts over the cooler air at the surface.

If the sky has mostly altocumulus and cirrus clouds accompanied by southerly or southwesterly winds, a cold front may be approaching. With the passage of the cold front, cumulonimbus clouds produce heavy rainfall. In addition, there is usually a rapid drop in temperature with gusty winds. This is followed by cold, clear conditions, particularly in winter, as the trailing high-pressure center produces subsidence over the region.

Atmospheric soundings

Twice a day across the United States, weather instruments are borne aloft by balloons—termed *radiosondes*. As they float skyward, they radio back information about the temperature, pressure,

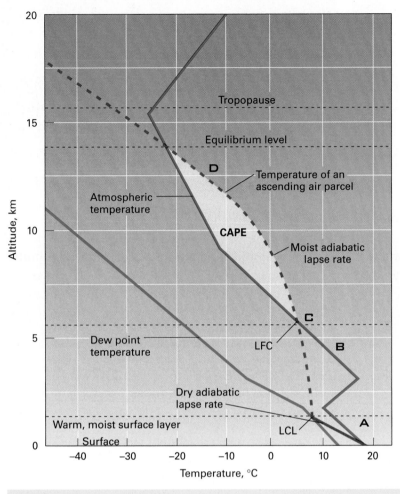

D The parcel will continue to rise as long as its temperature is warmer than the surrounding atmospheric temperature. Once its temperature matches the surrounding temperature, it will remain at that level, called the equilibrium level. This represents the cloud-top level. The area bounded by the two curves above the LFC and below the equilibrium level is related to the energy released during convection—the convective available potential energy (CAPE).

C If lifted high enough, the parcel's temperature will become warmer than the surrounding atmospheric temperature. At this point, it becomes buoyant and begins to rise on its own. The height at which the parcel's temperature is the same as the surrounding atmosphere marks the level of free convection (LFC).

B If the parcel is forced farther upward, its temperature will follow the moist adiabatic lapse rate, given by the dashed green line. As long as its temperature is cooler than the atmospheric temperature, the parcel will not rise on its own.

A The red line is the atmospheric sounding of temperature, and the blue line is the dew point temperature for this location. Starting at the surface, the temperature of a parcel that is forced to rise will first follow the dry adiabatic lapse rate, given by the green solid line. Where the parcel temperature reaches the dew point temperature, we find the lifting condensation level (LCL).

Atmospheric soundings FIGURE 15.5

Based on atmospheric soundings of temperature and dew point temperature, it is possible to determine the height of the lifting condensation level, the level at which free convection will occur, as well as the amount of energy that will be released during convection.

and dew point temperature of the atmosphere around them. In addition, by tracking their position visually or with global position systems (GPS), forecasters can get estimates of the wind speed and wind direction the balloons experience during their ascent.

This information can then be used to make local weather forecasts. For example, forecasters can use changes in wind direction and strength with height—*wind shear*—to signal the approach of cold air, characterized by counterclockwise turning—*backing*—of winds with height or the approach of warm air, characterized by clockwise turning—*veering*—of winds with height.

Profiles of temperature and dew point temperature also can provide information about the vertical stability

of the atmosphere, as discussed in FIGURE 15.5. Usually, air at the surface is stable—it requires some process to provide initial lift before it begins rising. This lift can be provided by the approach of a cold front or warm front, surface convergence associated with a mid-latitude cyclone, or flow over a mountain.

As the air begins to rise, its temperature will cool according to the dry adiabatic lapse rate. If it cools enough and reaches the dew point temperature, condensation begins—the height where this condensation occurs marks the *lifting condensation level* and the height of the cloud base. However, if the air parcel is still cooler than the surrounding air, it will remain stable unless it continues to be forced aloft. An air parcel that continues to be forced aloft will cool according to the moist adiabatic lapse rate.

If the air parcel cooling is less than the cooling of the surrounding atmosphere, at some point its temperature will become equal to and then greater than the temperature of the atmosphere around it. At this point, the parcel has reached the *level of free convection*. Only then will the parcel become unstable and continue to lift on its own. It will do so until its temperature cools below the temperature of the surrounding atmosphere, at which point convection will stop.

By looking at vertical profiles of the surrounding atmosphere in the local region, a weather forecaster can determine at what height clouds will form, how much lifting must occur in order for convection to begin, and then how long and how high the convection will go. All of this information will provide the weather forecaster with an indication of the likelihood and intensity of storms in the area.

Pressure tendency
A final local indicator of changing weather is the change in pressure with time,

called the *pressure tendency*. Generally, low-pressure centers move toward regions of falling pressure, while high-pressure centers move toward regions of rising pressures. Maps that show the pressure tendency rather than pressure are called **isallobaric maps**, as seen in FIGURE 15.6.

Local changes in pressure can also be used to forecast an approaching front. As a cold front approaches, pressures tend to fall steadily. After the passage of the front, pressures begin to rise again, suddenly at first, followed by a more gradual increase as the region comes under the influence of the colder air to the north. Pressures also tend to fall ahead of a warm front. After the passage of the warm front, however, pressures continue to fall as the cold front behind it approaches. In both cases, falling pressures are typically associated with an approaching front. Forecasters can determine whether it is an approaching warm front or cold front by the cloud cover and wind directions.

> **isallobaric map**
> Map of lines connecting places of equal change in pressure within a specified time period.

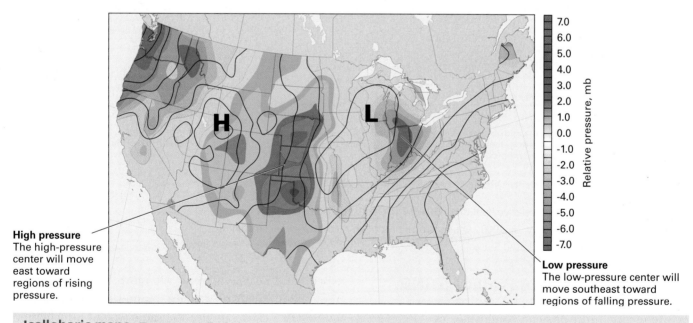

High pressure
The high-pressure center will move east toward regions of rising pressure.

Low pressure
The low-pressure center will move southeast toward regions of falling pressure.

Isallobaric maps FIGURE 15.6

This map shows the change in pressures over the previous 3 hours (in color) along with the surface pressures for January 8, 2008. Note the rapidly increasing pressures across the midwestern United States to the east of the high-pressure centers themselves, as well as the decreasing pressures south of the Great Lakes region. These indicate approaching anticyclones and cyclones, respectively.

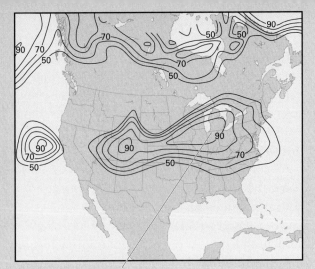

700 mb relative humidity
Relative humidity values of 90 percent are found across much of the central United States, indicating strong possibility of precipitation.

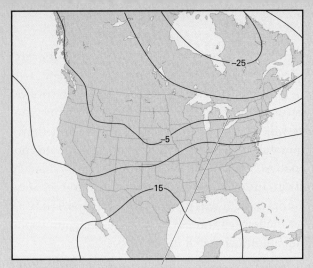

850 mb temperatures
With 850 mb temperatures below –5°C over the northern portion of this region, precipitation is expected to fall as snow, with rain farther to the south.

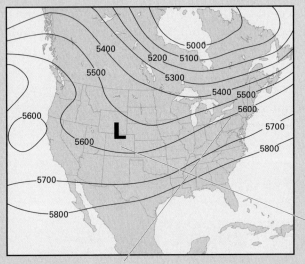

500 mb heights
Winds at 500 mb—the steering winds—follow the isobars, indicating that the storm will move northeast into New England and eastern Canada.

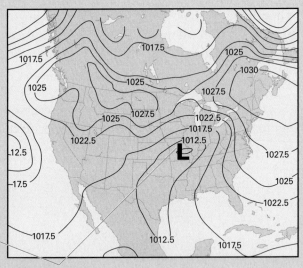

Sea-level pressures
The trough at 500 mb, which sits over Colorado, is situated to the west of the surface low-pressure center, indicating that this storm will intensify as it moves east.

Upper-air charts used for weather forecasting FIGURE 15.7

These maps show the 700 mb relative humidity, 850 mb temperatures, 500 mb heights, and sea-level pressures for February 18, 2000, when a storm produced record snowfall stretching from Nebraska to the eastern United States.

SYNOPTIC WEATHER FORECASTING

While local conditions can sometimes be used to make short-term forecasts, longer-term forecasts produced by forecasting agencies like the National Weather Service typically rely on synoptic and numerical weather forecasts. **Synoptic weather forecasts** are determined by looking at maps of the atmospheric circulation at many different levels to determine not only where

> **synoptic weather forecast** Weather forecast using synoptic maps to identify features at the surface and aloft that indicate how weather will develop over a specific time period.

storms and weather systems will move, but also whether they will intensify or weaken. Numerical weather forecasts can give the same type of information but are based on computer models that try to simulate how the atmosphere will behave over the next 6 to 72 hours (or longer).

Upper-air charts   As mentioned, synoptic weather forecasts usually rely on charts describing various aspects of the atmosphere at different levels (FIGURE 15.7).

We've already noted that the 500 mb wind field steers storms in a particular direction and can give the forecaster an idea of where storms will move over the coming time period.

Another useful weather map shows the 700 mb relative humidity fields. When the relative humidity is above 70 percent at this level, some type of cloud cover is usually present; as the relative humidity rises above 90 percent, precipitation usually develops. Temperature maps can then give an indication of whether the precipitation will fall as snow or as rain. For example, if the temperatures at 850 mb are below $-5°C$, then precipitation usually falls as snow, while regions with temperatures above $-5°C$ usually experience rain.

Finally, it is possible to compare upper-air pressure fields with low-level pressure fields to determine whether a storm is strengthening. Often, if there is a low-pressure center at 500 mb with a surface low to the east of it, then the storm is still developing and strengthening. In contrast, if the pressure center at 500 mb is situated over the surface low, the storm has occluded and is either fully mature or weakening.

hook echo Curve-shaped region seen in radar images when precipitation is drawn into a thunderstorm, indicating possible tornado formation.

Radar While upper-air weather charts can provide the forecaster with tools to predict how weather might develop over time, other tools have been developed to better estimate the extent and severity of the weather in certain locations. One of these is the use of ground-based radar to measure precipitation intensity.

This radar system, termed *Doppler radar*, constitutes a set of radar stations throughout the country that can be used to determine the intensity of rain, snow, sleet, and hail based on how strongly a radar signal emitted from the station is reflected back to a receiver at the same station. Radar can also provide information on the strength of the winds and whether they are blowing toward or away from the radar site. Although it cannot detect tornadoes directly, radar can also be used to detect strong wind shear—characterized by a **hook echo**—which can signal possible tornado development, as seen in **Figure 15.8**.

Information from radar stations situated across the United States provides forecasters with a detailed image of precipitation and wind speeds across the country. Combined with knowledge of how the storms and

Precipitation and winds from a radar image FIGURE 15.8

On May 8, 2003 a massive super-cell thunderstorm moved through Oklahoma City, accompanied by intense rainfall and an F4 tornado that caused extensive damage. Radar images like this show both the heavy precipitation as well as the signatures of tornado-producing storms. The colors give the reflectivity of the precipitation, with red representing heavy rainfall.

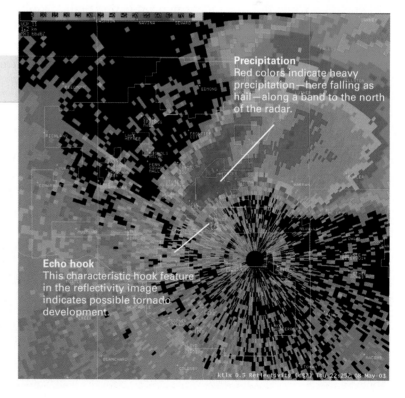

Precipitation Red colors indicate heavy precipitation—here falling as hail—along a band to the north of the radar.

Echo hook This characteristic hook feature in the reflectivity image indicates possible tornado development.

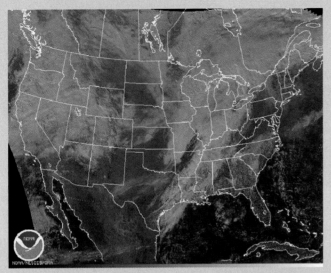

A Visible radiation is reflected from the cloud tops and underlying land surfaces.

B In this image of water vapor, reds and yellows indicate high water vapor content associated with a storm over the Mississippi Valley, while dry conditions—indicated by blacks and oranges—prevail over the southeastern United States.

Weather satellite images FIGURE 15.9

These satellite images show weather conditions on January 8, 2008.

fronts are moving, the forecaster can then provide cities and towns with updated forecasts warning of approaching severe weather.

Satellites   While Doppler radar can provide very high resolution images of local precipitation, satellites provide a way of obtaining a global-scale view of the Earth system, including weather systems and their movements.

Many satellites can see both visible light reflected by clouds as well as thermal infrared radiation emitted by clouds. Both of these radiation signals provide information about the location and height of cloud systems. Other satellites use sensors to track water vapor (Figure 15.9), which provides an image of atmospheric circulations in regions where there are no cloud formations to track. In combination with the visible and in-

frared tracks of clouds, these images can help the forecaster predict how intense storm systems are and where they are moving.

Still other sensors provide information about atmospheric temperature profiles. These can be used to track cold air masses, for example, and help predict where frost will develop. In addition, some satellite sensors can determine the intensity of winds at the surface of the ocean, which can reveal the intensity of midlatitude and tropical cyclones approaching coastal regions. Finally, new satellite sensors are being developed that, like Doppler radar, can estimate precipitation intensity and therefore help forecasters track severe weather as it moves across large areas. *What a Scientist Sees: Using Visual Information* explores the ways forecasters use these tools.

CONCEPT CHECK   STOP

What three methods are used for forecasting weather based on local conditions?

What types of maps and charts are used for synoptic weather forecasting?

How does a forecaster determine the movement of storms from upper-air wind fields?

What types of information can a weather forecaster get from radar?

Using Visual Information

The visual information provided by a combination of synoptic charts, ground-based radar, and space-based satellites are very powerful tools for tracking and forecasting weather across the nation. This set of pictures shows synoptic charts and images for the "Valentine's Day Storm" that developed over the central United States during February 13, 2007. The storm then moved up the east coast over the next several days, where it dropped over 750 mm (30 in.) of snow in some locations. In addition, severe ice storms caused extensive power outages, flight cancellations, and the closure of major interstate highways. By the time the storm passed, it had caused the loss of 37 lives.

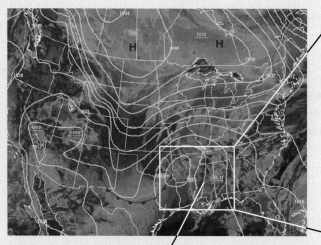

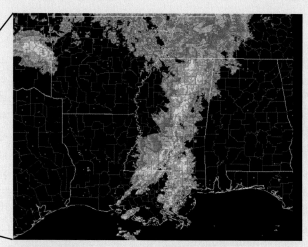

A 6:00 A.M. EST, February 13, 2007
The synoptic chart shows the development of a closed low-pressure center at the western end of a stationary front. A squall line, indicated by the dashed red line, lies to the east of the low. Infrared satellite images indicate significant cloud cover—in yellows and reds—associated with convective activity in this region. Elsewhere, skies are clear and appear green and blue.

B Radar image
Radar images for this time indicate very intense precipitation (in red) forming along the squall line.

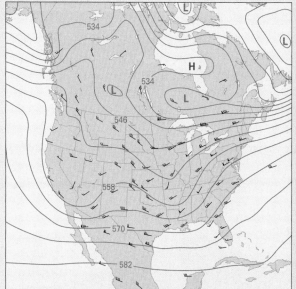

C Upper-air charts
The 500-mb heights and wind fields indicate that the steering winds will take the storm to the east and north at approximately 43–47 knots, which should bring it over the east coast (approximately 500 km away) within the next 24 hours.

D 6:00 A.M. EST, February 14, 2007
During the next 24 hours, the storm moved as predicted, bringing heavy precipitation—shown in this infrared satellite image by yellow and red colors—and strong winds to the northeastern United States.

Numerical Weather Forecasting

Describe how computers are used to produce numerical weather forecasts.

Identify the types of forecasts that can be generated by computer models.

Explain why forecasters use multiple models to produce numerical weather forecasts.

While many tools are available to the present-day weather forecaster, one of the most important is the computer, with its ability to solve complex mathematical equations that simulate the physical processes playing out in the real world.

Equations that simulate the forces in the atmosphere are contained in huge computer programs termed **atmospheric models**. The models include almost all of the physical equations that determine how the

atmospheric model A computer model designed to simulate and predict the behavior of the atmosphere.

atmosphere evolves over time. Based on numerical equations for the forces in the atmosphere—including the pressure gradient force, the Coriolis force, and friction—the model can determine how wind speed and direction will change over time. Similarly, the temperature at a particular location in the atmosphere will depend on the amount of shortwave or longwave radiation emitted and absorbed, the amount of sensible and latent heat released or absorbed, and the movement of cold or warm air into and out of the region. Atmo-

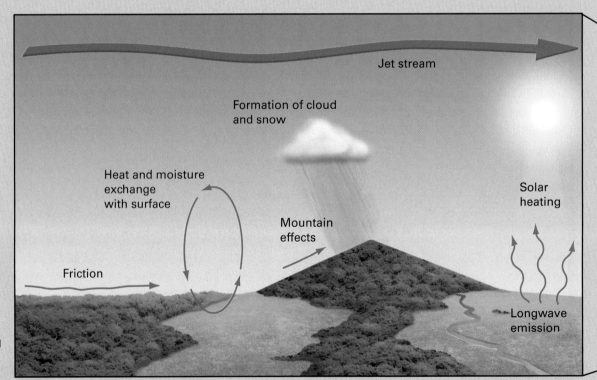

Surface atmosphere interactions
Models also have to account for heat and moisture provided by the Earth's surface, and energy provided by the Sun.

Jet stream

Formation of cloud and snow

Heat and moisture exchange with surface

Mountain effects

Solar heating

Friction

Longwave emission

spheric models contain equations for all of these processes and use them to determine how temperatures evolve over time.

The model has to solve these equations not only for the air near the surface, but also at all levels in the atmosphere. In addition, it has to solve these equations for every point in space, not only at one location. To do this, computer models typically break up the atmosphere into about 20–50 different vertical levels and 10,000–20,000 *grid points*, or global locations (FIGURE 15.10). These model grid points are typically spaced across the globe about 100 km apart, as shown in the figure. For finer-scale atmospheric models, which are run for a particular region or continent rather than the whole globe, the grid points can be as close as 10 km apart. Finally, in order to make a proper forecast, the model has to recompute these equations every 2–20 minutes of simulated time.

To make a 24-hour forecast of only one variable at one point and one level, the model typically has to make 100 calculations. Multiplied by 30 levels, 10,000 grid points, and 10 different variables—such as zonal and meridional winds, temperature, and humidity, as well as variables related to radiation and surface energy fluxes—this corresponds to at least 300 million separate calculations to make a single forecast of what the weather will be like in 24 hours. Only computers are able to make these computations more rapidly than the time it takes the actual weather changes to occur—that is, faster than "real time."

Model representation of the world
FIGURE 15.10

For numerical weather modeling, a computer divides the atmosphere into vertical and horizontal boxes. Within each box, the model must determine changes in winds, humidity, temperature, and pressure. From these changes, the model can determine where rainfall will occur, how air masses will move, and whether fronts will develop.

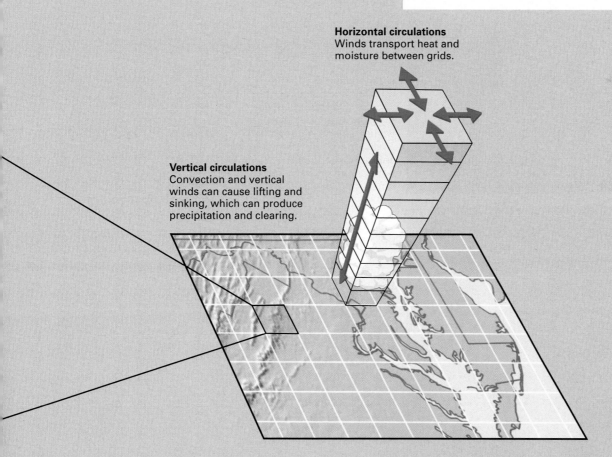

Horizontal circulations
Winds transport heat and moisture between grids.

Vertical circulations
Convection and vertical winds can cause lifting and sinking, which can produce precipitation and clearing.

VISUALIZATION

Once the model makes a forecast for a given time period, usually from 6 hours up to 10 days, the forecaster can use the model output as an estimate of what the weather will be like during that time. Although model data are usually output as numbers, for forecasting purposes it is easier to produce visual maps of the output, called **prognostic charts**. These charts, shown in FIGURE 15.11, are very similar to the synoptic charts discussed earlier. However, instead of showing present atmospheric conditions, they represent computer simulations of the atmospheric conditions at a given point in the future. Prognostic charts thus forecast the movement of storms, fronts, and precipitation.

In addition to producing prognostic charts, forecasters are interested in forecasting how conditions at a particular location might change. Using data from numerical models, they can produce plots of how variables like temperature, humidity, and precipitation at a fixed location change over time. These plots, called **meteograms**, allow the forecaster to

prognostic chart
A map or chart showing forecast pressure patterns and other meteorological variables for a specific time in the future.

meteogram A plot showing a numerical prediction of meteorological variables such as temperature, relative humidity, wind speed and direction, and pressure, at a given location over time.

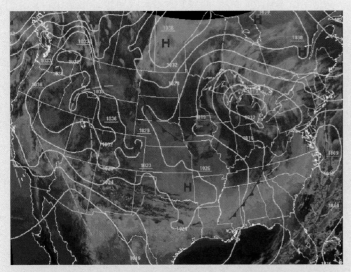

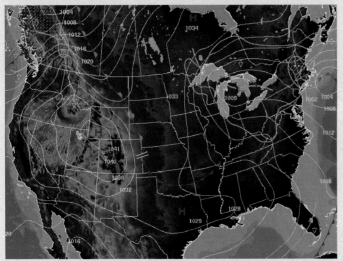

A Current conditions
This image shows the surface pressures, fronts, and cloud conditions for January 14, 2008. A storm system—indicated by the yellow and red colors in the infrared image—is situated over western Michigan and another is situated east of the mid-Atlantic states. High pressures are found over the northwestern United States, extending down to Texas, bringing mostly fair conditions to these regions indicated by the blue and green colors.

B 12-hour forecast
The 12-hour forecast indicates that the storm system over Michigan remains in place, while the one over the Atlantic will move north, bringing heavy snows to the northeastern United States. The high-pressure system will expand southward and eastward over the Rocky Mountains and Gulf Coast states, bringing clear conditions to these regions. Color indicates elevation, with green indicating low and brown high values.

Prognostic chart
FIGURE 15.11

Prognostic charts show the simulated evolution of weather patterns over time. They are plotted similar to synoptic charts but represent future, rather than current, conditions.

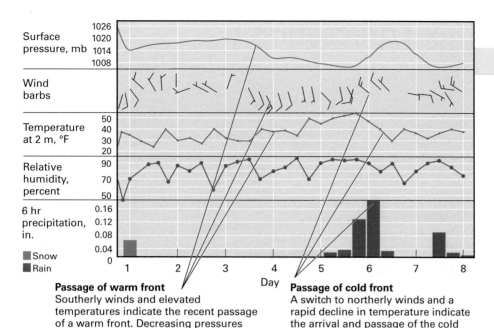

A meteogram takes the data from a numerical weather forecast and plots it for a specific location. This meteogram for Minneapolis shows the forecasted weather accompanying an approaching cyclone.

Passage of warm front
Southerly winds and elevated temperatures indicate the recent passage of a warm front. Decreasing pressures indicate the possible approach of a midlatitude cyclone and cold front.

Passage of cold front
A switch to northerly winds and a rapid decline in temperature indicate the arrival and passage of the cold front. Accompanying the cold front is heavy precipitation.

make a specific prediction of the weather for that location (**FIGURE 15.12**).

These forecasts represent estimates of what the weather will be like at certain times in the future. Because of differences in equations and models, as well as differences in the conditions used to start the models, called the *initial conditions*, model forecasts sometimes differ from one another. There are many model forecasts for the same time available from different atmospheric models. In the United States, the National Weather Service runs the Global Forecast System model (GFS), while the United Kingdom runs the United Kingdom Met Office model (UKMO), and European countries run the European Centre for Medium-Range Weather Forecast model (ECMWF). For regional forecasts, a forecaster can turn to the Weather Research and Forecasting model (WRF), developed and run by the National Weather Service, as well as others.

Often, it is difficult to determine which is the "better" model forecast until the weather event occurs, so a forecaster has to rely on experience with the models to determine which one to use. Forecasters can also take the average of the different forecasts to produce an

ensemble forecast A set of different weather forecasts for the same forecast time but generated from different models or from one model using different starting conditions.

ensemble forecast. These ensemble forecasts, which emphasize the features that the models agree on and deemphasize the ones they disagree on, typically represent the actual weather better than any one model in particular.

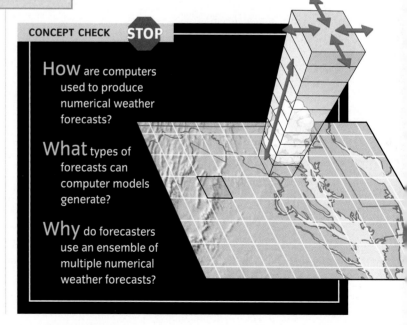

CONCEPT CHECK STOP

How are computers used to produce numerical weather forecasts?

What types of forecasts can computer models generate?

Why do forecasters use an ensemble of multiple numerical weather forecasts?

Operational Weather Forecasting

To see how a forecaster would develop a forecast for a specific region, we will look at a storm that hit Chicago, Illinois, on March 2–3, 2007. This storm brought heavy rainfall, thunderstorms, and high winds. It also produced tornadoes across a region stretching from southern Illinois to Alabama. Let's see how a forecaster might predict this storm's development.

USE OF SYNOPTIC CHARTS

Surface and upper-air synoptic charts provide the forecaster with valuable information about present weather conditions as well as those that may arise in the future. In the storm we're looking at, the surface pressure charts for February 28, 2007 (FIGURE 15.13) show generally light winds over the forecast region of

Current conditions FIGURE 15.13

A Surface pressures, fronts, and winds on February 28, 2007.

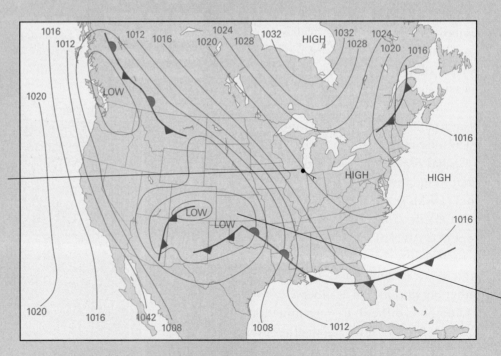

Current conditions
In the forecaster's region, winds are light—about 5 knots—but skies are cloudy. No apparent fronts are approaching.

Chicago, with no clear sign of strong frontal activity in the region. A simple **persistence forecast**, which predicts future conditions based on the present ones, would suggest little to worry about.

However, the upper-air chart provides three warning signs that bad weather may develop. First, the positioning of the upper-air trough to the *west* of the low centered in southern Colorado indicates that the storm may intensify over time. Second, the upper-air winds, which follow the isobars, show strong divergence over the region of the surface cyclone, which suggests rapid ascent of air leading to thick cloud cover and heavy precipitation. Last, the upper-air winds are positioned to steer a possible storm to the northeast, toward the Great Lakes. If the storm develops, it will move into the forecaster's region over the next 12–24 hours.

■ **persistence forecast** Forecast in which future weather conditions are predicted to be the same as present conditions.

■ **analog method** Method in which a forecast is made by examining the evolution of historical weather events that are similar to current events.

USE OF THE ANALOG METHOD

Once the potential for severe weather in a region has been identified, there are many ways a forecaster can go about deciding how that weather system will develop and how it might affect a particular region. One that has been in use for centuries is the **analog method**, which draws on a vast store of experience on the forecaster's part. Using this experience, the forecaster finds similarities between the synoptic charts for the current conditions and conditions in the past. Based on how the past conditions developed over time, the forecaster makes an educated estimate of how present conditions might develop over the next 24 hours. Unfortunately, while there may be similarities between the synoptic conditions today and those from previous days, the similarities are never exact.

B 500 mb heights and winds on February 28, 2007. Upper-air temperatures indicated by the red, dashed lines.

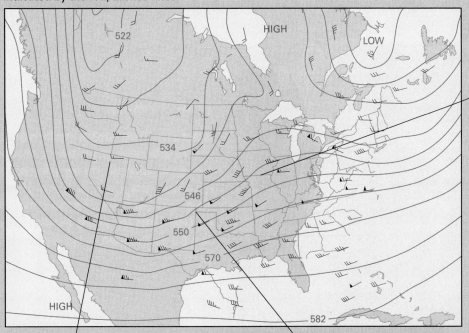

Steering winds
The 500 mb winds indicate that the midlatitude cyclone might move to the northeast and into the forecaster's region.

Position of surface and upper-air troughs
The positioning of the upper-air trough relative to surface pressure patterns indicates that the storm may intensify over time.

Upper-air divergence
The spreading of isolines, indicating upper-air divergence, leads to ascent of air from below that may produce condensation and heavy precipitation.

USE OF NUMERICAL FORECASTS

Instead of searching for analogs, the forecaster may turn to the prognostic charts produced by the numerical weather models. The forecaster first examines the individual forecasts from many different model runs. The models may all show the upper-air trough and the surface pressure system intensifying over the next 24 hours, just as the forecaster suspects.

However, each model may show a slightly different orientation and track for the storm and its fronts. For example, one model may indicate that the storm will track south of the forecaster's region, with only light precipitation accompanying the passage of the surface low pressure and no strong frontal precipitation. Another model, however, may show the storm passing directly over the forecaster's region, placing the region right in the path of the intensifying cold front, which could bring significant precipitation, strong winds, and possibly even lightning, thunderstorms, or tornadoes.

How does the forecaster determine which model to emphasize? In some cases, the forecaster will use the model that has performed the best in predicting similar storm development. Weather service providers, like the National Weather Service, keep detailed studies of how each model performs and what types of problems each has. The forecaster is aware of these studies and can make an informed decision based on the likelihood that one model will be better than another.

In addition, after many years of experience, the forecaster may simply have a preference for one model over another based on the best prior agreement with the weather patterns that eventually develop. Sometimes, the weather forecaster has a sense of how the weather will develop based on previous storms that have passed over the region and may decide to use the numerical forecast that shows a development similar to those previous situations. Finally, the forecaster can use an ensemble prediction, which is an average of all the different model forecasts available. Ultimately, the forecaster will make a decision based on all of these factors.

After selecting a particular model forecast, the forecaster then uses the additional data from the model to predict the timing and severity of the coming weather. Based on the model's atmospheric temperature and humidity profiles, for instance, the forecaster can determine whether the atmosphere is unstable enough to generate large-scale convection and possibly lightning and thunderstorms. These elements and others are ultimately incorporated into the forecaster's prediction of the weather over the next 24 hours (**FIGURE 15.14**).

24-hour forecast FIGURE 15.14

Based on multiple weather model forecasts and the forecasters' experience, a 24-hour forecast is produced. The low-pressure system shown here is forecast to intensify and move northeast, producing heavy snow over the Great Lakes and severe thunderstorms over the southeastern United States.

USE OF LOCAL OBSERVATIONS

Let's move forward 24 hours (FIGURE 15.15). The current synoptic charts and satellite images show that the weather system has developed into a storm with strong frontal activity, possibly producing severe weather along a front extending from Illinois to Texas. Just how severe is the weather associated with the front, and how might it affect Chicago? To answer this, the forecaster turns to local observations from the regions affected by the fronts.

Radar images show that conditions are presently clear at the forecaster's region of Chicago along the banks of the Great Lakes. However, very strong rainfall has developed in a narrow band in the vicinity of the front. Time-lapse images would show this band of rain slowly moving northeast, possibly bringing with it lightning and gusty, turbulent winds.

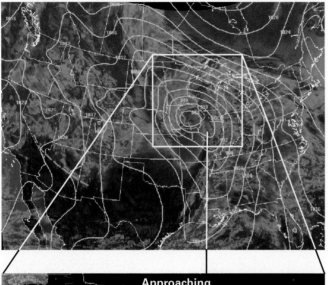

A Synoptic charts and satellite images
Infrared images from satellites indicate strong convection over many parts of the eastern United States—shown in red and yellow colors—while clear conditions—shown in blues—prevail behind the storm. Superimposed synoptic charts show the location of the low-pressure system and the warm, cold, and occluded fronts.

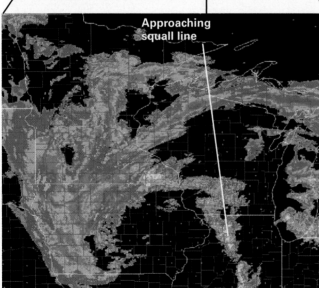

Approaching squall line

B Radar images
Local radar images indicate the formation of very intense precipitation along a squall line as the front approaches.

Satellite and radar images of an approaching storm FIGURE 15.15

WATCHES, WARNINGS, AND ADVISORIES

What can the forecaster do at this point? Typically, the forecaster will post a text-based **short-range forecast** that summarizes the predicted weather for the region over the next 48 hours or less. A forecaster can also produce a **nowcast** covering the next few hours, as well as an **extended-range forecast** that may go out as far as 5–10 days. These are part of the routine forecasts that weather forecasting services put out four times daily.

If the weather forecast is severe enough, the forecaster can put out watches, warnings, and advisories. Generally, watches are less dire than warnings, although both have distinct meanings

■ **short-range forecast** A weather forecast made for a time period of up to 48 hours.

■ **nowcast** A short-term weather forecast, generally for the next few hours.

■ **extended-range forecast** A forecast of weather conditions for a period extending beyond 3 days.

depending on the weather phenomenon that is being predicted. For example, a "severe thunderstorm watch" means that thunderstorms with winds higher than 92 km/hr (57 mph) and/or hail larger than 18.75 mm (3/4 in.) are possible, while a "severe thunderstorm warning" indicates that severe thunderstorms have been sighted or have been identified in the region on radar. In the case we have been discussing here, the weather service issued a severe thunderstorm watch for the area around Chicago. The text of this watch is provided in FIGURE 15.16.

Weather service providers can also issue special weather statements. These are not as formal as watches or warnings, but are intended to provide information to the public on the possible ap-

```
SEL7

    URGENT - IMMEDIATE BROADCAST REQUESTED
    SEVERE THUNDERSTORM WATCH NUMBER 47
    NWS STORM PREDICTION CENTER NORMAN OK
    1125 AM CST THU MAR 1 2007

    THE NWS STORM PREDICTION CENTER HAS ISSUED A
    SEVERE THUNDERSTORM WATCH FOR PORTIONS OF

         MUCH OF NORTHERN ILLINOIS
         NORTHERN INDIANA
         SOUTHWESTERN LOWER MICHIGAN
         LAKE MICHIGAN

    EFFECTIVE THIS THURSDAY MORNING AND EVENING FROM 1125 AM UNTIL 700 PM CST.

    HAIL TO 2 INCHES IN DIAMETER...THUNDERSTORM WIND
    GUSTS TO 70 MPH...AND DANGEROUS LIGHTNING ARE POSSIBLE IN THESE
    AREAS.

    THE SEVERE THUNDERSTORM WATCH AREA IS APPROXIMATELY ALONG AND 70 STATUTE
    MILES NORTH AND SOUTH OF A LINE FROM 55 MILES NORTH OF PEORIA
    ILLINOIS TO 40 MILES NORTH NORTHEAST OF FORT WAYNE INDIANA.  FOR A
    COMPLETE DEPICTION OF THE WATCH SEE THE ASSOCIATED WATCH OUTLINE
    UPDATE (WOUS64 KWNS WOU7).

    REMEMBER...A SEVERE THUNDERSTORM WATCH MEANS CONDITIONS ARE FAVORABLE FOR
    SEVERE THUNDERSTORMS IN AND CLOSE TO THE WATCH
    AREA. PERSONS IN THESE AREAS SHOULD BE ON THE LOOKOUT FOR
    THREATENING WEATHER CONDITIONS AND LISTEN FOR LATER STATEMENTS
    AND POSSIBLE WARNINGS. SEVERE THUNDERSTORMS CAN AND OCCASIONALLY
    DO PRODUCE TORNADOES.

    OTHER WATCH INFORMATION...CONTINUE...WW 42...WW 43...WW44...WW45...WW46...

    DISCUSSION...POWERFUL TROUGH AND DEEP UPPER LOW ROTATING TOWARD UPPER
    MS VALLEY. LN OF LOW TOPPED THUNDERSTORMS ARE DEVELOPING ALONG COLD FRONT
    WRN IL AND WILL MOVE RAPIDLY NEWD ACROSS WATCH AREA THIS AFTERNOON. PRIMARY
    SEVERE THREAT WILL BE DAMAGING WINDS AND HAIL ASSOCIATED WITH THE LINE AS IT
    MOVES ACROSS NRN IL INTO NRN IN AND SWRN MI.

    AVIATION...A FEW SEVERE THUNDERSTORMS WITH HAIL
    SURFACE AND ALOFT TO 2 INCHES. EXTREME TURBULENCE AND SURFACE
    WIND GUSTS TO 60 KNOTS. A FEW CUMULONIMBI WITH MAXIMUM TOPS TO
    400. MEAN STORM MOTION VECTOR 22040.


    ...HALES
```

Severe thunderstorm watch for March 1, 2007

FIGURE 15.16

This severe thunderstorm watch was issued for the storm approaching northern Illinois and the surrounding regions.

proach of severe weather. The statements usually accompany text that describes the possible weather changes, as well as reasons the statements are being issued and possible precautionary measures to take.

VALIDATION

Eventually, the **validation time** arrives—the time for which the forecast was made. The forecaster can now judge how good the forecasts for the region were. In

the case of the low-pressure system shown in **FIGURE 15.17**, the forecaster was right to be concerned. The storm tracked just northwest of Chicago, bringing with it heavy rain and hail, followed by damaging winds that knocked out power lines and toppled trees.

Overall, over a hundred severe weather reports were issued stretching from Missouri into Georgia. Tornadoes were reported across this region, along with severe thunderstorms and hail up to 19 mm (3/4 in.) across.

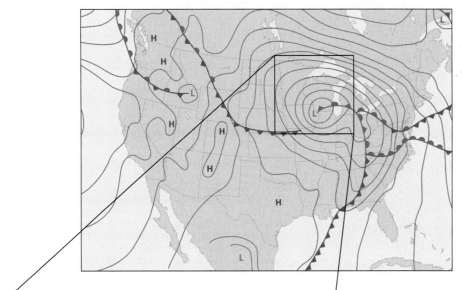

A The surface chart for March 2, 2007 indicates that the low-pressure center lies just west of Chicago.

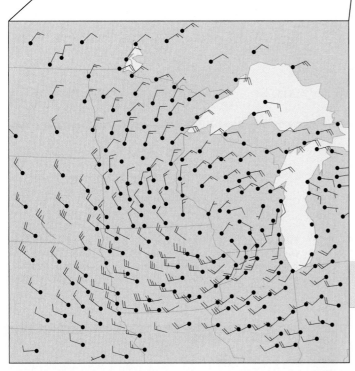

B Surface winds of up to 30 knots (56 km/hr) circling around the midlatitude cyclone disrupted power by downing power lines and affected flights into and out of local airports.

Surface weather chart and wind map detail
FIGURE 15.17

In this case, the forecast was correct. Not all forecasts turn out this well, though, and some can be very inaccurate. To determine how good a forecast is, the forecaster must validate it against the actual weather conditions. There are many ways of *validating*, or comparing, a forecast with the actual weather conditions. Usually, these are some measure of the forecast **skill**.

Over the last 50 years, weather forecasting has improved markedly with the advent of faster computers and improvement in our understanding of the atmosphere. However, despite improving technology, the skill of weather forecasts is limited after a certain point in time, due in part to **chaos** within the atmospheric system—a situation in which small disturbances can develop unpredictably.

Because the atmosphere is chaotic, slight differences in the starting conditions for model forecasts eventually grow over time, so that forecasts become significantly different from one another. It is impossible to determine which forecast is the correct one. The forecaster has reached the natural limit of how far into the future forecasts can be made. The role chaos plays in limiting the skill of weather forecasts is discussed more in FIGURE 15.18.

While there are natural limits to forecasting skill based on the chaotic nature of the atmosphere, there are also practical limits. One of these arises because not every aspect of the atmosphere can be modeled by the very powerful computers used to produce numerical forecasts. Some aspects of the atmospheric system—for example, cloud formation or winds over very mountainous terrain—occur on spatial scales that are too fine for a model to capture. Other processes—such as cloud condensation—are not understood well enough to program. These limitations of the models

skill A measure of the accuracy of a forecast based on how it compared with the actual weather conditions at the forecast time.

chaos In meteorology, a situation in which small differences in two weather forecasts grow over time, leading to significant differences past some point.

▼ Historical weather forecasting skill

This figure shows how forecasts of 500 mb heights over the northern hemisphere have improved with time. 100 percent would be a perfect score, while 0 percent would indicate no forecast skill. Generally, the skill of 3-day forecasts have improved over time. In addition, long-range forecasts, out to 7 days, have also improved, although long-range forecasts always have lower skill than short-range forecasts. This is a natural consequence of chaotic behavior in the atmosphere.

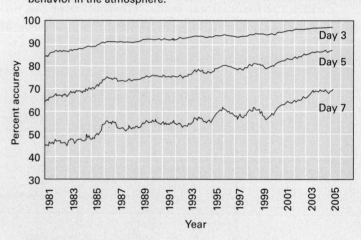

▼ Origins of chaos

Part of the reason long-term forecasts are less accurate than short-term forecasts has to do with the presence of chaos within the atmospheric system. For example, predicted changes in 500-mb heights at a given location from two model runs with slightly different initial conditions show how chaos can cause forecasts to deviate farther and farther from one another.

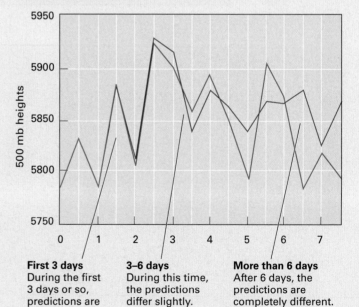

First 3 days
During the first 3 days or so, predictions are very similar.

3–6 days
During this time, the predictions differ slightly.

More than 6 days
After 6 days, the predictions are completely different.

Chaos is a situation in which a meteorological variable such as temperature or wind speed exhibits erratic behavior as very small changes in the initial state of the system rapidly grow over time, leading to unpredictable changes. Many systems experience chaotic behavior. It is a particularly important component of the atmosphere and limits how far into the future we can make weather forecasts.

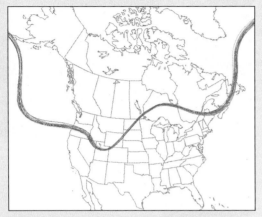

Initial conditions
Each model run is shown in a different color. Initially, all the models have almost identical jet streams.

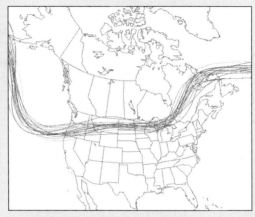

4-day forecast
The models still have very similar jet stream locations with only slight differences.

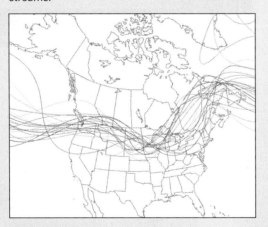

8-day forecast
By the eighth day of the forecast, a few models have very different jet stream locations, although there is still some similarity between many of them.

12-day forecast
Twelve days into the forecast, all the models show different jet stream locations. Each model will produce different weather for a specific location, making it impossible for a forecaster to make a prediction for that region.

▲ Chaos and the jet stream
By running the same computer forecast model many different times with only very slight differences, it is possible to see the effects of chaos. As the forecast time gets longer, chaotic behavior begins to appear in the forecasts, producing "spaghetti" plots in which each forecast produces a very different result from the others. Each of these simulations comes from the same model, with only slightly different initial conditions. The different results for the 12-day forecast make it impossible to decide which is the right one.

introduce uncertainties and differences between the model forecasts and the observations that, like chaotic differences, grow over time.

In addition to computational limitations of the models, there are also observational limitations. For computer models to run successfully, they need to be given data on the present atmospheric conditions. However, even the thousands of observations that are taken around the globe on a daily basis cannot completely describe the atmosphere. There are always characteristics of atmospheric circulation that are left out of the models but exist in the real atmosphere. Again, because the model's description of the atmosphere does not exactly match the real atmosphere, the model will produce forecasts of the atmospheric evolution that eventually will differ noticeably from the actual atmospheric evolution.

Because of these limitations, the skill of forecasts does not generally match the theoretical limit set by chaos alone. Thus, the practical limit of forecasting skill over the west coast of the United States is only 3–4 days. These regions have very active weather patterns in which slight variations in the atmospheric circulation can grow quickly over time. In contrast, in equatorial regions where there is no synoptic storm activity, forecast times are better. There, persistent patterns of rainfall or temperatures can last for quite a while, allowing a forecaster to make much more accurate forecasts farther into the future.

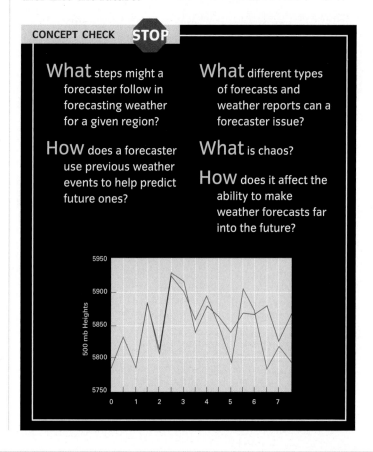

CONCEPT CHECK STOP

What steps might a forecaster follow in forecasting weather for a given region?

How does a forecaster use previous weather events to help predict future ones?

What different types of forecasts and weather reports can a forecaster issue?

What is chaos?

How does it affect the ability to make weather forecasts far into the future?

What is happening in this picture ?

This emergency responder is checking local weather conditions near an active forest fire. The data he is collecting will be used to predict the behavior of the fire over the next 20 minutes to 24 hours. Forecasts can be used to determine whether conditions are right for dropping firefighters into the middle of a blaze and whether to issue evacuation orders for nearby residents.

What type of weather conditions do you think he is trying to forecast?

SUMMARY

1 Local and Synoptic Weather Forecasting

1. Local weather forecasting relies on the use of local weather conditions to help make predictions of changes in weather over the coming 6–12 hours. Local conditions used for this type of forecasting include prevailing wind direction, cloud formations, atmospheric soundings, and changes in pressure over the previous 3–6 hours.

2. **Synoptic weather forecasting** relies on the use of **synoptic charts** depicting both local and regional weather conditions at the surface and aloft. In addition, synoptic weather forecasting makes use of local conditions provided by radar and large-scale conditions provided by satellites. In combination, these charts and maps allow the forecaster to make longer-range forecasts with better accuracy than those based on local conditions alone.

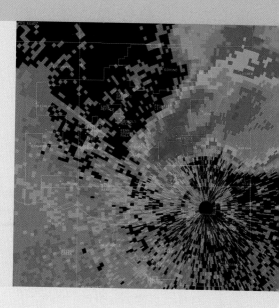

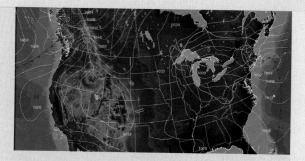

2 Numerical Weather Forecasting

1. **Numerical weather forecasts** rely on the use of atmospheric models to make predictions about the change in weather over the range of hours to days. Mathematical models simulate the physical processes occurring in the actual atmosphere and require high-speed computers to make the millions of calculations necessary for these types of predictions. Sometimes multiple models are all run at once, and the results are compared with one another to come up with a more reliable forecast.

2. Once numerical weather predictions are made using a computer, they can be visualized in different ways. Forecasters can print out **prognostic charts**, which are similar to synoptic weather charts except that they show maps of pressures, temperature, winds, and precipitation for the forecast period in the future. In addition, forecasters can produce **meteograms**, which show how pressure, temperature, winds, and precipitation will change over time at a given location.

3 Operational Weather Forecasting

1. To produce operational forecasts, a weather forecaster uses many different tools. Initial forecasts can be based on comparing current synoptic charts with previous charts. Using the **analog method**, the forecaster can then predict how current conditions might evolve based on how previous weather systems developed.

2. Weather forecasters can then turn to numerical weather forecasts. Because different models can produce different forecasts, and the same model can produce different forecasts given slightly different starting conditions, the forecaster must still rely on experience to determine which models to follow and which models to disregard.

3. If the forecaster feels that inclement weather is possible for a given region, the forecaster can then issue various types of forecasts. These include **nowcasts** (weather forecasts up to 6 hours in the future), **short-range forecasts** (up to 48 hours into the future) and **extended-range forecasts** (from 3 to 10 days into the future). In addition, the forecaster can issue warnings, watches, and advisories at any time based on how severe the weather is expected to be.

4. The time for which the forecast is made is called the **validation time**. At this point, a forecast can be compared with the actual weather to determine how accurate it was. Generally, forecasts have been improving over the decades with the advent of new technologies. However, there is a limit to how far ahead weather forecasts can be made, because the atmosphere is a chaotic system in which small disturbances can develop unpredictably.

KEY TERMS

CRITICAL AND CREATIVE THINKING QUESTIONS

1. If you wanted to predict the weather in your region but did not have access to a computer or any other source of information other than what you could measure or observe at your house, what types of measurements or observations would you make? What would these tell you about changes in the weather in your region?

2. What type of information about the weather can be obtained from ground-based radar? What type of information can be obtained from space-based satellites? Besides the different types of weather information, what are some other differences between the data gathered from satellites and from radar?

3. What information can weather forecasters gain by looking at the 500 mb winds? If they compare the 500 mb winds and pressures with surface winds and pressures, what additional information can they get?

4. What weather features can forecasters get from numerical models? What are the different ways that weather forecasters can visualize these data? If you were in Chicago and wanted to know how your weather compared with the weather in California, what type of forecast would you want? If you wanted to know how the weather at 11:00 A.M. compared with the forecasts at 2:00 P.M., what type of forecast would you want?

5. How does chaos in the atmospheric system affect our ability to make weather forecasts? What methods do forecasters use to improve their forecasts given that the atmosphere is chaotic?

SELF-TEST

1. Low-pressure centers always continue to travel in the same direction they traveled over the last 24 hours.
   a. True b. False

2. If you wanted to determine the possibility of cloud formation and precipitation in a given region, you would look at

   _____.
   a. sea-level pressure maps
   b. 500 mb wind maps
   c. 700 mb relative humidity maps
   d. 850 mb temperature maps

3. If you wanted to determine whether precipitation is likely to fall as snow or rain in a given region, you would look at

   _____.
   a. sea-level pressure maps
   b. 500 mb wind maps
   c. 700 mb relative humidity maps
   d. 850 mb temperature maps

4. Radar images can be used to measure

   _____.
   a. winds and precipitation
   b. temperature and precipitation
   c. temperature and pressure
   d. winds and pressure

5. Using the accompanying diagram, determine the following: (a) wind direction; (b) wind speed; (c) temperature; (d) dew point; (e) cloud cover; (f) sea-level pressure; (g) weather conditions.

6. Storms tend to travel _____ to the 500 mb winds at about _____ speed.
   a. perpendicular; twice the
   b. 45-degrees; the same
   c. parallel; half the
   d. opposite; the same

7. To make a numerical weather forecast requires approximately _____ of separate calculations.
   a. hundreds
   b. thousands
   c. tens of thousands
   d. tens of millions

8. To increase their confidence in numerical weather forecasts, forecasters sometimes run many forecasts for the same conditions, which is called a(n) _____.
   a. ensemble
   b. protectorate
   c. isollabar
   d. steerer

9. To show how forecasted weather conditions at a particular site will change over time, forecasters use a _____.
   a. prognostic chart
   b. synoptic chart
   c. meteogram
   d. wind barb

10. Using the accompanying diagram, determine: (a) the forecast day for the arrival of the warm front; (b) the forecast day for the arrival of the cold front; (c) the day with heaviest snowfall; (d) the day with coldest temperatures.

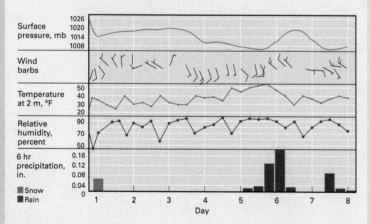

11. A method of weather forecasting based on the behavior of historical weather systems similar to the current one is called the

   _____.
   a. ensemble method
   b. persistence method
   c. analog method
   d. numerical method

12. A forecaster can issue _____.
    a. a severe thunderstorm watch
    b. a severe thunderstorm warning
    c. special weather statements
    d. all of the above

13. The time for which a forecast is made is called the _____.
    a. validation time
    b. expectation time
    c. prediction time
    d. assessment time

14. Generally, the skill of weather forecasts has _____ with time; however, long-term forecasts are still _____ than short-term forecasts.
    a. stayed the same; worse
    b. stayed the same; better
    c. improved; worse
    d. improved; better

15. On the accompanying diagram, indicate the following: (a) the forecast jet stream that would bring storms to northern California; (b) the forecast jet stream that would bring storms to the west coast of Canada; (c) the location over western North America where most models predict storms will go; (d) the jet stream that would bring storms the farthest south over eastern North America.

Human-Induced Climate Change and Climate Forecasting

Polar bears are perennial inhabitants of the far northern reaches of the globe. They can travel across arctic sea ice, water, and continental coastlines. During summer, when sea ice has broken up, polar bears are found over the continental margins of North America, Eurasia, and Greenland. In winter, as ice forms along the coastal margins and over the Arctic Ocean, polar bears can venture over great distances. Winter is when they do most of their hunting and feeding, but recent warm winters have meant shorter ice-covered seasons, as well as weaker sea ice. These conditions have limited the polar bears' hunting grounds during the winter, as well as their ability to hunt during the late winter and early spring, when they must find most of their food before the breeding season. As the sites of permanent snow cover decrease, the summertime habitat of the polar bears has also been shrinking.

Over the last 100 years, arctic temperatures have risen by 5°C (9°F), while the extent of the wintertime sea ice has decreased by 6 percent over the past 20 years. During the same time, the body mass of polar bears has declined by 15–20 percent. With projections of future climate change indicating a further erosion of wintertime sea ice of up to 50 percent over the next 100 years, there are now concerns that polar bears will run out of the wintertime hunting grounds needed to sustain wild populations.

A mother polar bear and her two-year-old cub on an iceberg in the open sea.

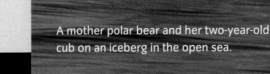

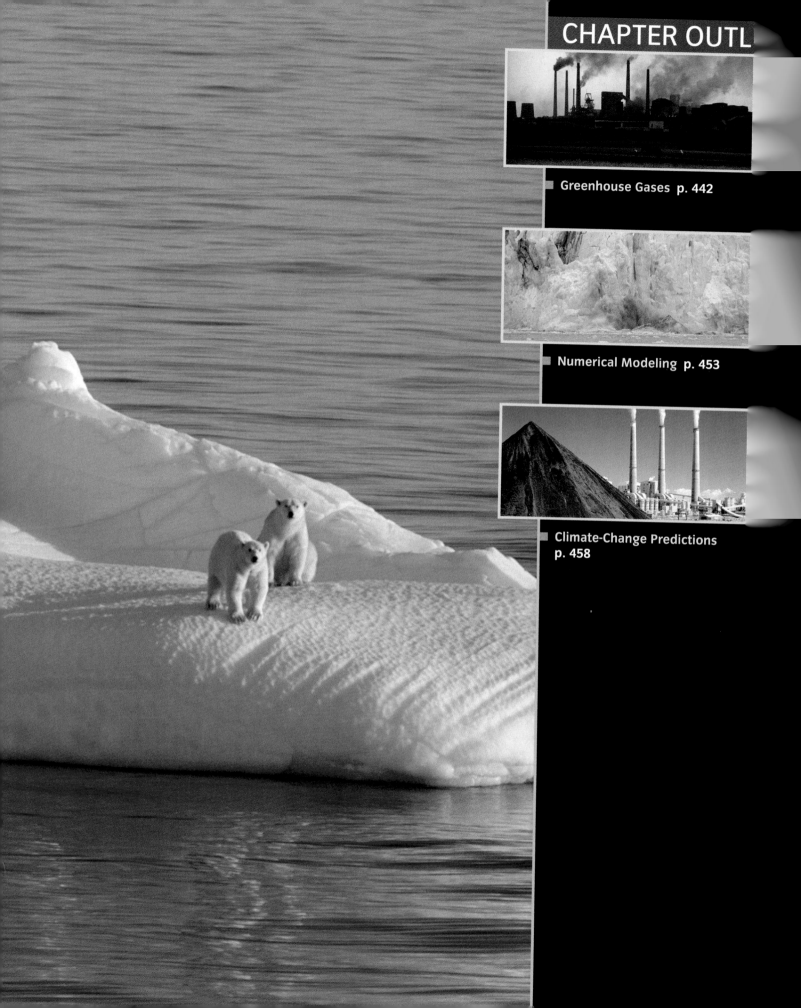

Greenhouse Gases

Identify the gases released through human activity that enhance the greenhouse effect.

Explain the relationship between emissions and concentrations of greenhouse gases.

Describe the factors to consider when determining the future emissions of greenhouse gases by humans.

uman activity has greatly affected both past and present climate. Humans have always endeavored to change the climate on a small scale to make life more hospitable and productive. For example, irrigation makes dry climates better for agriculture by importing water, making the climate more moist. Other human activities have also changed the climate in certain regions—the urban heat island effect is a prominent example. In that case, cities develop to support trade, commerce, and administration, but also change the land-surface characteristics that in turn change the climate of the overlying atmosphere. We use the term **anthropogenic climate forcing** to describe climate change induced by human activity.

anthropogenic climate forcing
Change in climate resulting from human activities at local to global scales.

Emissions of gases and aerosols FIGURE 16.1

Human activity around the globe has significantly altered the chemical composition of the atmosphere, adding gases and aerosols that remain for days to centuries.

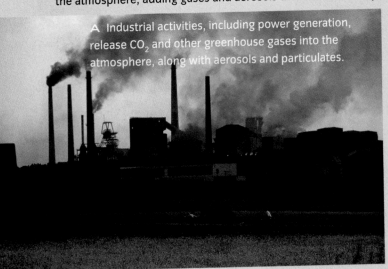

A Industrial activities, including power generation, release CO_2 and other greenhouse gases into the atmosphere, along with aerosols and particulates.

B Cars and trucks release CO_2 as well as smog-producing gases. Diesel engines also produce aerosols.

C Fires release CO_2 from storage in woody plant matter and also contribute aerosols.

D Cattle manure is a significant source of methane, an important greenhouse gas.

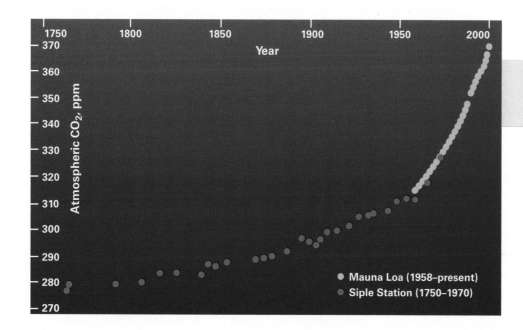

Atmospheric concentrations of carbon dioxide over the last 250 years, as estimated from ice cores (Siple Station) and direct measurements (Mauna Loa Observatory).

This chapter looks at the ways that human activity is changing the climate of the whole Earth, not only particular isolated regions. Humans are now responsible for a type of climate forcing, just like El Niños or volcanoes. Anthropogenic climate change is a side effect of nearly every human activity—driving to the grocery store, watching the Superbowl, growing food, even sleeping, particularly if your building is heated in the winter or cooled in the summer (FIGURE 16.1).

Unlike El Niño forcing, however, anthropogenic forcing occurs over longer time spans—decades to hundreds of years—and is based on the slow accumulation of human-produced gases in the atmosphere that eventually affect the natural greenhouse effect. These gases are referred to generically as *greenhouse gases* —also called *heat-trapping gases*. This chapter looks at where greenhouse gases come from and the effect they have on the climate.

CHANGING CHEMICAL COMPOSITION OF THE ATMOSPHERE

Although there is debate about the extent to which humans are changing the global climate, we have most certainly changed the **chemical composition** of the atmosphere.

chemical composition The amount and relative abundance of chemical elements and molecules within a given volume of air.

Before about 1800, the chemical constituents in the atmosphere were in equilibrium with the ocean/atmosphere/biosphere system, and their concentrations remained nearly constant over time. However, with the onset of industrial activity and the increase of agriculture, there has been a dramatic rise in certain constituents, such as CO_2.

Greenhouse gases The rise in certain chemical constituents is unequivocally tied to human activity. For example, the CO_2 (FIGURE 16.2) increase is documented in the records of emissions from fossil fuel burning and cement production, changes in the fraction of CO_2 *isotopes*—CO_2 molecules with different numbers of neutrons—in the atmosphere, and geographic distributions of CO_2 concentrations around the globe.

Carbon dioxide is not the only important greenhouse gas in the atmosphere. Methane (CH_4) occurs naturally and is emitted from saturated land surfaces, such as wetlands, and by animals, such as termites and cattle.

At present, these natural sources together account for about 20 percent of methane emissions. Today, most emissions of methane (about 60 percent) are related to managed agriculture and activities associated with it, including deforestation and landfills.

The other 20 percent of methane emissions is associated with fossil fuel burning and industrial activity.

Anthropogenic emissions of methane have increased atmospheric concentrations from 750 ppb (parts per billion) to about 1750 ppb over the last 100 years, which represents at least a doubling of the natural concentrations (FIGURE 16.3). Although its overall concentration is a thousand times smaller than the concentration of CO_2, methane is a very strong absorber of longwave radiation, making it a very efficient greenhouse gas. The increase in methane concentration over the last 100 years has added 0.5 W/m^2 of extra energy to the climate system, about a third of the amount added by the increase in CO_2 concentrations over the same time.

Another important chemical constituent is nitrous oxide (N_2O), which has also been increasing. Like methane, anthropogenic emissions of N_2O are mainly associated with agricultural activity, although some N_2O comes from industrial activity. Anthropogenic emissions have increased the concentrations of N_2O in the atmosphere by about 15 percent, while adding an extra 0.15 W/m^2 extra energy to the climate system—about one-tenth the amount added by CO_2 over the same time period.

Aerosols
Aerosols—liquid or solid particles suspended in the atmosphere—can also produce large-scale changes in the climate of the Earth (FIGURE 16.4). Natural and anthropogenic sources of aerosols include *dust* from deserts, dry lake beds, agricultural fields, and semiarid regions. Deforested regions and other areas in which the natural land cover has been disturbed can add to the dust content of the atmosphere, as can industrial dust from manufacturing and coal burning.

Another type of aerosol—*organic matter*—is released through inefficient burning of fossil fuels as well as the burning of vegetation after the clearing of forests—called *biomass burning*. Natural sources of these aerosols are the oxidation of plant debris and microbes such as bacteria, fungi, and algae.

Inefficient fossil fuel combustion and biomass burning also release another important aerosol—*black carbon*. These aerosols are almost entirely anthro-

pogenic, with few natural sources other than forest fires or large volcanic eruptions.

Finally, aerosols include *sulfate particles* that are formed when sulfur dioxide gas, SO_2, combines with liquid water and oxygen to form aerosol particles of

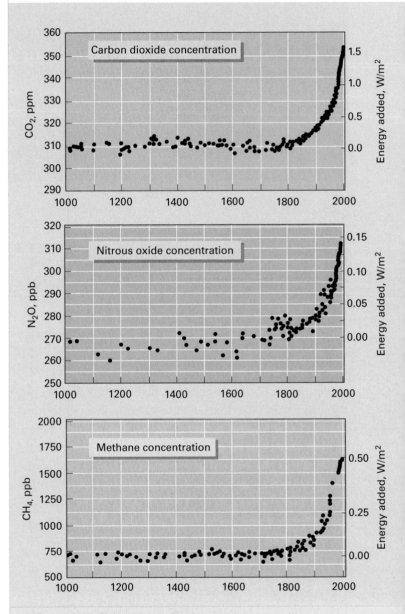

Other greenhouse gases FIGURE 16.3

Atmospheric concentrations of carbon dioxide, nitrous oxide, and methane over the last 1000 years. Shown on the right axis is the additional energy added to the climate system as a result of the increased gas concentrations.

A Dust In this picture, dust produced by agricultural activity appears white against the underlying fields. These aerosols reflect more sunlight, increasing the global albedo.

B Black carbon Inefficient combustion of fossil fuels, produced here during the refining of oil, results in aerosols that are dark and highly absorbing of incoming solar energy.

Aerosols and their impact FIGURE 16.4

C Cloud condensation nuclei Aerosols act as cloud condensation nuclei. Increases in these aerosols result in smaller, brighter cloud droplets. Here, aerosols emitted by ships produce tracks of clouds that can be seen from space.

liquid sulfuric acid. SO_2 is released by fossil fuel burning as well as by volcanic eruptions. Another source of SO_2 is marine plankton.

What role do aerosols play in changing the global climate? Some aerosols, like dust, organic carbon, and sulfate particles, cool the regional and global climate by increasing the albedo of the atmosphere, thereby reflecting more incoming solar radiation back to space.

This process occurs after large volcanic eruptions that inject SO_2 and sulfate aerosols high into the atmosphere.

Other aerosols, such as black carbon aerosols, have the opposite effect—they increase the absorption of incoming solar radiation within the atmosphere. The extra absorption increases the temperature of the atmosphere, which can then radiate more longwave energy to the surface, thereby warming the surface. By decreasing the overall albedo of the globe, black carbon aerosols tend to warm the planet.

A final impact involves aerosol interactions with clouds. When clouds form, the water condenses around *cloud condensation nuclei*—tiny centers of solid or liquid matter. Aerosols of all types can serve as cloud condensation nuclei. However, when more nuclei are present in the atmosphere, fewer water molecules condense on each individual cloud droplet.

These smaller cloud droplets tend to be brighter and more reflective than larger cloud droplets. If aerosols increase around the globe, clouds will be composed of smaller but brighter cloud droplets. The effect will be to raise the albedo of the clouds and thus the albedo of the Earth system. While the strength of this effect is still uncertain, it does appear to occur, particularly for liquid-water clouds in the low latitudes and near the Earth's surface.

SOURCES AND SINKS OF ANTHROPOGENIC CO₂

Atmospheric carbon dioxide makes up less than 2 percent of all the carbon in the Earth system. The atmo-

spheric pool is supplied by plant and animal respiration in the oceans and on the lands, outgassing volcanoes, and fossil fuel combustion. Plants remove carbon dioxide from the atmosphere by photosynthesis. On land, this carbon dioxide can accumulate as peat deposits under cold and wet conditions.

In the oceans, atmospheric carbon dioxide dissolves into ocean waters, where it is used in photosynthesis by phytoplankton to build skeletal structures of calcium carbonate. These skeletons settle to the ocean floor to accumulate as layers of sediment. The result is an enormous carbon storage pool, but it is not available to organisms until it is later released by rock weathering. Organic compounds synthesized by phytoplankton also settle to the ocean floor and eventually are transformed into the hydrocarbon compounds that make up petroleum and natural gas.

Humans burn petroleum, natural gas, and coal for energy, which alters the natural carbon cycle by releasing CO_2, as seen in FIGURE 16.5. The increase in carbon emissions started in the mid-1800s with the advent of the industrial revolution and large-scale use of coal as an energy source. Recently, however, emissions from burning petroleum-based fuels have superseded emissions from coal. During the 1990s, the average emissions from fossil fuel burning were 6.5 billion met-

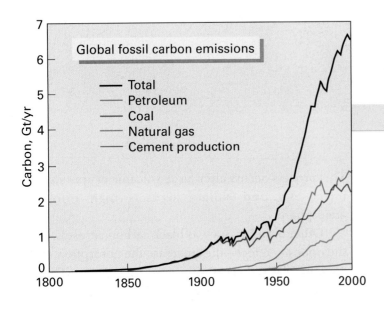

Sources of CO₂ emissions FIGURE 16.5

CO_2 emissions have increased exponentially since about 1850 with the start of the industrial revolution. Originally, most CO_2 emissions came from burning coal. Now, however, the burning of petroleum as liquid fuel is the major source of CO_2 emissions.

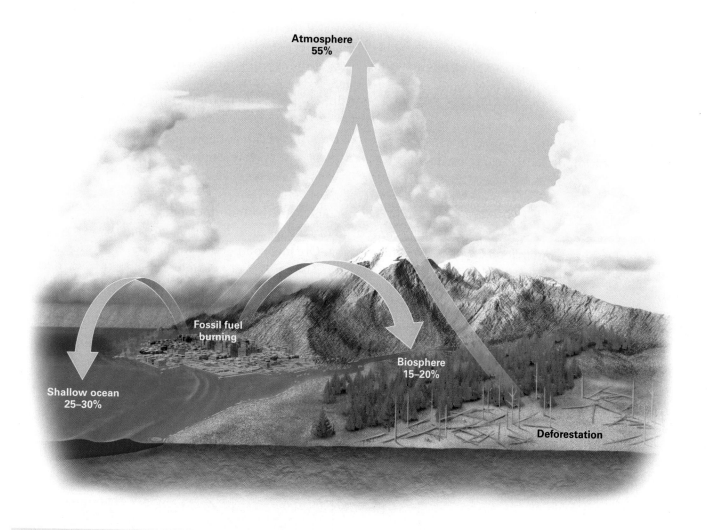

Fate of anthropogenic CO_2 emissions FIGURE 16.6

Humans emit approximately 6.5 billion metric tons of carbon per year through fossil fuel usage and another 2.5 billion metric tons per year through deforestation. Of this, about half—55 percent—stays in the atmosphere, while 15–20 percent is incorporated into the terrestrial biosphere through increased photosynthesis, and 25–30 percent is incorporated into the ocean through absorption and biological uptake.

ric tons of carbon per year (also called a gigaton of carbon or GtC).

Another important human effect on the carbon cycle lies in our alteration of the Earth's land cover—for example, by clearing forests or cultivating agricultural areas. Deforestation is estimated to have resulted in an average emission to the atmosphere of 2.5 billion metric tons of CO_2 per year.

What happens to the extra carbon that fossil fuel burning and deforestation liberate? About half remains in the atmosphere (FIGURE 16.6). A smaller portion (about 2.5 GtC/yr) is absorbed by the oceans, with the rest flowing into the biosphere—about 1.5 GtC/yr. The net impact of present human activity is to increase the amount of CO_2 in the atmosphere by about 5 billion metric tons each year.

RESIDENCE TIME

How long do anthropogenic emissions of greenhouse gases remain in the atmosphere? The **residence time** is the time it takes for an increment of additional emissions to be reduced or removed by other processes. The residence time of CO_2 is about 100–150 years, so today's concentration depends strongly on

the amounts released in prior years. It also means that even if the rate at which we release carbon into the atmosphere stops growing and remains steady, CO_2 concentrations will continue to grow. As concentrations continue to increase, so will the greenhouse heating of the climate system. FIGURE 16.7 describes in more detail the relation between greenhouse gas emissions and concentrations.

Process Diagram

Relation between emissions and concentrations FIGURE 16.7

VIEW THIS IN ACTION in your WileyPLUS course

Because CO_2 and other greenhouse gases are not removed immediately from the atmosphere, they tend to accumulate over time. The length of time a gas stays in the atmosphere is called its *residence time*. In the case of CO_2, the residence time is about 100–150 years, so the concentration of CO_2 in today's atmosphere depends on the emissions over the past 100–150 years.

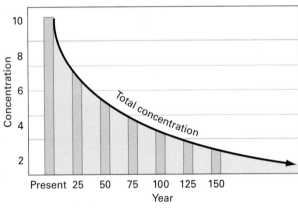

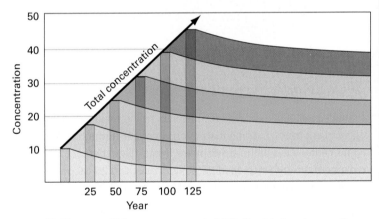

A If 10 units of CO_2 are emitted and nothing else is added, the atmospheric concentration of CO_2 will take around 100–150 years to decrease to the level at which it initially started.

B However, if the same amount of CO_2 is added each year, the concentration is the sum of the current emissions plus a fraction of all the CO_2 that has been added over the previous years.

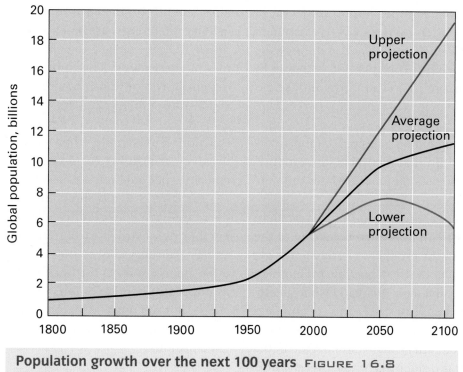

Population growth over the next 100 years FIGURE 16.8

In 1950, the population of the world began to increase at a faster rate. Rapid growth will continue into the next century. The rate of increase is expected to slow, however, as resources become limited and birthrates begin to decline.

FUTURE EMISSIONS

Next, we explore some important factors that will affect anthropogenic emissions of greenhouse gases over the next 100 years. These factors encompass social, economic, political, and technological developments during the 21st century.

Population growth   One of the primary factors influencing the amount of energy consumed and carbon released is population. Nearly all human activities, from transportation to entertainment to agriculture, involve energy use. However, not every individual uses the same amount of energy or releases the same amount of greenhouse gases to the atmosphere. While the average U.S. citizen consumes enough energy to emit 20 metric tons of carbon dioxide to the atmosphere each year, the average person in some African nations emits less than one metric ton. The United States, which has

about 5 percent of the global population, emits about 25 percent of the total CO_2.

Changes in population depend on the present population size, as well as the fertility rate and the mortality rate. Generally, the mortality rate has been falling with improvements in medicine, sanitation, and nutrition. The fertility rate, which is related to education, family planning, and economic conditions, has also fallen.

Presently, about six and a half billion people inhabit our planet. Depending on economic development and medical advances, it is estimated this will increase to 7–15 billion by 2100. FIGURE 16.8 shows the historical growth of the global population, along with estimates of population growth used to forecast future CO_2 emissions. Generally, these scenarios suggest that population growth will slow in the future. The most recent doubling of the world's population took approximately 40 years, while the average growth estimates predict that the next doubling will require 70 years.

Economic growth Another important factor in determining the rate of emission of CO_2 is the rate of economic growth. For each percentage point increase of economic growth, energy consumption increases by about half a percentage point, and greenhouse gas emissions increase by about a third of a percentage point. The increase in energy consumption is a function of many factors, including a desire for more products and services. Additional energy is subsequently consumed in the construction of more houses, automobiles, and refrigerators.

The forecasting of economic growth 100 years into the future is very difficult and introduces one of the largest uncertainties in the estimates of greenhouse gas emissions. **FIGURE 16.9** shows possible estimates of economic growth over the next 100 years. While all estimates show substantial increases, the range of estimates is large: 4–16 times present values. The average, however, suggests that there will be a 10-fold increase in the standard of living for the globe as a whole over the next 100 years. According to the rule of thumb given above, this translates into a tripling of the CO_2 emission rate through increases in economic growth alone.

Technology change Although increases in both population and standard of living tend to increase the emission rate of greenhouse gases, the emission rate

also depends on the **carbon intensity** of energy usage—how much CO_2 is emitted for each unit of energy used. This in turn depends primarily on *energy efficiency*—the amount of energy required to perform a given function. For example, by properly inflating your car's tires, you can get better fuel efficiency, and the carbon emissions per mile you travel will decrease.

FIGURE 16.10 shows that carbon intensity has decreased at a consistent rate over the last 100 years. Further decreases in carbon intensity will come from energy efficiency improvements and the substitution of other energy sources for fossil fuels. However, as long as population and economic growth continue to increase, it is still possible that the total amount of energy used by humans will increase, as will future greenhouse gas emissions.

Possible emissions and concentrations

The key to predicting how concentrations of CO_2 and other greenhouse gases will increase in the future is to predict how emissions will increase. To do that, scientists and economists develop a series of **emissions scenarios** that de-

> ■ **carbon intensity**
> The amount of carbon dioxide emitted for a given unit of energy produced.

> ■ **emissions scenarios**
> Estimates of how emissions of greenhouse gases may change based on estimates of changing global socioeconomic conditions.

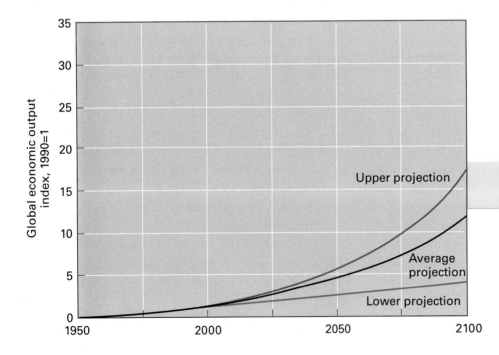

Economic growth over the next 100 years FIGURE 16.9

Economic growth is expected to increase over the next 100 years. Here, the 1990 value is equivalent to 1, and the numbers give the percentage change compared with 1990 (i.e., a value of 5 represents a 500 percent increase in the economic output of the world).

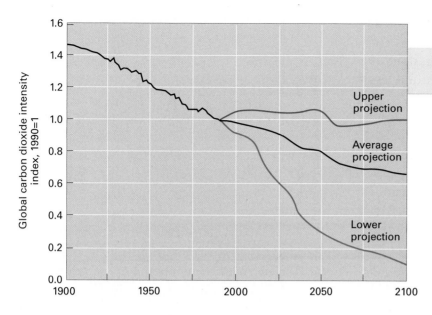

Historical and projected changes in CO_2 emissions per unit energy consumed. In this figure, the 1990 value is set to 1, and the other numbers give the fraction of 1990 emissions produced for the same amount of energy used. Decreasing values indicate more efficient energy usage with less CO_2 emitted.

pend on different assumptions about population, economic development, and technological change over the next 100 years. Using these scenarios, they then model the expected concentration of CO_2, as shown in FIGURE 16.11.

Regardless of the scenario, all emission forecasts predict that CO_2 concentrations will increase well above pre-industrial levels (about 286 ppm, or parts per million) by the year 2100. In addition, none of the scenarios shows concentration stabilizing before

Estimates of future CO_2 emissions and concentrations through 2100
FIGURE 16.11

Although the rate of emissions falls around 2040–2060 for some of the scenarios, CO_2 concentrations continue to increase under all scenarios. For reference, the A2 line represents a business-as-usual scenario. The A1B line shows the emissions needed to limit concentrations to a doubling of 1990 CO_2 by 2100. The B1 line represents the emissions needed to stabilize concentrations of CO_2 by 2100.

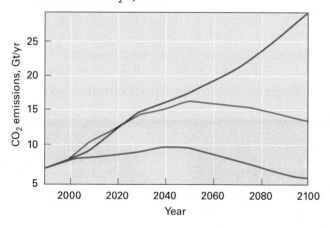
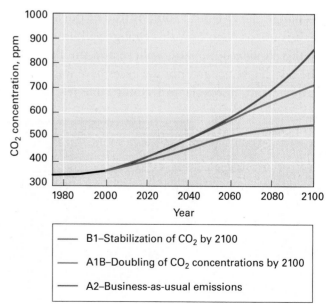

B1–Stabilization of CO_2 by 2100

A1B–Doubling of CO_2 concentrations by 2100

A2–Business-as-usual emissions

the year 2100. *Business-as-usual scenarios* (for instance, A2) produce a doubling of the pre-industrial concentration before 2070. This results in concentrations of about 800 ppm, 3 times greater than the natural levels, by 2100.

If emissions increases are scaled back by half by 2100 (line A1B), 1990 CO_2 concentrations will double to about 700 ppm by 2100. Because concentrations will continue to increase as long as emissions are increasing, the only way to achieve a constant CO_2 concentration is to keep constant or reduce CO_2 emissions (B1). However, even in this case, the final CO_2 concentrations of 550 ppm are still greater than the pre-industrial levels by about a factor of 2.

Overall, the general consensus is that over the next 100 years, CO_2 concentrations will double, and possibly triple above their natural levels of 280 ppm. In fact, to achieve concentrations less than a doubling requires an 80 percent decrease of emissions over the next 50 years by industrialized nations.

Concentrations of methane, halocarbons (CFCs), nitrous oxide, ozone, and aerosols will also change because of human activity. As with CO_2, their future concentrations depend on their residence times in the atmosphere. For example, N_2O has a residence time of 120 years, and therefore will behave similarly to CO_2. In contrast, CFCs have a residence time of thousands of years, and even small anthropogenic emissions will accumulate significantly over time.

Although it is possible to predict how human activity will change the chemical composition of the atmosphere, it is not as clear how this change will, in turn, change the climate of the atmosphere and the Earth. Over the last 250 years, greenhouse gases have warmed the planet, while aerosols and cloud changes have

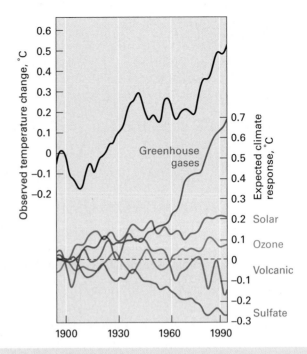

Contributions to historical increases in global temperatures FIGURE 16.12

Over the last 100 years, global temperatures have increased significantly. Anthropogenic emissions of greenhouse gases have contributed to this warming, as has increased solar energy and decreased stratospheric ozone. Increased aerosols—produced by volcanic eruptions as well as anthropogenic emissions of sulfates—have cooled the Earth-atmosphere system. In total, warming has strongly outweighed cooling.

cooled it (FIGURE 16.12). However, there are still large uncertainties in our projections, because different climate feedback processes can interact in very complex ways. Because of these uncertainties, forecasts of human-induced climate change use massive computer programs that are described in the next section.

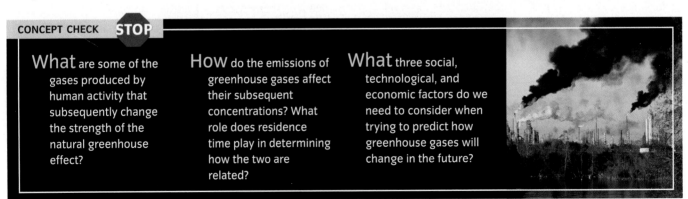

CONCEPT CHECK STOP

What are some of the gases produced by human activity that subsequently change the strength of the natural greenhouse effect?

How do the emissions of greenhouse gases affect their subsequent concentrations? What role does residence time play in determining how the two are related?

What three social, technological, and economic factors do we need to consider when trying to predict how greenhouse gases will change in the future?

Numerical Modeling

LEARNING OBJECTIVES

Identify the different types of models that make up a global climate model.

Describe some of the physical, chemical, and biological processes these different models are designed to represent.

Explain how scientists use global climate models to predict future climate variability.

The atmospheric models used to forecast climate—called *general circulation models*—are similar to the models used in day-to-day weather forecasting, but they generally have coarser spatial and temporal resolution—100–200 km between grid points and 6-hour time steps. Unlike weather prediction models, general circulation models must keep track of all the energy and water in the Earth–atmosphere system. For day-to-day forecasts, a loss or gain of energy or moisture does not amount to much, but when summed over years and decades, the loss and gain can result in much cooler and/or drier conditions. Hence, general circulation models use additional equations that ensure radiative equilibrium of the Earth–Sun system and the conservation of water.

A second limitation of numerical weather prediction models is their use of **parameterizations** to approximate the behavior of some atmospheric phenomena. Parameterizations allow the model to capture phenomena that the model cannot explicitly represent using physical laws. For example, some processes, like precipitation, occur at scales finer than the model grid resolution. Other processes, such as the exact flow of radiation through the atmosphere, are too time-consuming to compute. Still other processes, like condensation, are so complex that they are not completely understood.

parameterizations Estimates of certain atmospheric processes that approximate the actual behavior of the atmosphere.

The most difficult process to parameterize, at least with regard to climate simulations, is cloud formation, particularly in the upper troposphere (**FIGURE 16.13**). Problems in the simulation of clouds are important, because liquid and gaseous water in these regions have a very strong greenhouse effect. Models with exactly the same dynamics but different parameterizations for cloud formation can produce very different climate simulations, which is one of the greatest uncertainties in global climate forecasts at present.

Multilayered clouds FIGURE 16.13

Atmospheric climate models need to represent all important atmospheric processes, including cloud formation. In this image, we see the complexity of clouds, both in their spatial size and vertical extent.

OCEAN MODELS

Like atmospheric general circulation models, ocean dynamic models attempt to solve equations describing the movement, temperature, and salinity of the ocean. These equations are solved at grid points spanning the entire ocean, as well as at different levels of ocean depth. Many of the equations used in atmospheric models also apply to ocean models, because the same forces—such as the pressure gradient force, the Coriolis force, and the frictional force—are present.

However, some of the equations are different. For example, atmospheric models explicitly account for changes in humidity, while ocean models explicitly account for changes in salinity. Ocean models also incorporate biological activity, which can modify the ocean's albedo, the uptake and release of CO_2, and even the temperature structure in certain regions.

Ocean turbulent mixing associated with friction and density differences is much more complex than in the atmosphere, requiring additional parameterizations. As with clouds in the atmosphere, differences in how models treat these ocean processes can lead to substantially different climate simulations.

Ocean models also account for the formation and movement of sea ice (FIGURE 16.14). For climate purposes, the most important sea ice covers the Arctic Ocean and the fringes of Antarctica. The ice covers large areas and has a very high albedo compared with the underlying ocean waters. As this ice melts, there is a very large change in the amount of absorbed and reflected radiation.

To handle the formation and melting of sea ice, ocean models account not only for the temperature of the overlying atmosphere and the amount of snowfall, but also for the direction and temperature of the currents. For example, a change in the northward extent of the Gulf Stream can bring warmer water to the north and help erode sea ice from below. Also, a change in currents can float sea ice from the high latitudes to lower latitudes, where it will tend to melt. Compared to snow and ice on land, sea ice is a dynamic component of the climate system.

A final feature of ocean models is their spatial resolution. At midlatitudes, ocean models use the same horizontal resolution as the overlying atmosphere. However, at lower latitudes, the grid points have to be much closer together to simulate phenomena like

Calving iceberg FIGURE 16.14

Ocean climate models not only need to model changes in the temperature, salinity, and currents of the ocean, but they also need to model changes in the amount and extent of sea ice associated with glaciers and ice shelves. This glacier is crumbling, or "calving," into the Seno Eyre Fjord in Chile.

SENO EYRE FJORD, CHILE

Global Locator

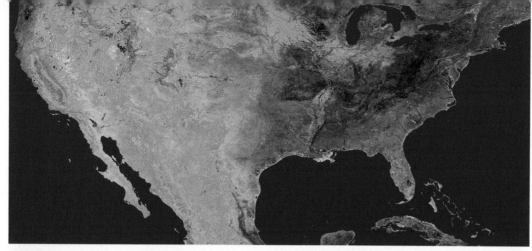

A In summer, the increased extent of green vegetation increases surface albedo and promotes increased evapotranspiration.

B As climate changes with the onset of winter, the amount and type of vegetation may change as well, changing albedo and soil moisture over large regions of the planet.

Vegetation and climate change FIGURE 16.15

Land-surface models need to model changes in soil moisture, stream flow, snow cover, and many other processes. This satellite image shows the change in vegetation cover between summer and winter over the United States.

El Niño and coastal upwelling. As more and more grid points are added to the model, the computation time required by the model increases.

LAND-SURFACE MODELS

In modeling the ocean and atmosphere, the dynamics—how fast the winds blow and where the currents go—are very important. In contrast, land-surface models focus on producing the correct exchange of heat and moisture between the land and atmosphere. This process requires many equations governing the flow of energy and water between different layers of the soil and between the soil and the atmosphere.

The flow of energy and water depends on the soil and vegetation type. While the soil type does not usually change much, vegetation is very sensitive to changes in climate, as seen in FIGURE 16.15. Land-surface models have now been developed that predict how the vegetation type will change with climate change. For example, will new species move into a region? What will happen to the number and size of fires as climate changes, and what effect will this have on vegetation? How does the change in vegetation affect climate—for example, by changing the albedo, the evaporation of water from the soil to the atmosphere, or the speed of the wind at the surface?

Another important concern of land-surface models for climate is the amount of snow and ice cover. Snow

Schematic diagram of global climate-change model processes FIGURE 16.16

Global climate models attempt to simulate numerous types of atmospheric, oceanic, and land-surface processes.

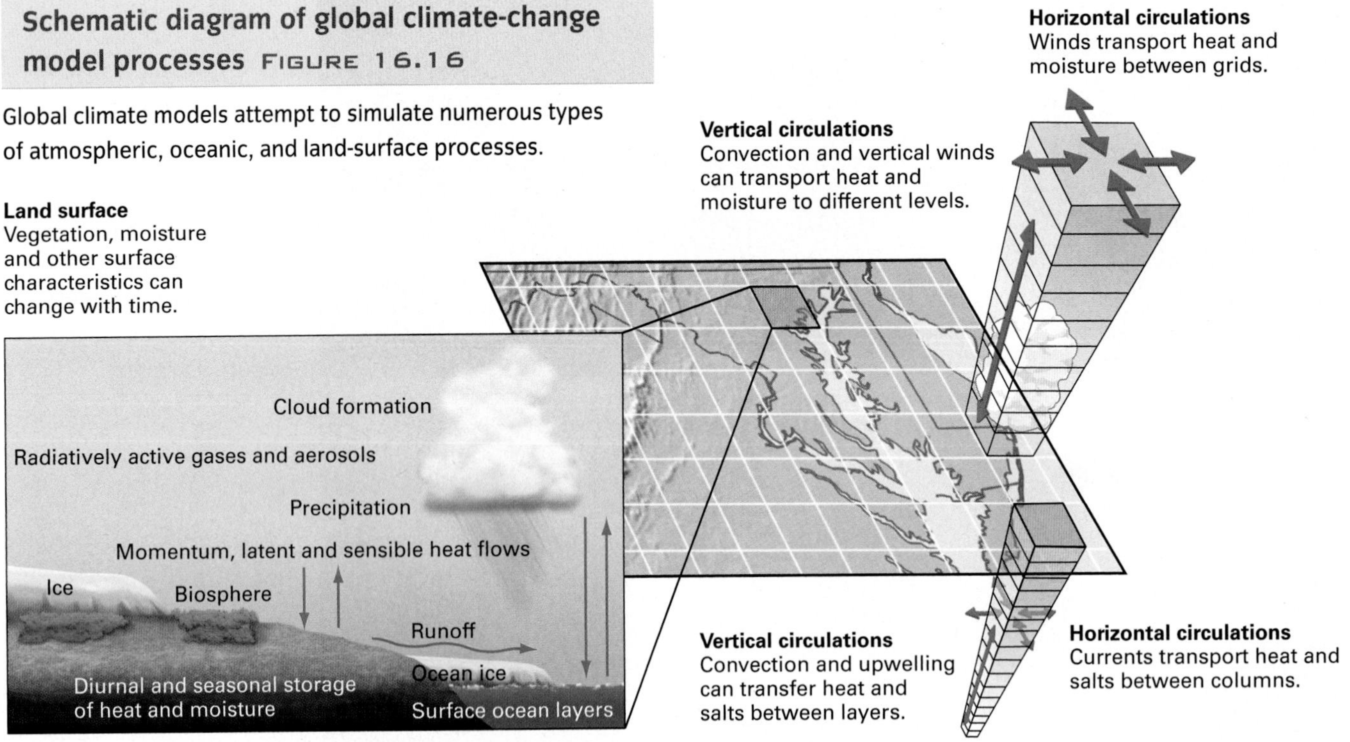

Land surface
Vegetation, moisture and other surface characteristics can change with time.

Horizontal circulations
Winds transport heat and moisture between grids.

Vertical circulations
Convection and vertical winds can transport heat and moisture to different levels.

Cloud formation

Radiatively active gases and aerosols

Precipitation

Momentum, latent and sensible heat flows

Ice   Biosphere

Runoff

Diurnal and seasonal storage of heat and moisture

Ocean ice

Surface ocean layers

Vertical circulations
Convection and upwelling can transfer heat and salts between layers.

Horizontal circulations
Currents transport heat and salts between columns.

and ice raise the albedo of the planet, so they have a significant impact on regional climate. Land-surface models therefore must determine how much snowfall remains on the ground and how much melts and when. In addition, the models have to keep track of glaciers and ice sheets. If glaciers grow or shrink, they can strongly affect the climate of higher-latitude regions.

MODELING "EXPERIMENTS"

To estimate the combined effect of all of the forcings and feedbacks within the climate system, scientists use **global simulation models (GSMs)** or **global climate models (GCMs)**. These models couple the atmosphere, ocean, and the land-surface models described above into a single, larger, and more complex model, as shown in FIGURE 16.16. By changing the chemical composition of the atmosphere and surface characteristics in the global climate model, scientists can predict how the climate might change over the coming years.

global simulation models and global climate models Computer models that combine atmosphere, ocean, and land-surface models into one model to simulate the entire coupled Earth system.

Global climate models are continuously evolving as we improve our knowledge of how the climate system works. Models can now simulate midlatitude storms and natural climate variations, such as El Niños. However, they cannot yet simulate tropical cyclones or regional-scale (less than 100 km) climate variations. However, given the global scale of human-induced changes in the chemical composition of the atmosphere, global climate models are one of the very few tools we have for making useful projections of future climates.

Before making projections, however, we need to see whether the climate model can capture observed changes within the climate system, including recent temperature changes, natural variability, and extreme events. FIGURE 16.17 shows that only by including human-induced changes in greenhouse gas concentrations can coupled models successfully simulate temperature changes over the last 100 years. This result is one of the strongest pieces of evidence that human activity has already changed the global climate.

If the global climate model accurately reproduces the observed changes in climate, the next step is to run the model forward, allowing the concentrations of greenhouse gases and aerosols to change over time. To do this, one set—or *ensemble*—of simulations, similar to the ensemble of simulations used in weather forecasting, is run with the present gas concentrations, and another ensemble of simulations is run with concentrations based on the various emissions scenarios discussed earlier. The difference in the simulated climate of the two sets of model runs indicates how climate may change due to emissions of human activity over the 21st century. The next section shows the projections in more detail.

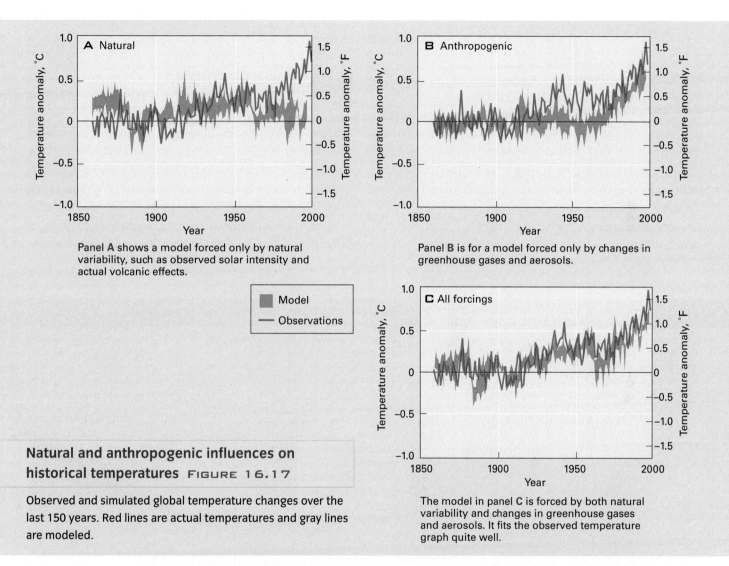

Panel **A** shows a model forced only by natural variability, such as observed solar intensity and actual volcanic effects.

Panel **B** is for a model forced only by changes in greenhouse gases and aerosols.

Natural and anthropogenic influences on historical temperatures FIGURE 16.17

Observed and simulated global temperature changes over the last 150 years. Red lines are actual temperatures and gray lines are modeled.

The model in panel **C** is forced by both natural variability and changes in greenhouse gases and aerosols. It fits the observed temperature graph quite well.

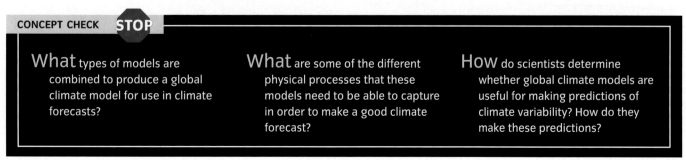

CONCEPT CHECK STOP

What types of models are combined to produce a global climate model for use in climate forecasts?

What are some of the different physical processes that these models need to be able to capture in order to make a good climate forecast?

How do scientists determine whether global climate models are useful for making predictions of climate variability? How do they make these predictions?

Climate-Change Predictions

Anthropogenic emissions of greenhouse gases over the 21st century can affect the climate system in many ways, as well as affect human societies and natural ecosystems. In this section, we examine the projected changes in temperature and precipitation as the human impact grows. In addition, we examine how anthropogenic climate change may affect sea levels, water resources, extreme weather events, and ocean circulation.

GLOBAL AND REGIONAL TEMPERATURE FORECASTS

The global average temperature in 2100 will depend on the emission rate of CO_2. FIGURE 16.18A depicts several possible scenarios. For a business-as-usual scenario, the 300 percent increase predicted in CO_2 concentrations will result in a 3–5.5°C (5.4–10°F) rise in global temperature. In comparison, the temperature changes during an ice age are about 6°C over 10,000 years.

Projections of global temperatures for the next 100 years FIGURE 16.18

These graphs and maps project changes in global average temperatures over the next 100 years.

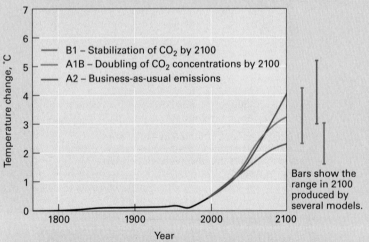

A Three scenarios of CO_2 emissions
Scenario A2 represents business as usual. Scenario B1 represents a stabilization of CO_2 by 2100, and scenario A1B represents a doubling of 1990 CO_2 concentrations by 2100. The bars on the right side are estimates of the range of temperature change produced by the different models.

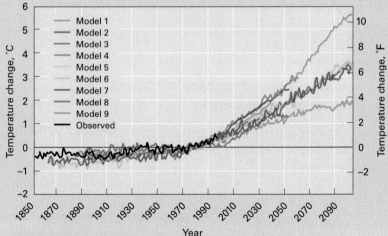

B One CO_2 scenario and variety of models
Change in temperature to 2100 for a doubling of 1990 CO_2 concentrations, as estimated by different climate models, is shown in this graph.

If 1990 CO_2 concentrations stabilize over the next 100 years (B1), the global temperature is still predicted to increase by 2–3°C (3.6–5.4°F).

The range in temperature predictions depends on uncertainties both in emission scenarios and in different models. To estimate the uncertainty produced by different models, we can look at the evolution of global mean temperatures for a subset of different model forecasts (FIGURE 16.18B). By 2100, the average global mean temperature change for these models is 2–5.5°C, (3.6–10°F) with an average of 3°C (5.5°F).

Most of the model differences arise from how the models treat clouds. Smaller temperature changes occur in models that have significant increases in low-level, highly reflecting clouds, which increase global albedo. Greater temperature changes occur in models that have significant increases in high-level clouds, which enhance the greenhouse effect. This uncertainty is why studies of cloud processes and cloud physics have gained such attention over the last few years.

In spite of uncertainties, all of the model forecasts predict increases in global temperatures with rising concentrations of greenhouse gases. We may be unsure of how much temperatures will increase by 2100, but we can safely say that temperatures will increase well beyond today's values.

Regional temperature changes can be significantly different from global-average values, as seen in FIGURE 16.18C. In the high-latitude regions, where there is strong ice–albedo feedback, temperatures for the business-as-usual scenario rise by more than 12°C (21.5°F), which would result in the melting of almost all high-latitude glaciers. Over the continental United States, the average temperature rise would be 4–5°C (7.2–9°F), principally during winter. Temperatures would rise least in the Southern Ocean, due to its large ocean expanse and moderating influence on climate change.

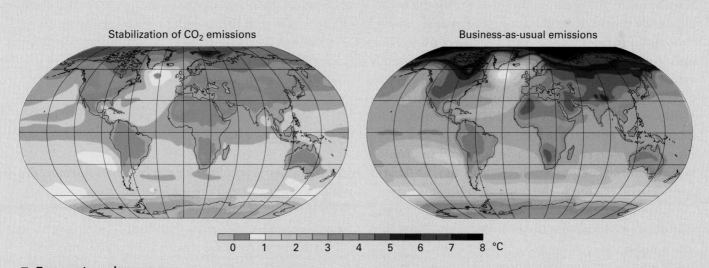

Stabilization of CO_2 emissions

Business-as-usual emissions

0   1   2   3   4   5   6   7   8   °C

C **Temperature-change maps**
These maps show change in temperature (in Celsius) over the next 100 years for the CO_2-stabilization scenario (B1) and the business-as-usual scenario (A2).

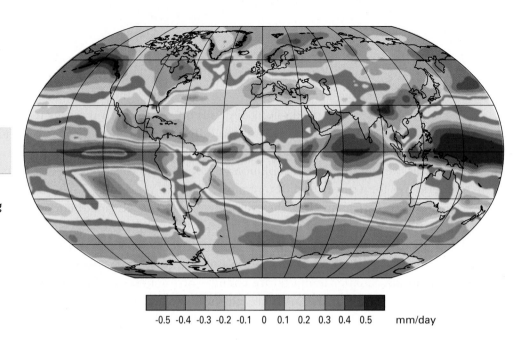

Change in rainfall over the next 100 years, given the CO$_2$-doubling scenario (A1B).

-0.5 -0.4 -0.3 -0.2 -0.1 0 0.1 0.2 0.3 0.4 0.5   mm/day

GLOBAL AND REGIONAL PRECIPITATION FORECASTS

Looking at precipitation, globally we expect higher temperatures to produce more evaporation, which then leads to more precipitation. In other words, the hydrologic cycle is expected to become more intense, making droughts and floods larger and more persistent. However, regional precipitation changes may differ greatly across the globe (FIGURE 16.19). In the ITCZ, precipitation is expected to intensify, stimulating the Hadley cell circulation. In turn, this will intensify subsidence in the poleward branch of the Hadley cell, leading to decreased rainfall in the tropics and subtropics. Wet equatorial regions will become wetter, while arid and semiarid regions in the tropics will become drier.

In addition, the Hadley cell may expand poleward, leading to a decrease in the environmental lapse rate and a more stable atmosphere. The expansion would result in drier conditions in the southern United States, southern Europe, and southern Australia. Over the continental United States, rainfall in the southern portions is expected to decrease, while rainfall in the northern portions is expected to increase.

For agriculture, **soil moisture** is affected by both rainfall and temperature, which increases evaporation.

soil moisture
Amount of liquid water stored in the soil.

Although rainfall over the United States is expected to increase, so will temperature and hence evaporation. Projections for the United States indicate that evaporation changes will be greater than precipitation changes, and summer soil moisture will decrease (FIGURE 16.20). Drier soil, in turn, will shift the growing regions for many crops. For example, suitable growing conditions for wheat will move from the Great Plains of the United States into Canada and even Alaska.

SEA-LEVEL RISE

As the upper level of the ocean becomes warmer, the water will expand and the sea level will rise accordingly. The melting of land-based glaciers will add to the ocean's volume, but not as significantly as the warming of the water itself, at least over the next 100 years.

As shown in FIGURE 16.21, sea level for the business-as-usual scenario is expected to increase about 50 cm by the year 2100. For the three emission scenarios considered here, sea level is projected to be between 38 and 55 cm (15–22 in.) higher than at present. As with temperature, the changes in future sea-level rise will not occur uniformly over the globe, and can be af-

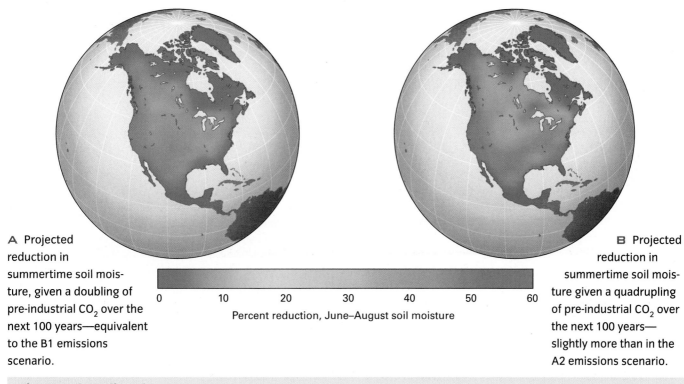

A Projected reduction in summertime soil moisture, given a doubling of pre-industrial CO_2 over the next 100 years—equivalent to the B1 emissions scenario.

Percent reduction, June–August soil moisture

B Projected reduction in summertime soil moisture given a quadrupling of pre-industrial CO_2 over the next 100 years— slightly more than in the A2 emissions scenario.

Changes in soil moisture FIGURE 16.20

fected by changing ocean circulations, wind patterns, and local ocean temperature changes.

Sea-level rise is a slow process. It is first influenced by changes in ocean temperature, but as glaciers and land-based ice sheets melt, they contribute substan-tially to the rise, even if temperatures stabilize. The speed of melting depends on how much higher future temperatures rise. Sea level 500 years from now may be 1.5 m (5 ft) higher than at present (FIGURE 16.21B).

Projections of sea-level rise FIGURE 16.21

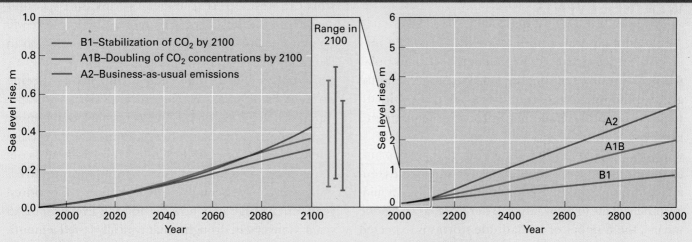

Next 100 years
Change in sea level through 2100, for different scenarios of CO_2 emissions. On the right side bars provide estimates of the range of sea-level rise produced by different models.

Next 1000 years
Even as global temperatures stabilize, land-based glaciers will continue to slowly melt, feeding further increases in sea level.

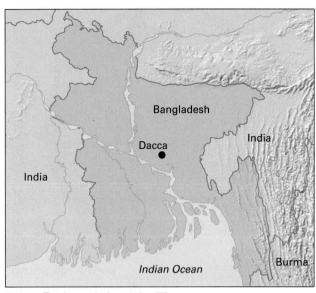
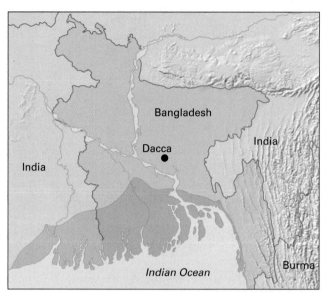

Today Total population: 112 million
Total land area: 134,000 km^2

1.5 m–Impact Total population affected: 17 million (15%)
Total land area affected: 22,000 km^2 (16%)

Sea-level rise and low-lying areas FIGURE 16.22

This figure shows the potential impact on Bangladesh of a 1.5-m (5-ft) change in sea level.

Although the overall rise of 1.5 m (5 ft) predicted by 2500 may not seem like much, it can inundate extensive low-lying coastal areas, as seen in FIGURE 16.22. During high tides or storms, waves will breach sea walls and flood coastal zones in other regions as well.

CHANGE IN EXTREME WEATHER EVENTS

As global temperatures increase, it is expected that equatorial temperatures will change less than polar temperatures, decreasing the latitudinal temperature gradient between these two regions. Because midlatitude storms derive their energy from this temperature gradient, storm activity will decrease. However, a warmer atmosphere will produce a higher tropopause as the air in the troposphere warms and expands, enabling pressure differences—and wind speeds—in midlatitude storms to grow larger. Therefore, as the globe warms, the number of midlatitude storms is expected to decrease, but the storms will be larger.

How will the number and intensity of tropical cyclones change? A warmer ocean can supply more energy to these storms, allowing them to grow larger in size and intensity. History confirms this trend. As the Atlantic has warmed over the last 40 years, the percentage of category 4 or 5 hurricanes more than doubled, and large storms are now the most prominent category of hurricane. However, changes in the upper-air winds can also affect hurricane formation, with stronger winds inhibiting the vertical growth necessary for a storm to develop. It is unclear at this time whether changes in the global circulation of the atmosphere will be more or less conducive to hurricane formation in the future.

Climate change may also bring more extreme droughts and floods. As temperatures increase, evaporation will increase, drying the soil, while adding more water vapor to the atmosphere and increasing the potential for heavier rainfall. This, in turn, will have consequences for infrastructure and planning. River systems that have been flooded only once in a hundred years on average may now be flooded every 50 or 25 years. Conversely, droughts that typically lasted a month or less may instead last whole seasons or even years, depending on the balance between evaporation and precipitation for the given region.

CHANGE IN THE THERMOHALINE CIRCULATION

Another important potential impact of increasing greenhouse-gas concentrations is a systematic change in the *thermohaline circulation*. This global-scale circulation of water through the world's oceans is driven by the sinking of high-density water in the high latitudes of the North and South Atlantic. The sinking brings warm equatorial water northward, warming the mid- and high-latitude climates. In some computer simulations, as the high latitudes of the North Atlantic warm, there is a melting of glaciers that produces fresher oceans. This effect is augmented by increased precipitation and freshwater runoff into the oceans produced by the acceleration of the hydrologic cycle.

Because the ocean water is fresher, it is less dense and does not sink as rapidly or as deeply as before. As the sinking of water in the North Atlantic weakens, there is less northward transport of warm water to replace it. The result is a weakening and southward shift of the Gulf Stream extension, which cools Greenland, northern Europe, and northeastern Canada. The cooling, however, would be superimposed upon the warming caused by the enhanced greenhouse and ice–albedo feedback effects in this region.

Why doesn't the cooling, which would make the water more dense, reinvigorate the thermohaline circulation? Because the density is much more sensitive to salinity than to temperature, small increases in the freshness of the water outweigh the effects of cooling. In addition, glacial ice can melt rapidly and produce a large infusion of fresh water into the ocean, but it takes centuries for glaciers to form again. Once the water becomes fresher, it takes many years of cold temperatures to return salinities to their former values. In this sense, the potential changes in the thermohaline circulation are sometimes referred to as *catastrophic* or *irreversible* changes, because they happen very quickly (over 10–15 years) but can be reversed only over a much longer time.

One contributor to the freshening of the high-latitude oceans is melting of land-based glaciers and sea ice. Recently, the disappearance of sea ice in the Arctic Ocean accelerated rapidly, outpacing the decreases predicted by most climate models, as discussed further in *What a Scientist Sees: Observed and Predicted Changes in Sea Ice*. This accelerated melting, enhanced by the positive ice–albedo feedback, could be an indicator of catastrophic changes that would have significant repercussions for ocean circulations in the arctic as well as around the globe.

Observed and Predicted Changes in Arctic Sea Ice

A Arctic sea ice, as recorded by satellites on September 14, 2007, when it reached its lowest recorded extent—25 percent lower than the previous record set in 2005 and almost 40 percent lower than the average amount since 1979.

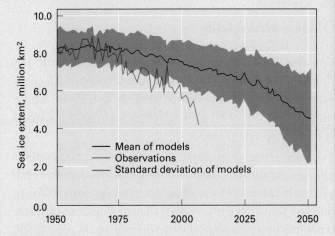

B Over the last 50 years, the observed sea ice extent in the Arctic Ocean has been decreasing. Recently, this decrease has been more rapid than predicted by most climate models. In 2007, sea ice extent hit a record low of 4.3 million km², a level not predicted to occur until 2035.

What a Scientist Sees

		0 — 1 — 2 — 3 — 4 — 5 (Global mean annual temperature change relative to 1980–1999, °C)
Water		Increased water availability in moist tropics and high latitudes – – – – – – – →
		Decreasing water availability and increasing drought in midlatitudes and semiarid low latitudes – – – →
		Hundreds of millions of people exposed to increased water stress – – – – – →
Ecosystems		Up to 30% of species at increasing risk of extinction ——— Significant extinctions around the globe →
		Increased coral bleaching —— Most corals bleached —— Widespread coral mortality – – – – →
		Terrestrial biosphere tends toward a net carbon source as:
	Increasing species range shifts and wildfire risk	15% ——— 40% of ecosystems affected – – – →
		Ecosystem changes due to weakening of the – – – → meridional overturning circulation
Food		Complex, localized negative impacts on small holders, subsistence farmers and fishers – – – →
		Tendencies for cereal productivity to decrease in low latitudes ——— Productivity of all cereals decreases in low latitudes – – →
		Tendencies for some cereal productivity to increase at mid- to high latitudes ——— Cereal productivity to decrease in some regions
Coasts		Increased damage from floods and storms – – – – – – – →
		About 30% of global wetlands lost – – →
		Millions more people could experience coastal flooding each year – – – – →
Health		Increasing burden from malnutrition, diarrheal, cardio-respiratory, and infectious diseases – – – →
		Increased morbidity and mortality from heat waves, floods, and droughts – – – – →
		Changed distribution of some disease vectors – – – – →
		Substantial burden on health services – – →

Summary of expected climate changes FIGURE 16.23

This table lists climate changes that are expected to occur by 2100, and places the text according to the approximate temperature at which effects will be felt. Dashed lines indicate continuing effects. Solid lines indicate effects that change over time.

OTHER IMPACTS

FIGURE 16.23 summarizes some of the predicted changes in temperature and precipitation produced by increasing concentrations of greenhouse gases. These changes will affect human activities and natural systems in many important ways:

Water resources. In many regions, more precipitation will fall as rain instead of snow, and snow that does fall will melt earlier. These changes are very important in regions that rely on meltwater from winter snowpack. With less water storage in the form of snow, runoff will be greater and occur in a shorter time period, which will place a strain on the dams and reservoirs that store the runoff for later use. If peak flows are too high, some water will be lost to the ocean. This change in timing and magnitude of runoff affects agriculture, power generation, and water availability during the summer, when it is needed most.

Agriculture and food security. In regions that are presently water-limited—such as the Sahel in Africa, the Imperial Valley in California, and parts of China and Australia—the combination of warmer summers and decreased rainfall will affect agriculture, particularly where irrigation is less prevalent. This loss of agriculture can lead to an increase in food scarcity and give rise to more frequent famines.

Terrestrial and freshwater ecosystems. Increasing temperatures may lead to a poleward migration of many plant and animal species. For some species that thrive in the high latitudes, increases in temperature will significantly decrease the size and range of their habitats. Animals such as polar bears, which rely on ice flows to increase their hunting grounds during the winter, will be especially affected. In arid or semi-arid grasslands and rangelands, climate change is likely to decrease soil moisture and reduce vegetation growth. Finally, increasing CO_2 concentration promotes vegetation growth, but different plants respond differently, with many weeds, such as poison ivy and kudzu, showing the most rapid response to increased CO_2 levels.

Coastal zones and marine ecosystems. The increases in sea-surface temperature and sea level, decreases in sea-ice cover, and changes in salinity and ocean circulation will affect all marine organisms, including plankton, fish, marine mammals, and seabirds. For example, as temperatures along the western coast of the United States recently increased, there was an 80 percent decline in phytoplankton and a rapid die-off of sea birds. As oceanic CO_2 concentrations and sea-surface temperatures rise above a certain threshold, coral rapidly die. Strong storms also tear corals from their beds. Both of these effects are seen during strong El Niño years and may become more prevalent as overall temperatures continue to rise.

Insurance and financial services. The cost of insuring damages resulting from extreme weather events has risen significantly in recent decades. Part of the increase comes from an increase in the number of people living in vulnerable areas and in the value of houses and buildings in these areas. Projected increases in the frequency and magnitude of storms, floods, and droughts have caused some insurance companies to refuse or cancel coverage in affected areas, such as the Gulf coast and eastern seaboard.

Human health. Human health is very sensitive to climate. During times of excess rainfall, mosquito-borne diseases can spread over vast areas. Rises in temperature, when sustained over time, produce heat waves that can result in serious illness and even death, particularly for older people and the poor. Finally, air quality in urban areas can decline substantially as temperatures rise and speed up the photochemistry that produces ozone.

MITIGATION AND ADAPTATION STRATEGIES

As we've shown, climate change produced by human activity will affect our environment, particularly in the incidence of extreme high-temperature events, storm activity, floods, and drought. At the same time, there is still uncertainty about how great an impact these events will have.

Policymakers are faced with responding to the risks posed by human-induced emissions of greenhouse gases in spite of significant scientific uncertainties. They are required to balance the costs of reducing greenhouse gas emissions (*mitigation*) with reducing the vulnerability of human and natural systems (*adaptation*). Mitigation strategies designed to reduce emissions of greenhouse gases are discussed in FIGURE 16.24 on pages 466–467.

Given the potential impact of continued fossil fuel emissions on the global climate, alternative energy sources are needed to meet future energy demands. What alternative energy sources can replace fossil fuels? Possibilities include nuclear energy, solar energy, hydropower (energy derived from the flow of water), and wind power (energy derived from the flow of air). Some of these—solar energy, wind power, and hydroelectric power—are *renewable energy resources* and have little environmental impact but cannot be produced in large enough quantities to fully replace fossil fuels.

Others, such as nuclear energy, can have a significant environmental impact. Nuclear fission generates radioactive waste that needs to be stored and monitored for thousands of years before it becomes inert or safe. At the same time, nuclear power can generate immense amounts of energy, if carefully managed.

Another option is to increase energy efficiency by developing and distributing technologies that capture more energy from fossil fuel combustion. Shifting from high-carbon fuels, such as coal and oil, to low-carbon fossil fuels, such as natural gas, can also enhance energy efficiency. Finally, technology is being developed that removes CO_2 from emissions before it enters the atmosphere. The captured CO_2 is liquefied and pumped into exhausted oil fields or deep ocean waters.

Mitigation of human-induced climate change depends on social, economic, and technological strategies that lessen the need for carbon-based fuels and reduce the greenhouse gas emissions produced by those fuels.

How to Cut Emissions

Scientists warn that current CO_2 emissions should be cut by at least half over the next 50 years to avert a future global warming disaster. Princeton researchers Robert Socolow and Stephen Pacala have described 15 "stabilization wedges" (far right) to realize that goal using existing technologies. Each carbon-cutting wedge would reduce emissions by a billion metric tons a year by 2057. Adopting any combination of these strategies that equals 12 wedges could lower emissions 50 percent.

Today
Global carbon emissions are estimated at 8 billion metric tons a year.

In the past 50 years
Rising carbon emissions

GLOBAL CARBON EMISSIONS

(billions of metric tons a year)
3.7 metric tons of CO_2 emissions contains a metric ton of carbon

1957   Today   2057

By 2057
Projected emissions of 16 billion metric tons of carbon a year

Three possible paths for future carbon emissions:

1 wedge

MAINTAIN current rate of increase

HOLD emissions at today's rate by **cutting 8 wedges** by 2057, then reduce further

REDUCE emissions by half over the next 50 years by **cutting 4 more wedges,** then reduce further

+9°F
Over 800 ppm

Consequences after 2057

Possible temperature rise and Atmospheric CO_2 concentration in parts per million (ppm)

+5.4°F
525 ppm

+3.6°F
450 ppm

New technologies

may be needed after 50 years to lower emissions further to reach a zero net level (CO_2 emissions minus CO_2 naturally absorbed by Earth's land and oceans).

ONE WEDGE AT A TIME

Each strategy listed below would, by 2057, reduce annual carbon emissions by a billion metric tons.

EFFICIENCY AND CONSERVATION

- Improve fuel economy of the two billion cars expected on the road by 2057 to 60 mpg from 30 mpg.
- Reduce miles traveled annually per car from 10,000 to 5000.
- Increase efficiency in heating, cooling, lighting, and appliances by 25 percent.
- Improve coal-fired power plant efficiency to 60 percent from 40 percent.

CARBON CAPTURE AND STORAGE

- Introduce systems to capture CO_2 and store it underground at 800 large coal-fired plants or 1600 natural-gas fired-plants.
- Use capture systems at coal-derived hydrogen plants producing fuel for a billion cars.
- Use capture systems in coal-derived synthetic fuel plants producing 30 million barrels a day.

LOW-CARBON FUELS

- Replace 1400 large coal-fired power plants with natural-gas-fired plants.
- Displace coal by increasing production of nuclear power to three times today's capacity.

RENEWABLES AND BIOSTORAGE

- Increase wind generated power to 25 times current capacity.
- Increase wind power to 50 times current capacity to make hydrogen for fuel-cell cars.
- Increase ethanol biofuel production to 50 times current capacity. About one-sixth of the world's cropland would be needed.
- Stop all deforestation.
- Expand conservation tillage to all cropland (normal plowing releases carbon by speeding decomposition of organic matter).

SOURCES: ROBERT H. SOCOLOW AND STEPHEN W. PACALA, PRINCETON UNIVERSITY (UPDATED REPORT); OAK RIDGE NATIONAL LABORATORY (GLOBAL CARBON EMISSIONS DATA); ICONS BY JONATHAN AVERY; GRAPHIC BY JUAN VELASCO, NGM ART

Efficiency and conservation

Replacing automobiles with public transport can reduce emissions and conserve fuel.

Carbon capture

The carbon in this coal will be released directly into the atmosphere as CO_2 unless it is captured and stored, either underground or in inert chemical compounds.

Low carbon fuels

Nuclear power generators release steam from their cooling towers.

Renewable energy

A hydroelectric plant generates energy as water flows past turbines.

CONCEPT CHECK STOP

How are global temperatures expected to change over the next 100 years given different emissions scenarios?

Why will certain regional temperature changes be different from the global temperature change?

What types of precipitation changes are expected over the next 100 years given a significant increase in greenhouse gases?

What other climate changes are predicted to occur over the next 100 years given a significant increase in greenhouse gases?

What is happening in this picture ?

This image is a composite of high-resolution satellite images of land and ocean from summer, 2001, with the cloud patterns of one summer day superimposed. Pictured are midlatitude cyclones, tropical cyclones, ice sheets, deserts, and tropical convection.

▪ How might these and other weather and climate processes change as global climate changes?

▪ Think about all aspects of the Earth system, including the atmosphere, ocean, ice, and land surfaces.

SUMMARY

1 Greenhouse Gases

1. **Anthropogenic climate forcing** occurs because of human activity. It includes local climate change induced by irrigation or the build-up of cities, but present concern centers on the emissions of heat-trapping gases—called **greenhouse gases**—released by almost all human activity.

2. The most immediate human influence on the atmosphere results from emissions of greenhouse gases by industry, agriculture, and transportation. These emissions have increased the concentrations of gases such as CO_2, N_2O, and methane well above the levels of the last 1000 years.

3. How our emissions affect overall concentrations depends both on the amount of gases

that are emitted and on how long the gases stay in the atmosphere, which is called the **residence time**. The longer the gases stay in the atmosphere, the longer will be their impact on climate. For example, CO_2 has a residence time of 100–150 years, so today's concentrations are based on the amount we emit now as well as the amounts emitted over the last 100–150 years.

4. Future concentrations depend on how our emissions change over time. Future emissions of greenhouse gases depend primarily on the overall population, the rate of economic growth, and the energy efficiency of our activities. How these three factors change over the next 100 years is as much a function of social and economic changes as it is of climate change.

2 Numerical Modeling

1. To produce forecasts of climate change over the next 100 years, scientists use complex computer models called **global climate models**. These are assembled from submodels that simulate the atmospheric, oceanic, and land-surface changes associated with changing climate. Each submodel includes a vast array of different physical components and processes and their interactions with other submodels.

2. To determine how future climate responds to changes in the chemical composition of the atmosphere, scientists first run global climate models many times using the present **chemical composition** of the atmosphere. They then run them again using concentrations associated with different **emissions scenarios**. The difference between the climate in the two simulations gives scientists an estimate of the changes that might occur in the real world.

3. Present global climate models can accurately reproduce the actual climate changes that have occurred over the last 100 years. To reproduce these climate changes, however, requires that the models include the present build-up of greenhouse gases, showing that recent change in global temperatures is already due to emissions of greenhouse gases arising from human activity.

3 Climate-Change Predictions

1. Different emissions scenarios yield different influences on global temperatures. For a business-as-usual scenario, which leads to a tripling of CO_2 in the next 100 years, global temperatures are expected to rise by 3–5°C. The uncertainty in predictions is due principally to differences in how the models treat changes in cloud cover.

2. In addition to global changes, there will be regional changes in both temperature and precipitation. Temperature increases are expected to be greatest over the high latitudes and least in the southern hemisphere tropics. Equatorial regions will receive even more rainfall than at present, while dry tropical and subtropical regions will receive less. In the midlatitudes, precipitation will increase, but so will evaporation due to increasing temperatures. Hence, some regions will have more **soil moisture** available for agriculture, while others will have less.

3. Other predicted climate changes include an increase in intensity of both midlatitude and tropical cyclones, more intense flooding and more persistent drought, and a slow but steady rise in sea level. Certain irreversible changes may also occur, including a slowdown of the global-scale thermohaline circulation and a significant melting of the northern and southern hemisphere glaciers, leading to increases in sea level of up to 3 m (10 ft).

KEY TERMS

CRITICAL AND CREATIVE THINKING QUESTIONS

1. Identify ways in which human activity may have influenced global air temperatures over the last 100 years.

2. What is the residence time of CO_2 in the atmosphere once it has been emitted via fossil fuel burning? Why is the residence time important when determining the concentrations of CO_2 based on these emissions? Would you expect the concentrations of CO_2 to be greater or less if the residence time of CO_2 were shorter?

3. What factors influence the future anthropogenic emissions of CO_2? Why can emissions continue to increase even if our technology improves and energy efficiency increases over the next 100 years?

4. What different model components make up a global climate model? How are these global climate models used to predict future climate change, and what do they tell us about the human impact on historical climate change?

5. Given the business-as-usual emissions scenario, which regions will experience the largest temperature increases? The smallest? Why do these regions have such large differences in their response to global climate change?

6. Why don't we plot global changes in precipitation the same way we plot temperature changes? Given significant increases in CO_2 and other greenhouse gases, which regions will experience significant increases in precipitation? Decreases in precipitation? What factors besides precipitation must we consider when we look at changes in water resources and availability?

7. Other than fossil fuels, what alternative energy sources can humans turn to?

SELF-TEST

1. _____ is not an anthropogenic source of CO_2.
   a. Coal burning
   b. Deforestation
   c. Rice cultivation
   d. Gasoline consumption

2. Increased aerosols can lead to _____.
   a. increased global temperatures
   b. decreased global temperatures
   c. both a and b
   d. no impact on global temperatures

3. _____ is considered a greenhouse gas.
   a. Sulfur dioxide
   b. Dust
   c. Oxygen
   d. Nitrous oxide

4. The residence time of excess CO_2 in the atmosphere is about _____.
   a. 5 years
   b. 25 years
   c. 100 years
   d. 500 years

5. On the accompanying diagram, indicate the fraction of anthropogenic CO_2 that stays in the atmosphere, enters the ocean, and enters the biosphere.

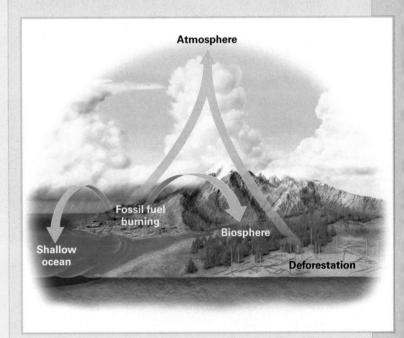

6. Future emissions of greenhouse gases depend on
_____.

   a. changes in population

   b. changes in fuel efficiency

   c. changes in world economic growth

   d. all of the above

7. On the accompanying diagram, indicate which CO_2 emissions scenario is associated with which future CO_2 concentrations.

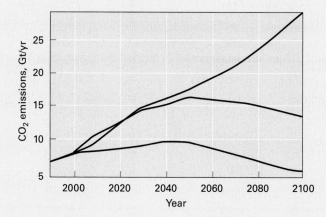

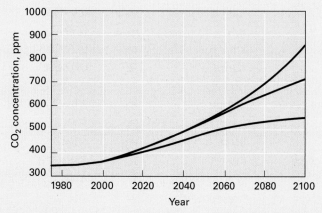

8. Global climate models rely on _____ to capture atmospheric phenomena the model cannot explicitly represent.

   a. interpolations

   b. parameterizations

   c. transformations

   d. gyrations

9. _____ is a process that ocean models do *not* have to represent.

   a. Calving of icebergs

   b. Changes in ocean currents

   c. Changes in salinity

   d. Changes in lake levels

10. For a doubling of CO_2 in the atmosphere, temperatures are expected to increase uniformly across the globe.

   a. True   b. False

11. The range of temperature increase expected for "business-as-usual" emissions is _____.

   a. 1–2.5°C   c. 3–5°C

   b. 2–3.5°C   d. 5–7°C

12. On the accompanying figure, indicate which climate models most likely have (a) large increases in low-level, highly-reflective clouds; (b) large increases in high-level, infrared-absorbing clouds; and (c) a mixture of low-level, highly-reflective clouds and high-level, infrared-absorbing clouds.

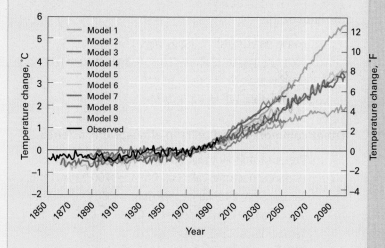

13. For a doubling of CO_2 in the atmosphere, _____ are expected to experience the largest temperature increases, while _____ are expected to experience the least.

   a. high-latitude oceans; high-latitude land surfaces

   b. high-latitude land surfaces; low-latitude oceans

   c. low-latitude oceans; high-latitude oceans

   d. low-latitude land surfaces; high-latitude land surfaces

14. For a doubling of CO_2 in the atmosphere, wet equatorial regions are expected to receive more rainfall, while dry tropical and subtropical regions are expected to receive less.

   a. True   b. False

15. The amount of soil moisture available for agriculture will _____ if precipitation _____ and evaporation _____.

   a. increase; increases; decreases

   b. increase; decreases; increases

   c. decrease; increases; decreases

   d. decrease; remains the same; decreases

Appendix A
Units of Measurement and Conversion

Units

Primary SI units (Système International d'Unités)

Length: meter (m)

Mass: kilogram (kg)

Temperature: kelvin (K)

Time: second (s)

Derived units

Force: newton (N): $= kg \times m/s^2$

Pressure: pascal (Pal): $= N/m^2 = kg/(m \times s^2)$

Energy: joule (J): $= kg \times m^2/s^2$

Power: watt (W): $= kg \times m^2/s^3$

Other commonly used units in atmospheric science

Pressure: millibar (mb): 1 mb $=$ 100 Pa $=$ 1 hPa

Pressure: hectopascal (hPa): 1 hPa $=$ 100 Pa $=$ 1 mb

Temperature interval: degree Celsius (°C): 1 °C $=$ 1 K

Temperature: Celsius degrees (°C): °C $=$ K $-$ 273.15

Temperature: Celsius degrees (°C):

°C $=$ (°F $-$ 32)/(1.8)

Temperature: Fahrenheit degrees (°F):

°F $=$ (°C $\times$ 1.8) $+$ 32

Wind Speed: knots (kts): 1 kt $=$ 0.51 m/s^2

Wind Speed: miles per hour (mph): 1 mph $=$ 0.48 m/s

Wind Speed: kilometers per hour (kph):

1 kph $=$ 0.28 m/s

Other useful conversions

Conversion	From	To	Multiply by
Length	inches	m	0.0254
	feet	m	0.3048
	miles	m	1609.344
	naut. mile	m	1853
Area	in^2	m^2	0.00064516
	ft^2	m^2	0.09290304
	mile^2	m^2	2.58999×10^6
Volume	in^3	m^3	16.3871×10^{-6}
	ft^3	m^3	0.0283168
	liter	m^3	0.001
Speed	knot	m/s^2	0.514791
	mph	m/s	0.44704
	ft/s	m/s	0.3048
	km/hr	m/s	0.277778
Mass	ounce	kg	0.0283495
	pound	kg	0.4535923
Pressure	in (Hg)	Pa	3386.39
	mb	Pa	100
	lb/in^2	Pa	6894.76
Energy	kW-hr	J	3.6×10^6
	kcal $=$ Food cal	J	4190
Power	horsepower	W	746

Appendix B
The U.S. Standard Atmosphere

Altitude [m]	Temperature [Celsius]	Pressure [mb]	Density [kg/m³]	Altitude [m]	Temperature [Celsius]	Pressure [mb]	Density [kg/m³]
0	15.00	1,013.25	1.23	16,000	−56.50	102.87	0.17
500	11.75	954.61	1.17	16,500	−56.50	95.08	0.15
1,000	8.50	898.75	1.11	17,000	−56.50	87.87	0.14
1,500	5.25	845.56	1.06	17,500	−56.50	81.21	0.13
2,000	2.00	794.95	1.01	18,000	−56.50	75.05	0.12
2,500	−1.25	746.83	0.96	18,500	−56.50	69.36	0.11
3,000	−4.50	701.09	0.91	19,000	−56.50	64.10	0.10
3,500	−7.75	657.64	0.86	19,500	−56.50	59.24	0.10
4,000	−11.00	616.40	0.82	20,000	−56.50	54.75	0.09
4,500	−14.25	577.28	0.78	20,500	−56.00	50.60	0.08
5,000	−17.50	540.20	0.74	21,000	−55.50	46.78	0.07
5,500	−20.75	505.07	0.70	21,500	−55.00	43.25	0.07
6,000	−24.00	471.81	0.66	22,000	−54.50	40.00	0.06
6,500	−27.25	440.35	0.62	22,500	−54.00	37.00	0.06
7,000	−30.50	410.61	0.59	23,000	−53.50	34.22	0.05
7,500	−33.75	382.51	0.56	23,500	−53.00	31.67	0.05
8,000	−37.00	356.00	0.53	24,000	−52.50	29.30	0.05
8,500	−40.25	330.99	0.50	24,500	−52.00	27.12	0.04
9,000	−43.50	307.42	0.47	25,000	−51.50	25.11	0.04
9,500	−46.75	285.24	0.44	25,500	−51.00	23.25	0.04
10,000	−50.00	264.36	0.41	26,000	−50.50	21.53	0.03
10,500	−53.25	244.74	0.39	26,500	−50.00	19.94	0.03
11,000	−56.50	226.32	0.36	27,000	−49.50	18.47	0.03
11,500	−56.50	209.16	0.34	27,500	−49.00	17.12	0.03
12,000	−56.50	193.30	0.31	28,000	−48.50	15.86	0.02
12,500	−56.50	178.65	0.29	28,500	−48.00	14.70	0.02
13,000	−56.50	165.10	0.27	29,000	−47.50	13.63	0.02
13,500	−56.50	152.59	0.25	29,500	−47.00	12.64	0.02
14,000	−56.50	141.02	0.23	30,000	−46.50	11.72	0.02
14,500	−56.50	130.33	0.21	30,500	−46.00	10.87	0.02
15,000	−56.50	120.45	0.19	31,000	−45.50	10.08	0.02
15,500	−56.50	111.31	0.18				

Appendix C
Weather Station and Map Symbols

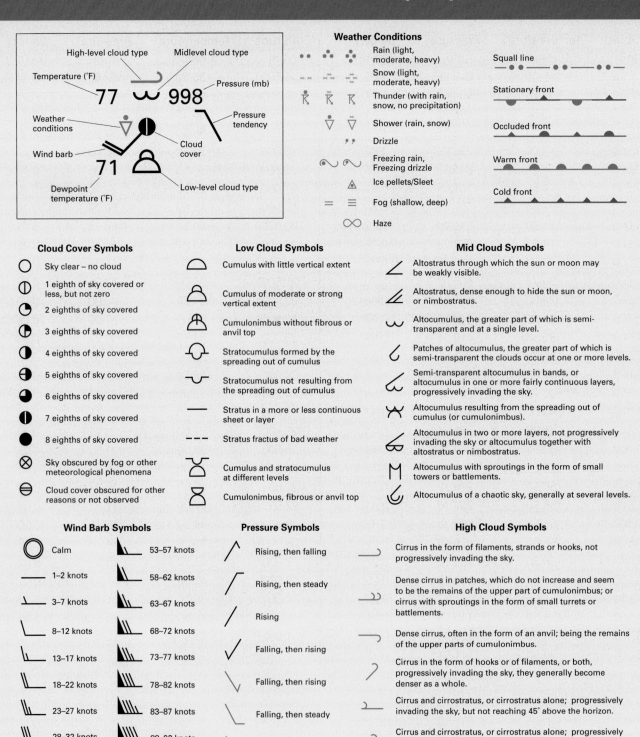

Weather Conditions

Rain (light, moderate, heavy)

Snow (light, moderate, heavy)

Thunder (with rain, snow, no precipitation)

Shower (rain, snow)

Drizzle

Freezing rain, Freezing drizzle

Ice pellets/Sleet

Fog (shallow, deep)

Haze

Squall line

Stationary front

Occluded front

Warm front

Cold front

Cloud Cover Symbols

- Sky clear – no cloud
- 1 eighth of sky covered or less, but not zero
- 2 eighths of sky covered
- 3 eighths of sky covered
- 4 eighths of sky covered
- 5 eighths of sky covered
- 6 eighths of sky covered
- 7 eighths of sky covered
- 8 eighths of sky covered
- Sky obscured by fog or other meteorological phenomena
- Cloud cover obscured for other reasons or not observed

Low Cloud Symbols

- Cumulus with little vertical extent
- Cumulus of moderate or strong vertical extent
- Cumulonimbus without fibrous or anvil top
- Stratocumulus formed by the spreading out of cumulus
- Stratocumulus not resulting from the spreading out of cumulus
- Stratus in a more or less continuous sheet or layer
- Stratus fractus of bad weather
- Cumulus and stratocumulus at different levels
- Cumulonimbus, fibrous or anvil top

Mid Cloud Symbols

- Altostratus through which the sun or moon may be weakly visible.
- Altostratus, dense enough to hide the sun or moon, or nimbostratus.
- Altocumulus, the greater part of which is semi-transparent and at a single level.
- Patches of altocumulus, the greater part of which is semi-transparent the clouds occur at one or more levels.
- Semi-transparent altocumulus in bands, or altocumulus in one or more fairly continuous layers, progressively invading the sky.
- Altocumulus resulting from the spreading out of cumulus (or cumulonimbus).
- Altocumulus in two or more layers, not progressively invading the sky or altocumulus together with altostratus or nimbostratus.
- Altocumulus with sproutings in the form of small towers or battlements.
- Altocumulus of a chaotic sky, generally at several levels.

Wind Barb Symbols

Calm	53–57 knots
1–2 knots	58–62 knots
3–7 knots	63–67 knots
8–12 knots	68–72 knots
13–17 knots	73–77 knots
18–22 knots	78–82 knots
23–27 knots	83–87 knots
28–32 knots	88–92 knots
33–37 knots	93–97 knots
38–42 knots	98–102 knots
43–47 knots	Wind direction variable
48–52 knots	Wind direction given but wind speed missing

Pressure Symbols

- Rising, then falling
- Rising, then steady
- Rising
- Falling, then rising
- Falling, then rising
- Falling, then steady
- Falling
- Rising, then falling

High Cloud Symbols

- Cirrus in the form of filaments, strands or hooks, not progressively invading the sky.
- Dense cirrus in patches, which do not increase and seem to be the remains of the upper part of cumulonimbus; or cirrus with sproutings in the form of small turrets or battlements.
- Dense cirrus, often in the form of an anvil; being the remains of the upper parts of cumulonimbus.
- Cirrus in the form of hooks or of filaments, or both, progressively invading the sky, they generally become denser as a whole.
- Cirrus and cirrostratus, or cirrostratus alone; progressively invading the sky, but not reaching 45° above the horizon.
- Cirrus and cirrostratus, or cirrostratus alone; progressively invading the sky, reaching more than 45° above the horizon, but without the sky being totally covered.
- Veil of cirrostratus covering the celestial dome.
- Cirrostratus not progressively invading the sky and not completely covering the celestial dome.
- Cirrocumulus alone, or cirrocumulus accompanied by cirrus or cirrostratus or both, but cirrocumulus is predominant.

Appendix D
Self-Test Answer Key

Chapter 1

1. a. constantly changing; 2. a. biosphere; 3. c. time cycle; 4. See Figure 1.3; 5. d. weather; 6. d. a month and longer; 7. c. Whether to put on a raincoat or shorts; 8. c. Formation of a tropical cyclone; 9. See Figure 1.8; 10. b. Change in wind speed from one airport to the next; 11. See *What a Scientist Sees*; 12. a. that they are flat; 13. c. a map projection; 14. d. polar, straight; 15. c. kilogram, mass

Chapter 2

1. a. nitrogen; 2. b. water vapor; 3. c. the stratosphere; 4. a. ultraviolet radiation; 5. c. chlorofluorocarbons (CFCs); 6. See Figure 2.9; 7. b. False; 8. c. a decrease of temperature with height; 9. b. the stratosphere; 10. b. False; 11. b. environmental temperature lapse rate; 12. a. 1013.2 mb; 13. a. True; 14. b. False; 15. c. ionosphere

Chapter 3

1. a. radiates much more energy than a cool object; 2. d. ultraviolet radiation; 3. See Figure 3.4; 4. a. the angle the Sun's rays make with the Earth's surface; 5. See Figure3.16; 6. d. conduction; 7. d. stored heat and as such cannot be directly measured with a thermometer; 8. a. scattering; 9. c. albedo; 10. b. upward, in all directions; 11. See Figure 3.20; 12. a. Carbon Dioxide; 13. b. would be much colder; 14. a. Net radiation; 15. b. oceanic currents, atmospheric currents

Chapter 4

1. See Figure 4.1; 2. c. perihelion, January 3; 3. d. 23 1/2 degrees from a perpendicular to the plane of the ecliptic; 4. c. circle of illumination; 5. a. North and South Poles; 6. See Figure 4.7; 7. a. a measure of the level of sensible heat of matter; 8. d. about one-half hour after sunrise; 9. b. False; 10. b. drier surfaces have less water to evaporate than do moist soils; 11. See Figure 4.16; 12. a. slowly; 13. d. Isotherms; 14. a. north; south; 15. $-2°F$

Chapter 5

1. a. latent heat; 2. See Figure 5.1; 3. a. True; 4. d. is the amount of water vapor in the air compared to the amount it could contain; 5. b. early afternoon; 6. c. specific humidity; 7. See Figure 5.6; 8. a. Adiabatic; 9. a. True; 10. d. moist adiabatic lapse rate; dry adiabatic lapse rate; 11. b. Advection; 12. b. Orographic; 13. a. warm, moist air and an unstable environmental lapse rate; 14. d. moist adiabatic; 15. b. Bergeron Process

Chapter 6

1. a. isobars; 2. a. True; 3. d. anemometer; 4. a. at nightfall, when the land cools below the surface temperature of the sea; 5. b. False; 6. b. False; 7. b. a result of the counterclockwise Earth's rotation around its axis; 8. a. True; 9. a. geostrophic wind; 10. See Figure 6.10; 11. b. False; 12. See Figure 6.11; 13. b. pressure gradient; 14. c. high-pressure system that rotates clockwise; 15. d. surface cyclones

Chapter 7

1. b. 1013.2; 2. a. polar fronts; 3. a. 0°; 4. c. subtropical high-pressure cells; 5. a. True; 6. c. subtropical high-pressure centers; 7. d. Wet and hot; dry and cool; 8. See Figure 7.5; 9. d. High; low; switch; 10. b. It is due to the temperature difference between oceans and land; 11. d. Polar fronts; 12. See Figure 7.14; 13. c. Low latitudes; high latitudes; 14. b. thermocline; 15. a. gyres

Chapter 8

1. d. air mass; 2. See Figure 8.2; 3. b. Continental polar; 4. a. True; 5. a. mE; 6. a. anticyclone; 7. c. cold front; 8. d. cyclones; anticyclones; 9. c. low-pressure trough; 10. c. midlatitude cyclone; 11. See Figure 8.9; 12. d. when a cold front has overtaken a warm front; 13. b. False; 14. d. a cold front with squalls; 15. See Figure 8.19

Chapter 9

1. c. convective in nature; 2. c. 5° to 30° N and S; 3. b. 10° to 20° N and S; 4. c. June through November; 5. b. South American; 6. d. All of the above; 7. a. True; 8. See Figure 9.7; 9. c. Warm, moist; latent heat; lower; 10. b. Presence of subsiding or sinking air; 11. a. storm surge; 12. d. Saffir-Simpson scale; 13. c. increased; warmer ocean temperatures; 14. a. Galveston, TX in 1900; 15. d. Rapid temperature drops

Chapter 10

1. c. warm, moist air and an unstable environmental lapse rate; 2. c. a microburst; 3. a. cumulus, mature, dissipating; 4. c. eye wall; 5. See Figure 10.3; 6. b. mT; cT; 7. d. charge separation; 8. See Figure 10.1; 9. a. Leader; 10. b. tornado; 11. b. False; 12. d. All of the above; 13. c. Spring and Summer; Southeastern and Central; 14. b. 100–450 m; 15. See Figure 10.15

Chapter 11

1. c. based on temperature and precipitation averages over decades; 2. c. latitude and coastal versus continental location; 3. c. the maximum precipitation is most likely to be in the colder months; 4. d. moderating the variation of temperature throughout the year; 5. See Figure 11.2; 6. c. warm temperatures; 7. c. annual precipitation; 8. See Figure 11.6; 9. d. stationary subtropical high-pressure cells; 10. b. no precipitation in any months; 11. See Figure 11.5; 12. c. cP; 13. a. polar front zone; 14. d. climograph; 15. a. no permanently flowing streams

Chapter 12

1. d. a very strong temperature cycle; 2. d. stationary subtropical cells of high pressure; 3. c. cP ; 4. b. moist subtropical; 5. d. mT; 6. a. polar-front zone; 7. b. mT; 8. a. True; 9. c. northern; 10. See Figure 12.16; 11. d. all of the above; 12. a. moist subtropical; 13. a. west sides; 14. c. Mediterranean; 15. a. climograph 1, b. climograph 3, c. climograph 2

Chapter 13

1. b. lower; aerosols; 2. c. wind flow patterns along the ITCZ; 3. d. location of jet streams; 4. See Figure 12.6; 5. d. all of the above; 6. d. all of the above; 7. d. decades or longer; 8. b. heliocity; 9. a. inter-glacial, 12,000; 10. c. 5°C; 11. c. decreases; decreased; moderating; 12. a. increases; increases; enhance; 13. a. decreased surface temperatures, b. increased surface temperatures; 14. b. jellyfish reproduction; 15. c. warm; acidic

Chapter 14

1. b. high, increases; 2. d. traveling anticyclones, cP; 3. c. mudflow, thunderstorm; 4. d. large; steep; 5. d. all of the above; 6. c. alongshore, onshore; 7. See Figure 14.10a; 8. b. False; 9. d. all of the above; 10. c. famine; 11. d. A mudflow arising from the passage of a thunderstorm; 12. See Figure 14.17; 13. c. nitric oxide; 14. a. increases, enhances; 15. c. evapotranspiration

Chapter 15

1. b. False; 2. c. 700 mb relative humidity maps; 3. d. 850 mb temperature maps; 4. a. Winds and precipitation; 5. See figure 15.3; 6. c. parallel; half the; 7. d. Tens of millions; 8. a. ensemble; 9. c. meteogram; 10. See Figure 15.12; 11. c. analog method; 12. d. All of the above; 13. a. validation time; 14. c. improved; worse; 15. See Figure 15.18

Chapter 16

1. c. Rice cultivation; 2. c. both a. and b.; 3. d. Nitrous oxide; 4. c. 100 years; 5. See Figure 16.6; 6. d. all of the above; 7. See Figure 16.11; 8. b. parameterizations; 9. d. changes in lake levels; 10. b. False; 11. c. 3–5°C; 12. See Figure 16.18; 13. b. high-latitude land surfaces; low-latitude oceans; 14. a. True; 15. a. increase; increases; decreases

This glossary contains definitions of terms shown in the text in italics or boldface.

absolute vorticity in meteorology, the sum of the planetary vorticity and relative vorticity

absolute zero the theoretical temperature at which all molecular motion ceases; defined as 0 K

absorption process in which electromagnetic energy is transferred to heat energy when radiation strikes molecules or particles in a gas, liquid, or solid

accretion growth of ice through collision with super-cooled water droplets

acid deposition the deposition of acid raindrops and/or dry acidic dust particles on vegetation and ground surfaces

adaptation (to climate change) process of modifying behavior in order to better prepare for consequences of expected change in climate

adiabatic lapse rate see *dry adiabatic lapse rate, moist adiabatic lapse rate*

adiabatic principle the physical principle that a gas cools as it expands and warms as it is compressed, provided that no heat flows into or out of the gas during the process

adiabatic process change of temperature within a gas because of compression or expansion, without gain or loss of heat from the outside

advection fog fog produced by condensation within a moist air layer moving over a cold land or water surface

aerosols tiny particles present in the atmosphere, so small and light that the slightest movements of air keep them aloft

agricultural drought condition in which crop growth is limited by available water, either due to decreased precipitation or increased evaporation

air a mixture of gases that surrounds the Earth

air density mass of air molecules found within a fixed volume of air

air mass extensive body of air in which temperature and moisture characteristics are fairly uniform over a large area

air-mass thunderstorm thunderstorm arising from daytime heating of the land surface, usually characterized by isolated cumulus and cumulonimbus clouds

air pollutant an unwanted substance injected into the atmosphere from the Earth's surface by either natural or human activities; includes aerosols, gases, and particulates

air temperature temperature of air, normally observed by a thermometer under standard conditions of shelter and height 1.2 m above the ground

albedo proportion of solar radiation reflected upward from a surface

alpine debris avalanche debris flood of steep mountain slopes, often laden with tree trunks, limbs, and large boulders

altocumulus cloud type in which patchy clouds or rolls are found near the middle of the troposphere

altostratus cloud type in which sheets or layers of clouds are found near the middle of the troposphere

analog method method in which a forecast is made by examining the evolution of historical weather events that are similar to current events

annual temperature range temperature difference between the warmest average daily temperature of the year and coldest average daily temperature of the year

anomaly the departure of a given field from its normal or average value

anthropogenic climate forcing change in climate resulting from human activities at local to global scales

anticyclone center of high atmospheric pressure

anvil cloud anvil shaped cloud that is found extending from the top of mature thunderstorms

aphelion point on the Earth's elliptical orbit when it is farthest from the Sun

arid (dry climate subtype) subtype of the dry climates that is extremely dry

and supports little or no vegetation cover

Atlantic multidecadal oscillation natural variations in sea surface temperatures over the north Atlantic spanning many tens of years

atmosphere layers of gases surrounding the Earth and bound to it by the Earth's gravity

atmospheric model a computer model designed to simulate and predict the behavior of the atmosphere

atmospheric pressure ressure exerted by the atmosphere because of the force of gravity acting upon the overlying column of air

aurora borealis natural display of lights occurring in high-latitudes of northern hemisphere, resulting from the interaction of charged particles from the sun with gases and ions in the ionsphere

backing (of winds) in northern hemisphere, a counterclockwise rotation of wind direction with height; in southern hemisphere, rotation is clockwise

baroclinic instability process in which horizontal disturbances in the jet stream grow over time, arising in the presence of large-scale meridional temperature gradients

barometer instrument that measures atmospheric pressure

Bergeron process process of ice-crystal growth in cold clouds as water vapor evaporates from super-cooled water droplets and deposits on ice crystals

biological pump a process in which CO_2 from the atmosphere is taken up by marine phytoplankton and is then removed to the deep ocean as the phytoplankton die and sink

biomass burning burning of organic matter

biosphere the network of all living organisms found on the Earth; also that portion of the Earth in which life can exist

black carbon (as aerosol) suspended solid particulates resulting from the incomplete burning of fossil fuels or organic matter

blackbody ideal object or surface that is a perfect radiator and absorber of energy; absorbs all radiation it intercepts and emits radiation perfectly according to physical theory

boreal forest climate ⑪ cold climate of the subarctic zone in the northern hemisphere with long, extremely severe winters and several consecutive months of frozen ground

business-as-usual scenario future projection of emissions of carbon dioxide and other greenhouse gases, which follows the same rate of increase as during the last 100 years (approximately 2% increase per year)

Canterbury northwester gusty, warm northwesterly wind caused as cross-mountain flow forces air to descend from the New Zealand Alps onto the Canterbury Plains of South Island, New Zealand

carbon intensity the amount of carbon dioxide emitted for a given unit of energy produced

cartographer person who studies the science and art of making maps

catastrophic (irreversible) changes a relatively rapid change in the climate system that subsequently takes substantially longer to revert back to its original conditions

Celsius scale temperature scale in which the freezing point of water at sea level pressure is 0° and the boiling point is 100°

centigrade see *Celsius scale*

centripetal force force required to keep object moving in a curved path

chaos in meteorology, a situation in which small differences in two weather forecasts grow over time, leading to significant differences past some point

charge separation process of cloud electrification in which convective transport of charged air parcels results in differences of charge between the top and bottom of a cloud

chemical composition the amount and relative abundance of chemical elements and molecules within a given volume of air

chemical weathering process in which exposed land surfaces undergo chemical changes as they interact with carbon dioxide, water, and oxygen

chinook wind a local wind occurring at certain times to the lee of the Rocky Mountains; a very dry wind with a high capacity to evaporate snow

chlorofluorocarbons (CFCs) synthetic chemical compounds containing chlorine, fluorine, and carbon atoms that are widely used as coolant fluids in refrigeration systems

circle of illumination the circle that separates the illuminated portion of the Earth from the portion that is not illuminated

circulation cells horizontal and vertical circulations found in the average wind patterns of the atmosphere; examples include the Hadley cell and the Walker cell

cirrocumulus cloud type in which thin cloud patches are found in the upper troposphere

cirrostratus cloud type in which thin, smooth layers of clouds are found in the upper troposphere

cirrus see *cirrostratus*; also refers to all clouds composed of small ice crystals, typically found in upper troposphere

climate the average or prevailing weather for a given region, characterized by temperature, moisture, precipitation, and winds

climograph graph on which two or more climate variables are plotted for each month of the year

cloud condensation nuclei suspended aerosols around which water vapor condenses to form cloud drops

cloud-to-ground lightning lightning discharge between cloud and ground

cold air outbreaks equatorward movement of continental polar (cP) air masses into the low-latitude regions of the subtropics

cold front moving weather front along which a cold air mass moves underneath a warm air mass, lifting the warm air mass

cold occluded front surface front marking the boundary between a relatively colder air mass behind the front and a cool air mass ahead of the front, with a warm air mass aloft separating the two

cold wave weather event in which temperature drops significantly over 24 hours and remains at least 8.33°C (15°F) below average

condensation nucleus a tiny bit of solid matter (aerosol) in the atmosphere on which water vapor condenses to form a tiny water droplet

conduction of heat transmission of sensible heat through matter by transfer of energy from one atom or molecule to the next in the direction of decreasing temperature

contour set of lines connecting points with constant values of a given variable, such as temperature or precipitation

contour map map in which a set of lines connect geographic locations with constant values of a given variable, such as temperature or precipitation

convection (atmospheric) air motion consisting of strong updrafts taking place within a convection cell

convective condensation level see *lifting condensation level*

convective precipitation precipitation induced when warm, moist air is heated at the ground surface, rises, cools, and condenses to form water droplets, raindrops and, eventually, rainfall

convergence the movement of atmospheric mass into a region

coral bleaching process in which changing environmental conditions cause coral animals to become stressed and to release the symbiotic algae that live within the coral tissue

Coriolis effect force produced by the Earth's rotation that appears to deflect a moving object on the Earth's surface to the right in the northern hemisphere and to the left in the southern hemisphere

counterradiation longwave atmospheric radiation moving downward toward the Earth's surface

cumuliform clouds clouds of globular shape, often with extended vertical development

cumulonimbus cloud large, dense cumuliform cloud yielding precipitation

cumulus cloud type consisting of low-lying, white cloud masses of globular shape well separated from one another

cumulus stage stage of air-mass thunderstorm development in which surface heating produces isolated regions of significant updrafts, leading to the formation of cumulus clouds and precipitation

cyclone center of low atmospheric pressure

cyclone families succession of cyclonic storms that follow one after the other along the same track

cyclonic storm intense weather disturbance within a traveling cyclone, generating strong winds, cloudiness, and precipitation

debris flood (debris flow) stream-like flow of muddy water heavily charged with sediment of a wide range of size grades, including boulders, generated by sporadic torrential rains upon steep mountain watersheds

declination latitude of the subsolar point at a moment in time

deep ocean region below the thermocline in which water is very cold. This region typically extends to the bottom of the ocean floor

deglaciation widespread recession of ice sheets during a period of warming global climate, leading to an interglaciation; see also *glaciation, interglaciation*

deposition (atmosphere) the change of state of a substance from a gas (water vapor) to a solid (ice)

dew-point temperature the temperature at which air with a given humidity reaches saturation when cooled without changing its pressure

diffuse radiation solar radiation that has been scattered (deflected or reflected) by minute dust particles or cloud particles in the atmosphere

discrete (data) set of separate data points from which intermediate values cannot be determined; opposite of continuous data

dissipating stage stage of air-mass thunderstorm development in which strong downdrafts throughout the air column inhibit the convection and latent heat release needed to sustain the thunderstorm

diurnal a type of process that takes place or changes over the course of a day

divergence the movement of atmospheric mass out of a region

Doppler radar radar system in which changes in frequency between the emitted signal and return signal can be used to determine the velocity of the reflecting object

Doppler shift the change in frequency of a signal upon reflection off an object moving towards or away from the receiver

downburst strong, damaging winds produced by the horizontal outflow of convective downdrafts as they encounter the ground

downvalley breeze downslope winds produced at night as cold air near the slope sinks in elevation towards the valley floor

drizzle see *mist*

drought a prolonged period of abnormally dry weather; see also *agricultural drought, hydrological drought, meteorological drought*

dry adiabatic lapse rate the rate at which rising air is cooled by expansion when no condensation is occurring; 10°C per 1000 m (5.5°F per 1000 ft)

dry climate climate in which evaporation is limited by low soil moisture

dry lightning lightning that occurs in the absence of precipitation

dry line boundary separating hot, dry air from warm, moist air along which thunderstorms tend to form

dry midlatitude climate ⑨ dry climate of the midlatitude zone with a strong annual temperature cycle and cold winters

dry subtropical climate ⑤ dry climate of the subtropical zone, transitional between the dry tropical climate and the dry midlatitude climate

dry tropical climate ④ climate of the tropical zone with high temperatures and low rainfall

dust solid particulates suspended in the atmosphere by vertical air motions

dynamic pressure gradient the difference in pressure between two locations, after removing the component that arises solely from differences in altitude

Earth-Sun geometry spatial relation of the Earth to the Sun as a function of time

easterly wave a traveling surface low-pressure system in the tropics that moves from east to west

eccentricity a measure of the elliptical shape of the Earth's orbit around the Sun

Ekman transport the total transport of mass within a layer of air or water produced by the Coriolis force and frictional force

El Niño phenomenon in which the waters of the eastern tropical Pacific warm significantly

El Niño/Southern Oscillation (ENSO) a coupled system of ocean and atmospheric circulation that produces cyclic changes in winds, currents, temperatures, and rainfall patterns in the equatorial Pacific region

electric charge a positive or negative imbalance of electricity within a body, determined by a loss or gain of electrons

electrical discharge flow of electricity through the atmosphere, resulting in the redistribution of electric charge from one location to another

electrical hygrometer a device that measures water vapor content of the atmosphere, based upon changes in flow of electricity through a conductor

electrical resistance thermometer a device for measuring the temperature of surrounding air, based upon the flow of electricity through a conductor in contact with the air

electromagnetic radiation (also *electromagnetic energy*) wave-like form of energy radiated by any substance possessing heat; it travels through space at the speed of light

electromagnetic spectrum the total wavelength range of electromagnetic energy

electron negatively charged subatomic particle

emission process in which bodies release radiation that subsequently travels through space

emissions scenarios estimates of how emissions of greenhouse gases may change based on estimates of changing global socioeconomic conditions

energy the capacity to do work, that is, to bring about a change in the state or motion of matter

energy balance (global) balance between shortwave solar radiation received by the Earth-atmosphere system and radiation lost to space by shortwave reflection and longwave emission from the Earth-atmosphere system

energy balance (of a surface) balance between the flows of energy reaching a surface and the flows of energy leaving it

energy efficiency amount of energy used to perform a given amount of work

energy system a collection of related energy flows; may refer to the global energy system of energy flows reaching and leaving land and ocean surfaces

enhanced Fujita intensity scale a measure of tornado intensity, based upon damage to structures and surrounding vegetation, introduced in 2007 as an improvement to the original Fujita intensity scale

ensemble forecast a set of different weather forecasts for the same forecast time but generated from different models or from one model using different starting conditions

entrainment process in which cool, dry environmental air mixes with warm, moist rising air within cumulus clouds

environmental lapse rate the rate at which the actual temperature at a particular location and time drops with increasing height

equatorial easterlies upper-level easterly air flow over the equatorial zone

equatorial trough atmospheric low-pressure trough centered more or less over the equator and situated between the two belts of trade winds

equinox time of year when the Earth's axis of rotation is neither pointed toward the Sun nor away from it

evaporation process in which water in liquid state or solid state passes into the vapor state

evapotranspiration the combined water loss to the atmosphere by evaporation from soil and transpiration from plants

extended-range forecast a forecast of weather conditions for a period extending beyond 3 days

extratropical cyclone see *midlatitude cyclone*

Fahrenheit scale temperature scale in which the freezing point of water at sea level pressure is 32° and the boiling point is 212°

fair weather system a traveling anticyclone, in which the descent of air suppresses clouds and precipitation and the weather is typically fair

famine a severe shortage of food, leading to malnutrition, starvation, or death for a large portion of a given population

flash flood flood in which heavy rainfall causes a stream or river to rise very rapidly

flood stream flow at a stream stage so high that it cannot be accommodated within the stream channel and must spread over the banks to inundate the adjacent floodplain

floodplain a broad belt of low, flat ground bordering a river channel that floods regularly

fog cloud layer in contact with land or sea surface, or very close to that surface; see also *advection fog, radiation fog*

forest fire (also *wildfire*) condition of uncontrolled burning in a forested or uncultivated region

frictional force force applied to atmospheric motions due to differences between the wind velocity and the velocity of the surface over which the wind moves

front surface of contact between two air masses with different temperature and moisture characteristics; see also *cold front, occluded front, polar front, warm front*

frontal precipitation precipitation arising from air motions associated with the approach or passage of a front

frostbite skin damage arising from exposure to below-freezing temperatures

Fujita intensity scale scale used to rate the intensity of a tornado by examining the damage caused to different types of human-made structures

general circulation average wind patterns around the globe

general circulation models computer-based models for predicting global-scale weather and climate phenomena

geographic grid complete network of parallels and meridians on the surface of the globe, used to fix the locations of surface points

geostrophic wind wind at high levels above the Earth's surface blowing parallel to the isobars

glaciation (1) general term for the total process of glacier growth and landform modification by glaciers. (2) single episode or time period in which ice sheets formed, spread, and disappeared

global energy balance see *energy balance (global)*

global radiation balance the energy flow process by which the Earth absorbs shortwave solar radiation and emits longwave radiation; in the long run, the two flows must balance

global scale scale at which we are concerned with the Earth as a whole, for example in considering Earth–Sun relationships

global simulation models and global climate models computer models that combine atmosphere, ocean, and land-surface models into one model to simulate the entire coupled Earth system

gradient change in the value of a given variable over a given distance

graph visual representation of relation between two or more variables

graupel suspended ice pellets formed when supercooled water drops freeze around central ice crystal

greenhouse effect accumulation of heat in the lower atmosphere and at the Earth's surface; produced through the absorption of longwave radiation by the atmosphere and the re-emission back to the surface

greenhouse gases gases which absorb longwave radiation and re-emit it back to the surface, producing an accumulation of heat in the lower atmosphere

grid points discrete locations within general circulation models at which values of winds, temperature, moisture and other weather and climate variables are computed

Gulf Stream extension the extension of the Gulf Stream off of the South Carolina coast and into the northern Atlantic and the seas near Greenland and Iceland

gyre circulation of large-scale currents around ocean basins bounded by continents

Hadley cell low-latitude atmospheric circulation cell with rising air over the equatorial regions and sinking air over the subtropical belts

hail form of precipitation consisting of pellets or spheres of ice with a concentric layered structure

heat capacity amount of heat needed to raise the temperature of one gram of a given substance by one degree Celsius

heat exhaustion elevated body temperature resulting in dizziness, fatigue, or heavy perspiration

heat index measure of apparent temperature based upon actual air temperature and relative humidity, designed to account for inability to remove heat through perspiration

heat island persistent region of higher air temperatures centered over a city

heat stroke uncontrolled increase in body temperature due to accumulation of heat, resulting in failure of organs and possible death

heat trapping gases see *greenhouse gases*

heat wave persistent period of significantly elevated temperatures, usually lasting 3–5 days

hectopascal measure of pressure, equivalent to 100 pascals or 1 millibar

heterosphere region of the atmosphere above about 100 km in which gas molecules tend to become increasingly sorted into layers by molecular weight and electric charge

high-level temperature inversion condition in which a high-level layer of warm air overlies a layer of cooler air, reversing the normal trend of cooling with altitude

high-pressure belt belt of persistent high atmospheric pressure centered approximately on latitudes 30° N and 30° S

homosphere the lower portion of the atmosphere, below about 100 km altitude, in which atmospheric gases are uniformly mixed

hook echo curve-shaped region seen in radar images when precipitation is drawn into a thunderstorm, indicating possible tornado formation

horizontal vortex circulation of air in which axis or rotation is parallel to the surface

humidity general term for the amount of moisture in the air

hurricane tropical cyclone of the western North Atlantic and Caribbean Sea

hydrologic cycle pathways of active movement of water between the ocean, atmosphere, and land surface

hydrological drought drought conditions in which surface and subsurface water supplies are substantially depleted, either through decreases in precipitation or increases in evaporation or runoff

hydrosphere total water realm of the Earth's surface, including the oceans, surface waters of the lands, ground water, and water held in the atmosphere

hygrometer instrument that measures the water vapor content of the atmosphere; some types measure relative humidity directly

hypothermia uncontrolled decrease in body temperature, resulting in reduced coordination, decreased blood flow and possible death

ice age geologic time period during which glaciations alternate with interglaciations in rhythm with cli-

mate changes; see also *glaciation, interglaciation*

ice sheet climate ⑬ severely cold climate found on the Greenland and Antarctic ice sheets

ice–albedo feedback positive feedback in which an increase in temperature results in a melting of ice and snow, thereby decreasing the albedo of the surface and subsequently increasing temperatures further

ideal gas law the law that describes the relationship between absolute temperature, pressure, and density of gases

infrared hygrometer instrument that measures the water vapor content of the atmosphere by measuring amount of emitted radiation absorbed by air between a source and receiver

infrared radiation electromagnetic energy in the wavelength range of 0.7 to about 200 μm

initial conditions quantitative description of initial state of atmospheric conditions used as input to general circulation model

insolation the flow of solar energy intercepted by an exposed surface, assuming a uniformly spherical Earth with no atmosphere

interglaciation within an ice age, a time interval of mild global climate in which continental ice sheets were largely absent or were limited to the Greenland and Antarctic ice sheets; the interval between two glaciations; see also *deglaciation, glaciation*

intertropical convergence zone (ITCZ) zone of convergence of air masses along the equatorial trough

inversion see *temperature inversion*

ion electrically charged atom or molecule

ionosphere layer of the upper atmosphere characterized by the presence of ions

isallobaric map map of lines connecting places of equal change in pressure within a specified time period

isobaric surface surface on which atmospheric pressures are constant

isobars lines on a map drawn through all points having the same atmospheric pressure

isohyets lines on a map connecting regions with equal amounts of rainfall

isolines lines on a map that connect locations with equal values of a given variable

isopleth line on a map or globe drawn through all points having the same value of a selected property or entity

isotherm line on a map drawn through all points with the same temperature

isotope form of an element with a unique atomic mass number

jet streak localized regions of very high winds embedded within the overall jet stream

jet stream high-speed air flow in narrow bands within the upper-air westerlies and along certain other global latitude zones at high levels

Joule unit of work or energy in the metric system; symbol, J

katabatic winds winds, usually cold, that flow from higher to lower regions under the direct influence of gravity

Kelvin scale (K) temperature scale on which the starting point is absolute zero, equivalent to –273°C

kinetic energy form of energy represented by matter (mass) in motion

knot measure of speed used in marine and aeronautical applications equal to one nautical mile per hour (1 kt = 0.514 m/s = 1.15 mph)

Köppen climate system classification of climates based upon annual temperatures and precipitation and their seasonality

La Niña situation in which the waters of the eastern tropical Pacific cool significantly

lahar rapid downslope or downvalley movement of a tongue-like mass of water-saturated tephra (volcanic ash) originating high up on a steep-sided volcanic cone; a variety of mudflow

land breeze local wind blowing from land to water during the night

land degradation degradation of the quality of plant cover and soil as a result of overuse by humans and their domesticated animals, especially during periods of drought

lapse rate the rate at which temperature drops with increasing height

latent heat heat absorbed or released as substances change from one phase to another

latent heat transfer flow of latent heat that results when water absorbs heat to change from a liquid or solid to a gas and then later releases that heat to new surroundings by condensation or deposition

latitude arc of a meridian between the equator and a given point on the globe

leader channel of ionized air extending from the cloud bottom to the ground that initiates the lightning stroke

lee-side trough low-pressure region found on the downwind (lee) side of a mountain chain, which can subsequently generate midlatitude cyclones

level of condensation see *lifting condensation level*

level of free convection Level at which a lifted air parcel's temperature, following first the dry adiabatic lapse rate until saturated and then the moist adiabatic lapse rate, becomes greater than surrounding environmental temperature

lifting condensation level level of the atmosphere to which an air parcel must be lifted before condensation starts to occur

lightning an electrical discharge produced by a thunderstorm, resulting in a flash of light extending from cloud to cloud or cloud to ground

lightning channel channel of ionized air which a lightning stroke follows

lightning flash the visible light produced by the rapid acceleration of electrons within the leader of the lightning stroke

lightning stroke the process of electrical discharge between a cloud and the ground that results in a visible lightning flash

line graph visual representation of relation between two or more variables, in which the measured values are connected by a line representing intermediate values

liquid-in-glass thermometer device for measuring temperature based upon the thermal expansion of a liquid within a closed glass container

lithosphere the solid portion of the Earth's surface, extending 100 km deep and comprising the ocean basins and continents

local weather forecasting the use of local atmospheric conditions to make a forecast of weather, usually within the next 6–12 hours

local winds general term for winds generated as direct or immediate effects of the local terrain

longitude arc of a parallel between the prime meridian and a given point on the globe

longwave radiation electromagnetic radiation in the range 3–30 mm; the type of radiation emitted by the Earth and atmosphere

longwave–temperature feedback negative feedback in which an increase in temperature of a surface is moderated by increasing longwave emission of energy

low-level temperature inversion atmospheric condition in which temperature near the ground increases, rather than decreases, with elevation

map a paper representation of space showing point, line, or area data

map projection a system of parallels and meridians representing the curvature of the Earth drawn on a flat surface

marine west-coast climate ⑧ cool moist climate of west coasts in the midlatitude zone, usually with abundant precipitation and a distinct winter precipitation maximum

mass convergence see *convergence*

mass divergence see *divergence*

mature stage stage of air-mass thunderstorm development in which strong updrafts and downdrafts are present, resulting in heavy rainfall, gusty winds, and lightning

mean annual temperature mean of (mean) daily air temperature for a given year or succession of years

mean daily temperature sum of daily maximum and minimum air temperature readings divided by two

mean monthly temperature mean of (mean) daily air temperature for a given calendar month

mechanical weathering see *physical weathering*

Mediterranean climate ⑦ climate type of the subtropical zone characterized by a very dry summer and a mild, rainy winter

Mercator projection map projection of horizontal parallels and vertical meridians, with the space between parallels increasing poleward

meridional transport flow of energy (heat) or matter (water) across the parallels of latitude, either poleward or equatorward

mesocyclone a vertical column of cyclonically rotating air that develops in the updraft of a severe thunderstorm cell

mesopause upper limit of the mesosphere

mesoscale convective complex a large complex of thunderstorms, generally round or oval-shaped, characterized by long-lived convection and heavy rainfall that persists through the night

mesoscale convective system a relatively long-lived, large, and intense convective cell or cluster of cells characterized by exceptionally strong updrafts

mesosphere atmospheric layer of upwardly diminishing temperature, situated above the stratopause and below the mesopause

meteogram a plot showing a numerical prediction of meteorological variables such as temperature, relative humidity, wind speed and direction, and pressure, at a given location over time

meteorological drought drought conditions characterized by below average precipitation, usually described as a percentage decrease of normal precipitation over a given time period

microburst a strong localized downdraft from a thunderstorm that produces rapid changes in surface wind speeds

midlatitude cyclone traveling cyclone of the midlatitudes involving interaction of cold and warm air masses along sharply defined fronts

midlatitude desert region of low precipitation and runoff, with low temperatures and high annual temperature range

midlatitude westerlies predominant west-to-east surface winds typically found in the latitudes 35–65N and 35–65S

Milanković cycle cyclical changes in the Earth's movement and orientation with respect to the Sun

millibar unit of atmospheric pressure; one-thousandth of a bar. Bar is a force of one million dynes per square centimeter

mist suspension of microscopic liquid or frozen water drops near the surface

mistral local drainage wind of cold air affecting the Rhone Valley of southern France

mitigation (of climate change) process of modifying behavior in order to reduce the expected change in climate associated with human activity

mixed layer upper portion of the ocean in which the water is well mixed by waves and wind, resulting in uniform temperatures and salinity

moist adiabatic lapse rate reduced rate at which rising air is cooled by expansion when condensation is occurring; ranges from 4 to 9°C per 1000 m (2.2–4.9°F per 1000 ft)

moist climate climate in which precipitation provides enough soil moisture to support evapotranspiration throughout the year

moist continental climate ⑩ moist climate of midlatitude zones with strongly defined winter and summer seasons and adequate precipitation throughout the year

moist subtropical climate ⑥ moist climate of the subtropical zone, characterized by a moderate to large annual

water surplus and a strong seasonal temperature cycle

monsoon a large seasonal shift in the wind direction and amount of rainfall over an extended area such as southeastern Asia

mountain sickness shortness of breath and possible nausea due to decrease in oxygen availability at high altitudes

mountain winds daytime movements of air up the gradient of valleys and mountain slopes; alternating with nocturnal valley winds

mudflow mixture of water and soil that flows rapidly downhill, following a stream channel

multiple vortex tornado tornado in which two or more suction vortices slowly rotate around larger vortex region

negative feedback in flow systems, a linkage between flow paths such that the flow in one pathway acts to reduce the flow in another pathway; see *positive feedback*

net radiation the difference between the amount of incoming radiation and the amount of outgoing radiation at a given location

Newton the SI unit of force required to accelerate 1 kg at the rate of 1 (m/s)/s; symbol, N

nimbostratus cloud type in which sheets or layers of clouds are found near the bottom of the troposphere, accompanied by precipitation

nimbus any cloud type in which precipitation is occurring

noon angle (of the Sun) angle of the Sun above the horizon at its highest point during the day

nor'easters storm system affecting the northeastern United States and eastern Canada in which northeasterly winds accompanying an off-shore cyclone bring cool, moist air on-shore, usually resulting in heavy snows and strong winds

nortes strong, northeasterly winds found in subtropical regions of North America, usually that accompany the intrusion of cold cP air from the north

North Atlantic oscillation (NAO) a large-scale climate signal over the mid-

and high latitudes of the North Atlantic characterized by changing pressure differences between polar (Iceland) and subtropical (Azores) surface pressures

North Pacific pressure pattern changes in sea surface temperatures and sea level pressures of the north Pacific occurring on seasonal to decadal time-scales

North Pole point at which the northern end of the Earth's axis of rotation intersects the Earth's surface

northeast trade winds surface winds of low latitudes that blow steadily from the northeast; see also **trade winds**

northern lights see **aurora borealis**

northes see **nortes**

nowcast a short-term weather forecast, generally for the next few hours

oblate ellipsoid geometric solid resembling a flattened sphere, with polar axis shorter than the equatorial diameter

obliquity a measure of the angle of tilt of the Earth's axis of rotation with respect to the plane of its orbit around the Sun

occluded front weather front along which a moving cold front has overtaken a warm front, forcing the warm air mass aloft

ocean current persistent, horizontal flow of ocean water

ocean upwelling process in which warm surface waters are replaced by colder waters from below

organic matter material derived from living organisms

orographic pertaining to mountains

orographic precipitation precipitation induced when moist air is forced to rise over a mountain barrier

overrunning situation in which a warmer, less dense air mass aloft moves relative to a colder, more dense air mass at the surface

overshooting top domelike cloud formation extending above an anvil cloud, resulting from rapid rise of unstable air past the equilibrium level

ozone form of oxygen with a molecule consisting of three atoms of oxygen, O_3

Pacific decadal oscillation slowly varying change in sea surface temperatures and sea level pressures of the north Pacific

Pampero strong, southerly or southwesterly winds found in Argentina and Uruguy that usually accompany the intrusion of cold mP air from the south

parameterizations estimates of certain atmospheric processes that approximate the actual behavior of the atmosphere

pascal metric unit of pressure, defined as a force of one newton per square meter ($1 N/m^2$); symbol, Pa; 100 Pa = 1 mb, 105 Pa = 1 bar

pathlength length of homogenous gas through which radiation passes

perihelion point on the Earth's elliptical orbit when it is closest to the Sun

permafrost soil and bedrock at a temperature below $0°C$ ($32°F$), found in cold climates of arctic, subarctic, and alpine regions

persistence forecast forecast in which future weather conditions are predicted to be the same as present conditions

phase the state of a substance: either solid, liquid, or gas

photosynthesis production of carbohydrate by the union of water with carbon dioxide while absorbing light energy

physical weathering breakup of massive rock (bedrock) into small particles through the action of physical forces acting at or near the Earth's surface; see also **weathering**

phytoplankton microscopic plants found largely in the uppermost layer of ocean or lake water

plane of the ecliptic imaginary plane in which the Earth's orbit lies

planetary vorticity rotation imparted to an object by the rotation of the Earth

plot see **graph**

plough winds see **straight-line winds**

polar cyclones see **polar lows**

polar easterlies system of easterly surface winds at high latitude, best developed in the southern hemisphere, over Antarctica

polar front boundary between cold polar air masses and warm tropical air masses

polar front jet stream jet stream found along the polar front, where cold polar air and warm tropical air are in contact

polar front zone broad zone in midlatitudes and higher latitudes, occupied by the shifting polar front

polar high persistent low-level center of high atmospheric pressure located over the polar zone of Antarctica

polar lows cyclone found in subpolar regions of the Pacific and Atlantic, usually in regions with large land-ocean temperature differences

polar outbreak tongue of cold polar air, preceded by a cold front, penetrating far into the tropical zone and often reaching the equatorial zone; it brings rain squalls and unusual cold

polar projection map projection centered on the Earth's North Pole or South Pole

Poles location on the Earth's surface through which the axis of rotation passes

poleward heat transport movement of heat from equatorial and tropical regions toward the poles, occurring as latent and sensible heat transfer

pollutants in air pollution studies, foreign matter injected into the lower atmosphere as particulates or as chemical pollutant gases

pollution dome broad, low dome-shaped layer of polluted air, formed over an urban area at times when winds are weak or calm prevails

pollution plume (1) the trace or path of pollutant substances, moving along the flow paths of ground water. (2) trail of polluted air carried downwind from a pollution source by strong winds

positive feedback in flow systems, a linkage between flow paths such that the flow in one pathway acts to increase the flow in another pathway; see also *negative feedback*

power rate at which energy is transferred, defined in the SI system by watts

precession change in the direction of the Earth's axis of rotation over time

precipitation particles of liquid water or ice that fall from the atmosphere and may reach the ground

pressure defined as the force applied perpendicular to the surface divided by the area of the surface

pressure gradient change of atmospheric pressure measured along a line at right angles to the isobars

pressure gradient force force acting horizontally, tending to move air in the direction of lower atmospheric pressure

pressure tendency change in pressure at a given location over the previous 3 hour period

prevailing westerly winds (westerlies) surface winds blowing from a generally westerly direction in the midlatitude zone, but varying greatly in direction and intensity

prognostic chart a map or chart showing forecast pressure patterns and other meteorological variables for a specific time in the future

proxy data information used for indirect measurement of past climate characteristics, such as the thickness of annual tree rings or the chemical composition of annual coral growth rings

radar an active remote sensing system in which a pulse of radiation is emitted by an instrument, and the strength of the echo of the pulse is recorded

radiation balance see *global radiation balance*

radiation fog fog produced by radiation cooling of the near-surface air layer

radiation see *electromagnetic radiation*

radiometer instrument designed to measure flux or intensity of electromagnetic radiation

radiosonde instrument or set of instruments born aloft by a balloon, designed to measure atmospheric conditions in the troposphere and lower stratosphere

rain form of precipitation consisting of falling water drops, usually 0.5 mm or larger in diameter

rain gauge instrument used to measure the amount of rain that has fallen

rain shadow belt of dry climate leeward of a mountain barrier, produced as a result of adiabatic warming of descending air

raindrop a drop of liquid water greater than 0.5 mm in diameter; in practice drops of liquid water between 0.2 mm and 0.5 mm are also considered raindrops

Rayleigh scattering scattering of electromagnetic radiation by particles much smaller than the wavelength of radiation itself

reflection outward scattering of radiation toward space by the atmosphere and/or Earth's surface

reflectivity ratio of reflected radiation flux to incident radiation flux

regional scale the scale of observation at which subcontinental regions are discernible

relative humidity the amount of water vapor in an air parcel as a fraction of the maximum amount it can contain based on its temperature

relative vorticity measure of local rotation with respect to an observer on the Earth's surface

remote sensing measurement of some property of an object or surface by means other than direct contact; usually refers to the gathering of scientific information about the Earth's surface from great heights and over broad areas, using instruments mounted on aircraft or orbiting space vehicles

remote sensor instrument or device measuring electromagnetic radiation reflected or emitted from a target body

renewable energy resources solar power, wind power, tidal power, biomass combustion, and other power sources that are continuously available and constantly renewed.

residence time the time it takes for an increase in the concentration of an atmospheric gas or aerosol to be effectively removed

respiration the oxidation of organic compounds by organisms that powers bodily functions

return stroke intense light that propagates upward from the Earth to the cloud base in the last phase of each lightning stroke

revolution motion of a planet in its orbit around the Sun, or of a planetary satellite around a planet

rhythm of the seasons progression of changes in temperature, precipitation, and other atmospheric conditions that accompany the passage of the Earth around the Sun

ridge on a map, a region of relatively high pressure at the surface or aloft in which the isobars do not form a closed high pressure center

roll cloud tube-shaped cloud whose axis is parallel to the ground, typically found above a gust front where outflowing air at the surface is in the opposite direction to the in-flowing air above it

rotation spinning of an object around an axis

runoff flow of water from continents to oceans by way of stream flow and ground water flow; a term in the water balance of the hydrologic cycle. In a more restricted sense, runoff refers to surface flow by overland flow and channel flow

Sahel (Sahelian zones) belt of wet–dry tropical and semiarid dry tropical climate in Africa in which precipitation is highly variable from year to year

salinity degree of "saltiness" of water; refers to the abundance of such ions as sodium, calcium, potassium, chloride, fluoride, sulfate, and carbonate

Santa Ana easterly wind, often hot and dry, that blows from the interior desert region of southern California and passes over the coastal mountain ranges to reach the Pacific Ocean

saturation the condition in which the specific humidity is equal to the saturation specific humidity

saturation specific humidity the maximum amount of water vapor an air parcel can contain based on its temperature

scale the magnitude of a phenomenon or system, as for example global scale or local scale

scattering process in which particles and molecules deflect incoming solar radiation in different directions; atmospheric scattering can redirect solar radiation back to space

sea level pressure observed (over oceans) or estimated (over land surfaces) pressure found at the mean level of the ocean

semiarid (steppe) dry climate subtype subtype of the dry climates exhibiting a short wet season supporting the growth of grasses and annual plants

sensible heat an indication of the intensity of kinetic energy of molecular motion within a substance; it is measured by a thermometer

sensible heat transfer flow of heat from one substance to another by direct contact

severe thunderstorm thunderstorm in which the surface winds are greater than 26 m/s (58 mph), hail is more than 19 mm (0.75 in.) in diameter, or there is an accompanying tornado

severe weather weather event that has the potential to cause substantial destruction or loss of life

sheet lightning lightning flash that occurs within a cloud or between clouds, and therefore appears as a diffuse flash to an observer on the ground

short-range forecast a weather forecast made for a time period of up to 48 hours

shortwave radiation electromagnetic energy in the range from 0.3 to 3 mm; shortwave radiation comes exclusively from the Sun

skill a measure of the accuracy of a forecast based on how it compared with the actual weather conditions at the forecast time

slash-and-burn agricultural system, practiced in the low-latitude rainforest, in which small areas are cleared and the trees burned, forming plots that can be cultivated for brief periods

sleet form of precipitation consisting of ice pellets, which may be frozen raindrops

sling psychrometer form of hygrometer consisting of a wet-bulb thermometer and a dry-bulb thermometer

smog mixture of aerosols and chemical pollutants in the lower atmosphere, usually found over urban areas

snow form of precipitation consisting of ice particles

snow pillow instrument designed to measure weight or pressure of overlying snowpack

soil moisture amount of liquid water stored in the soil

solar constant intensity of solar radiation falling upon a unit area of surface held at right angles to the Sun's rays at a point outside the Earth's atmosphere; equal to an energy flow of about 1400 W/m^2

solubility total amount of a chemical species that can be dissolved within a given liquid, for a given temperature and pressure

source region extensive land or ocean surface over which an air mass derives its temperature and moisture characteristics

South Pole point at which the southern end of the Earth's axis of rotation intersects the Earth's surface

southeast trade winds surface winds of low latitudes that blow steadily from the southeast; see also *trade winds*

Southern Oscillation episodic strengthening or weakening of prevailing barometric pressure differences between two regions, one centered on Darwin, Australia, in the eastern Indian Ocean, and the other on Tahiti in the western Pacific Ocean; a precursor to the occurrence of an El Niño event; see also *El Niño*

specific heat physical constant of a material that describes the amount of heat energy in joules required to raise the temperature of one gram of the material by one Celsius degree

specific humidity amount of water vapor (grams) contained within a kilogram of air

spectrum see *electromagnetic spectrum*

speed rate at which an object is moving, regardless of direction; magnitude of an object's velocity

squall line line of thunderstorms and strong winds that extends for several hundred miles

stable air mass air mass in which the environmental temperature lapse rate is less than the dry adiabatic lapse rate, inhibiting convective uplift and mixing

static pressure gradient the difference in pressure between two locations that arises solely from differences in altitude between the two locations

stationary front boundary between two differing air masses that has not moved over the last 3-6 hours, or that has moved relatively slowly

steering level level of the atmosphere at which the horizontal velocity has a direct relationship to the movement of pressure patterns and weather features at the surface

steering winds upper-air wind patterns associated with the jet stream that tend to steer the direction in which midlatitude cyclones travel

Stefan-Boltzmann equation equation that relates the total energy emitted by a body as a function of its temperature

steppe climate see *semiarid (steppe) dry climate subtype*

steppe semiarid grassland occurring largely in dry continental interiors

storm generation region regions in which midlatitude and tropical cyclones originate, generally due to some geographic feature that gives rise to disturbances in the atmospheric flow

storm surge rapid rise of the coastal water level accompanying the onshore arrival of a tropical or midlatitude cyclone

storm tracks common paths that cyclonic storms tend to follow, usually associated with the location of the jet stream

straight-line winds strong surface winds that do not have significant curvature to them, usually arising from downbursts spreading out at the surface

stratiform clouds clouds of layered, blanket-like form

stratopause upper limit of the stratosphere

stratosphere the layer of atmosphere directly above the troposphere; here temperature increases with altitude

stratus cloud type of the low-height family formed into a dense, dark gray layer

subgeostrophic winds winds in which the speed is less than the geostrophic wind speed associated with the surrounding horizontal pressure gradients

sublimation process of change of ice (solid state) to water vapor (gaseous state)

subsolar point the one point on Earth at any given time where the Sun is directly overhead

subtropical high-pressure belt belt of persistent high atmospheric pressure centered approximately on latitude 30° N and 30° S

subtropical jet stream jet stream of westerly winds forming at the tropopause, equatorward of the Hadley cell

subtropical zones latitude zones occupying the region of lat. 25° to 35° N and S (more or less) and lying between the tropical zones and the midlatitude zones

sulfate aerosols suspended liquid sulfuric acid drops, resulting from the oxidation of sulfur compounds emitted from volcanoes or fossil fuel combustion

summer (June) solstice moment in time when the North Pole is directed 23.5° toward the Sun

summer monsoon inflow of maritime air at low levels from surrounding oceans toward the low pressure centers found in continental interiors during the season of high Sun; associated with the rainy season of the wet–dry

tropical climate and the Asiatic monsoon climate

sunglint strong reflection of incident light upon water or snow, usually in a specific direction, producing a distinct bright spot upon a darker background

supercell thunderstorm strong, single-cell convective storm that persists for many hours and is usually accompanied by severe weather including downbursts, hail, and possible tornado formation

supercooled water water existing in the liquid state at a temperature lower than the normal freezing point

supergeostrophic winds winds in which the speed is greater than the geostrophic wind speed associated with the surrounding horizontal pressure gradients

surface chart chart showing the surface weather conditions, including sea-level pressures, wind speed and direction, cloud cover, fronts, precipitation and other weather features

surface energy balance equation equation expressing the balance among heat flows to and from a surface

surface pressure atmospheric pressure measured at the Earth's surface, usually at a specific location or locations

symbiosis form of positive interaction between species that is beneficial to one of the species and does not harm the other

synoptic chart map or chart showing meteorological data obtained simultaneously over a wide area

synoptic weather forecast weather forecast using synoptic maps to identify features at the surface and aloft that indicate how weather will develop over a specific time period

system (1) a collection of things that are somehow related or organized; (2) a scheme for naming, as in a classification system; (3) a flow system of matter and energy

teleconnections large-scale changes in the general circulation of the atmosphere that allow changes in atmospheric circulations in one region to

affect the climate of regions far from the original location

temperature a measure of the molecular energy within a given substance

temperature gradient rate of temperature change along a selected line or direction

temperature inversion upward reversal of the normal environmental temperature lapse rate, so that the air temperature increases upward; see also *high-level temperature inversion, low-level temperature inversion*

temperature regime distinctive type of annual temperature cycle

thermal infrared a portion of the infrared radiation wavelength band, from approximately 3 to 20 μm, in which objects at temperatures encountered on the Earth's surface (including fires) emit electromagnetic radiation

thermal radiation see *thermal infrared*

thermistor electronic device that measures (air) temperature

thermocline region below the mixed layer in which there is a rapid decrease in ocean temperature over a relatively short vertical distance

thermohaline circulation the global-scale, three-dimensional circulation of water through all of the ocean basins driven by the sinking of cold, dense water in the high latitudes of the Atlantic

thermometer shelter louvered wooden cabinet of standard construction used to hold thermometers and other weather-monitoring equipment

thermosphere atmospheric layer of upwardly increasing temperature, lying above the mesopause

thunder sound waves generated from the rapid expansion of superheated air around a lightning bolt

thunderstorm any storm in which vertical motions are sufficient to cause lightning and thunder

time cycle a regular alternation of a given system with time

time scale see *time cycle*

tornado a rapidly rotating column of air extending from the base of a thunderstorm that comes in contact with the ground

tornado alley qualitative description of region in central United States where tornado occurrences are most frequent

tornado emergency warning issued by National Weather Service only when a large tornado is predicted to impact populated regions

tornado vortex signature Doppler radar signature of tornado-producing circulations, characterized by opposing high velocity winds moving towards and away from radar receiver

tornado warning warning issued by National Weather Service when a tornado has been sited or the signature of possible tornado formation has been detected by radar

tornado watch a warning issued by the National Weather Service when weather conditions are forecasted that may give rise to tornado-producing severe thunderstorms

trade wind coast ocean coastal zone receiving easterly trade winds during some or all of the year, often with abundant convective precipitation.

trade winds (also trades) easterly winds found in the tropical regions north and south of the Equator

trade-wind coastal climate ② moist climate of low latitudes showing a rainfall peak in the high-Sun season and a short period of reduced rainfall in the middle of the low-Sun season

transmission process in which incoming solar radiation passes through the atmosphere without being absorbed or scattered

transpiration the process by which plants lose water to the atmosphere by evaporation through leaf pores

tropical cyclone intense traveling cyclone of tropical and subtropical latitudes, accompanied by high winds greater than 33 m/s (74 mph) and heavy rainfall

tropical cyclone track path of movement of a tropical cyclone, typically characterized by westward movement at low latitudes, followed by poleward

and eastward movement in higher latitudes

tropical depression tropical cyclone with a closed low and wind speeds lower than 17 m/s

tropical desert region of low precipitation and runoff, with high temperatures and low annual temperature range

tropical easterly jet stream upper-air jet stream of seasonal occurrence, running east to west at very high altitudes over Southeast Asia

tropical storm tropical cyclone with a closed low and wind speeds between 17–33 m/s

tropopause the level of the atmosphere between the troposphere and stratosphere, where temperatures stop decreasing with height and start increasing

troposphere the lowest layer of the atmosphere, in which temperature falls steadily with increasing height

trough on a map, a region of relatively low pressure at the surface or aloft in which the isobars do not form a closed low pressure center

tundra high-latitude climate region with no trees and usually underlaid by permafrost

tundra climate ⑫ cold climate of the arctic zone with eight or more months of frozen ground

typhoon tropical cyclone of the western North Pacific and coastal waters of Southeast Asia

unstable air air with substantial content of water vapor, capable of breaking into spontaneous convective activity leading to the development of heavy showers and thunderstorms

upper-air westerlies winds located at or slightly below the tropopause and poleward of 30° N and S, blowing from a generally westerly direction, but varying greatly in direction and intensity

upvalley breeze see *mountain winds*

upwelling upward motion of cold, nutrient-rich ocean waters, often associated with cool equatorward currents occurring along continental margins

urban heat island area at the center of a city that has a higher temperature than surrounding regions; see also *heat island*

validating process of determining the accuracy and skill of a weather forecast against the observed weather

validation time the time at which a weather forecast can be compared to the actual weather conditions

valley winds air movement at night down the gradient of valleys and the enclosing mountainsides; alternating with daytime mountain winds

variability measure of the variation in a series of observations that center around a mean

variable a characteristic of a system that can assume different values in space or time

veering (of winds) in northern hemisphere, a clockwise rotation of wind direction with height; in southern hemisphere, rotation is counterclockwise

velocity rate of change of position as a function of time, characterized by direction and speed

vertical winds velocity component of wind in the direction perpendicular to the local Earth's surface

visible light electromagnetic energy in the wavelength range of 0.4 to 0.7 μm

vorticity the rate of rotation of a fluid flow

warm front moving weather front along which a warm air mass slides over a cold air mass, leading to the production of stratiform clouds and precipitation

warm occluded front surface front marking the boundary between a cool air mass behind the front and a relatively colder air mass ahead of the front, with a warm air mass aloft separating the two

waste heat heat generated by human activity that is released to the environment

water vapor the gaseous state of water

water vapor–temperature feedback positive feedback in which an increase in temperature results in increased evaporation, thereby enhancing the natural greenhouse effect and subsequently increasing temperatures further

watt unit of power equal to the quantity of work done at the rate of one joule per second; symbol, W

wavelength the distance separating one wave crest from the next wave crest

weather the state of the atmosphere at a particular location and time, usually determined by temperature, moisture, precipitation, and winds

weather system recurring pattern of atmospheric circulation associated with characteristic weather, such as a cyclone or anticyclone

weathering total of all processes acting at or near the Earth's surface to cause physical disruption and chemical decomposition of rock; see also *chemical weathering, physical weathering*

weight measure of the gravitational force acting on an object

Wein's law law that relates the temperature of an object to the wavelength of the most intense radiation emitted by that object

westerlies see *prevailing westerly winds, upper-air westerlies*

western boundary currents narrow, fast-moving currents found along the western edge of gyre circulations

wet equatorial belt region at or near the Equator receiving abundant annual convective precipitation from intertropical convergence

wet equatorial climate ① moist climate of the equatorial zone with a large annual water surplus and uniformly warm temperatures throughout the year

wet-dry tropical climate ③ climate of the tropical zone characterized by a very wet season alternating with a very dry season

wildfire see *forest fire*

wind air motion relative to the Earth's surface, usually referring to the horizontal direction

wind shear a change in wind speed or direction with height

winter (December) solstice moment in time when the North Pole is directed 23.5° away from the Sun

x-axis the horizontal axis of a 2-dimensional plot or coordinate system

y-axis the vertical axis of a 2-dimensional plot or coordinate system

PHOTO CREDITS

Chapter 1

Pages 2–3: Jaques Descloitres/MODIS Land Rapid Response Team/NASA/Visible earth; page 2: (inset) NG Maps; page 4: Courtesy NASA; page 5: (top) Stacy Gold/NG Image Collection; (left) Todd Gipstein/NG Image Collection; (right) Norbert Rosing/NG Image Collection; (center) Nobert Rosing/NG Image Collection; page 8: Courtesy NOAA; page 9: (top) Courtesy NASA; (center) Todd Gipstein/NG Image Collection; (bottom) Peter Carsten/NG Image Collection; page 10: Christopher Knight/NG Image Collection; page 11: James Stanfield/NG Image Collection; page 12: (top left) Courtesy NASA; (center) Peter Carsten/NG Image Collection; (top right) Richard Nowitz/NG Image Collection; page 13: (left) Ira Block/NG Image Collection; (right) Ira Block/NG Image Collection; page 14: (top) Courtesy USGS; (center left) Courtesy NOAA; (center right) Richard Perry/The New York Times/Redux Pictures; page 15: Todd Gipstein/NG Image Collection; page 17: Bridgeman Art Library/Getty Images; page 20: (top) Courtesy NASA; page 24: (top) Courtesy NASA; (bottom) Bridgeman Art Library/Getty Images; page 25: Courtesy NASA

Chapter 2

Pages 28–29: ©AP/Wide World Photos; page 30: SPL/Photo Researchers, Inc.; page 31: Todd Gipstein/NG Image Collection; page 33: Jonathan Blair/NG Image Collection; page 34: Pritt Vesilind/NG Image Collection; page 35: (top left) David Edwards/NG Image Collection; (center left) Rich Reid/NG Image Collection; (bottom left) Sam Kittner/NG Image Collection; (top right) Rich Reid/NG Image Collection; (bottom right) Jonathan Blair/NG Image Collection; page 37: Courtesy NASA; page 38: Jodi Cobb/NG Image Collection; page 39: Rich Reid/NG Image Collection; page 41: Norbert Rossing/NG Image Collection; page 42: (top) Norbert Rossing/NG Image Collection; (bottom) Rich Reid/NG Image Collection; page 44: Jian Chen/The Stock Connection/Science Faction/Getty Images; page 45: Jimmy Chin/NG Image Collection; (inset) NG Maps; page 47: (top) Paul Zahl/NG Image Collection; (bottom left) Jodi Cobb/NG Image Collection; page 48: Jimmy Chin/NG Image Collection; page 49: (bottom) Norbert Rossing/NG Image Collection

Chapter 3

Pages 50–51: Courtesy NASA; page 52: (left) Steve Raymer/NG Image Collection; page 52: (right) Stephen J. Krasemann//DRK Photo; page 53: (top) Courtesy NOAA; (center left) Craig Aurness/Westlight/NG Image Collection; (center right) Tim Laman/NG Image Collection; (bottom left) Peter Carsten/NG Image Collection; (bottom left inset) NG Maps; (bottom right) Stephen J. Krasemann//DRK Photo; page 55: John Eastcott and Yva Momatiuk/NG Image Collection; page 56: NASA/SPL/Photo Researchers, Inc.; page 58: Gordon Wiltsie/NG Image Collection; page 59: Courtesy NOAA; page 64: Taylor S. Kennedy/NG Image Collection; page 65: (top) Sarah Leen/NG Image Collection; (bottom) Taylor S. Kennedy/NG Image Collection; page 67: (top) NASA/NG Image Collection; (bottom) John Dunn/Arctic Light/NG Image Collection; page 68: (top left) John Dunn/Arctic Light/NG Image Collection; (top right) Jeremy Woodhoue/Masterfile; (bottom) Jimmy Chin/NG Image Collection; page 69: (top) Courtesy NASA; (bottom) MODIS Land Rapid Response Team/Courtesy NASA; page 75: Jeremy Woodhoue/Masterfile; page 76: (top left) ©Corbis Digital Stock; (bottom) DAJ/Getty Images; page 77: (top left) Paul Mason/Getty Images; (top right) Courtesy NOAA; (bottom right) Photodisc/Getty Images; page 78: (top) Stephen Alvarez/NG Image Collection; (bottom) NASA/SPL/Photo Researchers, Inc.; page 80: John Eastcott and Yva Momatiuk/NG Image Collection

Chapter 4

Pages 82–83: Gerd Ludwig/NG Image Collection; page 82: (inset) NG Maps; page 85: (left) Medford Taylor/NG Image Collection; (right) Richard Nowitz/NG Image Collection; page 87: (center) Paul Nicklen/NG Image Collection; page 88: (top left) Raul Touzon/NG Image Collection; (top right) Mark Thiessen/NG Image Collection; (bottom left) Joyce Dale/NG Image Collection; (bottom right) Paul Nicklen/NG Image Collection; page 95: (top) Kike Calvo/V&W/The Image Works; (just below top) Richard Nowitz/NG Image Collection; (center) George Steinmetz/NG Image Collection; (bottom) Tim Laman/NG Image Collection; page 97: (top) Ira Block/NG Image Collection; (bottom) George F. Mobley/NG Image Collection; page 99: Joyce Dale/NG Image Collection; page 100: Michael Melford/NG Image Collection; page 101: Raymond Gehman/NG Image Collection; page 102: (top) Courtesy NASA/EPA. Provided by Dr. Dale Quattrochi, Marshall Space Flight Center.; (bottom left) Randy Olson/NG Image Collection; page 103: (top) Randy Olson/NG Image Collection; (bottom) Raymond Gehman/NG Image Collection; page 112: Courtesy Environmental Analytical Systems; page 116: (top) John Eastcott and Yva Momatiuk/NG Image Collection; (center) Michael Melford/NG Image Collection; (bottom) Richard Nowitz/NG Image Collection; page 117: Courtesy Environmental Analytical Systems

Chapter 5

Pages 120–121: YURI CORTEZ/AFP/Getty Images; page 120: (inset) NG Maps; page 123: (top left) NG Maps; (top right) Bill Curtsinger/NG Image Collection; (center right) Norbert Rosing/NG Image Collection; (bottom left) Dr. Maurice G. Hornocker/NG Image Collection; (bottom right) Jodi Cobb/NG Image Collection; page 129: Courtesy Arthur N. Strahler; page 133: Adalberto Rias Szalay/Sexto Sol/Photodisc/Getty Images; page 134: (center left) John Eastcott and Yva Momatiuk/NG Image Collection; (center right) Carsten Peter/NG Image Collection; (bottom left) John Eastcott and Yva Momatiuk/NG Image Collection; (bottom right) Todd Gipstein/NG Image Collection; page 135: Tim Laman/NG Image Collection; page 136: James A. Sugar/NG Image Collection; page 137: Courtesy NASA; (top right inset) NG Maps; page 141: Annie Griffiths Belt/NG Image Collection; page 144: Annie Griffiths Belt/NG Image Collection; page 146: (top) NG Maps; (bottom left) James P. Blair/NG Image Collection; (bottom right) Sarah Leen/NG Image Collection; page 147: (top left) ©Dick Blume/Syracuse Newspapers/AP/Wide World Photos; (top right) Nicholas Devore III/NG Image Collection; page 148: (bottom left) Courtesy USGS, Branch of Quality Systems; page 149: (top) Courtesy University of Wisconsin Space Science and Engineering Center; (bottom) Jodi Cobb/NG Image Collection; page 150: Tim Laman/NG Image Collection; page 151: ©Dick Blume/Syracuse Newspapers/AP/Wide World Photos

Chapter 6

Pages 154–155: Robb Kendrick/NG Image Collection; page 156: (bottom) ©AP/Wide World Photos; (inset) NG Maps; page 157: (top left) Courtesy Taylor Instrument Company and Wards Natural Science Establishment, Rochester, New York; (top right) Nick Caloyianis/NG Image Collection; page 158: Courtesy Taylor Instrument Company and Wards Natural Science Establishment, Rochester, New York; page 162: (top) Skip Brown/NG Image Collection; (bottom) JPL/Courtesy NASA; page 163: (right) Skip Brown/NG Image Collection; (bottom left) Gordon Wiltsie/NG Image Collection; page 164: David Hume/Reportage/Getty Images; page 165: Courtesy Janice Coen, National Center for Atmospheric Research; page 176: Courtesy NOAA; page 177: Courtesy NASA; page 178: (top) Nick Caloyianis/NG Image Collection; (bottom) Skip Brown/NG Image Collection; page 179: Courtesy NOAA; page 180: Courtesy Taylor Instrument Company and Wards Natural Science Establishment, Rochester, New York

Chapter 7

Pages 182–183: Christopher J. Morris/©Corbis; page 188: Steve Raymer/NG Image Collection; page 190: Steve Raymer/NG Image Collection; page 192: (left) Thomas J. Abercrombie/NG Image Collection; (right) Michael Nichols/NG Image Collection; (inset) NG Maps; page 194: Thomas J. Abercrombie/NG Image Collection; page 197: Courtesy NASA; page 198: Courtesy NASA; (inset) NGS Maps; page 203: Courtesy NOAA; page 204: (top) Courtesy NOAA; (center) Data courtesy of Rick Lumpkin (NOAA AOML). Joanna Gyory, Elizabeth Rowe, Arthur J. Mariano, Edward H. Ryan. Figure 2, The Florida Current, Ocean Surface Currents.; (bottom) Courtesy

NASA; page 206: Courtesy NASA; page 209: Courtesy Otis B. Brown, Robert Evans, and M. Carle, University of Miami, Rosenstiel School of Marine and Atmospheric Science, Florida, and NOAA/Satellite Data Services Division; page 210: (bottom left) NG Maps. Data from NOAA/NESDIS/NCDC/Satellite Data Services Division (SDSD) compiled by UNEP/GRID; (bottom right) Courtesy NASA; page 211: (top) Courtesy Otis B. Brown, Robert Evans, and M. Carle, University of Miami, Rosenstiel School of Marine and Atmospheric Science, Florida, and NOAA/Satellite Data Services Division; page 212: (top) Richard Nowitz/NG Image Collection; (top inset) NG Maps; (bottom left) Michael Nichols/NG Image Collection; (bottom right) Courtesy NASA; page 213: (left) Courtesy NASA; (right) Courtesy Department of Atmospheric Sciences, University of Illinois at Urbana-Champaign

Chapter 8

Pages 216–217: Patric McFeeley/NG Image Collection; page 218: (center left) David Doubilet/NG Image Collection; (center left inset) NG Maps; (right) NG Image Collection; (bottom left) John Dunn/NG Image Collection; (bottom left inset) NG Maps; (bottom right inset) NG Maps; page 219: David Doubilet/NG Image Collection; page 220: Courtesy NOAA; page 224: Courtesy NOAA; page 225: Courtesy Naval Research Laboratory; page 226: Courtesy NOAA; page 227: Courtesy NOAA; page 231: Courtesy David Dempsey, San Francisco State University; page 235: Raul Touzon/NG Image Collection; page 236: Courtesy NOAA; page 239: (top) Courtesy NASA; (bottom) Steve Pace/Envision; page 240: (top) John Dunn/NG Image Collection; (bottom) Courtesy NOAA; page 241: Courtesy NASA

Chapter 9

Pages 244–245: AFP/Getty Images; page 247: Courtesy NOAA; (inset) NG Maps; page 249: Courtesy NASA; pages 250–251: NG Image Collection; page 252: Courtesy NASA and National Hurricane Center. Image published on Wikipedia.; page 254: NG Maps; page 255: NG Image Collection; page 256: (top) Ty Harrington/FEMA News; (just below top) Andrea Booher/FEMA; (center) FEMA News Photo; (just below center) Mark Wolfe/FEMA; (bottom) FEMA News Photo; page 257: Annie Griffiths Belt/NG Image Collection; page 259: (top left) Courtesy NOAA; (top right) ©Chris Hanson/3D Nature LLC; (center right) ©Chris Hanson/3D Nature LLC; (bottom) Courtesy NOAA; page 260: Harold F. Pierce/NASA/NG Image Collection; page 261: Courtesy NOAA; page 262: (left) Jocelyn Augustino/FEMA; (right) ©AP/Wide World Photos; page 263: (center) NG Maps; (bottom) NG Maps; page 264: (top) Courtesy NOAA; (bottom) Courtesy NASA; page 265: (top) Courtesy NASA and National Hurricane Center. Image published on Wikipedia; (center) ©AP/Wide World Photos; page 267: NG Maps

Chapter 10

Pages 268–269: Gandee Vasan/Getty Images; page 270: Phil Shcermeister/NG Image Collection; page 271: Chris Johns/NG Image Collection; page 277: Dan Westergren/NG Image Collection; page 278: Courtesy NOAA; page 280: Jason Edwards/NG Image Collection; page 282: Jason Edwards/NG Image Collection; page 284: Richard Olsenius/NG Image Collection; page 285: Priit Vesilind/NG Image Collection; page 287: Courtesy NOAA; page 289: Courtesy NOAA; page 290: (top left) AFP PHOTO/YASSER AL-ZAYYAT/Getty Images; (top right) Peter Carsten/NG Image Collection; (bottom left) John McCombe/Getty Images; (bottom right) Priit Vesilind/NG Image Collection; page 291: (top) Win Henderson/FEMA; (bottom) Greg Henshal/FEMA; (bottom inset) NG Maps; page 293: (top) Courtesy NOAA; (bottom) Priit Vesilind/NG Image Collection; page 294: Phil Shcermeister/NG Image Collection; page 295: Greg Henshal/FEMA

Chapter 11

Pages 298–299: Brueghel, Pieter the Elder, *Hunters in the Snow*, January, 1565/Kunsthistorisches Museum, Vienna, Austria/The Bridgeman Art Library; page 300: (center left) John Dunn/NG Image Collection; (center right) Jim Richardson/NG Image Collection; (bottom left) George F. Mobley/NG Image Collection; (bottom right) Tim Laman/NG Image

Collection; page 301: (center left) John Dunn/NG Image Collection; (center right) Rich Reid/NG Image Collection; (bottom left) Peter Carsten/NG Image Collection; (bottom right) James P. Blair/NG Image Collection; page 306: Courtesy NASA; page 308: Courtesy NASA; page 312: (top) George Steinmetz/NG Image Collection; (center) Tim Laman/National Geographic Society; page 313: Barbara Summey, NASA/GSFC Visualization Analysis Laboratory/NG Image Collection; page 316: (top) Courtesy NASA; (bottom) Peter Carsten/NG Image Collection; page 317: (top left) James P. Blair/NG Image Collection; (top right) NG Maps

Chapter 12

Pages 320–321: Courtesy NASA; page 323: (left) Courtesy NASA; (right) Courtesy NASA; page 324: Raymond Gehman/NG Image Collection; page 326: ©AP/Wide World Photos; page 328: Andrew McConnell/Alamy; page 329: (top) Karsten Wrobel/Alamy; (bottom) Raymond Gehman/NG Image Collection; page 331: Richard Cummins/SUPERSTOCK; page 332: Raymond Gehman/NG Image Collection; page 335: Sisse Brimberg/NG Image Collection; page 336: David Alan Harvey/NG Image Collection; page 338: Andre Jenny/The Image Works; page 340: Brand X/SUPERSTOCK; page 342: Gerd Ludwig/NG Image Collection; page 343: Hinrich Baesemann/Landov LLC; page 344: Maria Stenzel/NG Image Collection; page 345: (center) Maria Stenzel/NG Image Collection; (bottom) Hinrich Baesemann/Landov LLC; page 346: (top left) Kari Niemelainen/Alamy; (center left) James P. Blair/NG Image Collection; (center right) Peter Essick/NG Image Collection; (bottom) Jim Richardson/NG Image Collection; page 347: (top left) Ted Spiegel/NG Image Collection; (top right) Will & Deni McIntyre/Photo Researchers, Inc.; (top right inset) NG Maps; (bottom left) Stephen Alvarez/NG Image Collection; (bottom right) Doug Cheeseman/Alamy; page 348: (top) Johnny Johnson/The Image Bank/Getty Images; (center) Raymond Gehman/NG Image Collection; (bottom) Karsten Wrobel/Alamy; page 349: Gerd Ludwig/NG Image Collection

Chapter 13

Pages 352–353: Tim Laman/NG Image Collection; page 355: (left) Peter Essick/National Geographic Society; (right) Taylor S. Kennedy/NG Image Collection; page 356: (top) NG Maps; (left) Peter Essick/NG Image Collection; (center right) NGM Art/NG Image Collection; page 357: Durieux/Sipa Press; page 358: Courtesy NOAA; page 359: Courtesy NOAA; page 360: (top left) Courtesy Gregory W. Shirah, NASA/GSFC/Scientific Visualization Studio/NG Maps; (top right) Courtesy NOAA; page 361: Courtesy NASA; page 363: Courtesy NASA; page 364: (bottom left) Nobert Rosing/NG Image Collection; (bottom right) Courtesy NASA; page 365: (left) Stephen Alvarez/NG Image Collection; (right) Ronen Zvulun/Reuters/Landov LLC; page 371: Courtesy NASA; page 372: Richard Nowitz/NG Image Collection; page 373: (center) Jack Stephens/Alamy; page 374: (left) James P. Blair/NG Image Collection; (right) John Eastcott and Yva Momatiuk/NG Image Collection; page 375: Courtesy NASA; page 376: (top) Todd Gipstein/NG Image Collection; (bottom) Courtesy National Oceanic and Atmospheric Administration Paleoclimatology Program/Department of Commerce; page 377: Michael Nichols/NG Image Collection; page 379: Karen Kasmauski/Getty Images; page 380: (top) Simon Fraser/SPL/Photo Researchers, Inc.; (center) Durieux/Sipa Press; page 381: Courtesy National Oceanic and Atmospheric Administration Paleoclimatology Program/Department of Commerce; page 382: Courtesy NOAA; page 383: (top) James P. Blair/NG Image Collection; (bottom) John Eastcott and Yva Momatiuk/NG Image Collection

Chapter 14

Pages 384–385: Albert Moldvay/NG Image Collection; page 386: Courtesy NASA; page 387: Courtesy NOAA; page 389: Courtesy NASA; page 390: (left) Skip Brown/NG Image Collection; (right) Cameron Davidson/NG Image Collection; page 391: (top) Jodi Cobb/NG Image Collection; (bottom) ©Arno Balzarini/epa/Corbis Images; page 392: (top) ©Arno Balzarini/epa/Corbis Images; (bottom) ©AP/Wide World Photos; page 394: ©AP/Wide

World Photos; page 395: John Eastcott and Yva Momatiuk/NG Image Collection; page 396: Alain Nogues/Corbis Sygma/©Corbis; (inset) NG Maps; page 397: Photo by John McColgan; Courtesy Bureau of Land Management, Alaska Fire Service; page 398: Courtesy NASA; page 399: Daniel Berehulak/Getty Images; page 404: (left) Gerd Ludwig/NG Image Collection; (right) Al Petteway/NG Image Collection; page 405: (top left) Michael Nichols/NG Image Collection; (top right) Vincent J. Musi/NG Image Collection; (bottom left) M. Collier/DRK Photo; (bottom right) Courtesy NASA; page 407: (top) M. Collier/DRK Photo; (bottom) James P. Blair/NG Image Collection; page 408: (top) ©AP/Wide World Photos; (center) Gerd Ludwig/NG Image Collection; (bottom left) Alain Nogues/Corbis Sygma/©Corbis

Chapter 15

Pages 412–413: Peter Carsten/NG Image Collection; page 414: Courtesy National Archives; page 415: Courtesy NOAA; page 417: (left) Annie Griffiths Belt/NG Image Collection; (right) James P. Blair/NG Image Collection; page 421: Courtesy NOAA; page 422: Courtesy NOAA; page 423: (top left) Courtesy NOAA; (top right) Courtesy WSI Corporation; (bottom right) Courtesy NOAA; page 426: Courtesy NOAA; page 430: Courtesy NOAA; page 431: (top)

Courtesy NOAA; (bottom) Courtesy WSI Corporation; page 435: Courtesy NOAA; page 436: ©AP/Wide World Photos; page 437: Courtesy NOAA; page 439: Courtesy NOAA

Chapter 16

Pages 440–441: Paul Nicklen/NG Image Collection; page 442: (center left) James P. Blair/NG Image Collection; (center right) Jodi Cobb/NG Image Collection; (bottom left) Roberto Campos/AFP/Getty Images; (bottom right) James P. Blair/NG Image Collection; page 445: (top left) Ira Block/NG Image Collection; (top right) Sam Kittner/NG Image Collection; (center) Courtesy NASA/GSFC MODIS Rapid Response Team; page 453: Courtesy NASA/GSFC/LaRC/JPL, MISR Team; page 454: Ira Block/National Geographic Society; (inset) NG Maps; page 455: Courtesy NASA/GSFC/University of Arizona; page 463: Courtesy NASA; page 466: Juan Velasco/National Geographic Society; page 467: (top) Karen Kasmauski/NG Image Collection; (center) Peter Krogh/National Geographic Society; (bottom) Dick Durrance II/National Geographic Society; (below top) ©Lester Lefkowitz/Corbis; page 468: (top) Courtesy NASA; (bottom left) James P. Blair/NG Image Collection; page 469: (bottom left) Ira Block/National Geographic Society; (top right) ©Lester Lefkowitz/Corbis

TEXT AND ILLUSTRATION CREDITS

Chapter 1

Figure 1.2: From Strahler, Alan and Arthur Strahler, *Physical Geography: Science and Systems of the Human Environment*, 3rd ed. Copyright © 2005 John Wiley & Sons, Inc. Reprinted with permission of John Wiley & Sons, Inc; Figure 1.3: Adapted from NOAA; Figure WASS1: Adapted from NOAA; Figure 1.6–El Niños: Courtesy NASA; Figure 1.6–Global Warming: Data from the Climatic Research Unit at the School of Environmental Sciences at the University of East Anglia. Used by permission; Figure 1.8: Courtesy NOAA; Figure 1.10: Courtesy NOAA; Figure 1.14: Courtesy U.S. Geological Survey.

Chapter 3

Figure 3.7: After W.D. Sellers, *Physical Climatology*, University of Chicago Press. Used by permission; Figure 3.9 Process Diagram: From A.N. Strahler, "The Life Layer," *Journal of Geography*, vol. 69, Figure 2.4. Used by permission; Figure 3.16: Copyright © A.N. Strahler. Used by permission.

Chapter 4

Figure 4.8: Copyright © A.N. Strahler. Used by permission; Figure 4.9: Copyright © A.N. Strahler. Used by permission; Figure 4.10: Copyright © A.N. Strahler. Used by permission; Figure 4.16: After EPA; Figure 4.21: Data courtesy of David H. Miller; Figure 4.22: Data compiled by John E. Oliver; Figure 4.25: Courtesy NOAA; Figure 4.26: Courtesy NOAA.

Chapter 5

Figure 5.3: Based on data of John R. Mather; Figure 5.4: Data of J. von Hann, R. Süring, and J. Szava-Kovats as shown in Haurwitz and Austin, *Climatology*; Figure 5.9: Copyright © A.N. Strahler. Used by permission; Figure 5.10: Copyright © A.N. Strahler. Used by permission; Figure 5.24: From R.H. Skaggs, Proc. Assoc. American Geographers, vol. 6, Figure 2. Used by permission.

Chapter 6

Figure 6.14: Adapted from NOAA.

Chapter 7

Global Locator, page 192: From Berg, Linda R. and Mary Catherine Hager, *Visualizing Environmental Science*. Copyright © 2007 John Wiley & Sons, Inc. Reprinted with permission of John Wiley & Sons, Inc; Figure 7.5b: Data

compiled by John E. Oliver; Figure 7.9: Data compiled by John E. Oliver; Figure 7.11: Copyright © A.N. Strahler. Used by permission; Figure 7.14: Copyright © A.N. Strahler. Used by permission; Figure 7.18: Copyright © A.N. Strahler. Used by permission; Figure 7.20: After A.J. Gordon, Nature 382: 399–400, August, 1996. Used by permission; Figure 7.21, page 211: From Murck, Barbara W., Brian J. Skinner, and Dana Mackenzie, *Visualizing Geology*. Copyright © 2008 John Wiley & Sons, Inc. Reprinted with permission of John Wiley & Sons, Inc.

Chapter 8

Figure 8.4: Data from U.S. Dept. of Commerce; Figure 8.5: Drawn by A.N. Strahler; Figure 8.6: Drawn by A.N. Strahler; Figure 8.7: Drawn by A.N. Strahler; Figure 8.20: Based on data of S. Pettersen, B. Haurwitz, and N.M. Austin, J. Namias, M.J. Rubin, and J-H. Chang; Figure 8.21: After M. A. Garbell.

Chapter 9

Chapter opener: From Warren Faidley, *Weatherwise*, vol. 45, #6. Used by permission; Figure 9.1: Data from H. Riehl, *Tropical Meteorology*, New York: McGraw-Hill; Figure 9.3: Redrawn from NOAA, National Weather Service; Figure 9.12: Courtesy NOAA.

Chapter 10

Figure 10.4: Based on data of John R. Mather; Figure 10.6: Adapted from diagrams by Research Applications Program, National Center for Atmospheric Research, Boulder, Colorado.

Chapter 11

Figure 11.2: Based on the Goode Base Map; Figure 11.5: Based on the Goode Base Map; Figure 11.3: From Strahler, Alan and Arthur Strahler, *Physical Geography: Science and Systems of the Human Environment*, 3rd ed. Copyright © 2005 John Wiley & Sons, Inc. Reprinted with permission of John Wiley & Sons, Inc; Figure 11.4: Simplified and modified from Plate 3, "World Climatology," Volume I, *The Times Atlas*, Editor John Bartholomew, The Times Publishing Company, Ltd., London, 1958; Figure 11.5: Based on the Goode Base Map; WASS1: Courtesy NOAA; Figure 11.8: Compiled from station data by A.N. Strahler.

Chapter 12

Figure 12.1: Based on the Goode Base Map; Figure 12.4: Based on the Goode Base Map; Figure 12.7: Based on the Goode Base Map; Figure 12.8: Compiled from station data by A.N. Strahler; Figure 12.12: Based on the Goode Base Map; Figure 12.18: Based on the Goode Base Map; Figure 12.22: Based on the Goode Base Map.

Chapter 13

Figure 13.1: Courtesy of Gordon C. Jacoby of the Tree-Ring Laboratory of the Lamont-Doherty Geological Observatory of Columbia University; Figure 13.2: IPCC 2001. Used by permission; Figure 13.6: Copyright © A.N. Strahler; Figure 13.7b: Data and image courtesy of Klaus Wolter at NOAA; Figure 13.8: National Weather Service; Figure 13.9: From Curtis, Scott., and Robert F. Adler, "Evolution of El Niño-precipitation relationships from satellites and gauges," *Journal of Geophysical Research*, 108(D4), 4153, doi:10.1029/2002 JD002690, 28 February 2003. Copyright © 2003 American Geophysical Union. Reproduced by permission of American Geophysical Union; Figure 13.11: Courtesy NOAA; Figure 13.14: Copyright © D. Reidel Pub. Co., 1984. Used by permission.

Chapter 14

Figure 14.3b: Data of U.S. Geological Survey; Figure 14.10: Adapted with the permission of Nelson Thornes from "The North Sea storm surge of 1 February 1953" from *Geography: An Integrated Approach*. David Waugh, ISBN 9780174447061, first printed in 2002; WASS: Courtesy of Sharon E. Nicholson, Department of Meteorology, Florida State University, Tallahassee; Figure 14.16: Adapted from Hugo Ahlenius, UNEP/GRID-Arendal. *http://maps.grida.no/go/graphic/climate_change_and_malaria_scenario_for_2050.* Based on data from Rogers, Randolph. *The Global Spread of Malaria in a Future, Warmer World.* Science (2000: 1763–1766). Used by permission.

Chapter 15

Figure 15.2: Courtesy NOAA; Figure 15.3 – PD01: Courtesy NOAA; Figure 15.7: Adapted from NOAA; WASS1: Adapted from NOAA; Figure 15.13: Adapted from NOAA; Figure 15.17a: Adapted from NOAA; Figure 15.17b: From The University Corporation for Atmospheric Research; Figure 15.18: "Chaos and the jet stream" figures adapted from NOAA.

Chapter 16

Figure 16.3: IPCC, 2001, Figure 2.1. Used by permission; Figure 16.5: Prepared by Robert A. Rohde for Global Warming Art. *http://www.globalwarmingart.com/wiki/Image:Global_Carbon_Emission_by_Type_png.* Used by permission; Figure 16.8a: Based on data from IPCC Special Report on Emissions Scenarios, Figure 2.4. Used by permission; Figure 16.9: Based on data from IPCC Special Report on Emissions Scenarios, Figure 2.5. Used by permission; Figure 16.10: Based on data from IPCC Special Report on Emissions Scenarios, Figure 2.11. Used by permission; Figure 16.11: Based on data from IPCC, 2001, Figures 17 & 18. Used by permission; Figure 16.12: Prepared by Robert A. Rohde for Global Warming Art. *http://www.globalwarmingart.com/wiki/Image:Climate_Change_Attribution_png.* Data based on Meehl, G.A., W.M. Washington, C.A. Ammann, J.M. Arblaster, T.M.L. Wigleym and C. Tebaldi (2004). "Combinations of Natural and Anthropogenic Forcings in Twentieth-Century Climate". *Journal of Climate* 17: 3721–3727 and Jones, P.D. and Moberg, A. (2003). "Hemispheric and large-scale surface air temperature variations: An extensive revision and an update to 2001". *Journal of Climate* 16: 206–223. Used by permission; Figure 16.17: IPCC, 2001, Figure 4. Used by permission; Figure 16.18a: Based on data from IPCC, 2001, Figure 22b. Used by permission; Figure 16.18b: IPCC, 2001, Figure 9.5a. Used by permission; Figure 16.18c: IPCC, 4th Assessment Report, Figure 10.8. Used by permission; Figure 16.19: IPCC, 4th Assessment Report, Figure 10.12. Used by permission; Figure 16.20: Courtsey NOAA; Figure 16.21: Based on data from IPCC, 2001, Figure 4.3. Used by permission; Figure 16.22: From UNEP/GRID-Arendal. *http://maps.grida.no/go/graphic/potential_impact_of_sea_level_rise_on_bangladesh.* Used by permission; WASS1b: From The University Corporation for Atmospheric Research; Figure 16.23a: Based on data from IPCC, 4th Assessment Report, Figure SPM2. Used by permission.

Line drawings in the following figures have been adapted from Strahler, Alan and Zeeya Merali, *Visualizing Physical Geography.* Copyright © 2008 John Wiley & Sons, Inc. Reprinted with permission of John Wiley & Sons, Inc.

Chapter 1: 1.9; 1.12; 1.13. **Chapter 2:** 2.2b; 2.3–PD; 2.9; 2.12a; 2.13. **Chapter 3:** 3.3; 3.10; 3.20; 3.23; 3.24. **Chapter 4:** 4.1; 4.3; 4.4; 4.5; 4.6; 4.7; 4.8; 4.11; 4.12; 4.13; 4.14–PD; 4.17; 4.19; 4.22; 4.24. **Chapter 5:** 5.1; 5.2a; 5.5; 5.6; 5.12a; 5.15–PD01; 5.16–PD02; 5.17; 5.18; 5.19; 5.21; 5.24; WHiP. **Chapter 6:** 6.4; 6.6–PD01; 6.7; 6.8; 6.9; 6.12. **Chapter 7:** 7.3; 7.4; 7.8; 7.10. **Chapter 8:** 8.2; 8.9; 8.10 PD01; 8.11; 8.14–PD02. **Chapter 9:** 9.6a; 9.7; 9.16. **Chapter 10:** 10.18. **Chapter 11:** 11.1; 11.6; 11.7; 11.9. **Chapter 12:** 12.2; 12.3; 12.5; 12.6; 12.9; 12.10; 12.11; 12.13; 12.14; 12.5; 12.16–PD01; 12.17; 12.19; 12.20–PD02; 12.21; 12.23; 12.24; 12.25. **Chapter 13:** 13.15–PD; 13.16. **Chapter 16:** 16.6.

Line drawings in the following figures have been adapted from Strahler, Alan and Arthur Strahler, *Introducing Physical Geography*, 4th ed. Copyright © 2006 John Wiley & Sons, Inc. Reprinted with permission of John Wiley & Sons, Inc.

Chapter 2: 2.11; 2.12b. **Chapter 3:** 3.22. **Chapter 4:** 4.18; 4.20. **Chapter 5:** 5.7. **Chapter 6:** 6.11a; 6.11b; 6.13. **Chapter 7:** 7.1; 7.2; 7.6; 7.16; 7.17; 7.21, page 210. **Chapter 14:** 14.17; 14.18; 14.19.

INDEX

circumpolar currents, ocean currents, 206–207
cirrocumulus clouds, 134–135
cirrostratus clouds, 134–135
cirrus clouds, 134–135
  along fronts, 224, 417
Clean Air Act of 1963, 38
climate. *See also* weather
  classification of, 309–315
  definition of, 7, 300
  factors affecting, 300–301
  relationship to weather, 7, 10–15
  severe changes, 395–400
climate changes. *See also* anthropogenic climate forcing; climate variability
  adaptation/mitigation strategies, 465–467
  global energy balance and, 74–77
  as norm, 354–355
  processes of, 11–12
  projections
    extreme weather, 462
    oceans and sea levels, 460–462, 463–465
    temperature and precipitation, 458–460
  severe climate, 395–400
climate controls, 300–301
climate extremes, in United States, 14–15, 24
climate forecasting
  climate-change projections, 458–465
  global models and simulations, 456–457
  land-surface models, 455–456
  ocean models, 454–455
climate regimes, changes in, 400
climate types. *See* climate, classification of; *specific climate type*
climate variability
  El Niño and La Niña, 358–362
  feedbacks, 371–379
  historical records and methods, 354–356
  millennial climate variations, 367–370
  oscillation patterns, 362–365
  thermohaline circulation, 366, 463
  volcanic eruption effects, 28, 41, 356–358
climographs, 310–311
  definition of, 310
cloud cover
  as climate feedback, 374–375
  temperature effects of, 67, 106, 374–375
cloud-to-ground lightning
  characteristics of, 281–283
  definition of, 281
clouds. *See also* cyclones; thunderstorms
  associated with fronts, 222–224, 417
  cloud cover and temperatures, 67, 106, 374–375
  forecasting difficulties, 453, 459

formation and types, 133–137, 142
  role in climate system, 77
coastal floods, characteristics of, 392–394
coastal regions
  climate change and, 464–465
  flooding in, 392–394
  storm damage in, 256–262
  temperatures in, 96–99, 100–101, 106, 108
cold air outbreaks, 238–239
  definition of, 238
cold emergencies, in cold waves, 389
cold fronts
  characteristics of, 222–224
  definition of, 222
cold waves
  characteristics of, 389–390
  definition of, 389
communications systems, ionospheric disturbances of, 42, 50
computers, in forecasting, 415, 424–427, 430, 434–436. *See also* numerical modeling
condensation
  in cloud formation, 131–133, 138–144
  saturation and, 127, 131–132
condensation nuclei, definition of, 133
conduction, in heat transfer, 64–65
continental (c) regions, air masses of, 219–221
continental antarctic (cAA) region
  air masses of, 219–221
  and climate classification, 310
continental arctic (cA) region
  air masses of, 219–221
  and climate classification, 310
  cold waves and, 388
continental climates, temperatures in, 96–99, 100–101, 106, 108
continental polar (cP) region
  air masses of, 219–221
  and climate classification, 310
  cold waves and, 388
continental tropical (cT) region
  air masses of, 219–221, 278–279
  and climate classification, 309–310
contour maps, 19–20
convection. *See also* convective precipitation
  and cold fronts, 222
  in cyclones, 252–253
  deforestation and, 406
  in heat transfer, 64–65
  and local forecasting, 418–419, 431
  over urban areas, 407
  in thunderstorms, 273, 274
  and tornadoes, 287
convection cells, in thunderstorms, 273
convective precipitation, 141–144
  definition of, 141
convergence, 175–176. *See also* global wind/pressure patterns
  in cyclones, 252–253
  definition of, 175

in midlatitudes, 187
  and upward air movement, 139
cooling degree days, commercial considerations of, 11, 114–115
coral growth, as climate feedback, 376–377
Coriolis effect, 166–171, 172–174
  definition of, 166
  on jet stream, 198
  on ocean currents, 202
Coriolis force
  characteristics of, 166–167
  and cyclonic wind speeds, 253, 284
  and oceanic circulation, 201
  rotation and, 85
  on upper-level winds, 196, 198
Coriolis, Gaspard-gustave de, and Coriolis effect, 166
counterradiation, 70–75
  definition of, 70
crops. *See* agriculture
cumuliform clouds, 134–135, 417
cumulonimbus clouds, 134–135
cumulus clouds, 134–135. *See also* thunderstorms
  along fronts, 222–223, 417
  formation of, 142
cumulus stage, definition of, 272
currents. *See* ocean currents
cyclone families, definition of, 238
cyclones. *See also* anticyclones; midlatitude cyclones; tropical cyclones
  characteristics of, 7, 8–9, 169–171, 173–176, 225–227, 246–251
  damage from, 244–245
  definition of, 169
  development of, 209–210, 252–255
  impacts from, 256–262
  trends in activity, 263, 462
  vorticity and, 232–236
cyclonic storms. *See also* cyclones
  definition of, 226
  forms of, 247
  storm tracks, 237–238, 253–255

D

daily cycles, of air temperature, 104–106
damage
  from cold weather, 239, 390
  from cyclones/hurricanes, 256–262
  from hail, 147
  from lightning, 268–269
  from tornadoes, 286, 287, 288–291
  "Valentine's Day Storm" of 2007, 423
day (solar), 84
debris floods
  characteristics of, 391–392
  deforestation and, 406
declination, definition of, 90
deep currents, of ocean, 206–208
deep ocean, definition of, 202
deforestation
  effects on weather/climate, 404–406
  and greenhouse gases, 447

density
  air. *See* air density
  of ocean water, 201, 202, 207, 208, 463
departures, temperature, 10
deposition
  acid, 404
  of water vapor, 122
deserts
  precipitation in, 146, 212, 303–306
  in subtropical high-pressure centers, 191–192
  vegetation in, 346
dew-point temperature
  definition of, 127
  fog and, 136
  saturation and, 127–128
  soundings measurements of, 417–419
diffuse radiation, 66
disease, climate change and, 400, 464–465
dissipating stage, definition of, 273
diurnal processes. *See also* daily cycles
  definition of, 85
divergence, 175–176
  and cyclone formation, 232–233, 235
  definition of, 175
  and thunderstorm formation, 274, 277
Doppler radar
  precipitation measurement and, 148, 421–422, 423
  tornadoes and, 288–289, 292
  wind measurement and, 158, 421–422, 423
downdrafts, 156
  in thunderstorms, 272–273, 274–276
downwelling, in El Niño events, 361
drainage winds, characteristics of, 164–165
droughts
  definition of, 395
  impact on human life, 395–397
  trends in, 14–15, 462
  vegetation and, 306
dry adiabatic lapse rate, 130–132, 142–144
  definition of, 131
  in thunderstorms, 271
dry climates
  definition of, 314
  vegetation in, 346
dry lines, 224, 278–279
  definition of, 278
dry midlatitude (9) climate
  characteristics of, 311–313, 338
  definition of, 338
dry seasons. *See* seasonality, of precipitation
dry subtropical (5) climate
  characteristics of, 311–313, 330–331
  definition of, 330
  satellite image of, 320–321
dry tropical (4) climate
  characteristics of, 311–313, 327–329

greenhouse effect *(cont.)*
  gases in, 31–33, 70–71, 74–77, 399, 443–448
  longwave radiation in, 59, 70–71
  and ocean warming, 399
  projections of, 458–459
greenhouse gases . *See* greenhouse effect, gases in; *specific gas*
ground water, as water reservoir, 123–124
growing degree days, in agriculture, 115
Gulf Stream, ocean current, 204–206, 208, 211
Gulf Stream extension
  and climate change, 463
  definition of, 366
gyres
  characteristics of, 203–205
  definition of, 202

H

Hadley cells. *See also* intertropical convergence zone (ITCZ)
  characteristics and effects, 184–188
  definition of, 184
  heat transfer and, 209–210
Hadley, George, and Hadley cells, 184
hail, characteristics of, 145, 147
Hawaiian high, high-pressure system, 188–189, 191, 194
heat capacity, definition of, 99
heat emergencies, in heat waves, 388
heat index, 113–114
  and heat waves, 387–388
heat transfer
  by global circulations, 208–211
  poleward, 63
  sensible and latent, 64–65, 101–102, 103, 208–211, 252
heat-trapping gases . *See* greenhouse effect
heat waves
  characteristics of, 386–388
  definition of, 386
heating degree days, commercial considerations of, 11, 114–115
hectopascals (hPa), 43
  definition of, 23
heterosphere, 40–41
high-latitude climate groups (Group III)
  characteristics of, 309–310, 312–314, 342
  climate descriptions, 342–347
high-latitude zones
  current climate changes in, 400
  heat redistribution in, 209–211
  North Atlantic oscillation and, 363–365
  oceanic circulation in, 206
  surface wind/pressure patterns in, 192–194
  upper-level wind/pressure patterns in, 196
high-pressure centers . *See* anticyclones
highland climate, 314

homosphere, 40–41
hook echo. *See also* radar
  definition of, 421
human activities. *See also* anthropogenic climate forcing; industrial processes
  climate change and . *See* anthropogenic climate forcing
  flooding effects, 390–394
  land-cover changes, 323, 404–407, 447
  pollution and, 401–404
  rapid climate change and, 395–400
  severe weather effects on, 388, 390–394, 400
Humboldt current, ocean current, 205, 206
humidity, 126–130
  definition of, 126
  relative, 114, 128–130, 387
  specific, 23, 126–128
hurricanes
  Andrew (1992), 260, 263
  intensity and impacts, 256–262
  Katrina (2005), 14, 261–262, 398
  Mitch (1998), 120–121, 257
  monitoring, 250–251, 255
  Rita (2005), 250–251, 262
  storm surge and, 392–393
  trends in, 14–15, 263
  U.S. season of 2004, 260
  U.S. season of 2005, 398–399
  Wilma (2005), 254–255, 398
hydrologic cycle, 122–125
  definition of, 124
  projected intensity of, 460–461
hydrosphere
  definition of, 6, 122
  and hydrologic cycle, 122–125
hygrometers, 129–130

I

ice
  crystals and lightning, 281
  in ice storms, 145
  ocean models and, 454
  as water reservoir, 122–123
ice ages
  as climate-change process, 12, 354–355, 366
  definition of, 367
ice-albedo feedback
  and climate change projections, 459, 463
  definition of, 373
ice cores, temperature records and, 355–356, 366
ice sheet (13) climate
  characteristics of, 312–314, 344–345
  definition of, 344
  temperature increases in, 440–441
ice storms
  characteristics of, 145
  "Valentine's Day Storm" of 2007, 423
ideal gas law, definition of, 46
indexes, temperature, 113–115

industrial processes
  aerosols and, 33, 35, 401–404
  affected by severe weather, 390
  air pollution and, 401–403
  CFCs in, 37–38
  climate change and, 400
  indexes used in, 114–115
infrared radiation, 54–55, 56–61
insolation. *See also* solar radiation
  definition of, 62
  factors affecting, 62–63
  net radiation and, 104–105, 107
  variations in, 90–96, 367–370
  in world latitude zones, 95–96
instruments . *See specific weather feature*
interior locations . *See* continental climates
intertropical convergence zone (ITCZ). *See also* low-latitude climate groups (Group I)
  circulation in, 184–185
  definition of, 185
  monsoon circulation and, 188–191
  precipitation and, 192
  seasonal shifts in, 188–189
ionosphere
  characteristics of, 41–42
  definition of, 41
ions
  definition of, 41
  and ionosphere, 41–42
iron, role in biological pump, 375–376
irrigation, effects on weather/climate, 406–407
isallobaric map, definition of, 419
isobars
  definition of, 159
  and winds, 159–161
isohyets, 140–141, 304–305
  definition of, 140, 304
isolines
  on contour maps, 19–20
  definition of, 19
isotherms, definition of, 109
ITCZ . *See* intertropical convergence zone (ITCZ)

J

jet streak
  and cyclone formation, 231–233, 235–236
  definition of, 231
jet stream
  chaos and, 435
  characteristics of, 182, 197–198
  and cyclone formation, 231–233
  definition of, 197
  disturbances in, 199–201, 208–211, 231–233, 277, 375–376
  in El Niño events, 362, 398–399
  and heat and moisture transfer, 208–211
  and North Atlantic oscillation, 364–365
joules (J), definition of, 23

K

Kamchatka current, ocean current, 206
katabatic winds, 165
Katrina, U.S. hurricane (2005), 14, 261–262, 398
kelvin (K), 21–22
Kelvin scale, 21–22
Köppen climate classification system, 314–315
  definition of, 315
Köppen, Vladimir, classification system of, 314–315
Krakatoa eruption (1883), as climate-change event, 357
Kuroshio current, ocean current, 205, 208

L

La Niña
  and climate changes, 362, 398
  definition of, 360
Labrador current, ocean current, 206
lahars, 391
land. *See also* soil
  land-biomass feedback, 377
  land-cover changes, 404–407, 447
  land-surface models, 455–456
  solar heating of, 96–99
  temperature cycles and, 108–111
  in water balance, 124–125
land-biomass feedback, 377
land breezes, characteristics of, 162, 164
land-cover changes, 404–407, 447
land-surface models, in climate forecasting, 455–456
lapse rates
  adiabatic, 131–132, 142–144, 271
  environmental, 39–40, 132, 142–144, 254, 271
latent heat, 64–65
  and atmospheric moisture, 122, 131, 132, 144, 372
  definition of, 64
  in global energy budget, 72–74
  in heat transfer, 208–211
  release in cyclones, 248, 252–253
  temperatures and, 104, 372
  transpiration and, 101–102, 103
latitude
  as climate factor, 300, 309–310, 337
  effects on temperatures, 100, 109–111
  and insolation, 62–63, 90–96
  parallels and, 17–20
  world latitude zones, 95–96
leaders, definition of, 282
lee-side trough
  and cyclone formation, 235–236
  definition of, 236
lifting condensation levels
  and cloud formation, 131–132, 142–144
  definition of, 132
  and local forecasting, 418–419
light, in electromagnetic spectrum, 54

oceans (cont.)
  temperature increases in, 398–399
  temperature layers of, 202
  in water balance, 124–125
  as water reservoirs, 122–123
operational weather forecasting
  formulation of forecast, 428–431
  statements and validation, 432–436
orbit (Earth), 85–90
  and climate variations, 368–370
orographic precipitation, 139–141,
    146
  definition of, 139
  and moist climate types, 314
oxygen
  absorption of radiation by, 57, 67
  as atmospheric component, 30–31
ozone
  absorption of radiation by, 57,
    66–67
  as atmospheric component, 36–38,
    464–465
  definition of, 37
  pollution and, 137, 401
ozone hole, 37–38

P
Pacific Ocean, oscillation patterns,
    358–363
parallels, and latitude, 17–20
parameterizations, definition of, 453
pascals (Pa), 43
  definition of, 23
perihelion, definition of, 86
persistence forecast, definition of, 429
Peru current, ocean current, 205, 206
phase, definition of, 64
photosynthesis, as carbon dioxide
    sink, 446–447
phytoplankton, and biological pump,
    375
Pinatubo, Mount, eruptions of,
    28–29, 357
Pineapple Express, weather system,
    220
planetary vorticity, definition of, 234
polar deserts, precipitation region,
    304–306
polar easterlies, winds, 186–187
polar front
  definition of, 186
  and Group II climates, 310
  high latitude circulation and,
    186–187
  and jet stream, 197–198
polar jet stream
  characteristics of, 197–198
  and North Atlantic oscillation,
    364–365
polar projection, definition of, 19
polar zones
  characteristics of, 95, 96
  surface wind/pressure patterns in,
    186–187, 192–194
  temperature projections for, 462
  upper-level wind/pressure patterns
    in, 196

poles
  definition of, 84
  and seasons, 87–91
pollution
  increase during heat waves, 388
  ozone in, 38
  weather and, 401–403
population growth, and greenhouse
    emissions, 449
power, definition of, 23
precession, definition of, 369
precipitation
  in cyclones, 230, 248–249
  definition of, 124
  deforestation effects on, 404–406
  El Niño changes in, 12
  formation in clouds, 138–139
  global annual, 304–305
  global and regional forecasts,
    460–461, 464–465
  and hydrologic cycle, 124–125
  measurement of, 148
  orographic and convective
    processes, 139–144
  radar and, 421–422, 423
  regimes, 303–308, 400
  seasonality of, 306–308, 312–313
  types of, 144–147
  world climate distribution, 312–313
precipitation regimes, 303–308, 400
prediction, of weather . See weather
    forecasting
pressure
  atmospheric . See atmospheric
    pressure
  definition of, 23
pressure gradients
  definition of, 160
  jet stream and, 197
  at surface, 172
  and temperature gradients,
    160–161, 196, 287
  in tornadoes, 284
  at upper levels, 195–201
  winds and, 159–165, 172–176,
    194–201
pressure tendency, and local
    forecasting, 419
prevailing currents, as climate factor,
    301
prime meridian, 17
prognostic charts
  definition of, 426
  in forecasting, 426–427
projections . See map projections
proxy data, 355–356
  definition of, 355

R
radar
  history of, 412–414
  precipitation measurement and,
    148, 421–422, 423
  tornado measurement and,
    288–289, 292
  wind measurement and, 158,
    421–422, 423

radiation . See electromagnetic
    radiation
radiation fog, 136
radiometers, 112
radiosondes, 157, 417–418
rain. See also rainfall
  acid rain, 404
  characteristics of, 144–147
rain gauges, 148
rain shadow, 140–141
  definition of, 140
  effect on precipitation, 304
rainfall. See also drought; specific
    climate type
  in cyclones, 248–249, 252, 257
  mean annual, 146
  measurement of, 148
  severe weather changes and,
    398–399
rainforests
  and agriculture, 323
  in subtropical high-pressure
    centers, 191–192
Rayleigh scattering, and blue skies, 58
realms, of Earth, 5–6
reflection, of radiation, 66–69
regimes
  changes in, 400
  temperature and precipitation,
    302–308
relative humidity, 114, 128–130
  definition of, 128
  heat index and, 114
  and heat waves, 387
relative vorticity, definition of, 234
renewable energy resources, 465–467
reservoirs, of hydrosphere, 122–124
residence time, definition of, 448
return strokes, definition of, 282
revolution (Earth), 86–91
  and climate variations, 367–370
ridge, high-pressure disturbance, 231
Rita, U.S. hurricane (2005), 250–251,
    262
river floods, characteristics of,
    390–391
rotation (Earth), 84–85
  and climate variations, 367–370
  seasons and, 87–91
rural areas, temperature in, 101–103,
    106

S
Saffir-simpson scale, of hurricane
    intensity, 256
Sahel, drought and famine in, 396
salinity
  climate changes in, 463
  of ocean water, 201, 202, 207, 454
salt
  as condensation nuclei, 133
  salinity of ocean water, 201, 202,
    207, 454
Santa Ana winds
  characteristics of, 162, 164–165
  wildfires and, 397
satellites, in forecasting, 149,
    412–415, 422–423

saturation, definition of, 127
saturation specific humidity, 127–128
  definition of, 127
scales, of temperature, 21–22
scattering
  albedo and, 67–69
  atmospheric, 58, 59, 61, 66
  definition of, 58
sea breezes, characteristics of, 162,
    164
sea ice
  melting and sea-freshening, 463
  ocean models and, 454
sea level
  projected rise in, 460–462
  sea-level pressure, 159–160
  storm surges and, 258–259, 262
sea-level pressure, adjustment for,
    159–160
seasonality, of precipitation, 306–308,
    312–313
seasons. See also insolation
  as climate-change process, 12
  cyclone/hurricane, 253
  Earth motion variations and,
    369–370
  rhythms of, 6, 88
  seasonality, 302–303, 306–308
  and weather changes, 13–15,
    188–194, 306–308
semiarid/steppe climate subtype (s),
    304–305, 314
  satellite image of, 320–321
  vegetation in, 346
sensible heat, 64–65
  definition of, 64
  in global energy budget, 72–74
  in heat transfer, 208–211
  temperatures and, 104
severe thunderstorms, 274–276
  definition of, 274
  and tornado formation, 285
severe weather
  cold waves, 389–390
  definition of, 386
  flooding, 390–394, 462
  heat waves, 386–388
  increases in, 14–15, 398–399
  projections for, 462
short-range forecasts, definition of,
    432
shortwave radiation
  absorption and scattering of, 66–73
  definition of, 58
  flow in energy budgets, 71–75
  in global energy balance, 54–55,
    57–61
SI (Système International)
    measurements, 21–23
Siberian high, high-pressure system,
    188
Sidr, cyclone (2007), 244
skill
  chaos and, 434–436
  definition of, 434
sleet, characteristics of, 145, 147
sling psychrometer, 129

tropical cyclones (cont.)
  definition of, 247
  development and movement of,
      252–255
  and Group I climates, 310
  impacts from, 256–262
  structure of, 248
  trends in activity, 263
  as weather systems, 7, 9
tropical depressions, 247
tropical deserts, precipitation region,
      304–305
tropical easterly jet stream, 198
tropical rainforests, in subtropical
      high-pressure centers, 191–192
tropical storms, 247
  and climate change, 398–399
tropical zones
  characteristics of, 95, 96
  precipitation in, 144, 303–304, 305,
      308
  surface wind/pressure patterns in,
      191–192
  upper-level wind/pressure patterns
      in, 196
tropopause, 40–41, 198
  definition of, 41
troposphere
  characteristics of, 40–41
  definition of, 40
  ozone in, 38
trough, low-pressure disturbance, 231
tundra (12) climate
  characteristics of, 312–314, 343–344
  definition of, 343
typhoons, 247
  tracks of, 254

U
ultraviolet radiation
  in electromagnetic spectrum, 54
  ozone and, 37–38, 66
United Kingdom Met Office model
      (UKMO), 427
units, of measurement, 21–23
unstable air, convection in, 142–144
updrafts, 156
  in thunderstorms, 272–273
upper-air charts, 420–421
upper-air disturbances . See upper
      atmosphere, disturbances and
      storm formation
upper atmosphere. See also jet stream
  disturbances and storm formation,
      231–236, 238–239, 274, 277
  in forecasting, 420–421
  ozone in, 37–38
  upper-level winds, 166–171,
      195–201
upper-level winds, 166–171, 195–201

upwelling, 203–206, 454–455
  in El Niño events, 358–361
urban areas
  climate change and, 464–465
  effects on weather/climate,
      406–407
  flooding in, 391
  temperature in, 101–103, 106
urban heat island
  characteristics of, 101–103, 405
  definition of, 102
  effects on weather/climate, 405,
      406–407
U.S. Weather Service, statistics
      compiled by, 113

V
validation, of forecasts, 433–436
validation time, definition of, 433
valley breezes, characteristics of, 163,
      164
variability, of climate . See climate
      variability
veering, of winds, 418
vegetation
  climate benefits of, 101–103
  climate change and, 464–465
  and climate classification, 314–315,
      346–347
  droughts and, 306
  and global energy flows, 76
  land-cover and albedo, 455
  temperature effects of, 101–103,
      106
  wildfires and, 397
velocity
  definition of, 22
  wind, 157–158
vertical winds, 156
visible light, in electromagnetic
      spectrum, 54
visualization tools, 15–20. See also
      specific tool
  graphs, 15–16
  maps, 17–20
volatile organic compounds (VOCs),
      air pollution and, 401
volcanoes
  atmospheric effects of, 28, 41,
      356–357
  and global temperatures, 357
  vortexes, of tornadoes, 285–286
vorticity
  and cyclones, 232–236
  and tornadoes, 285–286

W
warm fronts, definition of, 223
warnings, weather, 432–433

Washington, Mount, observatory,
      182–183
watches/warnings, 432–433
  for tornadoes, 292
water. See also oceanic circulation;
      oceans; precipitation
  as atmospheric component, 31,
      123–125
  climate change and, 464–465
  energy absorption by, 71. See also
      latent heat
  and lightning formation, 281
  reservoirs, 122–124
  solar heating of, 96–99
  states of, 122
  temperature cycles and, 96–99,
      100–101, 108–111
water balance, global, 124–125
water vapor. See also atmospheric
      circulation; cyclones;
      thunderstorms
  absorption of radiation by, 57, 67,
      70–71
  as atmospheric component, 31
  as climate feedback, 372
  in forecasting, 422
  as water reservoir, 123–124
water vapor–temperature feedback,
      definition of, 372
watts (W), definition of, 23
wavelength, definition of, 52
weather. See also climate
  definition of, 7
  relationship to climate, 7, 10–15
weather forecasting
  history of, 412–415
  local forecasting, 414–419, 431, 436
  numerical (computer) forecasting,
      424–427, 430
  operational forecasting, 428–436
  skill and chaos in, 434–436
  synoptic forecasting, 420–423,
      428–429
  of tornadoes, 292
weather systems. See also specific system
  definition of, 218
  differences between regions, 246
  examples of, 7
  seasonal change examples, 13–14
weathering, as climate feedback,
      378–379
Wein's Law, and radiation, 56
western boundary currents
  characteristics of, 203–206
  definition of, 203
western hemisphere, cyclone
      formation in, 254–255
wet-dry tropical (3) climate
  characteristics of, 311–313, 314,
      325–326

climate changes in, 400
  definition of, 325
  satellite image of, 320–321
  vegetation in, 346
  wildfires in, 397
wet equatorial (1) climate
  characteristics of, 310, 312–313,
      322–323
  definition of, 322
  vegetation in, 347
wet equatorial belt, precipitation
      region, 304–305
wet seasons . See seasonality, of
      precipitation
wildfires, drought and, 397
Wilma, U.S. hurricane (2005),
      254–255, 398
wind barbs
  explanation of, 170
  and surface weather charts, 416
wind chill index, 113
wind direction, and local forecasting,
      415–418, 436
wind shear
  in forecasting, 418
  in storms, 275, 285
  and tropical storm formation, 399
wind speeds . See specific wind type
winds
  characteristics of, 156–157
  commercial considerations of, 11
  in cyclones, 248, 252–253
  definition of, 156
  flooding and, 392–394
  in forecasting, 415–416
  jet stream and disturbances,
      195–201, 364–365
  local, 161–165
  measurement of, 157–158, 421–422,
      423
  ocean currents driven by, 202–206
  pressure gradients and, 159–165,
      172–176, 194–201
  steering winds, 237
  surface, 172–176, 184–187
  upper-level, 166–171, 195–201
  wind shear, 275, 285, 399
  wind speed record, 182
winter (December) solstice. See also
      solstices
  definition of, 87
world latitude zones, insolation in,
      95–96
world patterns . See global climate
      patterns; specific feature

X
X-rays
  atmospheric absorption of, 66
  in electromagnetic spectrum, 54